Functional Metallosupramolecular Materials

RSC Smart Materials

Series Editors:
Professor Hans-Jörg Schneider, *Saarland University, Germany*
Professor Mohsen Shahinpoor, *University of Maine, USA*

Titles in this Series:
1: Janus Particle Synthesis, Self-Assembly and Applications
2: Smart Materials for Drug Delivery: Volume 1
3: Smart Materials for Drug Delivery: Volume 2
4: Materials Design Inspired by Nature
5: Responsive Photonic Nanostructures: Smart Nanoscale Optical Materials
6: Magnetorheology: Advances and Applications
7: Functional Nanometer-Sized Clusters of Transition Metals: Synthesis, Properties and Applications
8: Mechanochromic Fluorescent Materials: Phenomena, Materials and Applications
9: Cell Surface Engineering: Fabrication of Functional Nanoshells
10: Biointerfaces: Where Material Meets Biology
11: Semiconductor Nanowires: From Next-Generation Electronics to Sustainable Energy
12: Supramolecular Materials for Opto-Electronics
13: Photocured Materials
14: Chemoresponsive Materials: Stimulation by Chemical and Biological Signals
15: Functional Metallosupramolecular Materials

How to obtain future titles on publication:
A standing order plan is available for this series. A standing order will bring delivery of each new volume immediately on publication.

For further information please contact:
Book Sales Department, Royal Society of Chemistry, Thomas Graham House, Science Park, Milton Road, Cambridge, CB4 0WF, UK
Telephone: +44 (0)1223 420066, Fax: +44 (0)1223 420247
Email: booksales@rsc.org
Visit our website at www.rsc.org/books

Functional Metallosupramolecular Materials

Edited by

John G. Hardy
Queen's University Belfast, Northern Ireland, UK
Email: john.hardy@qub.ac.uk

and

Felix H. Schacher
Friedrich-Schiller-University, Jena, Germany
Email: felix.schacher@uni-jena.de

ROYAL SOCIETY OF CHEMISTRY

THE QUEEN'S AWARDS FOR ENTERPRISE: INTERNATIONAL TRADE 2013

RSC Smart Materials No. 15

Print ISBN: 978-1-78262-022-8
PDF eISBN: 978-1-78262-267-3
ISSN: 2046-0066

A catalogue record for this book is available from the British Library

Published by the Royal Society of Chemistry,
Thomas Graham House, Science Park, Milton Road,
Cambridge CB4 0WF, UK

Registered Charity Number 207890

For further information see our web site at www.rsc.org

Preface

The field of supramolecular chemistry is clearly one of the most vibrant areas of chemistry. Here, inorganic, organic, macromolecular, physical and materials chemistry meet and interdisciplinary research projects cover all aspects from truly fundamental studies to application-driven approaches. Within the field of supramolecular chemistry, metal-ion-directed assembly is a common feature exploited for the creation of functional supramolecular entities of both natural and synthetic origins. Examples are well-defined assemblies composed of a discrete number of ligands and metal ions, including metalloproteins, helical structures (*e.g.* G-quartets and helicates), grids, racks, squares/rectangles, metallamacrocycles, capsules and cages, and moreover, entities composed of a non-discrete number of ligands and metal ions, such as metallosupramolecular polymers.

The first chapter of the book focuses on naturally-occurring functional metallosupramolecular materials formed from metal-containing proteins with a particular focus on the wide range of roles metal ions fulfil within these structures. The subsequent chapters cover numerous potential applications of such materials in a systematic fashion. This includes DNA-mediated assembly, constitutionally dynamic materials capable of adapting shape or properties in response to external stimuli, metallopolymer-based materials (*e.g.* ferrocene-containing polymers), metallogels, metal–organic frameworks, surface-anchored assemblies, molecular computing, spin-crossover, light harvesting and photocatalysis.

We hope that this compilation provides the reader with a relatively comprehensive overview of metallosupramolecular materials at a very exciting and important time in their history. We believe that this class of materials has arrived at a stage where products will be translated from the laboratory to the marketplace.

RSC Smart Materials No. 15
Functional Metallosupramolecular Materials
Edited by John G. Hardy and Felix H. Schacher

Published by the Royal Society of Chemistry, www.rsc.org

We would like to thank Hans-Jörg Schneider and Mohsen Shahinpoor for the invitation to contemplate a book project on the topic of *Functional Metallosupramolecular Materials* for the Royal Society of Chemistry series on Smart Materials, and Leanne Marle and Alice Toby-Brant at the Royal Society of Chemistry for their support throughout the production of the book. We wholeheartedly and unreservedly thank the authors of the chapters for their thoughtful contributions on a variety of highly interdisciplinary topics, and hope that the readers find the chapters as interesting, accessible, educational and enjoyable as we have.

We dedicate this book to Jean-Marie Lehn, Jack Harrowfield and Ulrich S. Schubert on the occasions of their 75th, 70th and 45th birthdays, respectively. We warmly thank them for insightful discussions, inspiration and mentorship.

Felix Schacher and John Hardy

Contents

RSC Smart Materials No. 15
Functional Metallosupramolecular Materials
Edited by John G. Hardy and Felix H. Schacher

Published by the Royal Society of Chemistry, www.rsc.org

CHAPTER 1

Interaction of Metal Ions with Proteins as a Source of Inspiration for Biomimetic Materials

ANDREW M. SMITH*[a]

[a]School of Materials and Manchester Institute of Biotechnology, The University of Manchester, Oxford Road, Manchester, M13 9PL, UK
*E-mail: Andrew.smith@manchester.ac.uk

1.1 Introduction

Nature utilises proteins to fulfil a range of functions from catalysing reactions, storage and transfer of energy to acting as structural supports. Proteins are, at the most basic level, polymers made up of a specific sequence of amino acids; this is known as the primary structure. These polymer chains have the potential to form highly complex structures where the individual amino acids interact to form specific folds, such as α-helices and β-sheets, known as the protein's secondary structure. These individual secondary structure units combine to give the protein fold, or tertiary structure. Individual proteins can interact with other proteins and molecules to give the supramolecular structure, which is the quaternary structure. The information that determines how the protein folds and interacts is encoded in the

RSC Smart Materials No. 15
Functional Metallosupramolecular Materials
Edited by John G. Hardy and Felix H. Schacher

Published by the Royal Society of Chemistry, www.rsc.org

amino acid side chains and these are also responsible for the catalytic function of enzymes. The side chains of the twenty naturally occurring (canonical) amino acids cover a wide swathe of interactions including hydrophobic, hydrophilic, ionic and covalent bonds, however, there is a limit to what can be done using the canonical amino acids; as a consequence, nature has evolved to use inorganic components within the organic proteins to extend their functionality and facilitate reactions that would not be feasible using amino acids alone. For some functions, proteins have evolved to incorporate cofactors to increase the variety or efficiency of their functions. These cofactors can be organic or inorganic; here we are interested in the latter. Inorganic cofactors can be as simple as metal ions that are coordinated by amino acid side chains to complex organic non-protein groups that coordinate the metal ions. These metal ions are used for a variety of functional reasons, from aiding in the formation of the correct protein fold to forming the active site of an enzymatic reaction. The variety of metals incorporated in proteins is also very broad with a range of valences of metals such as iron, copper, zinc, magnesium, nickel, molybdenum, and manganese being seen in humans alone. The degree of variety is also increased when organic groups such as haems are included as well as cases of multiple metal ion clusters.

The first thing to bear in mind with the involvement of metal ions with proteins is that the concentration of each metal ion is maintained in cells and organelles within a specific range dependent on the cells.[1] If metal ions exceed their normal levels then they can bind to additional sites as well as cause the displacement of other metal ions, resulting either in a change in specificity and reactivity or deactivation of the protein. The concentration ranges of free divalent metal ions is roughly based on the binding affinities of the metal ions, with Zn(II) being the strongest binder, and if sufficiently high concentrations are used, it will bind to most proteins. For the purpose of the discussions here, the interactions of physiologically relevant levels will be discussed.

Metal ions are involved in an extremely wide range of proteins, with over half of all proteins being metallo-proteins[2] including all types of enzymatic reactions as well as being involved in electron transport chains. They are also involved in the formation of extracorporeal structures such as holdfast mechanisms, silks, and glues. Additionally, metal ions are also involved in disease states, such as neurodegenerative diseases involving the formation of amyloid plaques. Due to the extremely wide breadth of the involvement of metal ions, this chapter will give a general overview and pick a few examples to go into more detail.

First, we will look at metal ions in the active site of enzymes covering a few examples and cover different metal ions as well as multiple metal ion clusters. We will then move on to the use of metal-containing organic cofactors, such as haem, and the variety of functions such a structure can fulfil. We will then discuss the use of metal ions in structural proteins both in the formation of the structure and also in its interactions. Finally, we will have a look at the influence of metal ions on protein folding and misfolding and its involvement in neurodegenerative diseases as well as anti-microbial peptides.

1.2 Enzymes

Enzymes are the molecular machines of life as they catalyse an extremely wide range of reactions with varying efficiencies and specificities. There are an extremely wide range of proteins that contain metal ions; these can generally be split into two main groups, those which have inorganic cofactors, where the metal ion is directly bound by the protein side chains and those that have an organic cofactor, a non-protein group for example a haem group that coordinates the metal ions.

1.2.1 Single Metal Ion Enzymes

1.2.1.1 Zinc Dependent Enzymes

Zinc was the first of the metal ions to be recognised as essential for life[3,4] and in 1939 the first enzyme, erythrocyte carbonic anhydrase, was discovered where zinc was essential for activity.[5] There are now in the order of 3000 zinc proteins and they have functionalities across all classes of enzymes.[6] One of the major discoveries was the "zinc finger" which was first identified in the *Xenopus laevis* transcription factor IIIA, which contains nine repetitive sequences of cysteine and histidine residues that coordinate nine zinc ions in the protein.[7] The characteristic pattern of cysteine and histidine residues separated by spacers of amino acids (X) $CX_aCX_bHX_cH$ which can coordinate metal ions has allowed the identification of these small domains in many proteins.[8] The sequence forms a small elongated domain (Figure 1.1) where the Zn(II) acts as the pinning point at the base of the loop that forms the finger, hence its generic name of zinc finger. In the case of the initially identified protein, it can interact with the major groove of DNA.[9] While the initial zinc finger discovery was for DNA binding, and many of the related proteins also have this role, it has also been shown to enable binding of proteins and lipids.[10,11]

In the case of zinc finger proteins, the zinc does not have a catalytic function but a purely structural role which is nonetheless essential for the binding and hence activity of such proteins. Zn(II) is the tightest binding metal ion to proteins and it is generally accepted that there is no activation of Zn(II) enzymes, rather that they are controlled through inhibition mechanisms. Zinc is also known to bind and inhibit a number of enzymes, including zinc metalloproteases. For example, there is a secondary non-active zinc binding site in carboxypeptidase A[12] that is 3.3 Å from the zinc active site and has a K_i of 0.7 μM.[13] In the case of the human receptor protein tyrosine phosphatase β, Zn(II) has a K_i of 21 pM,[14] while for bovine dimethylarginine dimethylaminohydrolase (isoform 1), it has a K_i of 4 nM.[15] Thus enzymes such as these may be inhibited by Zn(II) at physiological conditions and only activated by removal of the Zn(II). However, the identification of such enzymes is difficult as the vast majority of enzyme studies utilise enzymes that are purified from a heterologous system rather than their host tissues.[16] As a large number of

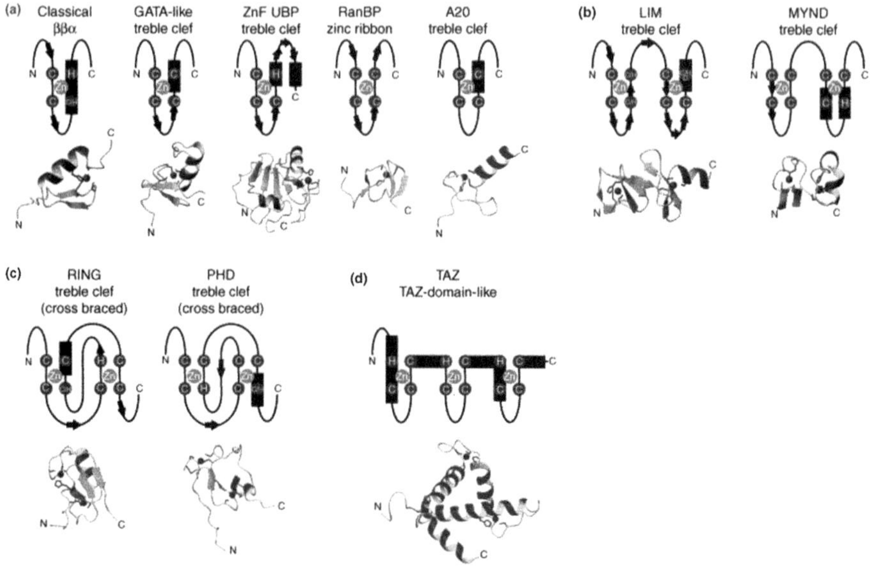

Figure 1.1 Topology and structures of zinc finger domains. Examples of the various types of zinc finger domains: (a) classical (PDB code 1FU9), GATA (1GAT), ZnF UBP (2G43), RanBP (1Q5W), A20 (2FID). (b) LIM (1A7I) and MYND (1FV6). (c) RING (1CHC) and PHD (1XWH). (d) TAZ (1R8U). Reprinted from ref. 11. Copyright 2007, with permission from Elsevier.

enzymes have catalytic active sites that incorporate cysteine, histidine, glutamic/aspartic acid, it is feasible that the active site can be occupied by a metal ion. The zinc level within cells is kept remarkably low at around 1 nM, but it is known to fluctuate within cells[17,18] due to a whole range of conditions dependent on cell type[1] and zinc ions are stored and released both intra- and inter-cellularly.[19] In pancreatic islet β cells, the zinc(II) ions participate in the formation of insulin into granules for storage and are secreted along with insulin.[20] Thus the concentrations of zinc can have a great effect on cellular activities. One cellular component that is known to be involved in the regulation of zinc is thionein or metallothionein which binds up to seven Zn(II) ions.[21] It has been shown that this thionein can efficiently remove the Zn(II) ions from inhibited enzymes such as glyceraldehyde 3-phosphate dehydrogenase, which recovers activity from inhibition within seconds of addition of thionein.[22] While it has been shown for multiple enzymes that thionein does not remove the Zn(II) ions tightly bound in the catalytic sites of zinc metalloproteins,[23,24] for example, with a 20-fold excess of thionein, alcohol dehydrogenase loses only marginal activity.[22]

Thus we can see that zinc has a wide range of functions within proteins and can act to form specific structures and as both an inhibitor and the catalytic centre of enzymes. This breadth of functions for zinc is not unique. Rather than go through each possible metal ion, we will now look at specific reactions or functions and how the metal ions are utilised by proteins.

1.2.1.2 Non-Haem Oxygenases

There are several classes of non-haem oxygenases that utilise iron in the catalytic core of the enzyme, many of which utilise additional cofactors in the reaction. One class of these enzymes are the aromatic amino acid hydroxylases (AAAHs), which have a tetrahydropterin cofactor and are involved in the synthesis of the aromatic amino acids. The original enzyme appears to have developed in prokaryotes as a phenylalanine hydroxylase, which has then evolved in eukaryotes as hydroxylases for the other two aromatic amino acids.[25] All the AAAHs feature a facial triad of two histidine residues and a glutamic acid residue, a common feature of Fe(II) non-haem oxygenases;[26] the other three ligands to Fe(II) are water, due to it being solvent exposed in mammalian AAAHs. Mutation of any of these described conserved residues to an alanine residue results in the Fe(II) being absent thus inactivating the enzyme. A study into tryptophan hydroxylase involved mutating the histidine residues to glutamate and glutamine, and the glutamate residues to histidine and glutamine respectively. This experiment showed that while in some cases the Fe(II) binding is retained, the activity of the enzyme is removed in all cases.[27] The enzymes require that all three catalytic components—pterin, amino acid substrate and dioxygen—are bound prior to the start of the catalytic reaction (Figure 1.2). The exact order of binding is unclear, but it appears that O_2 binds reversibly as the last binding step prior to reaction.[28,29]

Another non-haem iron oxygenase is the 1-aminocyclopropane-1-carboxylic acid (ACC) oxygenase, which is the last step in the synthesis of ethylene in plants. This is an important enzyme as the levels of ethylene produced by plants as a result of stress can produce deleterious effects in the plants.[30] As with the AAAHs, the Fe(II) is coordinated by two histidines and in this case an aspartate instead of a glutamate, with ascorbate and CO_2 or bicarbonate needed alongside ACC.[31–33] In this case it is proposed that ascorbic acid or ACC binds first, followed by dioxygen binding as the second step in binding.[31] The exact role of ascorbate in this reaction is unclear and it has been proposed that it binds away from the Fe(II) centre with other amino acids in the catalytic site; it has also been classed as a co-substrate and effector.[32,33] This highlights that with some enzymatic reactions, the sequence of binding of substrate, cofactors and ligands is sometimes hard to determine but is still important for the catalytic reaction.

1.2.2 Binuclear Metallohydrolases

Binuclear metal enzymes contain two metal ions that are either coordinated together or through a bridging ligand. In some cases these enzymes also contain additional metal ions or even additional metal ion active sites. We are interested in the many cases where the close proximity of the two metal ions is utilised in the enzymatic reaction that they catalyse. While it is possible that the two metal ions are the same, it is also possible to have metal centres containing different valence metal ions and also completely different metal ions.

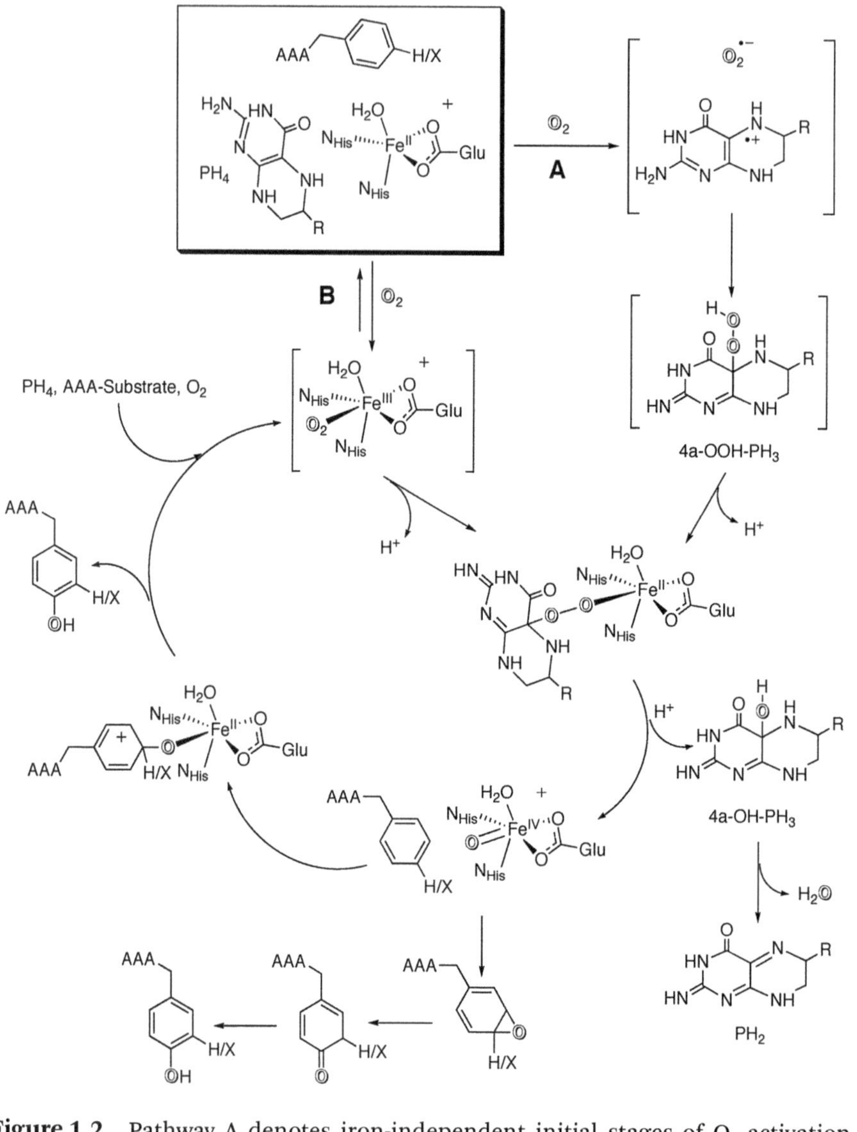

Figure 1.2 Pathway A denotes iron-independent initial stages of O_2 activation. AAA is short for aromatic amino acid, and X represents isotopically labeled hydrogen (D or T), CH_3, Br, or Cl. Reprinted with permission from ref. 25. Copyright 2005 American Chemical Society.

As a consequence of having multiple metal ions, often in different states, the catalytic reaction can become quite complex; this is further complicated by the fact that both metal sites are not always occupied throughout the catalytic cycle. There has been extensive research into various binuclear metallohydrolase groups and here we will present two examples, however, there is a large range of additional examples that we will not cover for the sake of brevity.[34]

1.2.2.1 *Purple Acid Phosphatases*

Purple acid phosphatases (PAPs) are the only examples of binuclear hydrolases where it has been shown that a heterovalent catalytic site is necessary for activity. These phosphatases have earned their name due to the characteristic colour of the enzymes. This colour comes from a conserved tyrosine ligand to a ferric ion in the active site of the protein.[35–37] These enzymes have been found in mammals,[38–46] plants,[47–52] and a limited number of micro-organisms.[53] Mammalian PAPs are approximately 35 kDa and are highly conserved with 85% identity.[54] They are proposed to have roles in iron transport, the generation of reactive oxygen species and bone resorption.[55] On the other hand, plant PAPs are mostly homodimers with a subunit mass of approximately 55 kDa, but some heterodimers have also been reported.[50,51] The homodimeric PAPs share roughly 65% identity but the identity between mammalian and plant PAPs is low.[54] In plants, PAPs are proposed to have roles in the generation of reactive oxygen species and phosphate metabolism.

The active site of PAPs consists of two metal binding sites, a trivalent and a divalent metal ion site. The trivalent site is typically occupied by Fe(III),[56] but in at least one case, that of the PAP from sweet potato, this can be replaced with Mn(II), resulting in a loss of the purple colour.[57] The trivalent metal ion is coordinated to a conserved deprotonated tyrosine residue, as well as a histidine residue and two aspartic acid residues, one of which also coordinates the other metal site. The divalent metal ion is coordinated by two histidines, an asparagine and the shared aspartate residue.[56] In mammals, the divalent metal site is Fe(II) which is essential for the functioning of the enzyme.[58,59] In plants, the divalent site is usually Zn(II) or Mn(II)[47–49,60,61] and in at least one case Fe(II), a recombinant isoform from sweet potatoes.[62] There are additional ligands to the metal ions but these vary depending on the system and the study, however, a general reaction mechanism for the PAPs has been developed from a range of spectroscopic and crystal structure studies.[63] This reaction mechanism is illustrated in Figure 1.3. The first step is substrate binding creating a hydrogen bonding network, followed by the rearrangement of the substrate so that it is coordinated to the divalent ion so that it can undergo a nucleophilic attack by a solvent-derived hydroxyl ion, or the metal ion bridging hydroxide (μ-OH), it is not clear which. Following the nucleophilic attack and hydrolysis of the substrate ester bond, the alcohol leaving group is released. The secondary product of the hydrolysis of the phosphate group remains bound to the active site. The regeneration of the active site is not clearly understood as it requires the removal of the phosphate group and potentially the regeneration of the μ-OH bridging the metal ions. One suggested mechanism, illustrated in Figure 1.3, is the coordination of a water molecule to the Fe(III) centre and then the regeneration of the μ-OH bridge, which would weaken the bonding of the phosphate leading to its release and return to the resting state.[64] There are two nucleophiles in the overall reaction as determined by the sequential cleavage of the ester bonds in a diester substrate[65] so that both are used but it is not clear in which order.

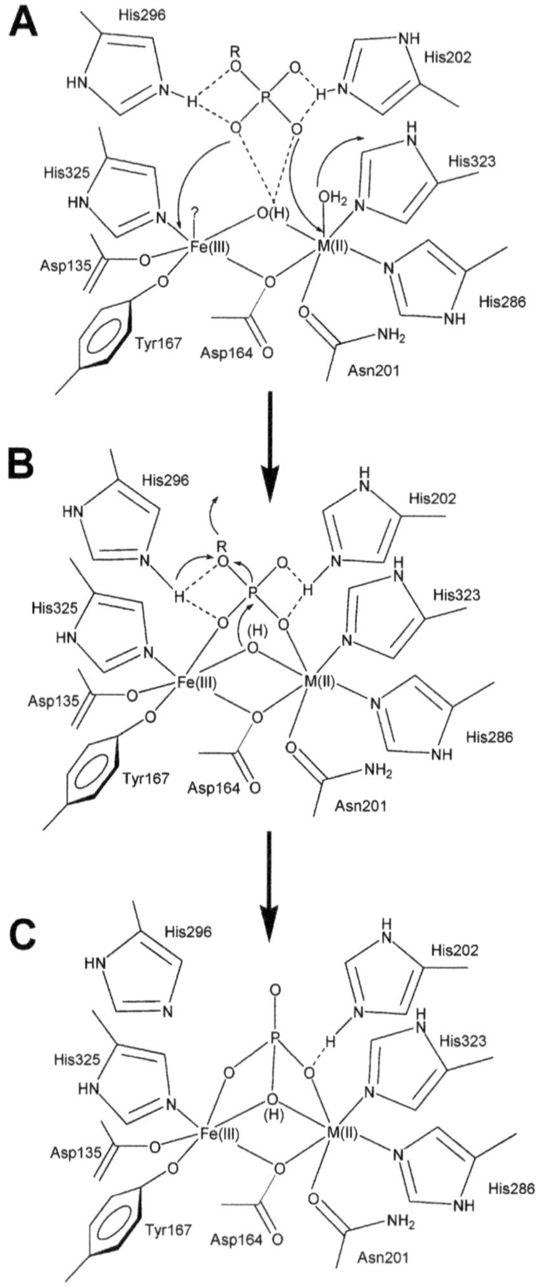

Figure 1.3 Proposed mechanism of hydrolysis by PAPs. (A) Substrate binding; (B) nucleophilic attack and release of leaving group; (C) regeneration of the active site; phosphate is replaced by two or three water molecules. Reprinted with permission from ref. 34. Copyright 2006 American Chemical Society.

1.2.2.2 *Metallo-β-lactamases*

β-Lactams are one of the most important groups of antibiotics that are currently available to deal with bacterial infections. However, over time bacteria have evolved that possess resistance to this class of antibiotics. This resistance has been achieved through a number of methods, the most important of which has been through the evolution of enzymes that can hydrolyse these antibiotics, the β-lactamases. Three of the four classes of β-lactamases, possess an essential serine residue in the catalytic centre. However, based on sequence analysis, the fourth class, class B β-lactamases, have an ion centre that requires either one or two Zn(II) ions bound in the active site to be functional and a αββα fold.[66] These are termed the metallo-β-lactamases (MβLs); they share a common fold and a small amount of sequence homology, but have been split up into three subclasses.[67] The MβLs are important as within their group they are resistant to all β-lactamase inhibitors[68] and are capable of cleaving all β-lactam antibiotics including the carbapenems which are regarded as the antibiotic of last resort.[69,70] B1 is the largest sub-class and can be coded either on chromosomes or on mobile elements that enable them to disseminate easily.[71] β-Lactamases in class B1 and B3 have very broad specificity being able to inactivate all bicyclic β-lactams,[72–74] while B2 β-lactamases are all known to be carbapenemases. B1 and B3 type MβLs have three histidine residues coordinating the first Zn(II) site, Zn1, while the B2 type MβLs have replaced one of the histidine residues with an asparagine (Figure 1.4). The second metal binding site is conserved between the B1 and B2 type MβLs consisting of an aspartate, histidine and a cysteine residue, while the B3 type MβLs replace the cysteine with another histidine residue.[67] In the case of the B2 type MβLs, the second metal site is not occupied in the catalytic form of the protein[75] (Figure 1.4). However, it is not clear whether this is also the case for the other types of MβLs.[76] The role of the second zinc ion in the reactivity of the MβLs is also not clear to the extent that researchers are uncertain

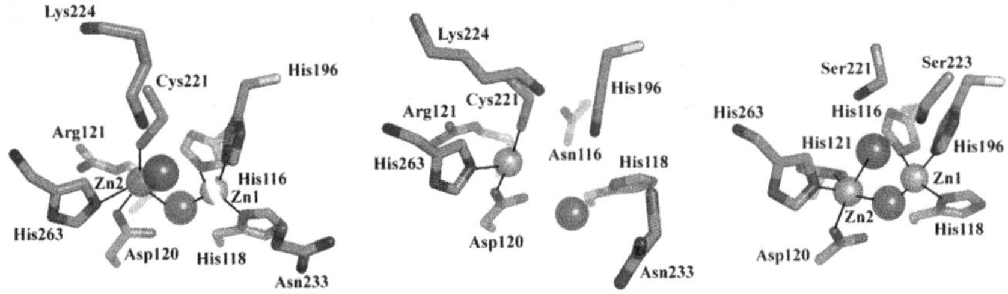

Figure 1.4 Metallo-β-lactamase active sites: *B. cereus* BcII (B1, left), *A. hydrophila* CphA (B2, centre), and *S. maltophilia* L1 (B3, right). Zinc ions are rendered as light grey spheres, and water molecules are rendered as dark grey spheres. Coordination bonds are shown as solid lines. Adapted with permission from ref. 66. Copyright 2006 American Chemical Society.

whether it is completely necessary or not. As there are differences in the metal centres of the subtypes of these enzymes, it is not completely clear how the exact reactive mechanism of these enzymes functions and several slightly different mechanisms for the reactivity of these proteins have been suggested.[76] The enzymes are a subclass of the metallo-proteases and thus their reaction mechanisms are all centred around the cleavage of the C–N bond in the β-lactam.[77,78] The mechanisms described for these enzymes have looked at both mono- and bi-nuclear enzymatic reactions. For B1 class MβLs, it has been proposed that with a single Zn(II), the reaction mechanism is similar to single Zn(II) peptidases where, after substrate binding, a deprotonated zinc-bound water molecule attacks the carbonyl centre forming a negatively charged tetrahedral intermediate, which is stabilised through interactions with the Zn(II). The aspartate can then donate a hydrogen to the nitrogen resulting in the cleavage of the C–N bond.[79] The exact order of these last steps in the breakage of the C–N bond is not clear, and it may be that the bond is cleaved before the hydrogen is donated by the aspartate residue. Whether the second zinc ion is necessary is also unclear; some studies have shown that at physiological conditions, the second Zn site is only partially occupied and also that it is not always necessary for the highest reactivity,[80] but this is dependent on the study.[81,82]

1.2.3 Metal-Containing Organic Cofactors

A number of proteins utilise metal ions within an organic chelator such as a haem group. This permits a wider range of electronic transitions for the metal ions and is important for forming electron transport chains, many components of which contain metal ions. Obvious examples are the haem groups used in the transport of oxygen molecules around most animals and also cytochrome c, where they act as electron transporters. A similar system is also present in plants for the photosynthetic system to convert light into chemical energy.

1.2.3.1 Oxygen-Carrying Haem-Based Proteins

Animals require a system for the transport of oxygen around their bodies due to the slow diffusion of oxygen through tissue and its poor solubility in blood plasma. As such, organisms have evolved a few oxygen-carrying proteins. In mammals, these are myoglobin and haemoglobin, which utilise bound Fe(II) as the oxygen transporter. Some invertebrates utilise haemocyanin, a Cu(II)-based protein, which is blue when it binds oxygen. For this discussion, we will concentrate on the mammalian systems of myoglobin and haemoglobin. Myoglobin and each of the subunits of haemoglobin contain a non-covalently bound haem group. The haem group in these systems is protoporphyrin IX which is a heterocyclic ring system containing four pyrrole rings linked by methane bridges, with four methyl, two propionate, and two vinyl substituents on the outer ring. The central Fe(II) is coordinated by the four nitrogen atoms of the pyrroles and a nitrogen from the side chain of a histidine residue, which form a square pyramid arrangement. O_2 binds to the Fe(II) on the opposite side to the histidine residue making the Fe(II) octahedrally coordinated. While the oxidation state of the Fe(II) remains the same on oxygen binding, its electronic state changes resulting in a change in absorbance maximum from a dark purple to a scarlet colour on oxygen binding. While O_2 is the desired binding partner for the Fe(II) centre, within the porphyrin ring, it can be replaced with a range of other small molecules such as CO_2, NO and H_2S, which is why these compounds are highly toxic as they prevent the transport of oxygen.

While myoglobin and the 4 subunit haemoglobin have the same porphyrin ring, their oxygen binding properties are quite distinct. Myoglobin has a hyperbolic binding curve and binds oxygen at a lower partial pressure than haemoglobin, which also means that oxygen does not dissociate from myoglobin within the normal pO_2 pressures found in both arterial and venous blood. In comparison, haemoglobin has a sigmoidal binding curve for oxygen, which over the variation in oxygen content of arterial and venous blood binds and then releases the oxygen, respectively. As a consequence, the two proteins act in sync, ensuring oxygen uptake in the lungs and release in low oxygen areas of the body where it is needed. The sigmoidal binding curve for haemoglobin is an indicator of the cooperative binding of oxygen by the four subunits of haemoglobin.

While it is the porphyrin group that binds the oxygen, the surrounding protein is key to its properties, as implied by the difference in binding properties of the two proteins, as the side chains surrounding the ring influence the binding. The porphyrin ring alone can bind oxygen, however, this is not reversible as the oxidation state of the Fe(II) shifts to Fe(III) through an intermediate with an O_2 molecule bridging two iron centres from two porphyrins. This can be prevented in synthetic porphyrins by creating a picket fence of chemical groups which prevents the two porphyrin rings from approaching close enough together to form the bridging intermediate. Thus, in nature, the proteins act to hold the porphyrins apart (this is especially true with haemoglobin as it contains four porphyrin groups) and make O_2 binding reversible.

Myoglobin has eight helices that are mostly α-helical with the haem group in a hydrophobic pocket. Upon binding oxygen, the only structural change is the binding of a histidine residue to the opposite side of the oxygen molecule to the Fe(II).

Haemoglobin, on the other hand, consists of two copies of two subunits (α and β), which are symmetrically related by a twofold rotation; the subunits are structurally related to one another and myoglobin. The subunits are arranged in a tetrahedron with extensive interactions between dissimilar subunits *i.e.* α_1–β_1, most of which are hydrophobic in nature; the few interactions between similar subunits *i.e.* β_1–β_1 are mostly polar. Binding of oxygen to one of the subunits causes extensive rearrangement of the subunit, and there is evidence of two distinct states t and r, the deoxy and oxy versions of the subunit, T and R being the state of the whole protein. In the t state,

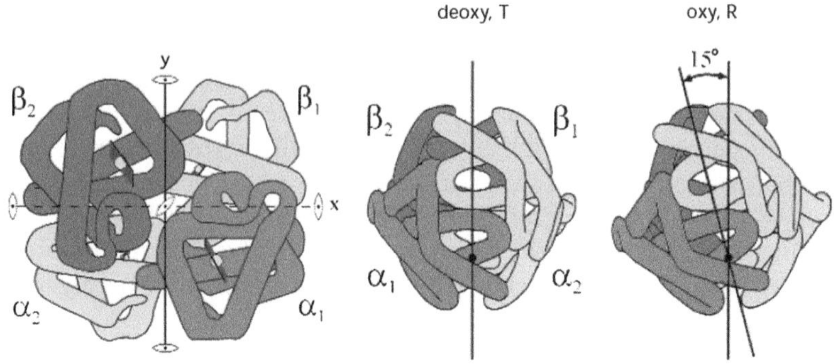

Figure 1.5 Schematic structures of haemoglobin. The two views on the right are looking down the pseudo two-fold x axis. The quaternary conformational change consists of a rotation of the symmetrically related ab dimers by approx. 15° relative to each other and a translation of approx. 0.1 nm along the rotation axis. Modified from W. A. Eaton, E. R. Henry, J. Hofrichter and A. Mozzarelli, *Nat. Struct. Biol.*, 1999, **6**, 351–358.[83]

the proximal histidine causes the porphyrin structure to adopt a pyramidal structure. Upon binding O_2, the pyramidal structure is flattened and the proximal histidine residue moves; with the β subunits, one of the residues in the fifth helix must move aside to allow O_2 to bind, which all result in a shift in the structure upon binding. As mentioned, in both the t and the r states, the subunits have tight interactions of structural components to each other, which means that the subunits exist as only one or the other state, but all the subunits must be in the same state. It is proposed that when the t state binds O_2 the subunit comes under strain as it cannot switch unless the other subunits also switch. As the t state has lower affinity for O_2 than the r state, binding is slow until the protein comes under enough strain for the entire protein to switch to the R state which has a 15° rotation in the overall structure (Figure 1.5). The r state subunits that have not already bound oxygen have a higher affinity for oxygen, which helps explain the sigmoidal binding curve of haemoglobin.[83]

Thus, the sigmoidal binding curve of haemoglobin is due to two hyperbolic curves—one of poor oxygen binding: the T state, and one of good oxygen binding: the R state. The T state is the more favourable in the absence of oxygen due to the formation of salt bridges between the subunits, while the R state is more favourable when all the iron centres have bound oxygen.

1.2.3.2 *Cytochromes*

Cytochromes are redox-active proteins that contain haem groups that can reversible alternate between Fe(II) and Fe(III) oxidation states; in this way, they can act as components in electron transport systems. As such they occur in the vast majority of organisms except for a few obligate anaerobes. In eukaryotes, there are several cytochromes within the mitochondrial electron transport system, cytochrome a and b are part of the membrane bound Complex III and IV respectively while cytochrome c is a peripheral membrane protein that is on the outer surface of the inner mitochondrial membrane. There are a large number of other proteins and cofactors involved in the mitochondrial electron transport system and its full description goes beyond the remit of this chapter.[84] However, the important factor is that the iron centres of the haem groups of the different cytochromes have different reduction potentials, which is essential if as much useful energy as possible is to be harvested from this energy system.

C-type cytochromes are a class of enzyme that contain a covalently-bound haem group. The haem group contains an iron ion which can be either as Fe(II) or Fe(III), which is coordinated by four nitrogen atoms of the protoporphyrin IX ring and the nitrogen of a histidine and the sulfur of a methionine. The protoporphyrin structure is covalently linked to the protein by two thioether bonds linking two of what were the haem vinyl groups to two cysteine residues. Thus, in cytochrome c of eukaryotes, these residues are invariant: Cys 14, Cys 17, His 18 and Met 80.[85] Additionally, there are a series of highly conserved residues that line the pocket of the haem binding site, as well as a ring of lysine residues around the exposed edge of the haem group. The latter acts as an interaction site for the two reaction partners, cytochrome c reductase and cytochrome c oxidase, which it shuttles between as part of the mitochondrial electron transport system.[85] The protein structure of cytochrome c acts as a cage to prevent the highly reactive reduced haem unit from reacting non-specifically. The exact process by which electrons are transferred to and from the iron centre is still not fully elucidated due to the fact that the proteins involved commonly contain multiple metal centres,[86] for example in the case of cytochrome c oxidase there are two copper centres as well as two haem groups within the protein.[86,87]

While prokaryotes do not possess cytochrome c they do have c-type cytochromes that bind haem in the same fashion but have differing primary sequences as well as variations in the secondary structure of the proteins. Genetic studies have shown that there are hundreds of genes that encode proteins that are potentially type-c cytochromes and that some organisms have multiple examples; these proteins can also have multiple haem sites.[88] Bacterial multi-haem cytochrome c are involved in a wide range of reactions related to energy metabolism, and as bacteria utilise a variety of energy sources, these proteins have evolved to use a wide range of substrates. There are many examples that utilise either nitrogen- or sulfur-based compounds.[89]

Another major group of cytochromes are the P450 enzymes, which catalyse a diverse range of oxidative reactions and are present in most organisms and over 14 000 genomic sequences have been identified.[90] These enzymes utilise O_2 and two electrons from either nicotinamide adenine dinucleotide (NADH) or nicotinamide adenine dinucleotide phosphate (NADPH) to catalyse their reactions with various cofactor mediators. Cytochrome P450s contain an iron protoporphyrin IX which when reduced has a strong affinity for O_2 but a stronger affinity for CO which results in the characteristic absorbance at 450 nm.[91]

The iron centre is coordinated by the four nitrogens of the protoporphyrin ring as well as a deprotonated cysteine residue[92] while the final coordination position is occupied by a displaceable water molecule.[93] Cytochrome P450 acts to generate a single oxygen atom from O_2 to be inserted into the substrate of the catalytic reaction. The general reaction process is based on bacterial enzymes but mammalian enzymes are believed to react in a similar fashion. The resting state of the reaction has Fe(III) in the haem, in the first reactive step the substrate binds, shifting the valence state to Fe(II) with NAD(P)H supplying an electron *via* cofactors.[94,95] In the second step, O_2 is bound to the Fe(II) transfer. In the third step, a second electron is supplied and a supernucleophilic dinegatively charged intermediate is formed. One of these negative charges is on the O_2, the second is delocalised over the cysteine ligand. In step four, a proton is added to the O_2 to reduce it, leaving the delocalized negative charge over the cysteine ligand. In the fifth step a second proton is added to the bound O_2 to form a transient Fe(II)–O–H_2O adduct which immediately forms H_2O and a porphyrin Fe(IV)O intermediate.[92] In the sixth step this activated oxygen is transferred to the substrate forming the mono-oxygenated product and regenerating the Fe(III) resting state. It is possible for this reaction to become decoupled and instead of adding an oxygen atom to a substrate, a superoxide anion can be formed, which can then become H_2O_2 and then also hydroxyl radicals. The range of reactions undertaken by cytochrome P450s is extensive and beyond the scope of this chapter.[96]

1.2.4 Applications of Metallo-Enzymes

The enzymes mentioned in the previous sections are all described in the context of their natural environment and function. Obviously, the catalytic capabilities of enzymes make them highly interesting candidates for use in the production of organic compounds[97,98] and as sensors for the detection of compounds.[99–101] The modification of enzymes to use different substrates, or the combination of different enzymes to create a catalytic pathway to make new organic molecules, is also of great interest with a wide range of enzymes having been studied in this context.[102,103] There are also many examples of metallo-enzymes being modified—the most common is the catalytic site of cytochrome P450 enzymes to improve their stability or alter their selectivity.[104] Also there has been an interest in altering the binding properties of zinc finger proteins to different DNA sequences to allow more control in the manipulation of DNA sequences.[105] For the development of applications it has often been necessary to immobilise the enzymes onto surfaces so that they can be utilised more efficiently.[100,106–108] For example, cytochrome P450 3A4/5 which is involved in drug metabolism has been adsorbed to indium tin oxide film electrodes to act as a sensor for drug molecule metabolism.[109] Additionally, it has been discovered that the immobilisation of enzymes tends to alter their stability to degradation as well as alter, usually an improvement, their catalytic properties.[110]

1.3 Structural Proteins

1.3.1 Silks

Silks are extracorporeal proteinaceous fibres that are extruded by a number of organisms, mostly from the phylum arthropoda. The ability to produce silk has evolved multiple times within this phylum;[111] as a result, the proteins utilised to form the silk can be quite diverse. However, the vast majority rely on the formation of β-sheets and in all except the underwater silks, the removal of water from the forming silk to make a hardened composite polymer. There has been interest in silk for thousands of years due to the extraordinary properties of silks, some of which compete with modern man-made fibres. As silks have evolved multiple times and involve different proteins we will only concentrate on the most heavily characterised systems, namely the silk from *Bombyx mori*, more commonly referred to as the silkworm and the silk produced by spiders. *B. mori* are moths where the pupae produce a silk cocoon that they use as protection while they metamorphose into the moth. This species of moth has been harvested and cultivated over thousands of years[112] for the production of silk in China and is the source of most of the silk used in clothes and is now being developed as a biomaterial.[113,114] As a consequence, the species is highly reliant on leaves from mulberry trees as a food source, and effectively reliant on humans for species survival due to farming. Due to this, there have been extensive studies into this silk to understand how it is produced and how the constituent proteins convert from a solution to a hard dry fibre.

Silk from *B. mori* mostly consists of two proteins: the heavy (391 kD) and the light chain (28 kD) that are produced in a fixed ratio with 85% of degummed silk being the heavy chain.[115] The heavy chain consists of hydrophobic repetitive motif sequences rich in glycine and alanine; there are also 11 long hydrophilic sequences which are regularly interspersed between the hydrophobic motifs.[115]

Silk threads are produced from specific glands where a spinning dope is stored as a concentrated solution to be converted to the silk thread on demand. As such the proteins must be stored as a solution and be converted to a dry material *via* a highly controlled process. Silk from *B. mori* exists in two states: silk I, which is helical in structure and silk II which is rich in β-sheets, this latter form being the dominant structural form in threads of silk.

Another class of silks that have been extensively studied are those produced by spiders, which can produce up to seven different types of silk dependent on species and sex; these different silks have various properties related to the disparate functions of the materials. Most research on spider silk has concentrated on major ampullate (MA) silk which is also known as dragline silk and is produced by spiders as a safety line behind them should they fall off of a surface. As a result it is relatively easy to harvest this silk and characterise it, which has shown that it is an extremely strong material with properties that challenge man-made fibres. Most spider silk proteins consist of a central repetitive core region flanked by α-helical N and C-terminal domains. The central repetitive core is made up of short motifs that vary between silk proteins. In MA silk there are primarily motifs of A_8, GGX, GPGXX where the amino acids X tend to be charged or hydrophilic amino acids, with a repeat being made up of the A_8 motif followed by several copies of the other two motifs.

A factor that has to be borne in mind with the physical properties of silks is that while being heavily dependent on the proteins involved, they can also be affected by a wide range of factors from temperature and humidity, to reeling speed, and diet of the organism producing the silk.[116–122] In the case of both silkworm silk and MA silk from spiders, these factors include, but are not limited to, shear stress, removal of water, extensional flow, heating,[123] changes in pH[124] and metal ions.[125–128] These factors allow the proteins in the solution to switch state and rapidly convert from a free flowing solution to a hardened material. In the case of spiders it has been shown that the pH and ion content of the spinning dope changes composition as it passes along the spinning duct, thus it is highly likely that metal ions also have an impact on the spinning process.[116] In the case of silkworms, analysis of the spinning dope at various points along the spinning gland shows that potassium and calcium are the most abundant metal ions present, with potassium, sodium, magnesium, copper, and zinc increasing in quantity from the posterior to the anterior of the spinning gland, the direction of secretion, with copper and zinc increasing sharply at the end of the gland.[127] Raman and NMR spectroscopy studies of silk from *B. mori* show that addition of Cu(II) results in a shift of the spectra from a silk I to a silk II structure with increasing amounts.[127,129] It was initially proposed that a section of the heavy chain which contains a histidine residue is the most likely binding site for Cu(II),[129] however, more recent quantitative measurements indicate that 6 or 7 Cu(II) ions can be coordinated by a single heavy chain and therefore coordination is possibly through histidine, tryptophan, tyrosine and serine residues.[130] The coordination of metal ions is through pH-sensitive residues such as the nitrogen within the imidazole ring of histidine, thus it is not surprising that metal ion binding is pH-dependent as in the case of silk. A study into Cu(II), Co(II), and Cd(II) binding to two silk fibroins from two different species of silkworm showed that metal ion binding was stronger towards pH 7, but slightly different between the different metal ions. Interestingly, the silk fibroin that had a higher content of histidine and tryptophan bound more metal ions supporting the notion that these residues are important in metal binding for silks.[131]

1.3.2 Mussel Byssus

A number of aquatic organisms live on the sea floor and in tidal zones of seas and survive by attaching themselves to rocks on the sea floor with anchors so that the tide does not cause them to be washed away. With the advent of man-made structures from other materials, such as wood, metal and fibre glass, these organisms also successfully attach themselves to these surfaces as well. One such group of organisms are mussels, which attach to the surface by forming a byssus (Figure 1.6). These are protein fibres that are moulded by the foot of the mussel with an anchor point on the desired surface and a long protein strand that attaches to their body. The thread has an outer protective layer surrounding bundles of collagen-like fibres which are embedded in a protein matrix. The thread has two distinct regions: a straight distal region and a crumpled proximal region.[132] The collagen fibres are made up

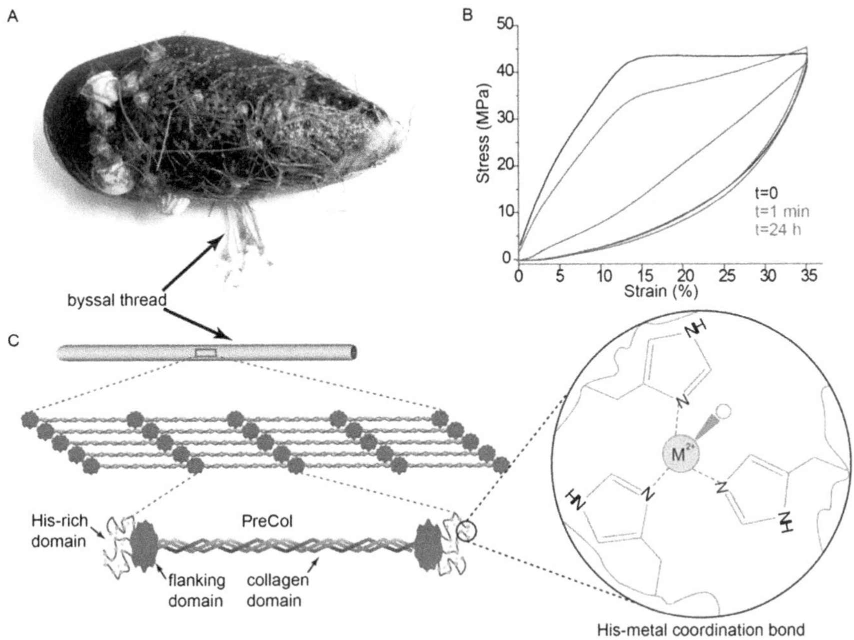

Figure 1.6 Mussel byssus structure and mechanics. (A) Mussels use a proteinaceous attachment holdfast, the byssus, to remain sessile in marine environments. (B) Distal region of the byssal thread (outside the shell) is stiff and capable of 70% hysteresis during cyclic loading. Damage caused by material yield can be healed over time. (C) Threads are composed of a semi-crystalline arrangement of protein building blocks called preCols with a primary structure reminiscent of a block copolymer possessing a large central triple helical collagen domain, a flanking domain composed of tightly-folded domains with β-sheet conformation, and at each terminus, domains with approx. 20 mol% histidine. The histidine residues are believed to be coordinated to metal ions such as Zn(II), Cu(II), and Ni(II), which behave as reversible sacrificial bonds under load-bearing conditions contributing to self-healing behaviour. The structure and geometry of the His–M(II) complex in the byssus is as of yet undetermined; thus, the structure presented in the schematic is based on that determined for marine worm mandibles with a similar protein sequence.[140] Reprinted with permission from ref. 137. Copyright 2014 American Chemical Society.

of three variant proteins called preCols, these are the distal (D), proximal (P) and non-gradient (NG) preCols. PreCol D and P are dominant in the distal and proximal ends of the thread respectively, and decrease gradually along the thread towards the opposite end. The third preCol NG is present throughout the length of the thread.[133] The preCols have a central collagen region which contains various deletions in the canonical repeat. The collagen region is flanked by regions that contain repetitive motifs seen in other proteins (Figure 1.6C). PreCol D has flanking regions that resemble spider dragline silk sequences, preCol P has flanking regions that resemble the sequence of elastin, and preCol NG has sequences that are glycine rich and have been proposed to be similar to plant cell wall proteins.[132] The mechanical properties of the flanking regions of preCols D and P are proposed to impart strength and elasticity, respectively, to the thread based on the properties of the related proteins. The gradient of proteins and properties related to them are important as when a hard and soft material are attached to one another, it is necessary to be able to dissipate any forces applied along the thread gradually otherwise a sharp change in the materials' properties will result in breakage at that point.[133] This is very important for organisms such as mussels in the tidal zones where they regularly experience strong forces applied by wave motion.[134] Both ends of the preCols after the flanks have histidine rich regions containing DOPA (L-3,4-dihydroxyphenylalanine) residues, which are a translational modification of tyrosine.[135,136] The histidine rich regions have been proposed to coordinate Zn(II) and Cu(II) ions based on the fact that groupings of histidine residues are known to bind these ions.[132,137] It, however, has also been shown that mussels utilise whichever divalent metal ions they have absorbed into their bodies to form the metal interactions within the byssus.[138] As the preCols form aligned collagen bundles, the coordination of metal ions by the histidine residues at the ends of each bundle ensure the linkage along the thread structure (Figure 1.6C). This has been shown by treatment of threads with EDTA to remove the metal ions from the structure; as a consequence, the threads had 45% lower Young's modulus.[139] It is assumed that the metal ions are coordinated by three histidine residues and a chloride ion as seen for the high zinc content in the jaws of marine worms of the *Nereis* species.[140] The presence of DOPA residues in these regions would indicate that covalent crosslinks are formed between these highly reactive modified tyrosine residues.[135,136] The prevalent use of non-covalent crosslinking of these proteins in combination with covalent crosslinking is proposed to impart the important self-healing properties of these threads. Applying repeated forces to these threads to just below breaking point steadily reduces the modulus of the threads presumably due to the steady breakage of non-covalent bonds.[137,141] As the non-covalent bonds are weaker, they will be the initial breakage points which can act to reduce the stress being applied to the covalent bonds, and due to the number of non-covalent bonds, they can maintain the integrity of the thread. If the threads are then left for a period of time and the modulus of the threads is measured again, they are found to have recovered most if not all of the original modulus due to the non-covalent bonds reforming over time (Figure 1.6B).[137] Thus the reliance on non-covalent bonds between proteins within the thread allows the byssus to be reused after each tidal event, in the case of mussels, without the need to waste resources on making new holdfasts.

While the preCols form the core of the thread, there are a number of additional proteins that form a matrix around the preCol bundles and also form the plaque that attaches to a surface.[142] There are multiple proteins in the mussel foot and attachment plaque, with the exact function of many unknown, however, it is clear that there is a high prevalence of the DOPA residue in each of these proteins.[143] One of the proteins in the plaque is mussel foot protein 1 (Mfp-1) which consists mostly of tandem repeats of a positively charged decapeptide [AKPSYO*OTY*K], where O, O*, and Y* are *trans*-4-hydroxyproline, *trans*-2,3, *cis*-3,4-dihydroxyproline, and DOPA, respectively.[144] The DOPA side chains can form tris- and bis-catecholate complexes in solution with Fe(III).[145,146] There are a number of other mussel foot proteins; Mfp-3 and Mfp-5 are the two that appear to be predominate in the plaque region of the foot, which have in the region of 20–30 mol% of DOPA residues in their sequences.[147] Analysis of protein precursor extracts that would form the plaque indicate that the hardening of the plaque is reliant on the presence of Fe(III) with other metal ions having little or no effect.[148] Over short time scales of below about 100 minutes, the Fe(III) acts as a non-covalent linker, but over longer periods it contributes to the oxidation of the DOPA side chains to one another to form covalent cross links.[149] While both Mfp-1 and Mfp-3 contain relatively high levels of DOPA, it appears that Mfp-1 can crosslink to itself as shown by examining the forces preventing separation of two mica sheets coated with the respective protein.[147] This highlights that these complexes can be between proteins rather than within proteins; as a result, the presence of these metal binding sites acts to glue the proteins together and hardens them to form the plaque which is in contact with a hard surface anchoring the mussel to the surface.

1.4 Metal-Induced Protein Misfolding

As described in the previous sections, metal ions are important for the function of a wide range of proteins and are essential elements for life.[3] While metal ions are essential for some proteins, they can also cause problems in other cases as they have been identified to be involved in the misfolding of proteins in a number of diseases.

Amyloid plaques are a component of a number of neurodegenerative diseases and consist of dense networks of fibres formed from misfolded proteins. The development and role of these amyloid plaques are still being studied but it is clear that the components along the pathway to the fibrils, including cleavage products, misfolded monomers and oligomers, may be responsible for the deleterious effects of the disease and that the amyloid plaques may be a storage solution to prevent damage. It is also proposed that normally the amyloid fibres that form the plaques would be recycled *via* various degradation pathways. As a consequence, amyloid-based diseases are highly complex protein diseases that can involve protein mutations in both the amyloid protein as well as the ancillary proteins linked to its processing. Also it is clear that these diseases tend to be age-related which brings with it a host of possible issues related to long term build-up of components and systems that are failing.[150]

In the case of Alzheimer's and the Aβ protein, it has become clear that metal ions are involved in this process. The exact roles that metal ions play, and which metal ions are specifically involved in the oligomerisation process is still not completely clear, but it is accepted that some do have an effect. Some research suggests that Cu(II) has a role as it acts to sequester the amyloid into plaques preventing the build-up of toxic species,[151,152] while others suggest that it triggers the formation of toxic species,[153,154] however, the main issue here is the difficulty in identifying the toxic species.[155]

Cu(II) ions have been shown to promote the misfolding of the Aβ as they bind to the N-terminal end of the Aβ protein residues 1–16[155–158] where it is coordinated by at least two histidine residues. It has also been shown that the protein can bind Zn(II) and Cu(II) through a tyrosine and two glutamic acid residues,[156,159,160] to a lesser degree.[153] It has been proposed that the binding of metal ions occurs in the monomeric protein causing it to adopt a more stable structure rather than acting as a bridging molecule to promote the dimerization of the proteins.[153,161] As well as Cu(II) ions, other metal ions such as Zn(II),[153,162–167] Fe(II) and (III),[164–167] have also been identified as present in amyloid plaques and interacting with the Aβ protein.

Several studies have looked into the effect of metal ions on the degradation of the formed amyloid fibres and found that different metal ions offer differing protease resistance. It has been shown that fibres formed in the presence of Zn(II) have resistance to the proteases that are capable of degrading amyloid fibrils while those formed in the presence of Cu(II) are still susceptible to degradation.[168] This further highlights the complexity of the system with clearly different binding and folding properties imparted by Zn(II) and Cu(II).

Metal ions are clearly involved in the direct formation of amyloid, but it has also been proposed that it has an additional indirect effect. As has been highlighted throughout this chapter, metal ions are important in the catalytic centres of a wide range of enzymes including proteases, and metal ions, especially Zn(II), can act as inhibitors to enzymes as well. It has been shown that there are elevated levels of both Cu(II) and Zn(II) up to 400 μM and 1 mM respectively, in amyloid deposits.[169] These high concentrations of metal ions are likely to have a direct effect on the function of any proteases involved in the degradation of amyloid plaques as replacement of the ion in the catalytic centre of proteases with another is known to alter its activity.[170] Additionally, as discussed previously, metal ions can act as an inhibitor to enzymes. Studies with insulin degrading enzyme (IDE), a Zn(II) protease which can degrade amyloid,[171] have shown that Cu(II) can displace the Zn(II) at the catalytic centre causing the inhibition of the reaction.[172,173]

While amyloid structures are related to disease states, the structures have inspired the development of materials.[174] The ease with which it is possible to make similar structures from the self-assembly of related proteins to, often more defined, fibrous structures *in vitro* has allowed a wide range of designed sequences to incorporate functional groups. For example, it has been possible to decorate amyloid fibres with cytochrome P450 enzymes,[175] as well as use them as a scaffold to create nanowires through the decoration of the fibres with metal ions.[176–178]

1.5 Antimicrobial Peptides

There is a wide range of naturally occurring peptides that are produced by organisms for the purpose of defence against pathogens. Higher organisms have evolved a selection of antimicrobial peptides (AMPs) as part of their innate immune system. As such there has been interest in exploiting and further promoting these peptides for their intended role of preventing bacterial infections. One source of these peptides is saliva, which contains over 200 different proteins and peptides, many of which come about due to proteolysis and can be split up into a number of classes. One class is the metal binding histatins. These peptides are histidine rich (18–28 mol%) and are present at the micromolar concentration range in saliva.[179,180] There are three main histatins: -1, -3,[181] and -5[182] which contain 38, 32 and 34 amino acids, respectively. Histatins 4–12 are proteolytic cleavage products of histatin-3 with histatin-5 being the major cleavage product.[183] Both histatin-3 and -5 also present high anti-fungal properties[184–186] and thus have been extensively characterised. Structural studies indicate that histatin-5 adopts a helical structure[187,188] (Figure 1.7) and as most of the histatins are related in sequence, it is likely that they also adopt a similar structure. As histatins are rich in histidine, it is not surprising that there is the potential for these peptides to bind metal ions. Saliva itself contains high levels of calcium[189] but also lower but variable levels of other divalent ions such as zinc, magnesium and copper.[190] Structural studies of histatin-5 have shown that there are two metal binding sites. The first is at the amino terminus and coordinates both Cu(II) and Ni(II) through the amino terminus, a histidine residue at position 3 and two backbone amines.[191] This is referred to as an ATCUN binding site as it is seen in a number of other proteins including albumins, where it is proposed to act as a specific Cu(II) and Ni(II) transporter.[192] The second binding site is a Zn(II) binding site with the motif HEXXH, a motif seen in Zn(II) proteases, and is related to the ability of the histatin to fuse negatively charged vesicle membranes. In the absence of Zn(II) or in the presence of other divalent cations, histatin-5 cannot fuse vesicles, while in the presence of Zn(II), there is rapid fusion of vesicles.[193] Circular dichroism studies show that while it has a random coil structure in water, when specifically in the presence of Zn(II) and negatively charged vesicles, histatin-5 switches to a helical structure.[193] While it may be expected that the anti-microbial activity of these peptides would be related to their ability to insert into membranes and form pores,

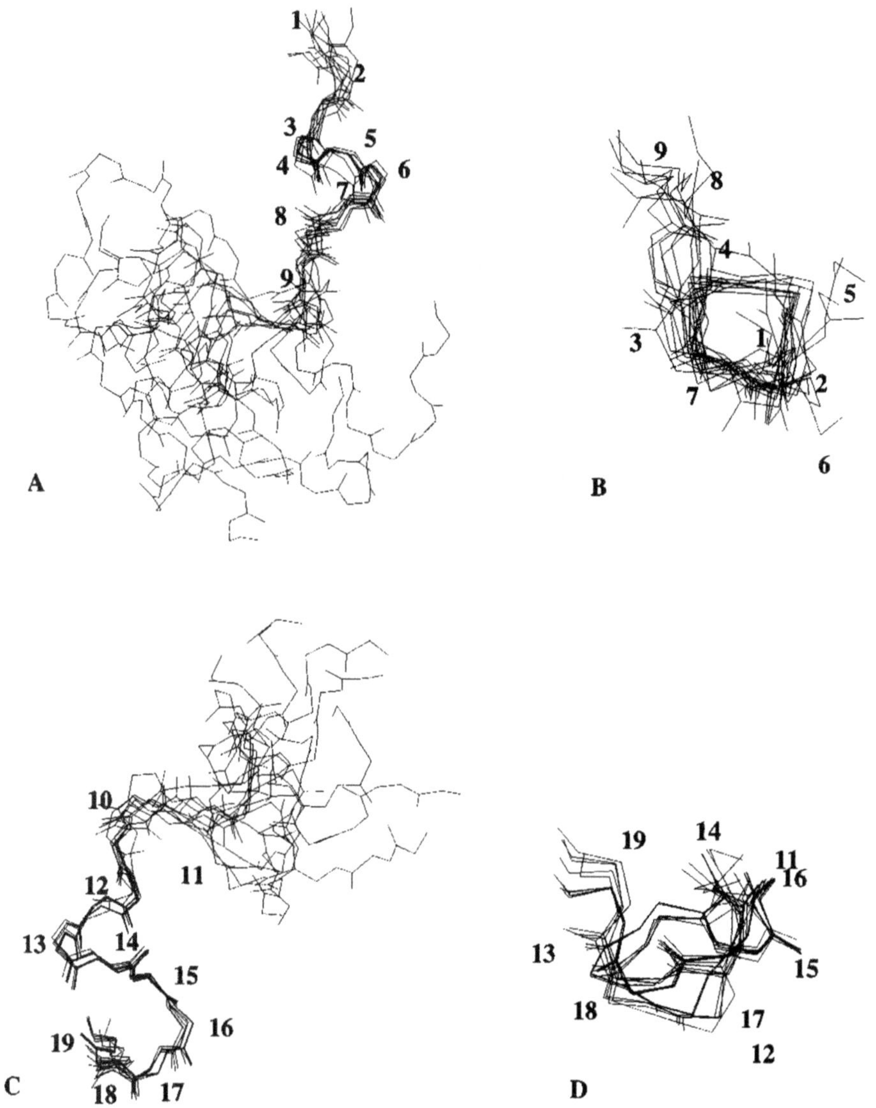

Figure 1.7 Superposition of converged backbone conformations of histatin-5 (RMSD ≤ Å), deduced from restrained molecular simulation studies, using NOE data. Family of eight conformations superimposed from residues 2 to 9 of histatin-5 is reported in (A) and (B) (edge view and top view, respectively); family of 10 conformations superimposed from residue 10 to residue 19 is reported in (C) and (D) (edge view and top view, respectively). Reprinted with permission from ref. 193. Copyright 1999 American Chemical Society.

these peptides do not possess an amphipathic nature like most pore forming helical proteins; as a consequence, it is still not clear how exactly these peptides function to bring about their anti-microbial properties.

While the full action of the histatins is still not clear, analogues based on the known sequences have been shown to have broad spectrum antibacterial effects[194,195] and also appear to promote osteoblast differentiation when immobilised on titanium implants, the exact reason for this being unclear in the study.[196]

1.6 Conclusions

In this chapter, we have presented a brief and broad overview of the wide range of roles that metal ions fulfil within proteins. It is hopefully clear that their role is extremely diverse and varies from forming catalytic centres within enzymes, acting as electron transporters in energy conversion to forming structural interactions for macroscopic holdfast systems. It is also clear that a complete understanding of all these metal binding properties and functions has still not been achieved. One important aspect is to highlight how difficult it is to study metal binding, where even very low levels of a metal ion can have a dramatic effect on the activity or specificity of an enzyme.

While initially it might be expected that all metal binding sites would be fairly promiscuous, binding any divalent metal ion, it is clear that for a lot of proteins it is important to be able to be highly selective, limiting a binding motif to only one specific metal ion. Thus these proteins have evolved binding sites with specific spatial restrictions on the binding site preventing other metal ions from binding. Nature has also taken advantage of its ability to synthesise specific small molecules to bind metal ions and act as a mediator between the protein and the metal ion, *e.g.* to alter the binding capacity of the structure as seen by the different binding capabilities of haemoglobin and myoglobin for oxygen. Also, advantage is taken from the fact that metal ions can coordinate proteins and also act to eventually convert non-covalent bonds to covalent bonds over long periods in the case of the DOPA residues in the mussel byssus.

As such, nature provides a wide range of inspiration of what can be done with metal ions and gives plenty of examples of how this can be achieved. While nature is limited to the canonical amino acids, we have the advantage of using many different methods to incorporate other additional amino acids into proteins to further expand the possibilities for interactions between the protein and metal ions. Thus, there is great potential to exploit metal ion binding as a way of creating functional systems be it with proteins or other polymers.

Acknowledgements

I would like to thank Dr Mohamed Elsawy and Dr Victoria Workman for advice and proofreading.

References

1. W. Maret, *Metallomics*, 2010, **2**, 117.
2. Y. Lu, N. Yeung, N. Sieracki and N. M. Marshall, *Nature*, 2009, **460**, 855.
3. A. S. Prasad, J. A. Halsted and M. Nadimi, *Am. J. Med.*, 1961, **31**, 532.
4. W. R. Todd, C. A. Elvehjem and E. B. Hart, *Am. J. Physiol.*, 1934, **107**, 146.
5. D. Keilin and T. Mann, *Nature*, 1939, **144**, 442.
6. B. L. Vallee and A. Galdes, *Adv. Enzymol. Relat. Areas Mol. Biol.*, 1984, **56**, 283.
7. J. Miller, A. D. McLachlan and A. Klug, *EMBO J.*, 1985, **4**, 1609.
8. A. Muller, R. M. MacCallum and M. J. E. Sternberg, *Genome Res.*, 2002, **12**, 1625.
9. L. Fairall, D. Rhodes and A. Klug, *J. Mol. Biol.*, 1986, **192**, 577.
10. J. H. Laity, B. M. Lee and P. E. Wright, *Curr. Opin. Struct. Biol.*, 2001, **11**, 39.
11. R. Gamsjaeger, C. K. Liew, F. E. Loughlin, M. Crossley and J. P. Mackay, *Trends Biochem. Sci.*, 2007, **32**, 63.
12. J. E. Coleman and B. L. Vallee, *J. Biol. Chem.*, 1960, **235**, 390.
13. K. S. Larsen and D. S. Auld, *Biochemistry*, 1989, **28**, 9620.
14. M. Wilson, C. Hogstrand and W. Maret, *J. Biol. Chem.*, 2012, **287**, 9322.
15. M. Knipp, J. M. Charnock, C. D. Garner and M. Vasak, *J. Biol. Chem.*, 2001, **276**, 40449.
16. W. Maret, *BioMetals*, 2013, **26**, 197.
17. A. Krezel and W. Maret, *J. Biol. Inorg. Chem.*, 2006, **11**, 1049.
18. Y. Li and W. Maret, *Exp. Cell Res.*, 2009, **315**, 2463.
19. K. M. Taylor, S. Hiscox, R. I. Nicholson, C. Hogstrand and P. Kille, *Sci. Signaling*, 2012, **5**, ra11.
20. E. A. Bellomo, G. Meur and G. A. Rutter, *J. Biol. Chem.*, 2011, **286**, 25778.
21. B. L. Vallee and K. H. Falchuk, *Physiol. Rev.*, 1993, **73**, 79.
22. W. Maret, C. Jacob, B. L. Vallee and E. H. Fischer, *Proc. Natl. Acad. Sci. U. S. A.*, 1999, **96**, 1936.
23. C. Jacob, W. Maret and B. L. Vallee, *Proc. Natl. Acad. Sci. U. S. A.*, 1998, **95**, 3489.
24. T. Y. Li, A. J. Kraker, C. F. Shaw and D. H. Petering, *Proc. Natl. Acad. Sci. U. S. A.*, 1980, **77**, 6334.
25. M. M. Abu-Omar, A. Loaiza and N. Hontzeas, *Chem. Rev.*, 2005, **105**, 2227.
26. L. Que, *Nat. Struct. Biol.*, 2000, **7**, 182.
27. P. F. Fitzpatrick, E. C. Ralph, H. R. Ellis, O. J. Willmon and S. C. Daubner, *Biochemistry*, 2003, **42**, 2081.
28. E. I. Solomon, A. Decker and N. Lehnert, *Proc. Natl. Acad. Sci. U. S. A.*, 2003, **100**, 3589.
29. O. A. Andersen, T. Flatmark and E. Hough, *J. Mol. Biol.*, 2002, **320**, 1095.
30. N. Hontzeas, S. S. Saleh and B. R. Glick, *Mol. Plant-Microbe Interact.*, 2004, **17**, 865.
31. J. S. Thrower, R. Blalock and J. P. Klinman, *Biochemistry*, 2001, **40**, 9717.

32. A. M. Rocklin, K. Kato, H. W. Liu, L. Que and J. D. Lipscomb, *J. Biol. Inorg. Chem.*, 2004, **9**, 171.
33. J. Zhou, A. M. Rocklin, J. D. Lipscomb, L. Que and E. I. Solomon, *J. Am. Chem. Soc.*, 2002, **124**, 4602.
34. N. Mitic, S. J. Smith, A. Neves, L. W. Guddat, L. R. Gahan and G. Schenk, *Chem. Rev.*, 2006, **106**, 3338.
35. B. C. Antanaitis, P. Aisen and H. R. Lilienthal, *J. Biol. Chem.*, 1983, **258**, 3166.
36. B. A. Averill, J. C. Davis, S. Burman, T. Zirino, J. Sanders-Loehr, T. M. Loehr, J. T. Sage and P. G. Debrunner, *J. Am. Chem. Soc.*, 1987, **109**, 3760.
37. Y. S. Yang, J. M. McCormick and E. I. Solomon, *J. Am. Chem. Soc.*, 1997, **119**, 11832.
38. H. D. Campbell, D. A. Dionysius, D. T. Keough, B. E. Wilson, J. D. Jersey and B. Zerner, *Biochem. Biophys. Res. Commun.*, 1978, **82**, 615.
39. J. Chen, Y.-M. Lee, K. M. Davis, X. Wu, M. S. Seo, K.-B. Cho, H. Yoon, Y. J. Park, S. Fukuzumi, Y. N. Pushkar and W. Nam, *J. Am. Chem. Soc.*, 2013, **135**, 6388.
40. S. H. Allen, P. R. Nuttleman, C. M. Ketcham and R. M. Roberts, *J. Bone Miner. Res.*, 1989, **4**, 47.
41. A. R. Hayman, M. J. Warburton, J. A. S. Pringle, B. Coles and T. J. Chambers, *Biochem. J.*, 1989, **261**, 601.
42. A. J. Janckila, T. A. Woodford, K. W. Lam, C. Y. Li and L. T. Tam, *Leukemia*, 1992, **6**, 199.
43. D. B. Robinson and R. H. Glew, *J. Biol. Chem.*, 1980, **255**, 5864.
44. E. G. Funhoff, C. H. W. Klaassen, B. Samyn, J. Van Beeumen and B. A. Averill, *ChemBioChem*, 2001, **2**, 355.
45. C. M. Ketcham, G. A. Baumbach, F. W. Bazer and R. M. Roberts, *J. Biol. Chem.*, 1985, **260**, 5768.
46. B. Ekrylander, P. Bill, M. Norgard, S. Nilsson and G. Andersson, *J. Biol. Chem.*, 1991, **266**, 24684.
47. J. L. Beck, L. A. McConachie, A. C. Summors, W. N. Arnold, J. Dejersey and B. Zerner, *Biochim. Biophys. Acta*, 1986, **869**, 61.
48. A. Durmus, C. Eicken, B. H. Sift, A. Kratel, R. Kappl, J. Huttermann and B. Krebs, *Eur. J. Biochem.*, 1999, **260**, 709.
49. G. Schenk, Y. B. Ge, L. E. Carrington, C. J. Wynne, I. R. Searle, B. J. Carroll, S. Hamilton and J. de Jersey, *Arch. Biochem. Biophys.*, 1999, **370**, 183.
50. G. G. Bozzo, K. G. Raghothama and W. C. Plaxton, *Eur. J. Biochem.*, 2002, **269**, 6278.
51. G. G. Bozzo, K. G. Raghothama and W. C. Plaxton, *Biochem. J.*, 2004, **377**, 419.
52. H. Nakazato, T. Okamoto, M. Nishikoori, K. Washio, N. Morita, K. Haraguchi, G. A. Thompson and H. Okuyama, *Plant Physiol.*, 1998, **118**, 1015.
53. G. Schenk, M. L. J. Korsinczky, D. A. Hume, S. Hamilton and J. DeJersey, *Gene*, 2000, **255**, 419.

54. G. Schenk, L. T. Guddat, Y. Ge, L. E. Carrington, D. A. Hume, S. Hamilton and J. de Jersey, *Gene*, 2000, **250**, 117.
55. G. W. Oddie, G. Schenk, N. Z. Angel, N. Walsh, L. W. Guddat, J. De Jersey, A. I. Cassady, S. E. Hamilton and D. A. Hume, *Bone*, 2000, **27**, 575.
56. D. E. Wilcox, *Chem. Rev.*, 1996, **96**, 2435.
57. N. Mitic, C. J. Noble, L. R. Gahan, G. R. Hanson and G. Schenk, *J. Am. Chem. Soc.*, 2009, **131**, 8173.
58. P. V. Bernhardt, G. Schenk and G. J. Wilson, *Biochemistry*, 2004, **43**, 10387.
59. D. L. Wang, R. C. Holz, S. S. David, L. Que and M. T. Stankovich, *Biochemistry*, 1991, **30**, 8187.
60. J. L. Beck, J. Dejersey, B. Zerner, M. P. Hendrich and P. G. Debrunner, *J. Am. Chem. Soc.*, 1988, **110**, 3317.
61. G. Schenk, C. L. Boutchard, L. E. Carrington, C. J. Noble, B. Moubaraki, K. S. Murray, J. de Jersey, G. R. Hanson and S. Hamilton, *J. Biol. Chem.*, 2001, **276**, 19084.
62. T. Waratrujiwong, B. Krebs, F. Spener and P. Visoottiviseth, *FEBS J.*, 2006, **273**, 1649.
63. G. Schenk, T. W. Elliott, E. Leung, L. E. Carrington, N. Mitic, L. R. Gahan and L. W. Guddat, *BMC Struct. Biol.*, 2008, **8**, 6.
64. S. J. Smith, A. Casellato, K. S. Hadler, N. Mitic, M. J. Riley, A. J. Bortoluzzi, B. Szpoganicz, G. Schenk, A. Neves and L. R. Gahan, *J. Biol. Inorg. Chem.*, 2007, **12**, 1207.
65. R. S. Cox, G. Schenk, N. Mitic, L. R. Gahan and A. C. Hengge, *J. Am. Chem. Soc.*, 2007, **129**, 9550.
66. M. W. Crowder, J. Spencer and A. J. Vila, *Acc. Chem. Res.*, 2006, **39**, 721.
67. M. Galleni, J. Lamotte-Brasseur, G. M. Rossolini, J. Spencer, O. Dideberg, J. M. Frere and The Metallo-Beta-Lactamase Working Group, *Antimicrob. Agents Chemother.*, 2001, **45**, 660.
68. F. Jose Perez-Llarena and G. Bou, *Curr. Med. Chem.*, 2009, **16**, 3740.
69. K. M. Papp-Wallace, A. Endimiani, M. A. Taracila and R. A. Bonomo, *Antimicrob. Agents Chemother.*, 2011, **55**, 4943.
70. J. M. Frere, *Mol. Microbiol.*, 1995, **16**, 385.
71. D. J. Payne, *J. Med. Microbiol.*, 1993, **39**, 93.
72. A. Felici and G. Amicosante, *Antimicrob. Agents Chemother.*, 1995, **39**, 192.
73. A. Felici, G. Amicosante, A. Oratore, R. Strom, P. Ledent, B. Joris, L. Fanuel and J. M. Frere, *Biochem. J.*, 1993, **291**, 151.
74. P. S. Mercuri, F. Bouillenne, L. Boschi, J. Lamote-Brasseur, G. Amicosante, B. Devreese, J. Van Beeumen, J. M. Frere, G. M. Rossolini and M. Galleni, *Antimicrob. Agents Chemother.*, 2001, **45**, 1254.
75. G. Garau, C. Bebrone, C. Anne, M. Galleni, J. M. Frere and O. Dideberg, *J. Mol. Biol.*, 2005, **345**, 785.
76. M. I. Page and A. Badarau, *Bioinorg. Chem. Appl.*, 2008, 576297.
77. M. W. Crowder, T. R. Walsh, L. Banovic, M. Pettit and J. Spencer, *Antimicrob. Agents Chemother.*, 1998, **42**, 921.
78. J. Weston, *Chem. Rev.*, 2005, **105**, 2151.

79. A. Carfi, S. Pares, E. Duee, M. Galleni, C. Duez, J. M. Frere and O. Dideberg, *EMBO J.*, 1995, **14**, 4914.
80. N. V. Kaminskaia, B. Spingler and S. J. Lippard, *J. Am. Chem. Soc.*, 2000, **122**, 6411.
81. M. Umayal and G. Mugesh, *Inorg. Chim. Acta*, 2011, **372**, 353.
82. H. Yang, M. Aitha, A. M. Hetrick, T. K. Richmond, D. L. Tierney and M. W. Crowder, *Biochemistry*, 2012, **51**, 3839.
83. W. A. Eaton, E. R. Henry, J. Hofrichter and A. Mozzarelli, *Nat. Struct. Biol.*, 1996, **6**, 351–358.
84. R. Baradaran, J. M. Berrisford, G. S. Minhas and L. A. Sazanov, *Nature*, 2013, **494**, 443.
85. F. Scott Mathews, *Prog. Biophys. Mol. Biol.*, 1985, **45**, 1.
86. D. Alvarez-Paggi, U. Zitare and D. H. Murgida, *Biochim. Biophys. Acta, Bioenerg.*, 2014, **1837**, 1196.
87. C. M. Paquete and R. O. Louro, *Acc. Chem. Res.*, 2014, **47**, 56.
88. S. Sharma, G. Cavallaro and A. Rosato, *J. Biol. Inorg. Chem.*, 2010, **15**, 559.
89. J. Simon, M. Kern, B. Hermann, O. Einsle and J. N. Butt, *Biochem. Soc. Trans.*, 2011, **39**, 1864.
90. D. R. Nelson, T. Kamataki, D. J. Waxman, F. P. Guengerich, R. W. Estabrook, R. Feyereisen, F. J. Gonzalez, M. J. Coon, I. C. Gunsalus, O. Gotoh, K. Okuda and D. W. Nebert, *DNA Cell Biol.*, 1993, **12**, 1.
91. T. Omura and R. Sato, *J. Biol. Chem.*, 1964, **239**, 2370.
92. M. Sono, M. P. Roach, E. D. Coulter and J. H. Dawson, *Chem. Rev.*, 1996, **96**, 2841.
93. C. A. Hasemann, R. G. Kurumbail, S. S. Boddupalli, J. A. Peterson and J. Deisenhofer, *Structure*, 1995, **3**, 41.
94. D. F. V. Lewis and J. M. Pratt, *Drug Metab. Rev.*, 1998, **30**, 739.
95. F. P. Guengerich and W. W. Johnson, *Biochemistry*, 1997, **36**, 14741.
96. E. G. Hrycay and S. M. Bandiera, *Arch. Biochem. Biophys.*, 2012, **522**, 71.
97. F. Lopez-Gallego and C. Schmidt-Dannert, *Curr. Opin. Chem. Biol.*, 2010, **14**, 174.
98. G. M. Whitesides and C. H. Wong, *Angew. Chem., Int. Ed.*, 1985, **24**, 617.
99. Y. Liu, M. K. Wang, F. Zhao, Z. A. Xu and S. J. Dong, *Biosens. Bioelectron.*, 2005, **21**, 984.
100. G. Chang, Y. Mori, S. Mori, T. Irie, H. Nagai, T. Goto, Y. Tatsu, H. Imaishi and K. Morigaki, *Anal. Chem.*, 2012, **84**, 5292.
101. Y. Sun, L. Wang and H. Liu, *Anal. Methods*, 2012, **4**, 3358.
102. Y. Li and P. C. Cirino, *Biotechnol. Bioeng.*, 2014, **111**, 1273.
103. V. B. Urlacher and M. Girhard, *Trends Biotechnol.*, 2012, **30**, 26.
104. R. Bernhardt and V. B. Urlacher, *Appl. Microbiol. Biotechnol.*, 2014, **98**, 6185.
105. C. A. Gersbach, T. Gaj and C. F. Barbas, III, *Acc. Chem. Res.*, 2014, **47**, 2309.
106. E. Kanemura, T. Goto, Y. Tatsu, H. Imaishi and K. Morigaki, *Anal. Methods*, 2014, **6**, 2117.

107. F. Jia, B. Narasimhan and S. Mallapragada, *Biotechnol. Bioeng.*, 2014, **111**, 209.
108. C. Dai, Y. Ding, M. Li and J. Fei, *Microchim. Acta*, 2012, **176**, 397.
109. K. Yoshioka, D. Kato, T. Kamata and O. Niwa, *Anal. Chem.*, 2013, **85**, 9996.
110. R. A. Sheldon, *Adv. Synth. Catal.*, 2007, **349**, 1289.
111. T. D. Sutherland, J. H. Young, S. Weisman, C. Y. Hayashi and D. J. Merritt, *Annu. Rev. Entomol.*, 2010, **55**, 171.
112. S.-Y. Yang, M.-J. Han, L.-F. Kang, Z.-W. Li, Y.-H. Shen and Z. Zhang, *BMC Evol. Biol.*, 2014, **14**, 185.
113. G. H. Altman, F. Diaz, C. Jakuba, T. Calabro, R. L. Horan, J. S. Chen, H. Lu, J. Richmond and D. L. Kaplan, *Biomaterials*, 2003, **24**, 401.
114. F. G. Omenetto and D. L. Kaplan, *Science*, 2010, **329**, 528.
115. C. Z. Zhou, F. Confalonieri, N. Medina, Y. Zivanovic, C. Esnault, T. Yang, M. Jacquet, J. Janin, M. Duguet, R. Perasso and Z. G. Li, *Nucleic Acids Res.*, 2000, **28**, 2413.
116. F. Vollrath and D. P. Knight, *Nature*, 2001, **410**, 541.
117. H. J. Jin and D. L. Kaplan, *Nature*, 2003, **424**, 1057.
118. Y. Liu, Z. Z. Shao and F. Vollrath, *Nat. Mater.*, 2005, **4**, 901.
119. B. Madsen and F. Vollrath, *Naturwissenschaften*, 2000, **87**, 148.
120. F. Vollrath, *Int. J. Biol. Macromol.*, 1999, **24**, 81.
121. R. W. Work, *Text. Res. J.*, 1977, **47**, 650.
122. D. B. Zax, D. E. Armanios, S. Horak, C. Malowniak and Z. T. Yang, *Biomacromolecules*, 2004, **5**, 732.
123. L. F. Drummy, D. M. Phillips, M. O. Stone, B. L. Farmer and R. R. Naik, *Biomacromolecules*, 2005, **6**, 332.
124. A. E. Terry, D. P. Knight, D. Porter and F. Vollrath, *Biomacromolecules*, 2004, **5**, 768.
125. Q.-X. Ruan and P. Zhou, *J. Mol. Struct.*, 2008, **883**, 85.
126. Q.-X. Ruan, P. Zhou, B.-W. Hu and D. Ji, *FEBS J.*, 2008, **275**, 219.
127. L. Zhou, X. Chen, Z. Z. Shao, Y. F. Huang and D. P. Knight, *J. Phys. Chem. B*, 2005, **109**, 16937.
128. P. Zhou, X. Xie, F. Deng, Z. Ping, X. Xun and D. Feng, *Biochemistry*, 2004, **43**, 11302.
129. X. H. Zong, P. Zhou, Z. Z. Shao, S. M. Chen, X. Chen, B. W. Hu, F. Deng and W. H. Yao, *Biochemistry*, 2004, **43**, 11932.
130. L. Zhou, X. Chen, W. L. Dai and Z. Z. Shao, *Biopolymers*, 2006, **82**, 144.
131. R. Rajkhowa, R. Naik, L. Wang, S. V. Smith and X. Wang, *J. Appl. Polym. Sci.*, 2011, **119**, 3630.
132. J. H. Waite, X. X. Qin and K. J. Coyne, *Matrix Biol.*, 1998, **17**, 93.
133. J. H. Waite, E. Vaccaro, C. J. Sun and J. M. Lucas, *Philos. Trans. R. Soc., B*, 2002, **357**, 143.
134. C. M. Yonge, *J. Mar. Biol. Assoc. U. K.*, 1962, **42**, 113.
135. L. A. Burzio and J. H. Waite, *Protein Sci.*, 2001, **10**, 735.
136. B. Liu, L. Burdine and T. Kodadek, *J. Am. Chem. Soc.*, 2006, **128**, 15228.
137. S. Schmidt, A. Reinecke, F. Wojcik, D. Pussak, L. Hartmann and M. J. Harrington, *Biomacromolecules*, 2014, **15**, 1644.

138. T. L. Coombs and P. J. Keller, *Aquat. Toxicol.*, 1981, **1**, 291.
139. E. Vaccaro and J. H. Waite, *Biomacromolecules*, 2001, **2**, 906.
140. H. C. Lichtenegger, T. Schoberl, J. T. Ruokolainen, J. O. Cross, S. M. Heald, H. Birkedal, J. H. Waite and G. D. Stucky, *Proc. Natl. Acad. Sci. U. S. A.*, 2003, **100**, 9144.
141. M. J. Harrington and J. H. Waite, *J. Exp. Biol.*, 2007, **210**, 4307.
142. H. Zhao and J. H. Waite, *Biochemistry*, 2006, **45**, 14223.
143. Q. Lin, D. Gourdon, C. Sun, N. Holten-Andersen, T. H. Anderson, J. H. Waite and J. N. Israelachvili, *Proc. Natl. Acad. Sci. U. S. A.*, 2007, **104**, 3782.
144. S. W. Taylor, J. H. Waite, M. M. Ross, J. Shabanowitz and D. F. Hunt, *J. Am. Chem. Soc.*, 1994, **116**, 10803.
145. S. W. Taylor, D. B. Chase, M. H. Emptage, M. J. Nelson and J. H. Waite, *Inorg. Chem.*, 1996, **35**, 7572.
146. S. W. Taylor, G. W. Luther and J. H. Waite, *Inorg. Chem.*, 1994, **33**, 5819.
147. D. S. Hwang, H. Zeng, A. Masic, M. J. Harrington, J. N. Israelachvili and J. H. Waite, *J. Biol. Chem.*, 2010, **285**, 25850.
148. J. Monahan and J. J. Wilker, *Chem. Commun.*, 2003, 1672.
149. H. Zeng, D. S. Hwang, J. N. Israelachvili and J. H. Waite, *Proc. Natl. Acad. Sci. U. S. A.*, 2010, **107**, 12850.
150. F. Chiti and C. M. Dobson, *Annu. Rev. Biochem.*, 2006, **75**, 333.
151. L. M. Hou and M. G. Zagorski, *J. Am. Chem. Soc.*, 2006, **128**, 9260.
152. J. Zou, K. Kajita and N. Sugimoto, *Angew. Chem., Int. Ed.*, 2001, **40**, 2274.
153. K. H. Lim, Y. K. Kim and Y. T. Chang, *Biochemistry*, 2007, **46**, 13523.
154. C. J. Sarell, S. R. Wilkinson and J. H. Viles, *J. Biol. Chem.*, 2010, **285**, 41533.
155. C. S. Atwood, R. D. Moir, X. D. Huang, R. C. Scarpa, N. M. E. Bacarra, D. M. Romano, M. K. Hartshorn, R. E. Tanzi and A. I. Bush, *J. Biol. Chem.*, 1998, **273**, 12817.
156. J. W. Karr, H. Akintoye, L. J. Kaupp and V. A. Szalai, *Biochemistry*, 2005, **44**, 5478.
157. B. K. Shin and S. Saxena, *Biochemistry*, 2008, **47**, 9117.
158. C. D. Syme, R. C. Nadal, S. E. J. Rigby and J. H. Viles, *J. Biol. Chem.*, 2004, **279**, 18169.
159. C. C. Curtain, F. Ali, I. Volitakis, R. A. Cherny, R. S. Norton, K. Beyreuther, C. J. Barrow, C. L. Masters, A. I. Bush and K. J. Barnham, *J. Biol. Chem.*, 2001, **276**, 20466.
160. S. Zirah, S. A. Kozin, A. K. Mazur, A. Blond, M. Cheminant, I. Segalas-Milazzo, P. Debey and S. Rebuffat, *J. Biol. Chem.*, 2006, **281**, 2151.
161. C. Talmard, L. Guilloreau, Y. Coppel, H. Mazarguil and P. Faller, *ChemBioChem*, 2007, **8**, 163.
162. C. A. Damante, K. Osz, Z. Nagy, G. Pappalardo, G. Grasso, G. Impellizzeri, E. Rizzarelli and I. Sovago, *Inorg. Chem.*, 2009, **48**, 10405.
163. P. Faller and C. Hureau, *Dalton Trans.*, 2009, 1080.
164. W. P. Hu, G. L. Chang, S. J. Chen and Y. M. Kuo, *J. Neurosci. Methods*, 2006, **154**, 190.

165. X. D. Huang, C. S. Atwood, R. D. Moir, M. A. Hartshorn, R. E. Tanzi and A. I. Bush, *J. Biol. Inorg. Chem.*, 2004, **9**, 954.
166. G. Klug, D. Losic, S. S. Subasinghe, M. I. Aguilar, L. L. Martin and D. H. Small, *Eur. J. Biochem.*, 2003, **270**, 4282.
167. J. Ryu, K. Girigoswami, C. Ha, S. H. Ku and C. B. Park, *Biochemistry*, 2008, **47**, 5328.
168. P. J. Crouch, D. J. Tew, T. Du, D. N. Nguyen, A. Caragounis, G. Filiz, R. E. Blake, I. A. Trounce, C. P. W. Soon, K. Laughton, K. A. Perez, Q.-X. Li, R. A. Cherny, C. L. Masters, K. J. Barnham and A. R. White, *J. Neurochem.*, 2009, **108**, 1198.
169. M. A. Lovell, J. D. Robertson, W. J. Teesdale, J. L. Campbell and W. R. Markesbery, *J. Neurol. Sci.*, 1998, **158**, 47.
170. J. Aqvist and A. Warshel, *J. Am. Chem. Soc.*, 1990, **112**, 2860.
171. W. Q. Qiu and M. F. Folstein, *Neurobiol. Aging*, 2006, **27**, 190.
172. G. Grasso, A. Pietropaolo, G. Spoto, G. Pappalardo, G. R. Tundo, C. Ciaccio, M. Coletta and E. Rizzarelli, *Chem.–Eur. J.*, 2011, **17**, 2752.
173. G. Grasso, F. Salomone, G. R. Tundo, G. Pappalardo, C. Ciaccio, G. Spoto, A. Pietropaolo, M. Coletta and E. Rizzarelli, *J. Inorg. Biochem.*, 2012, **117**, 351.
174. D. Hamada, I. Yanagihara and K. Tsumoto, *Trends Biotechnol.*, 2004, **22**, 93.
175. A. J. Baldwin, R. Bader, J. Christodoulou, C. E. MacPhee, C. M. Dobson and P. D. Barker, *J. Am. Chem. Soc.*, 2006, **128**, 2162.
176. M. Hamedi, A. Herland, R. H. Karlsson and O. Inganas, *Nano Lett.*, 2008, **8**, 1736.
177. E. Kasotakis, E. Mossou, L. Adler-Abramovich, E. P. Mitchell, V. T. Forsyth, E. Gazit and A. Mitraki, *Biopolymers*, 2009, **92**, 164.
178. T. Scheibel, R. Parthasarathy, G. Sawicki, X. M. Lin, H. Jaeger and S. L. Lindquist, *Proc. Natl. Acad. Sci. U. S. A.*, 2003, **100**, 4527.
179. M. Campese, X. L. Sun, J. A. Bosch, F. G. Oppenheim and E. J. Helmerhorst, *Arch. Oral Biol.*, 2009, **54**, 345.
180. X. L. Sun, E. Salih, F. G. Oppenheim and E. J. Helmerhorst, *FASEB J.*, 2009, **23**, 2691.
181. L. M. Sabatini and E. A. Azen, *Biochem. Biophys. Res. Commun.*, 1989, **160**, 495.
182. P. A. Raj, M. Edgerton and M. J. Levine, *J. Biol. Chem.*, 1990, **265**, 3898.
183. M. Castagnola, R. Inzitari, D. V. Rossetti, C. Olmi, T. Cabras, V. Piras, P. Nicolussi, M. T. Sanna, M. Pellegrini, B. Giardina and I. Messana, *J. Biol. Chem.*, 2004, **279**, 41436.
184. M. Edgerton, S. E. Koshlukova, T. E. Lo, B. G. Chrzan, R. M. Straubinger and P. A. Raj, *J. Biol. Chem.*, 1998, **273**, 20438.
185. E. J. Helmerhorst, R. F. Troxler and F. G. Oppenheim, *Proc. Natl. Acad. Sci. U. S. A.*, 2001, **98**, 14637.
186. Y. Y. Xu, I. Ambudkar, H. Yamagishi, W. Swaim, T. J. Walsh and B. C. O'Connell, *Antimicrob. Agents Chemother.*, 1999, **43**, 2256.
187. P. A. Raj, E. Marcus and D. K. Sukumaran, *Biopolymers*, 1998, **45**, 51.

188. P. A. Raj, S. D. Soni and M. J. Levine, *J. Biol. Chem.*, 1994, **269**, 9610.
189. S. Matsuo and F. Lagerlof, *Arch. Oral Biol.*, 1991, **36**, 525.
190. C. W. Bales, J. H. Freelandgraves, S. Askey, F. Behmardi, R. S. Pobocik, J. J. Fickel and P. Greenlee, *Am. J. Clin. Nutr.*, 1990, **51**, 462.
191. H. Gusman, U. Lendenmann, J. Grogan, R. F. Troxler and F. G. Oppenheim, *Biochim. Biophys. Acta, Protein Struct. Mol. Enzymol.*, 2001, **1545**, 86.
192. C. Harford and B. Sarkar, *Acc. Chem. Res.*, 1997, **30**, 123.
193. S. Melino, S. Rufini, M. Sette, R. Morero, A. Grottesi, M. Paci and R. Petruzzelli, *Biochemistry*, 1999, **38**, 9626.
194. E. J. Helmerhorst, W. VantHof, E. C. I. Veerman, I. SimoonsSmit and A. V. N. Amerongen, *Biochem. J.*, 1997, **326**, 39.
195. M. M. Welling, C. Brouwer, W. van 't Hof, E. C. I. Veerman and A. V. N. Amerongen, *Antimicrob. Agents Chemother.*, 2007, **51**, 3416.
196. S. Makihira, H. Nikawa, T. Shuto, M. Nishimura, Y. Mine, K. Tsuji, K. Okamoto, Y. Sakai, M. Sakai, N. Imari, S. Iwata, M. Takeda and F. Suehiro, *J. Mater. Sci.: Mater. Med.*, 2011, **22**, 2765.

CHAPTER 2

DNA-Based Metallosupramolecular Materials

JANANE RAHBANI[†a], KIMBERLY METERA[†a] AND HANADI F. SLEIMAN*[a]

[a]Department of Chemistry and Center for Self-Assembled Chemical Structures, McGill University, 801 Sherbrooke St. W., Montreal, QC H3A 0B8, Canada
*E-mail: Hanadi.sleiman@mcgill.ca

2.1 Introduction

One of the goals of metallosupramolecular chemistry is the precise organization of metal complexes for applications in nanoelectronics, logic computation, optics, light harvesting or catalysis.[1,2] Few materials are better suited for controlled nanoscale self-assembly than nucleic acid biopolymers such as DNA. In this chapter, we highlight the use of DNA for the programmable organization of metal ions and complexes. The positioning of metal nanoparticles using DNA scaffolds has been covered in a number of reviews,[3,4] and will not be discussed further here.

Structurally, with a persistence length of about 50 nm, rigid rod character over ~10 nm and a width of 2 nm, the typical B-DNA double helix has

[†]J. R. and K. M. are equally contributing authors.

RSC Smart Materials No. 15
Functional Metallosupramolecular Materials
Edited by John G. Hardy and Felix H. Schacher

Published by the Royal Society of Chemistry, www.rsc.org

appropriate physical dimensions for nanoscale construction. The highly selective base-pairing interactions of adenine–thymine and cytosine–guanine and cooperativity of association of two single strands into a double helical complex yield predictable binding between nucleic acid strands and the potential construction of "smart" materials. The ability to generate a near infinite number of different DNA sequences, each capable of binding only its complementary strand allows DNA strands to program the patterning of different components into any isotropic or anisotropic arrangement. The stability of a DNA double helix is influenced by the specific DNA sequence, length, and solution conditions, and combined with mechanisms such as strand displacement reactions, it is possible to control the dynamic character of a DNA nanostructure. Furthermore, native DNA can be readily modified by enzymes that can replicate, elongate or site-specifically cleave it, allowing additional control over the individual building blocks and the final self-assembled construct. In addition to the typical double helical structure of DNA, triplex structures are possible with appropriate base-pairing, and tetrameric G-quadruplex and i-motif structures can be formed from G-rich and C-rich sequences, respectively. Perhaps most important for the development of DNA nanotechnology is that DNA can be readily synthesized *via* automated methods, making it a relatively accessible and cost-effective material. Additional synthetic analogues of DNA such as peptide nucleic acids (PNA) or locked nucleic acids (LNA) have been developed that offer unique properties, and these can be hybridized with or incorporated into native DNA to generate systems with new properties. The combination of ready availability, highly predictable interactions, and versatile modifications makes DNA an excellent construction material for the assembly of metal ions and metal complexes.

2.2 DNA Nanotechnology

The double helix is composed of two hybridized complementary DNA chains. Adenine (A)–thymine (T) and cytosine (C)–guanine (G) base pairs interact by hydrogen bonding and provide the specificity of the interactions of one strand with another. A combination of other forces such as π-stacking of the base pairs, electrostatic and hydrophobic interactions cooperatively contribute to duplex stability. Other factors that affect the stability of a duplex are the specific base sequence (G–C pairs have three hydrogen bonds and a donor–donor–acceptor hydrogen bonding pattern, A–T pairs have two and a donor–acceptor pattern), chain length (longer is typically more stable), and environmental conditions such as temperature and ion concentration.

The design of an immobile branched DNA junction by Seeman was a significant innovation that initiated the field of DNA nanotechnology (Figure 2.1). This branched DNA structure, inspired by the naturally existing Holliday junction, motivated the Seeman group and other researchers to design and assemble more complex motifs, such as double and triple crossovers to form planar tiles, branched junctions consisting of multiple DNA arms radiating from a focal point, helix bundles and DNA origami structures.[5] By blending together DNA building blocks with synthetic organic molecules, researchers have expanded the field of DNA nanotechnology into new supramolecular 2D- and 3D-structures with dynamic and switchable structural features (Figure 2.1). These structural motifs, with their varied forms and physical properties, have allowed the extension of DNA scaffolds into two and three dimensions. The significant control over the final DNA scaffold makes it an excellent platform for controlled placement of metal ions.

Adding metal ions to DNA nanoassemblies can contribute important functionalities to otherwise passive templates. Metal complexes incorporated into DNA can impart added stability, photo- or redox activity and addressability, and magnetic or catalytic properties to the hybrid constructs. They can introduce new energy or charge transport pathways into the DNA double helix. Structurally, the diverse coordination geometries of transition metal complexes (tetrahedral, octahedral, square planar and trigonal-bipyramidal) can provide a straightforward route to introduce linker angles and architectures that are difficult, or simply inaccessible, to DNA itself.

There are some challenges, however, to combining metals and DNA, including the non-specific binding of metals with the nitrogen-rich DNA bases or negatively charged phosphate backbone, the activity of some metal complexes towards DNA degradation, and the degradation of metal complexes in the reaction conditions of automated DNA synthesis and the aqueous buffer

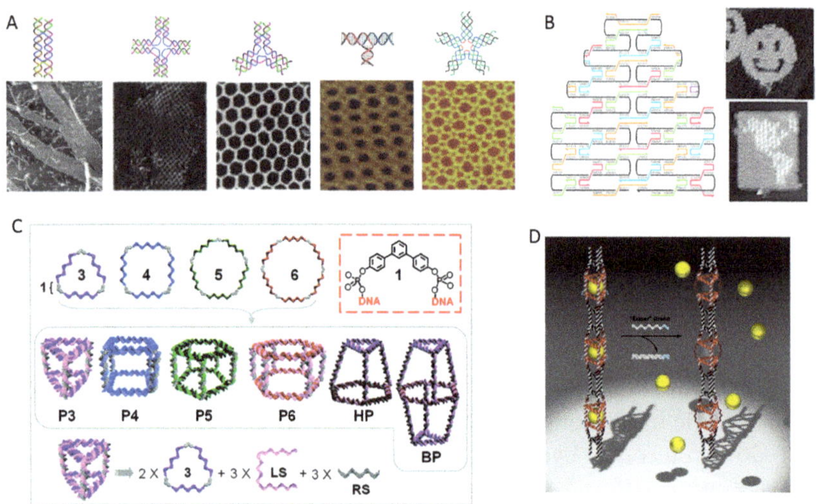

Figure 2.1 (A) DNA tiles, assembled into periodic 2D arrays from (left to right) double crossover (DX),[6] cross-motif (CM),[7] three-point-star,[8] T-junction[9] and 5-point star motifs.[10] (B) DNA origami.[11] (C) Geometrically well-defined single-stranded DNA templates with organic vertices are used to organize 3D prisms (reprinted with permission from ref. 12. Copyright 2007 American Chemical Society). (D) DNA nanotubes encapsulate gold nanoparticles and release them with an added DNA strand.[13]

solutions of DNA assemblies. This review will describe the various synthetic strategies that have been used to overcome these challenges and create metal–DNA nanostructures.

DNA fulfills the criteria for nanoscale construction. Here, we describe the approaches and examples of programmable metal ion placement onto a variety of DNA constructs. Some applications of these assemblies are discussed, and finally, the potential of DNA-like variants to function in metallo-supramolecular assemblies is described.

2.3 Methods to Incorporate Metal Centers into DNA

Approaches to site-specifically incorporate metal ions into DNA can be divided into three categories: metal binding to nucleobases (either natural or artificial) in the interior of the DNA double helix (Figure 2.2A), attachment of complete metal complexes to DNA (Figure 2.2B), and incorporation of metal-binding ligands to DNA for subsequent metal coordination (Figure 2.2C).

2.4 Metal Binding to Natural DNA Nucleobases

Metal-mediated base pairs represent an interesting approach to extend the "genetic alphabet", by replacing the hydrogen bonding between the natural base pairs G:C (guanine and cytosine) and A:T (adenine and thymine) with coordination bonds between the bases and various transition metals (Figure 2.2A). In this section, we discuss metal-mediated base pairs between natural nucleobases, and describe their characterization and their effect on the structure of nucleic acids.

The coordination of metal ions to DNA base pairs enables the incorporation of these metal ions along the helical axis, thus generating new chemical and physical properties of the resulting metal-modified nucleic acid. The interaction of Hg^{2+} with DNA was first reported by Katz in the 1950s.[15] Forty years later, an NMR study on the selective interaction of mercury(II) with a DNA hairpin was published.[16] It has been found that DNA duplexes with T–T mismatches were stabilized by Hg^{2+} binding to the two thymine residues (Figure 2.3).[17]

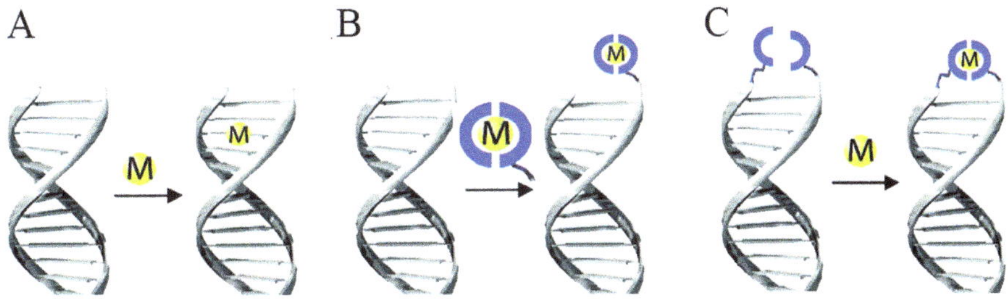

Figure 2.2 Site-specific incorporation of metal ions to DNA. (A) Metal binding to unmodified DNA. (B) Attachment of metal complexes to DNA. (C) Coordination of metals to ligand-modified DNA.[14]

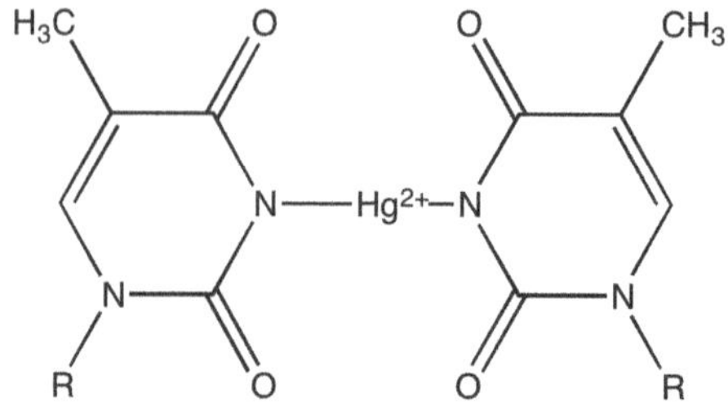

Figure 2.3 Mercury(II)-mediated base pair with thymine residues.

Ono and coworkers reported an increase of 3 °C of the melting temperature of a DNA duplex coordinated to mercury(II) compared to its corresponding regular A:T base pair duplex.[17] They used ^{1}H NMR spectroscopy to support the replacement of the imino protons by Hg^{2+} cations during the binding process. The binding constant of the T–Hg^{2+}–T base pair was measured using isothermal titration calorimetry and was reported to be around 5.7×10^5 M^{-1}.[18]

The incorporation of metal-mediated base pairs by DNA acting enzymes is a significant achievement. The first example reported by Urata and coworkers examined the site-selective incorporation of a thymine base opposite to another thymine in the presence of Hg^{2+}. After the successful insertion of thymine, DNA polymerases were able to further extend the primers.[19] In the presence of Hg^{2+}, a thymidine triphosphate was inserted by the enzyme into the opposite position of a thymine residue, instead of adenine. This observation revealed the size similarity between the regular base pairs and the T–Hg^{2+}–T base pair, since the latter motif was able to fit into the enzyme active site. The base pair U–Hg^{2+}–U (where U is uracil) has been studied by Sigel and coworkers.[20] Owing to the A-conformation of RNA, it was concluded from ^{15}N NMR that mercury(II) is bound less tightly to uracil compared to the T–Hg^{2+}–T base pair, even though they share a similar coordination pattern. Recently, Ono *et al.* presented the first crystal structure of a B-DNA duplex including two successive T–Hg^{2+}–T base pairs (Figure 2.4).[21] The results showed that in the absence of Hg^{2+}, the duplex adopted a distorted conformation, whereas after metal addition the duplex switched back to the B-form. The structure of the double stranded DNA containing two consecutive T–Hg^{2+}–T base pairs was also resolved in solution using NMR spectroscopy.[22] Both methods proposed that the short distance between the two mercury ions at the center of the duplex may be due to a metallophilic attraction and that the overall structure of the two T–Hg^{2+}–T base pairs was analogous to the normal base pairs.[21,22]

Another stable metal-mediated natural base pair is C–Ag^+–C. Similar to the T–Hg^{2+}–T base pair, a DNA duplex containing a C:C mismatch is stabilized by the incorporation of silver(I). The thermal denaturation temperature of an Ag^+ mediated DNA duplex increases 8 °C with respect to a double stranded DNA containing solely the C:C mismatch[23] and its binding constant was calculated to be 4.2×10^5 M^{-1}.[24] The thermal denaturation temperature of a DNA

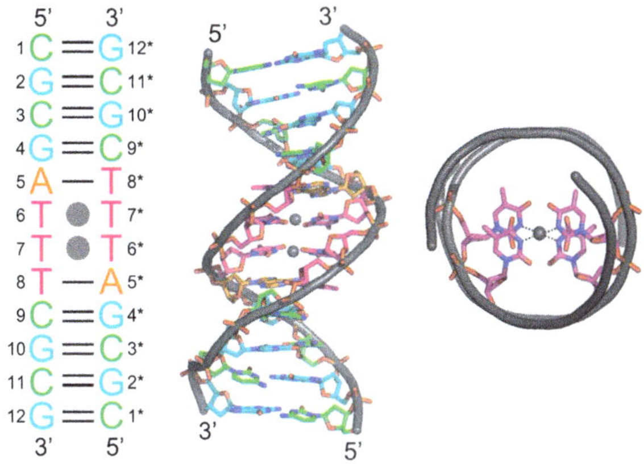

Figure 2.4 Crystal structure of a DNA duplex containing T–T mismatches in the presence of Hg^{2+} (side view in the middle and top view to the right). The metal ions are displayed as gray spheres.[21]

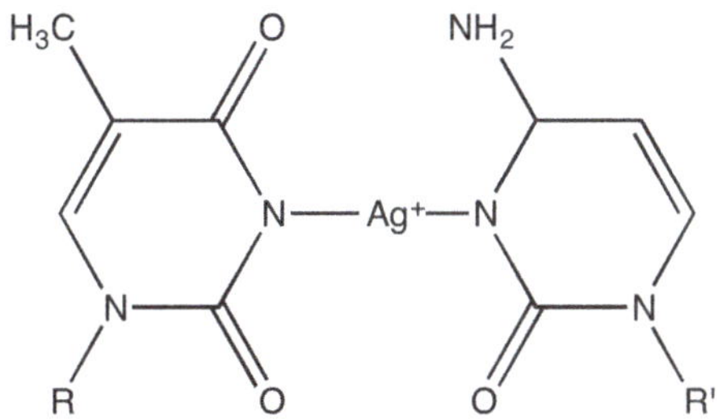

Figure 2.5 Silver(I)-mediated base pair between thymine and cytosine residues.

duplex containing two pyrimidine mismatched units C:T was also reported to increase by 9.4 °C upon binding to silver(I) (Figure 2.5).[25]

Metal mediated natural nucleobases such as T–Hg^{2+}–T have been used to detect Hg^{2+}, which is an environmental pollutant in drinking water. Liu *et al.* engineered a DNA-based enzyme (DNAzyme) with a T–T mismatched loop; in the presence of Hg^{2+} the DNAzyme folds into its active structure and can cleave an RNA strand, which separates an attached fluorescence dye from a quencher, thus turning on luminescence with a mercury detection limit of 2.4 nM.[26] Willner *et al.* designed an electrochemical detection system in which binding to Hg^{2+} stabilizes a DNA duplex, thus bringing together an enzyme with a ferrocene relay molecule near an electrode, causing enzymatic signal amplification of DNA detection.[27] A simple system that does not involve DNA labeling is based on a minor groove interacting molecule, which binds to double stranded DNA with a concomitant increase in its fluorescence. Upon mercury(II) binding, a DNA duplex forms *via* T–Hg^{2+}–T linkages, causing enhancement of the dye fluorescence signal.[28] The detection of Hg^{2+} *via* DNA functionalized gold nanoparticles (AuNPs) has been studied by Liu and coworkers.[29] This method was based on the high sensitivity of AuNPs

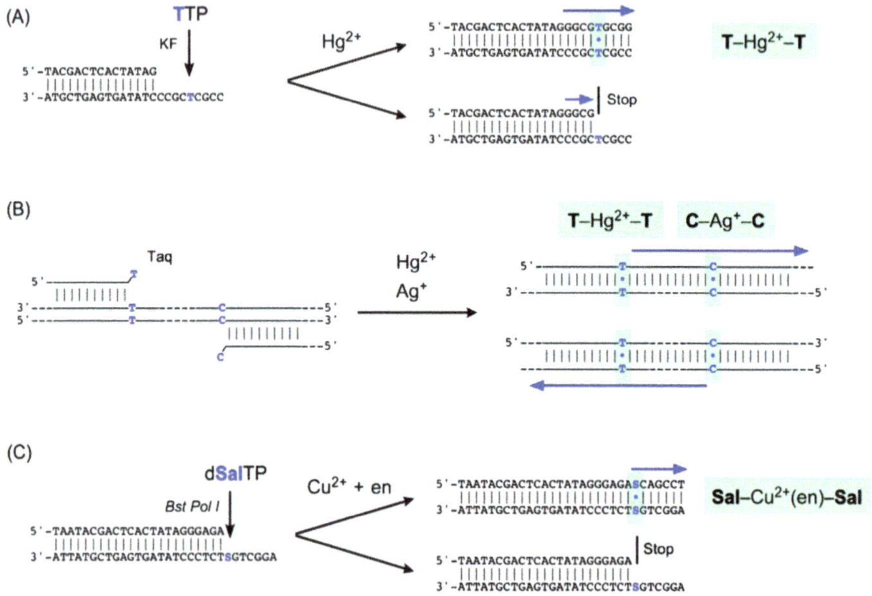

Figure 2.6 DNA polymerase reaction *via* metal-mediated artificial base pairs: (A) T–Hg–T nucleobase incorporation *via* Klenow fragment (KF), (B) depiction of an AND logic gate using both T–Hg–T and C–Ag–C base pairs, and (C) enzymatic incorporation by Bst polymerase I of S–Cu(en)–S base pairs (reprinted with permission from ref. 31. Copyright 2012 American Chemical Society).

(due to the high extinction coefficient of their surface plasmon) and their quenching effect. The conformational changes of DNA upon binding to Hg^{2+} have been investigated by applying surface plasmon resonance (SPR), by coupling these changes with the formation of a guanine quadruplex and binding of a porphyrin unit (hemin).[30] Note that by modifying the oligonucleotide sequences, other metal ions can be detected using the same principles.

The construction of a molecular logic gate (Figure 2.6) was reported, where the presence of both Hg^{2+} and Ag^{+} triggered the activation of DNA polymerases in the presence of T:T and C:C mismatches (input) and proceeded with the PCR amplification (output).[32] These promising results have encouraged chemists to use metal-ion coordinated base pairing to activate endonucleases,[33] ligases,[34] and exonucleases,[35] in the context of metallo–DNA base pairs.

In the last two decades, programmable molecular machines based on the self-assembly of DNA have been demonstrated, owing to the specificity of Watson–Crick base pairing. In this perspective, metal-mediated nucleic acids can play an important role in the construction of reversible and controllable mechanical nanodevices. New molecular beacon probes, based on a DNA-hairpin structure with a fluorophore and a quencher at opposite ends, have been built to detect small molecules (Figure 2.7A).[36] The DNA-hairpin was formed by replacing hydrogen bonding between canonical base pairs with metal-mediated

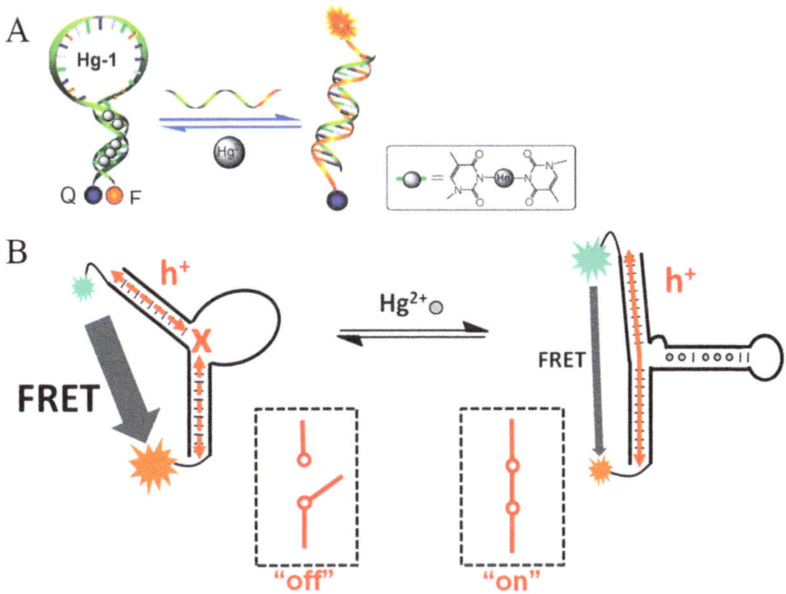

Figure 2.7 (A) Representation of a T–Hg–T based molecular beacon where Q is a quencher and F is a fluorophore.[36] (B) Depiction of a three-way junction mechano-electronic switch driven by Hg(II).[38]

nucleobases such as T–Hg^{2+}–T and C–Ag^{+}–C. Although these molecular beacons require a double labeling synthetic procedure, they constitute the starting point towards applications in biological systems *via* a label-free strategy.[37] Another type of switching device based on the coordination of Hg^{2+} to thymine base pairs was developed by Sen and coworkers.[38] In this example, a three-way junction exhibited reversible structural change in the presence of Hg^{2+} ions that facilitated charge transport through the DNA duplex (Figure 2.7B).

2.5 Binding of Pre-Formed Metal Complexes to DNA

Pre-formed metal complexes can be incorporated into unmodified DNA duplexes or other assemblies after their synthesis. In addition, stable, premade metal complexes can be derivatized for direct attachment to DNA during the DNA synthesis. In the following sections, we will outline the metallation and functionalization of DNA by intercalation, the association of metallosupramolecular structures with DNA, and some examples of metal complexes attached to DNA as structural building blocks.

2.5.1 Metallointercalators

Metallointercalators are an interesting class of metal complexes that can be arranged using DNA templates. Intercalators are molecules that insert into the DNA double helix by π-stacking with the base pairs. They should be

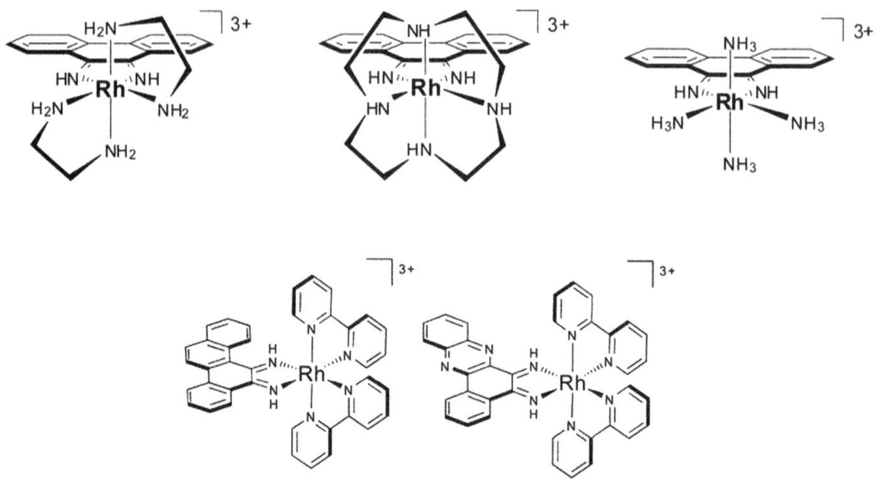

Figure 2.8 Top: rhodium intercalators targeting 5′-CG-3′ sequences. From left to right: Δ-$[Rh(en)_2phi]^{3+}$, $[Rh[12]aneN_4phi]^{3+}$, and $[Rh(NH_3)_4phi]^{3+}$ (reprinted with permission from ref. 41. Copyright 1993 American Chemical Society). Bottom: intercalators targeting DNA mismatches: left, Δ-$[Rh(bpy)_2chyrsi]^{3+}$ (reprinted with permission from ref. 43. Copyright 1997 American Chemical Society), right, Δ-$[Rh(bpy)_2phzi]^{3+}$.[43]

substitutionally inert, so as not to coordinate with the bases within the DNA base stack. They may be planar, such as the square planar Pt^{2+} complexes reported by Lippard,[39] or octahedral and possess a planar aromatic ligand with an extended π-surface that can π-stack with the bases inside the double helix. Intercalation results in a degree of unwinding of the duplex and modifications to the distances between stacked base pairs.

Typical intercalators can insert anywhere along the DNA duplex with little sequence specificity, resulting in non-specific binding to DNA. Here, we will briefly discuss some requirements and examples of *site-specific* intercalation, where metal complexes have been designed to intercalate into specific locations along a DNA double helix. This has been achieved in some cases by tethering a metal complex to the DNA so its intercalation site is pre-programmed, but there are also examples of site-specific intercalators that are not covalently bound to the DNA. Extensive reviews on metallointercalators have been published;[40–43] some illustrative examples will be described here.

Site-specific intercalation has most commonly been achieved with octahedral complexes. These allow a greater opportunity to match the 3D shape and chirality of the complex to the DNA binding site. With octahedral complexes, site-specific intercalation can be enhanced by incorporating functionalities on the ligands that can interact with the specific bases adjacent to the intended binding location. For example, 5′-CG-3′ sequences can be targeted by rhodium complexes Δ-$[Rh(en)_2phi]^{3+}$, $[Rh[12]aneN_4phi]^{3+}$ and $[Rh(NH_3)_4phi]^{3+}$ (Figure 2.8, top). These complexes bind in the major groove through hydrogen bonding between the axial amines of the complex and the

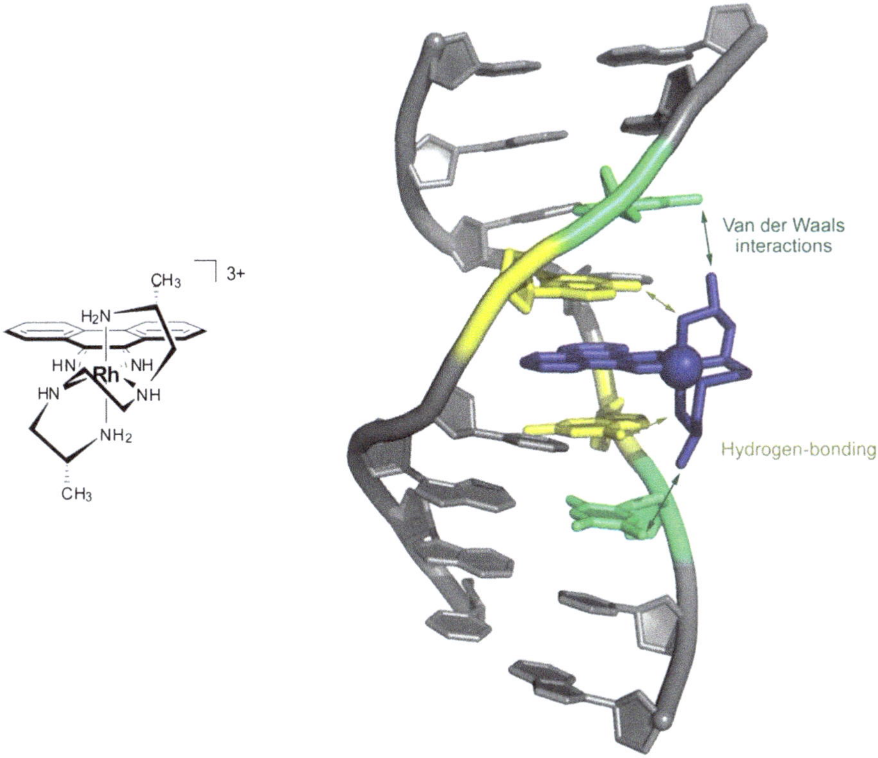

Figure 2.9 Δ-α-[Rh[(*R*,*R*)-Me_2trien]phi]$^{3+}$ complex and interactions supporting its site-specific intercalation into 5′-TGCA-3′ sequences.[43]

O6 of the guanine.[41] The complex Δ-α-[Rh[(*R*,*R*)-Me_2trien]phi]$^{3+}$ was designed to recognize 5′-TGCA-3′ sequences, again *via* hydrogen bonding between the axial amines of the complex and the guanine O6, as well as van der Waals interactions between the pendant methyl group on the complex and the methyl groups on the flanking thymines (Figure 2.9).[41]

Other intercalators preferentially target DNA mismatches (Figure 2.8, bottom). For example, in the complexes Rh(bpy)$_2$(chrysi)$^{3+}$ (chrysi = chrysene-5,6-quinone diimine) and Rh(bpy)$_2$(phzi)$^{3+}$ (phzi = benzo[*a*]-phenazin-5,6-quinone diimine), the chrysi and phzi ligands are too wide to properly intercalate between well-matched bases in a duplex.[43]

Instead, they insert into a DNA stack containing mismatched base pairs (thermodynamically destabilized sites) *via* the minor groove, and eject both mismatched bases out of the double helix.[43] In this intercalation mode, the bulky chrysi or phzi ligand replaces a destabilized base pair. The chrysi complex was shown to recognize a single mismatch within a 2725 base-pair plasmid.[44] In this particular example, the metal complex cleaves the DNA at its intercalation site. In terms of programmed placement of metal complexes into DNA strands, intentionally designed DNA mismatches could allow, in well-designed cases, site-specific intercalation of metal complexes into DNA nanostructures.

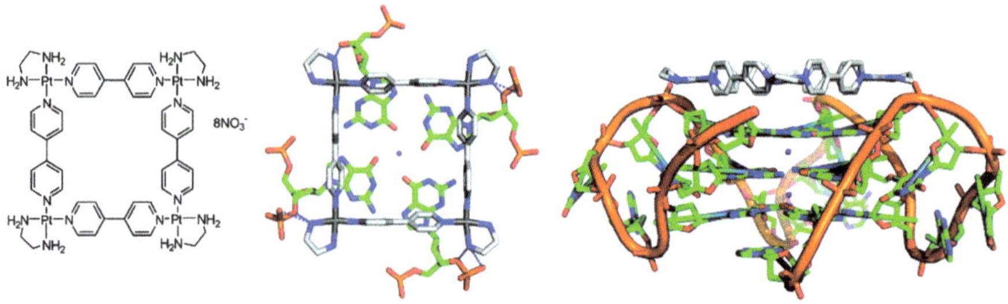

Figure 2.10 Platinum supramolecular square that selectively binds to guanine quadruplexes (reprinted with permission from ref. 45. Copyright 2008 American Chemical Society).

Platinum complexes with large aromatic ligands have been shown to target guanine quadruplexes. These higher-order DNA structures are brought together through hydrogen bonding of guanine units into stacked tetramers. Recent studies have shown that targeting these structures can bring about antitumor effects through a number of mechanisms, including inactivation of oncogene promoters. Thus platinum complexes that bind G-quadruplexes are being developed as a new class of selective antitumor therapeutics. Our group discovered that a platinum self-assembled square complex is not only ideally suited to bind G-quadruplexes, but results in significant inhibition of the cancer-specific enzyme telomerase (Figure 2.10).[45]

Some interesting functionalities have been demonstrated for non-covalent DNA intercalators. Among the most well-known examples is the molecular "light switch" complex $Ru(phen)_2(dppz)^{2+}$ (dppz = dipyrido[3,2-*a*:2′,3′-*c*] phenazine) developed by Barton.[43] This compound is luminescent in organic solutions but not in aqueous solutions because hydrogen bonding between water and the endocyclic nitrogens on the dppz ligand deactivates the excited state. Upon intercalating into DNA, however, the complex becomes luminescent as the intercalating ligand is shielded from the aqueous solution.

Other examples of functional intercalated complexes are catalytic Cu^{2+} complexes developed by Feringa and Roelfes.[46,47] In this work, non-chiral ligands are brought into close proximity to (chiral) DNA by intercalation. This allows the chirality of the DNA to be transferred to the reaction, resulting in enantioselective Diels–Alder and Friedel–Crafts reactions, among others.

Metal complexes have also been covalently attached to DNA strands so that they may intercalate in pre-determined positions. One of the most significant outcomes of site-specific intercalation has been the discovery that DNA is, under specific conditions, able to mediate charge transfer between one intercalated metal complex and another, or between an intercalated complex and an electrode.

The Barton group has extensively studied DNA-mediated charge transport using metallointercalators. The photophysical and electrochemical properties of metal complexes make them excellent redox-active probes. In one example,[40] an emissive $Ru(phen)_2dppz^{2+}$ donor and a $Rh(phi)_2phen^{3+}$

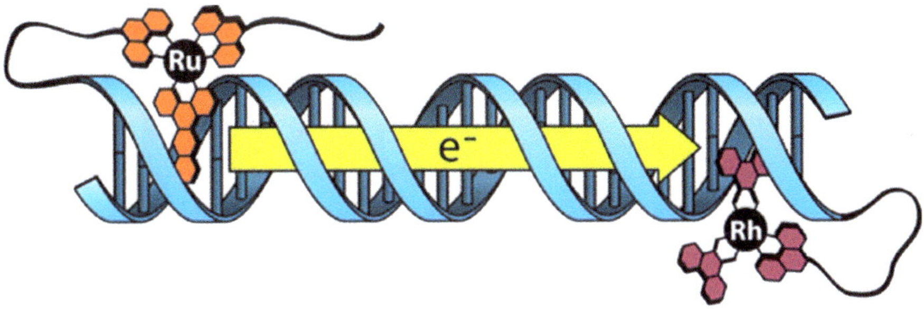

Figure 2.11 Charge transport mediated by DNA.[40]

acceptor (quencher) were designed for covalent attachment to complementary strands of a DNA oligomer (Figure 2.11). This approach allowed for site-specific intercalation of the complexes, so the distance between the donor and acceptor was well defined and diffusion-controlled quenching was eliminated. In the absence of the rhodium complex, emission from the intercalated $Ru(phen)_2dppz^{2+}$ complex was observed as expected. Incorporating the $Rh(phi)_2phen'^{3+}$ acceptor resulted in emission quenching. Quenching occurred over a 35 Å distance, and more recently was shown to occur over 34 nm, or 100 base pairs.[48] Control experiments confirmed that quenching was by charge transfer mediated by DNA, and not through energy transfer. These experiments and others that incorporated base mismatches or nicked sugar–phosphate backbones have shown that charge transport occurs through the π-stacked bases, and not *via* the DNA backbone. In the presence of DNA base mismatches, charge transfer is interrupted. Thus, photo- and electroactive metallointercalators that fit well and couple efficiently with a properly hybridized base stack can interact with each other, mediated by DNA.

Tethered and thus precisely located metallointercalators, such as $Rh(phi)_2bpy'^{3+}$, (phi = phenanthrenequinone; bpy′ = 4-butyric acid-4′-methylbipyridine) have also been used to show that charge transfer can occur through junctions such as the four-way- and double crossover motifs that can be used to form higher-order DNA constructs.[49] In these experiments, the less rigid four-way junction was a poorer conduit for charge transfer than the more rigid double crossover motif. Such structure–function relationships should be considered when designing higher-order DNA constructs with incorporated redox-active metals.

2.5.2 Non-Covalent Binding of Supramolecular Assemblies to DNA

The majority of the work discussed thus far in this chapter has involved, or will involve, the combination of metal ions or complexes with DNA to yield controlled positioning of the metals or to form functional DNA nanoassemblies. Another interesting facet of DNA–metallosupramolecular chemistry is the binding of larger, preassembled metallosupramolecular structures to

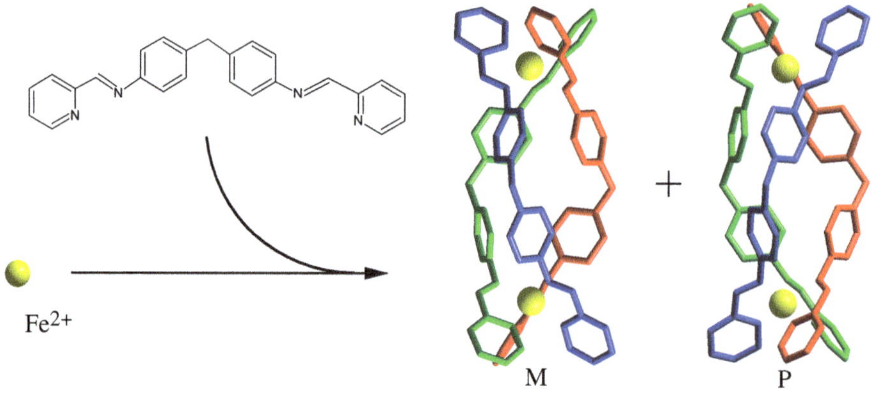

Figure 2.12 Formation of bimetallic helicate structures.[54]

DNA. As an example, the Hannon and Rodger groups have generated bi-metallic helicate structures as synthetic DNA binders (Figure 2.12). These helicates are similar in size (~2 nm long × ~1 nm diameter) and shape to natural biomolecule DNA binders such as the zinc finger motifs of proteins, and unlike many small molecules, bind DNA only in the major groove.

The bimetallic helices can be constructed with Fe^{2+} to form $Fe_2L_3^{4+}$, where L is a variation of the pyridyl-imine ligand shown in Figure 2.12, but related structures have also been formed with Ru^{2+}, Cu^+ and Ag^+ ions (Cu^+ and Ag^+ helicates are dicationic). The positive charge on the complex aids in its non-covalent binding to DNA. Variation of the metal results in different interactions of the helices with the DNA. For example, the Fe complexes induce significant intramolecular coiling of DNA and DNA unwinding. Related $Ru_2L_3^{4+}$ complexes exhibit similar binding properties, but also cleave DNA upon UV or visible radiation.[50] Furthermore, unsaturated ruthenium-based helicates were made with vacant coordination sites to allow direct coordination with DNA, and those complexes showed promising activity against two breast cancer cell lines.[51] In the presence of hydrogen peroxide, Cu^+ analogue $Cu_2L_2^{2+}$ can act as an artificial nuclease.[52]

The activity of the helicates on the DNA has also been explored by varying the structure of the helicate to include heteroatom bridging linkers,[53] make the structure wider[54] or longer,[55] or to attach short peptides to the ends of the helicate to improve sequence-specificity of binding to DNA.[52]

Another interesting feature of these helical assemblies is their ability to bind to unusual DNA structures. Compared to recognition of certain sequences, shape recognition is an interesting alternative to gaining specificity. Fe^{2+} helicates were shown to bind to three-way DNA junctions (Figure 2.13).[56] Neither the helicate structure nor the DNA junction was altered upon binding, and crystal structures were obtained showing that the helical compound fit well into the central hollow space of the three-way junction. The binding was facilitated by the electrostatic dipole–dipole interactions between the positively charged helicate and the negatively charged DNA backbone, and by π-stacking

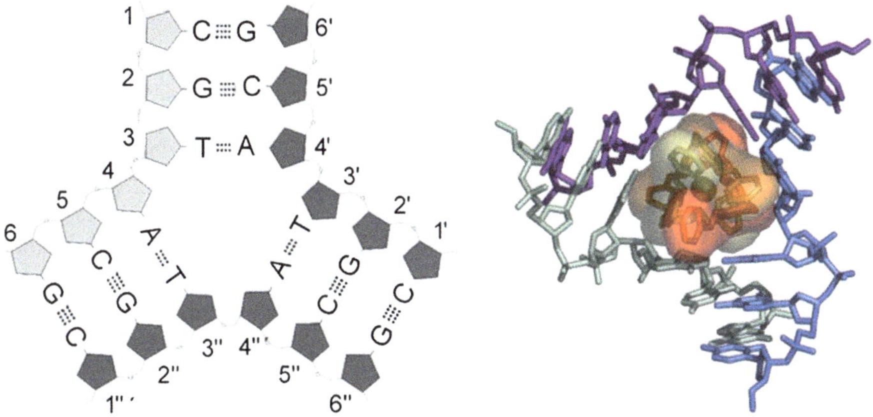

Figure 2.13 Three-way DNA junction and 3D structure of $Fe_2L_3^{4+}$–DNA complex.[56]

of the phenyl rings at the center of the helicate with the bases at the junction. Significantly, in terms of DNA nanotechnology, the DNA sequence used to form the three-way junction was a palindromic sequence that could have formed a variety of structures (dsDNA, three-way- or four-way junction, *etc.*), but the final DNA structure was dictated by the binding of the helicate. More recently, the recognition by Fe^{2+} helicates of DNA bulges was demonstrated.[57] DNA bulges are formed from one or more unpaired nucleotides in an otherwise complementary duplex structure. They occur during a number of natural biochemical processes, but can also be programmed into a DNA assembly. Association with the helicates increased the melting temperatures (and thus, the thermal stability) of bulge-containing oligomers, and it was proposed that the enhanced affinity of the helicates for the bulges is the result of unique structural features of the bulges not found in fully complementary DNA.

The binding of large-preassembled metallosupramolecular structures to DNA may have significant implications for designing functional metal–DNA structures that can start to mimic natural biomolecule–DNA interactions.

2.5.3 DNA Covalently Modified with Metal Complexes as a Building Block for Higher-Order Assemblies

Metal complexes can be incorporated into DNA as structural components in DNA assembly. These complexes can be integrated into DNA during DNA synthesis, however, they must be robust enough to withstand the harsh conditions of automated synthesis. The first examples include Sleiman's formation of $Ru(bpy)_2(imidazole)_2^{2+}$ and $Ru(bpy)_3^{2+}$ (Figure 2.14) complexes as junctions between DNA arms.[58,59] Both assemblies hybridized complementary DNA, and in the case of the $Ru(bpy)_3^{2+}$ complex, we demonstrated the specific consequence of the metal complex geometry on assembly of cyclic DNA nanostructures. In both these cases, the ruthenium complexes were stable enough to be incorporated using automated synthesis.

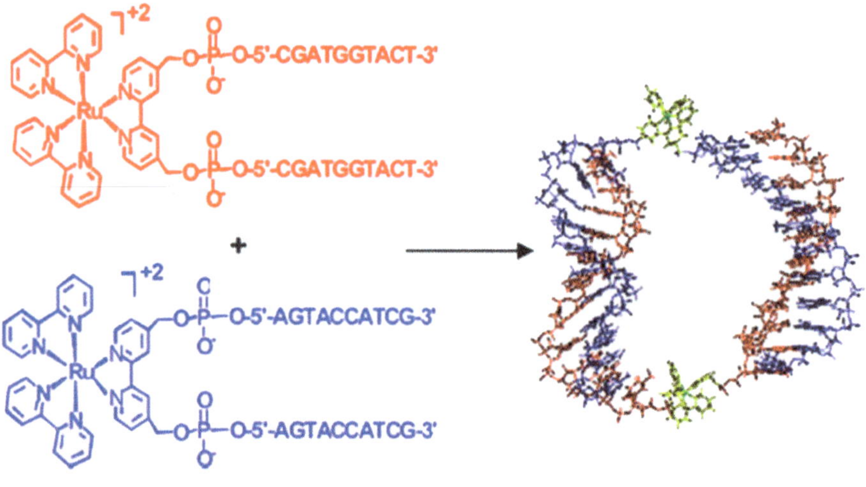

Figure 2.14 Self-assembly of a cyclic DNA nanostructure with $Ru(bpy)_3^{2+}$ complexes.[59]

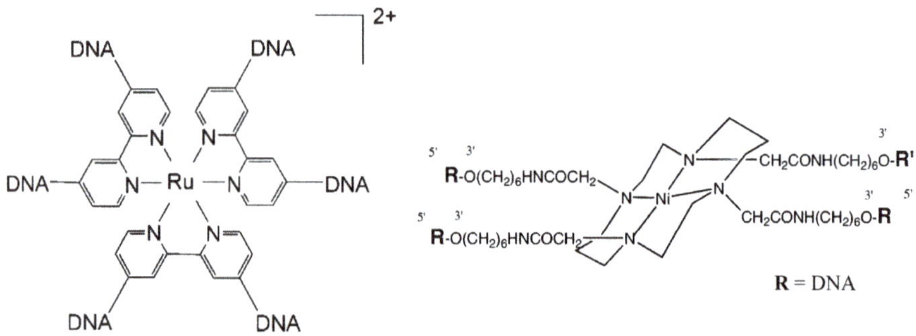

Figure 2.15 Six-arm[61] and four-arm[60] building blocks (reprinted with permission from ref. 60. Copyright 2004 American Chemical Society).

Other examples include McLaughlin's four-arm[60] and six-arm[61] building blocks of DNA radiating from Ni^{2+} cyclam and $Ru(bpy)_3^{2+}$ cores, respectively, that were able to assemble into higher order structures (Figure 2.15). From the Ni^{2+} cyclam building block a number of higher-order constructs were identified, but the six-arm building block generally formed large structures that were difficult to define.

2.6 DNA Modified with Ligands

There are limitations for attaching a complete metal complex to DNA. Complexes must be robust enough to withstand the conditions of automated DNA synthesis, or must be stable, soluble, and compatible with aqueous conjugation conditions to DNA (typically less efficient than coupling during automated DNA synthesis, which is carried out in organic solvents). Complexes that bind to DNA *via* intercalation must have the proper ligands to allow sufficient π-stacking with the DNA base pairs, and this mode of binding is not appropriate for the generation of higher-order DNA assemblies. An alternative approach is to first attach a metal-binding ligand to DNA, and then add the metal (Figure 2.2C). One advantage of this method is that a large variety of individual ligands are generally compatible with DNA synthesis. This in turn can allow metal complexes of varying robustness and functionality to be attached to the DNA. Another advantage of this approach is that it is relatively straightforward to incorporate ligands into any position of a DNA sequence. The most common methods to incorporate ligands into DNA are the synthesis of metal base pairs or the attachment of ligands to the DNA backbone.

2.6.1 Artificial Metallo–Base Pairs

Incorporating metal-binding ligands as artificial bases within the DNA base stack is an excellent way to site-specifically place metal ions inside a DNA duplex. Since Shionoya's first example of a Pd^{2+} ion coordinated to a pair of amine-modified bases[62] (Figure 2.16, **3**),[63] there have been many examples of

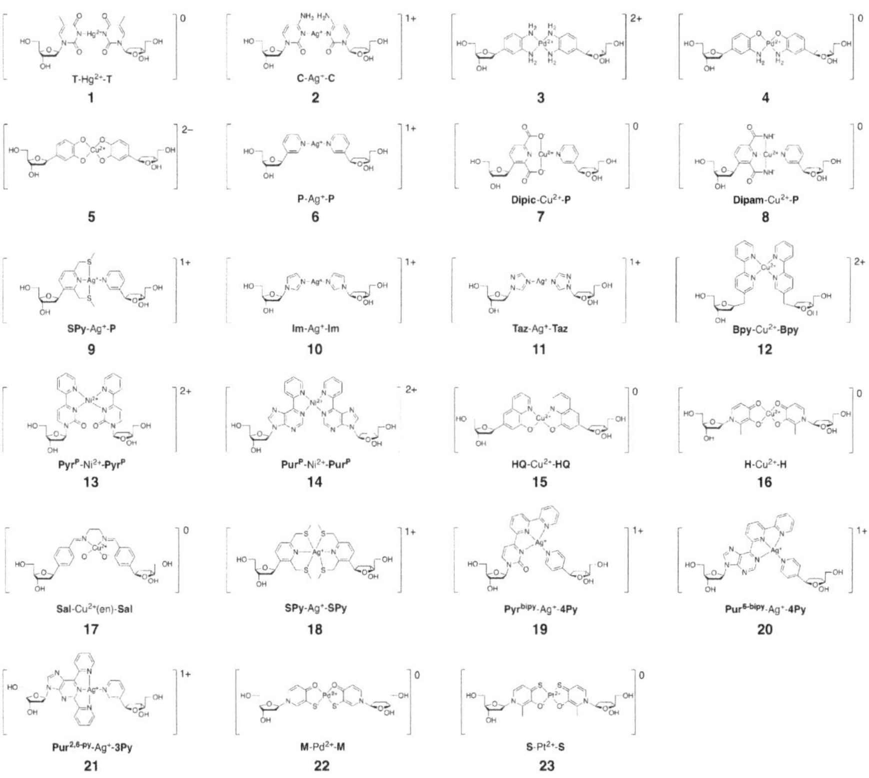

Figure 2.16 Selection of metal-mediated base pairs (reprinted with permission from ref. 63. Copyright 2013 American Chemical Society).

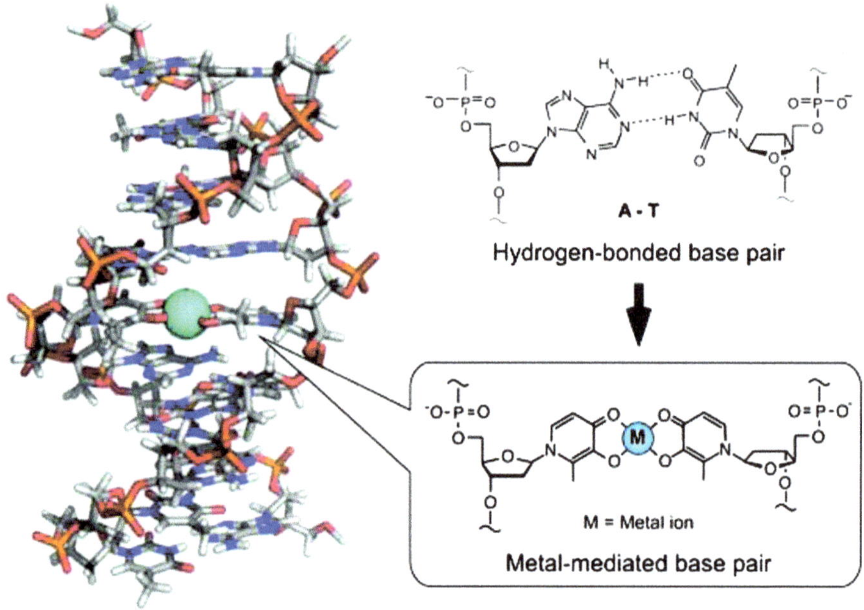

Figure 2.17 Computer model of an artificial metal-mediated base pair (reprinted with permission from ref. 31. Copyright 2012 American Chemical Society).

metal-coordinating base pairs and their incorporation into DNA. Such constructs are anticipated to apply to generating artificial systems capable of information storage, creating molecular-scale magnets, or facilitating long-distance electron transfer. Some thorough reviews on the subject have been published recently,[31] and thus here we aim to give a brief overview of the field.

A selection of metallo–base pairs is shown in Figure 2.16. For a metallo–base pair to fit well into the DNA duplex, it must have appropriate size, planarity, and charge. Figure 2.17 shows a computer model of a metal-mediated base pairing system, and how metal ions might be placed inside the DNA double helix.[31] Linear and square planar complexes are the most commonly employed. The constructs are most commonly characterized by thermal denaturation (melting temperature, T_m) and circular dichroism (CD) experiments, and in some cases detailed NMR experiments[64] and crystallographic techniques[65,66] have been used to elucidate the structures of the incorporated metallo–base pairs and their effects on the DNA duplex.

Linear metallo–base pairs have been generated using pyridine, imidazole, or triazole, such as the imidazole–Ag^+–imidazole (Im–Ag^+–Im, Figure 2.16, **10**) complexes reported by the Müller group.[67] The binding affinity of Ag^+ in these complexes was ~10^6 M^{-1}, which is stronger than the binding of Ag^+ to natural base pairs in the C–Ag^+–C mismatch. While the inclusion of the non-metallated artificial bases results in a slight destabilization of the duplex as measured by melting temperatures, coordination of Ag^+ ions results in duplex stability comparable to unmodified DNA. In further work, up to three

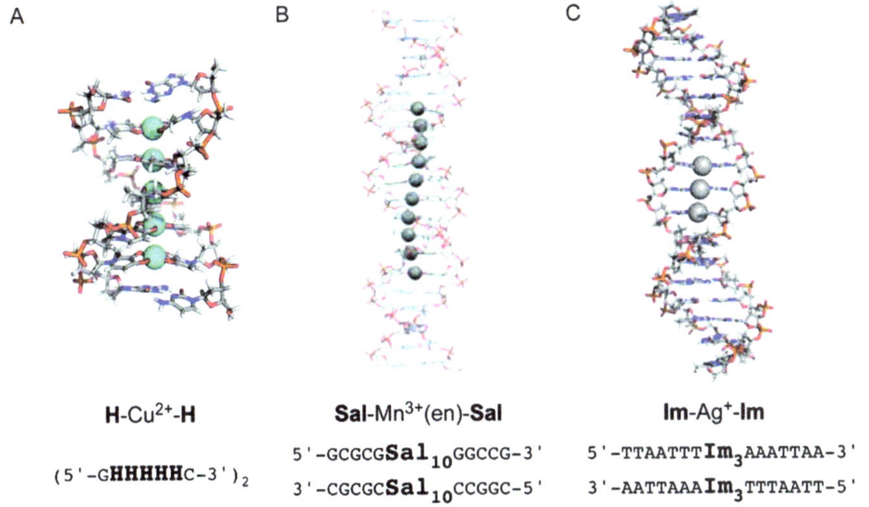

Figure 2.18 Systems with multiple metallo-base pairs. (A) Five H–Cu^{2+}–H base pairs. (B) Ten Sal–Mn^{3+}(en)–Sal base pairs. (C) Three Im–Ag^{+}–Im base pairs. (Reprinted with permission from ref. 31. Copyright 2012 American Chemical Society).

Im–Ag^{+}–Im were incorporated sequentially into a duplex (Figure 2.18C) and detailed NMR studies were conducted to obtain a solution structure of the metallo-DNA construct.[64]

Shionoya has extensively studied metallo-base pairs formed by Cu^{2+} ions and hydroxypyridone (H) (H–Cu^{2+}–H, Figure 2.16, **16**).[68] When the hydroxypyridone base pair was programmed into the middle of a 15-base oligonucleotide, the non-metallated artificial base pair caused a slight destabilization of the duplex (T_m = 37 °C, compared to unmodified DNA, T_m = 44.2 °C). When Cu^{2+} was added, however, the resulting duplex was even more stable than the unmodified DNA (T_m = 50.1 °C). Electron paramagnetic resonance (EPR) confirmed the presence of the bound Cu^{2+} ions, and CD measurements showed the characteristic features of typical B-DNA. In subsequent work, when up to five H–Cu^{2+}–H base pairs were arranged into a short oligonucleotide it was found that the Cu^{2+}–Cu^{2+} distance was ~3.7 ± 0.1 Å, similar to the distance between natural base pairs in DNA. Interestingly, neighboring metallo-base pairs coupled with each other ferromagnetically (Figure 2.18A).[69] More recently, oligonucleotides (15 bases long) were connected to single walled carbon nanotube break junctions deposited onto surfaces. It was found that oligonucleotides with one H–Cu^{2+}–H base pair exhibited conductance parallel to duplex DNA. Upon addition of EDTA to remove the Cu^{2+}, the conductance decreased significantly, and upon addition of fresh copper(II), the conductance increased again. Three H–Cu^{2+}–H base pairs in a duplex exhibited higher conductance than a single metallo-base pair.[70]

The Carell group has studied metal–salen complexes. In these base pairs, salicylic aldehyde-containing bases combine with added ethylene diamine (en) to create a salen coordination sphere (Sal) to which metal ions can be added (Figure 2.16, **17**). The ethylene diamine acts in a similar way to a cross-linker and imparts added stability to the metallo–base pairs. For example, for 15-mer oligonucleotides with a salen base pair in the center, incorporation of just the salicylic aldehyde bases resulted in destabilization of the duplex (T_m measurements), but adding excess ethylene diamine had a stabilizing effect (although not as high as an unmodified duplex). However, adding Cu^{2+} resulted in significant stabilization (~30 °C).[71]

In contrast to the ferromagnetic coupling of neighboring Cu^{2+} complexes by Shionoya, antiferromagnetic coupling was observed for two Cu^{2+} ions in the salen system, attributed to subtle differences in the two ligand environments.[72] Multiple metal ions could be incorporated, and up to 10 Mn^{2+} ions were added to oligomers in which the Sal–Mn^{2+}(en)–Sal base pairs were flanked by five C–G base pairs (Figure 2.18B).[73] The ethylene diamine reaction with the salicylic aldehyde at pH 7 in water is reversible, which could allow opening and closing of the salen coordination environment for more dynamic structures. In very interesting work, oligonucleotides containing a Sal–Cu^{2+}(en)–Sal base pair were replicated by the polymerase *Bst Pol 1*, and could be amplified by the polymerase chain reaction (PCR).[65] Also in this study, three crystal structures were obtained showing the fit of the Sal–Cu^{2+}(en)–Sal-containing duplex inside the active *Bst Pol 1* fragment, as well as of the salicylic aldehyde-containing primer strand for replication in complex with the polymerase.

There are many other examples of metallo–base pairs in addition to the select examples described here. For instance, Seela developed metallo–base pairs that coordinate two Ag^{+} ions per base pair, resulting in duplex stabilization,[74] and three of these artificial base pairs were incorporated into duplexes for the coordination of six Ag^{+} ions.[75] Müller recently developed 1,2,3-triazole-based Click nucleosides,[76] Switzer generated purine- and pyrimidine-like bases that coordinate to Ni^{2+},[77,78] and Carell organized five Cu^{2+} ions using a novel pyrazole metal-binding base pair and significantly stabilized 15-mer duplexes.[79] Shionoya used the hard-soft-acid-base rules to coordinate the softer Pd^{2+} and Pt^{2+} ions to mercaptopyridine- and hydroxypyridinethione-bearing nucleosides.[80]

The precise positioning of multiple, but *different*, metals within the DNA duplex *via* artificial metallo–base pairs was achieved by Shionoya and Carell (Figure 2.19, top). Hydroxypyridone-based (H) and pyridine-based (P) oligonuleosides were programmed into DNA during automated synthesis. Site-specific coordination of Cu^{2+} and Hg^{2+} to form H–Cu^{2+}–H and P–Hg^{2+}–P resulted in the organization of the metals into their predetermined patterns.[81] In another example, salen (Sal) and thymine (T) base pairs were programmed to organize up to 10 Cu^{2+} and Hg^{2+} ions (Figure 2.19, bottom).[31] In these experiments, the judicious choice of metals and metal-binding ligands is crucial for the site-selectivity of metal binding. This work shows the potential to align metal ions in precise arrangements using DNA strands.

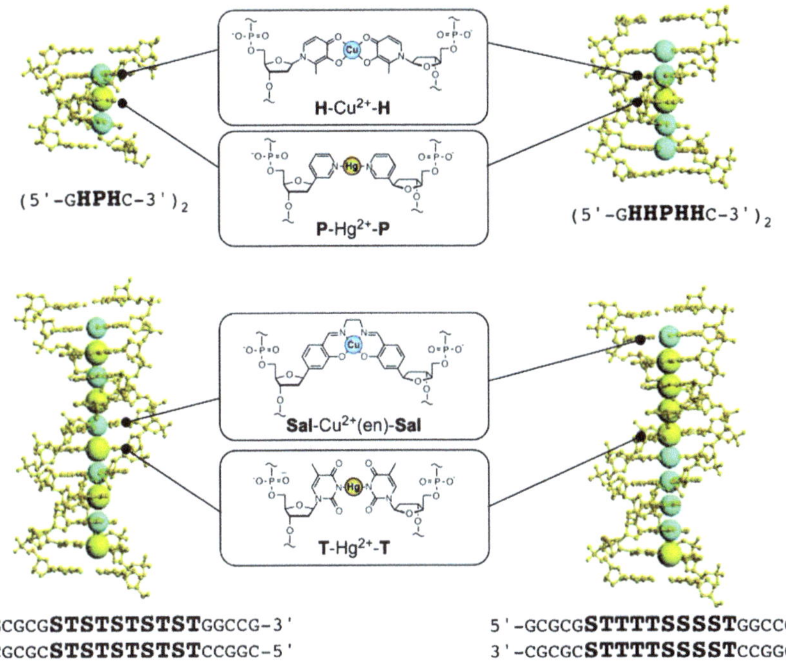

Figure 2.19 Arrangement of different metal ions *via* metallo-base pairs (reprinted with permission from ref. 31. Copyright 2012 American Chemical Society).

2.6.2 Attaching Metal-Binding Ligands to the DNA Backbone

2.6.2.1 Salen Complexes (Salen: N,N′-Bis(salicylidene) ethylenediamine)

Metallosalen complexes are interesting candidates for the organization of metal complexes in DNA. Many metallosalens are compatible with aqueous conditions and have demonstrated enantioselective catalytic activity. Salen complexes are suitable for a wide range of metal ions, as was noted earlier in the formation of salen-containing metallo-base pairs. Czlapinski and Sheppard used a template strategy to generate Mn^{2+} and Ni^{2+} metallosalen complexes in the middle of DNA strands (Figure 2.20A),[82,83] or at the end of a DNA duplex.[84] Further work demonstrated the potential functionality of these complexes, as under certain conditions Ni^{2+} metallosalens could site-specifically cleave DNA.[85]

Gothelf created DNA/salen conjugates as modules for higher-order assembly (Figure 2.20B). Two or three salicylaldehyde groups were incorporated within the same molecule, depending on the module, with DNA arms attached to bring the modules together in a preprogrammed alignment. Using bi-functionalized modules, addition of Mn^{2+} and ethylene diamine

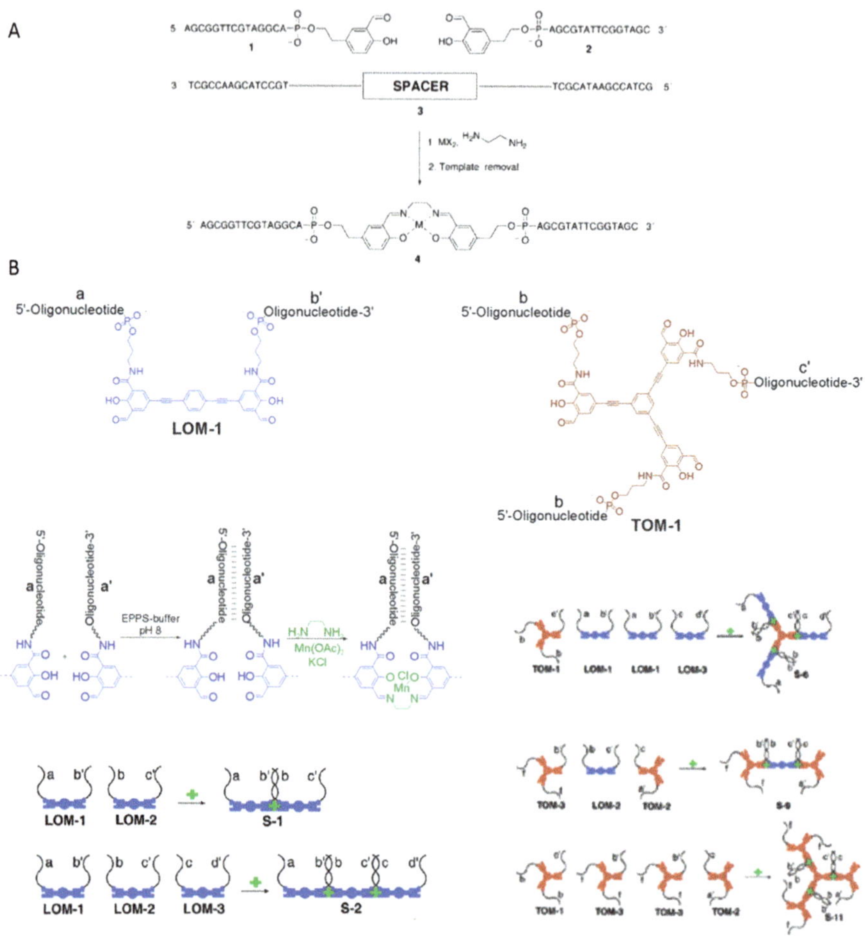

Figure 2.20 DNA functionalized with metallosalen complexes. (A) Reprinted with permission from ref. 82. Copyright 2001 American Chemical Society; (B) reprinted with permission from ref. 86. Copyright 2004 American Chemical Society.

formed linear Mn–salen complexes that were locked in a coplanar geometry. Three-armed modules could also be assembled into branching architectures, or the linear and three-armed modules could be combined into any desired pattern. In this design, no additional external DNA template was required, and a variety of final assemblies could be made depending on the modular building blocks used (Figure 2.20B).[86] Modules with disulfide linkers between the salicylaldehyde moieties and the DNA were connected with Al^{3+} ions.[87] In this design, DNA was again used to align the organic molecules for metallation, but the DNA component could later be removed to leave functionalized organic arrays in pre-determined arrangements.

2.6.2.2 Porphyrins

DNA–porphyrin conjugates have been intensely studied. Porphyrins are synthetically accessible, planar molecules with well-developed chemistry. They have tunable photophysical properties and can participate in electron- and energy transfer reactions. Porphyrins have been studied for applications including light-harvesting, electronic and photonic wires, logic gates, optoelectronics and biomolecule sensing. For these applications the precise ordering of the porphyrin molecules is key for obtaining controlled and predictable electron and energy transfer between array components.

Porphyrins can bind to DNA either non-covalently (π-stacking or hydrogen-bonding) or covalently. π-Stacking of porphyrin molecules with DNA motifs may be a simpler, lower-cost approach, but results in less controlled porphyrin arrangements. The literature regarding porphyrin–DNA is extensive; thus our discussion will be mainly limited to examples of metal-containing porphyrins covalently conjugated to DNA.

The Stulz group has attached porphyrins to DNA bases *via* semi-rigid linkers, forcing the porphyrin to protrude through the major groove of the double helix. This avoids intercalation and ensures minimal interactions between the DNA and the porphyrin. In one example, they generated two different porphyrins for incorporation into a DNA-mediated array (Figure 2.21).[88] Multiple porphyrins attached on alternating positions to both DNA strands can π-stack in the major groove of the double helix, thus creating a "zipper-like" motif that significantly stabilizes the DNA duplex. Furthermore, it is important to note that in general, it is difficult to form mixed-metal porphyrin–DNA conjugates because any metal ion bound inside the porphyrin coordination site is lost as a result of the conditions for automated DNA synthesis. Stulz approached this by re-introducing metal ions (Zn^{2+}, Co^{2+} or Cu^{2+}) to single stranded porphyrin–DNA conjugates. Hybridization of a metal-containing strand to a non-metallated strand resulted in the formation of a duplex DNA–porphyrin array containing 12 alternating free base–metal porphyrins.[89]

Balaz has demonstrated that conjugates of Zn^{2+}-porphyrins and DNA can form dimers, depending on the linker between the DNA and the porphyrin and the ionic strength of the solution (Figure 2.22A),[90] and that porphyrins conjugated to the 5′ end of DNA can promote the formation of a dA–dG antiparallel homoduplex (Figure 2.22B). The porphyrin also serves as a spectroscopic reporter of the DNA structure formed by circular dichroism.[91] More recently, the group organized free-base-, Ni^{2+}- and Zn^{2+} porphyrins onto a single stranded DNA template using hydrogen bonding interactions (Figure 2.23).[92] If the porphyrin was free-base or contained Ni^{2+}, chiral arrays were formed with helicity induced by the hydrogen bonding porphyrins. High-resolution TEM showed that the Ni^{2+}-containing arrays further aggregated to form nanowires and fibrils that were a few hundred nanometres to several micrometres in length, and 15–16 nm wide. In this example of using single

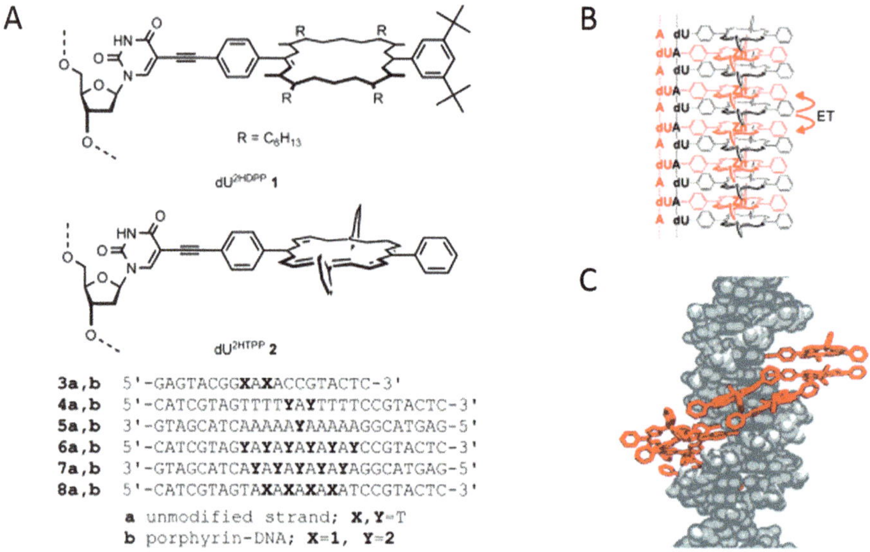

Figure 2.21 (A) Structures of porphyrins incorporated into DNA strands; (B) alignment of the mixed-porphyrin array; (C) molecular modeling of porphyrin array structure.[88]

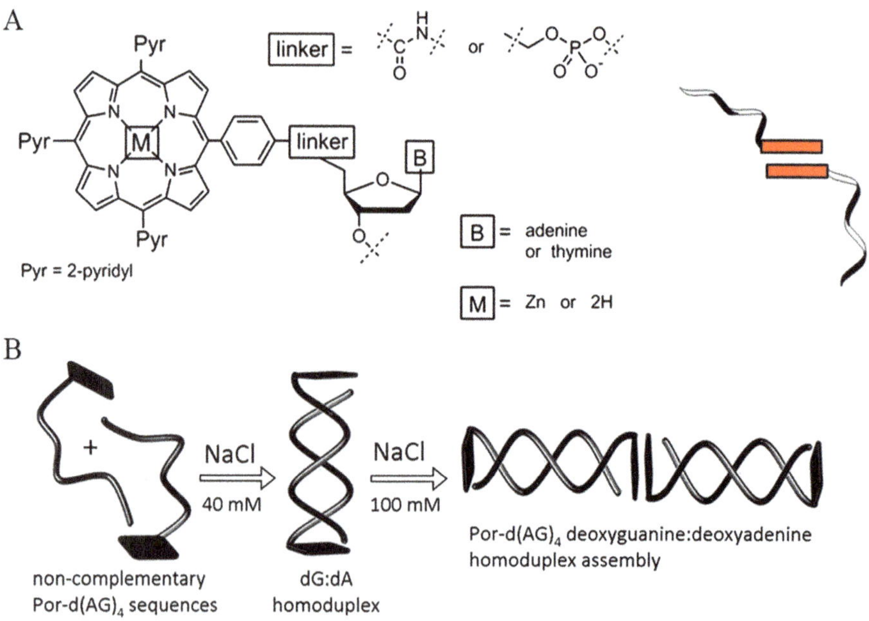

Figure 2.22 (A) DNA–porphyrin conjugates forming dimers.[90] (B) dA:dG duplexes stabilized by porphyrin interactions.[91]

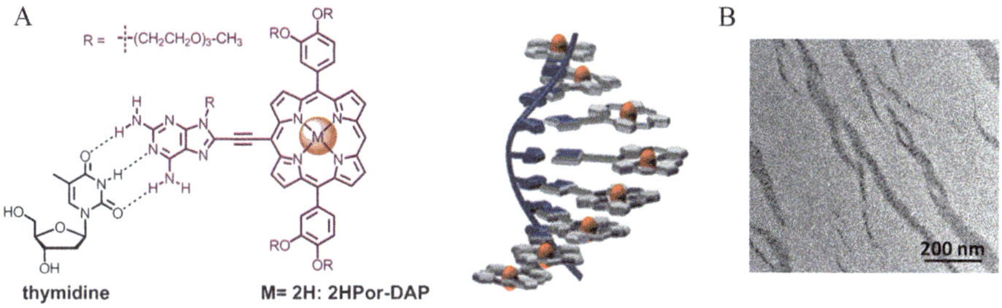

Figure 2.23 (A) Alignment of porphyrin molecules on single stranded DNA by hydrogen bonding, and (B) TEM image of porphyrin–DNA assemblies.[92]

stranded DNA as the scaffold for porphyrin arrangement, the metal ion used and the annealing rate strongly affected the final assembly. Zn^{2+}-porphyrins did not π-stack efficiently to form chiral structures, and this is believed to be a result of the five-coordinate Zn^{2+} ion not readily surrendering its axially coordinated water molecule. Ni^{2+} on the other hand could more easily adopt a four-coordinate conformation, and helices of single strand DNA/Ni^{2+}-porphyrin retained chirality upon heating up to 85 °C. Furthermore, fast annealing of the DNA with the porphyrins favored left-handed helices and slow annealing favored right-handed helices.

In a recent study by Wagenknecht,[93] porphyrins metallated with Cu^{2+} or Au^{3+} were attached to DNA by either copper-mediated alkyne–azide coupling ('click' chemistry) or *via* a Diels–Alder reaction. Click chemistry functionalization involves copper ions, and as a result, Cu^{2+} ions were automatically incorporated into the porphyrins. If Au^{3+}-containing porphyrins were desired, the Au^{3+} had to be added to the porphyrins prior to the post-synthesis modification. The Diels–Alder modification is a metal-free method and is thus more flexible with respect to metal insertion.

Majima used DNA hybridization to organize two different porphyrin molecules, one containing no metal and one containing Zn^{2+}, by cross-linking four DNA duplexes (Figure 2.24).[94] The arrangement of the porphyrins was programmed by the sequences of the complementary strands. Majima and Seeman combined a porphyrin-based four-way junction with a DNA tile system known to form 2D arrays. This porphyrin junction induced the tile to roll into circular tetramers and led to rolled up nanotubes rather than the flat sheets normally observed.[95]

The Albinsson group used porphyrins as anchors to attach DNA to lipid membranes (Figure 2.25).[96,97] One of the goals of this work is the generation of light-harvesting systems through the precise positioning of components onto surfaces *via* DNA hybridization. Various features of the DNA–porphyrin binding to liposomes were examined, including the length of the DNA and spacer separating the DNA from the porphyrin anchor. One anchored single-stranded DNA–porphyrin conjugate could be hybridized to either an

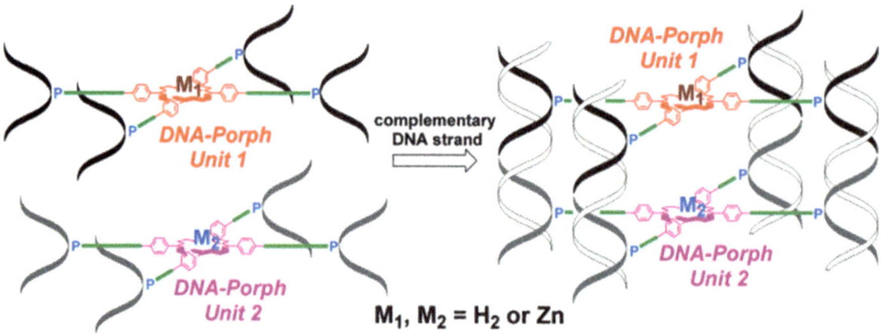

Figure 2.24 Four-duplexes assembled with two porphyrin derivatives (reprinted with permission from ref. 94. Copyright 2005 American Chemical Society).

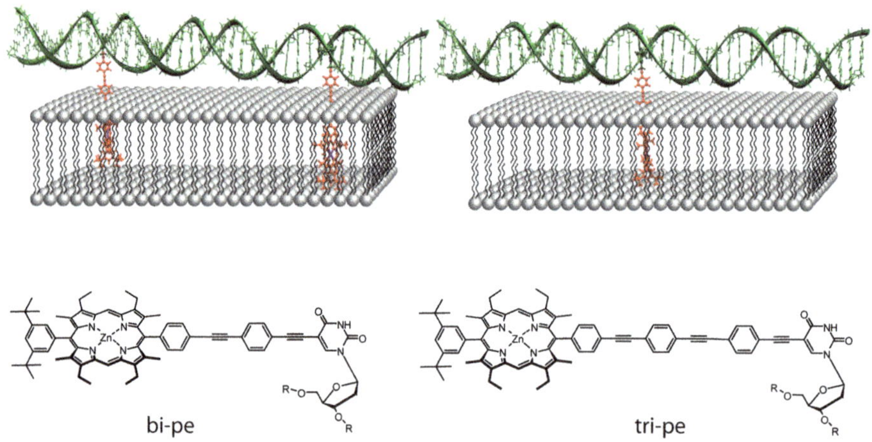

Figure 2.25 Anchoring of Zn^{2+}-porphyrin–DNA conjugates to lipid surfaces (reprinted with permission from ref. 96. Copyright 2010 American Chemical Society).

unmodified DNA strand or to another porphyrin-modified DNA to yield a duplex anchored by two porphyrin moieties.

2.6.2.3 DNA 2D- and 3D-Metallosupramolecular Structures with Polypyridine Ligands

Incorporation of bipyridine-, phenanthroline-, and terpyridine derivatives into DNA assemblies is a promising method of adding structural diversity and functionality to DNA assemblies. These ligands are generally tolerant of the conditions of automated DNA synthesis and thus their site-specific incorporation into DNA is relatively straightforward. Careful selection of the combinations of such ligands can create ligand environments suitable for a

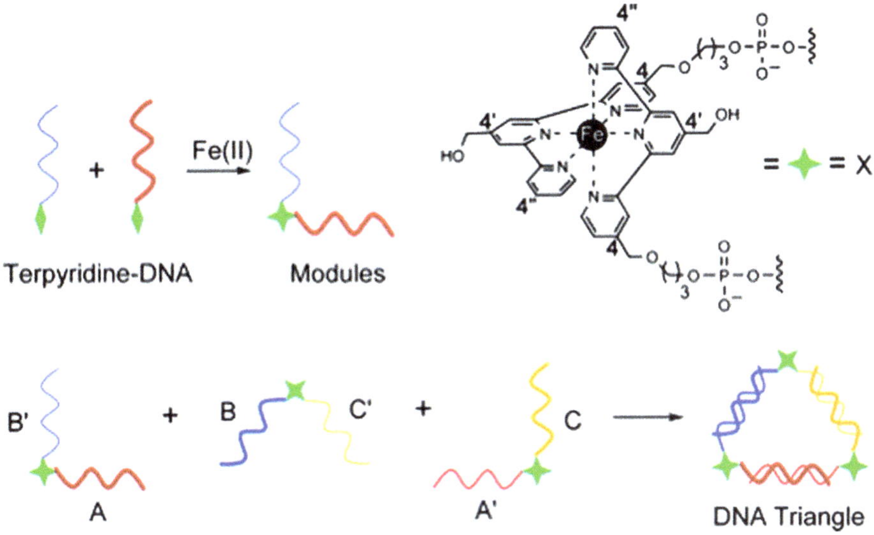

Figure 2.26 Fe^{2+}-templated building blocks for assembly into discrete triangle (reprinted with permission from ref. 98. Copyright 2004 American Chemical Society).

wide variety of metal ions, which can be added after DNA synthesis. These ligands are well suited to function as junctions for DNA assembly, and conversely, DNA assembly can position these complexes in well-defined arrangements. The end results are generally DNA assemblies with controlled 2D and 3D structures and site-specific placement of metal complexes. Such predictable metal arrangement is expected to have implications in nanoelectronics, nanooptics, catalysis, data storage, and artificial photosynthesis.

In one early report, Han functionalized DNA with terpyridine ligands. The addition of Fe^{2+} resulted in the generation of building blocks, which could further assemble into discrete triangles by DNA hybridization (Figure 2.26).[98]

Kramer demonstrated the power of metal ion coordination to affect DNA assembly (Figure 2.27A).[99] Terpyridine moieties were conjugated to a DNA strand at both the 5′- and 3′ ends. The addition of one equivalent of Zn^{2+} or Fe^{2+} caused the formation of a cyclic DNA structure, even if the DNA had been hybridized to a partially complementary strand. Addition of excess Zn^{2+}, but not Fe^{2+}, could cause the cycle to reopen by saturating the two terpyridine sites. Haner generated oligonucleotides functionalized with bipyridine (bpy) or phenanthroline (phen) that, when brought together by hybridization with a third strand, formed a ligand environment for a metal ion (Figure 2.27B).[100] Ligand combinations of $(bpy)_2$, bpy:phen, and $(phen)_2$ generally stabilized the duplex even without metal coordination, but addition of Cu^{+} or Cu^{2+} resulted in additional stabilization. Zn^{2+} ions had no significant effect.

The Sleiman group used DNA organization to create optimized ligand environments for the site-selective incorporation of different metal ions into assemblies.[101] DNA strands were synthesized that were terminated with either

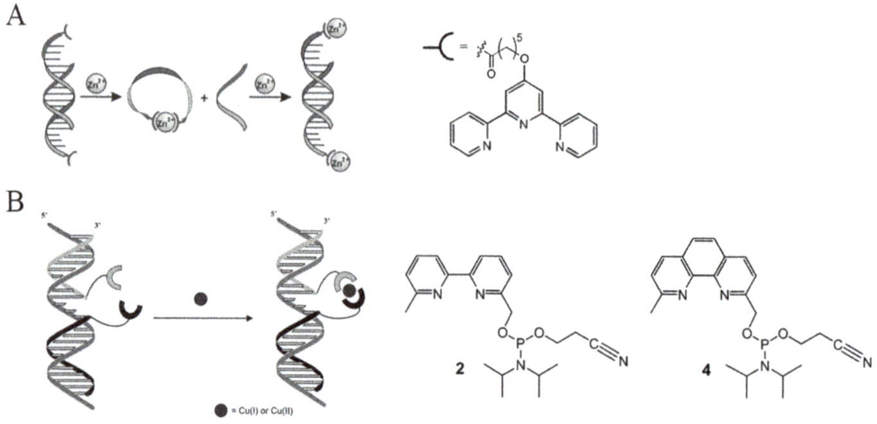

Figure 2.27 (A) Cyclization and hybridization control of terpyridine-modified DNA by added Zn^{2+} (reprinted with permission from ref. 99. Copyright 2005 American Chemical Society). (B) Copper complex-assisted DNA hybridization (reprinted with permission from ref. 100. Copyright 2006 American Chemical Society).

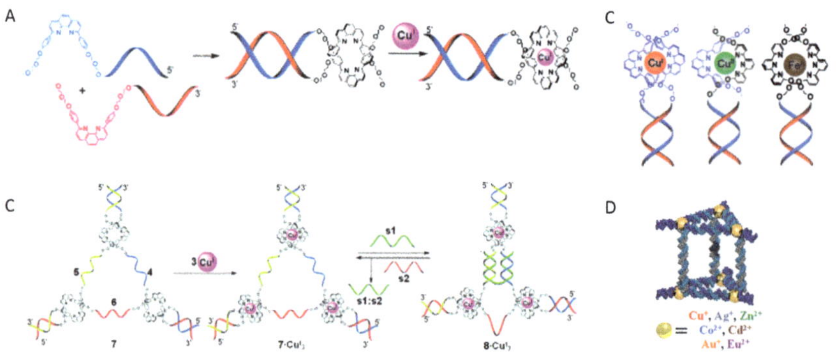

Figure 2.28 (A and B) Assembly of metal-specific metal environments;[101] (C) formation of metal–DNA triangle with variable metal–metal distances;[102] (D) metalated 3D DNA cage.[103]

terpyridine (tpy)- or diphenylphenanthroline (dpp) derivatives. The DNA strands were hybridized together to create three specific ligand environments: $(tpy)_2$, tpy:dpp, or $(dpp)_2$. Various metals were incorporated into these constructs, with significant stabilization of the duplex observed for metals added to their preferred ligand spheres (Figure 2.28A and B). Co^{2+} and Fe^{2+} preferred the octahedral $(tpy)_2$, and Fe^{2+} in particular resulted in one of the highest reported melting temperature increases for a metal-duplex construct. Tetrahedral Cu^+ provided the most stabilization to the $(dpp)_2$ complex, and Cu^{2+} preferred the five-coordinate tpy:dpp environment. Furthermore, the specificity of the complexes towards certain metal ions allowed "error correction" if a metal was put into the "wrong" environment. For example, Cu^+ spontaneously oxidized to Cu^{2+} if added to the tpy:dpp duplex, and Cu^{2+} was spontaneously reduced

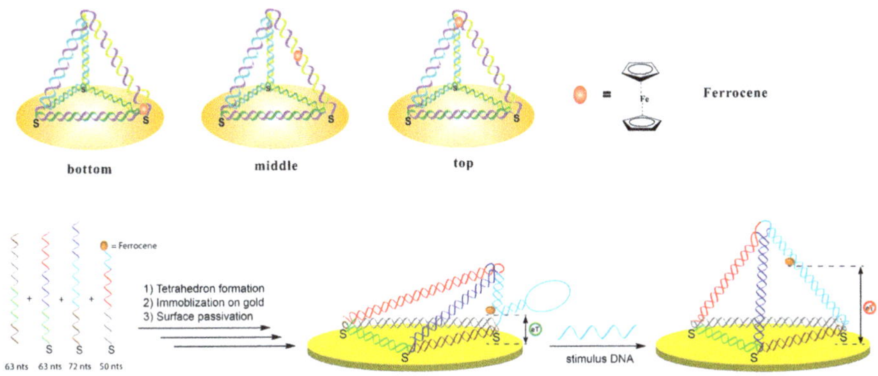

Figure 2.29 Self-assembled 3D DNA tetrahedral bound to gold electrodes. Top: DNA-controlled positioning of ferrocene (reprinted with permission from ref. 104. Copyright 2012 American Chemical Society). Bottom: structurally switchable tetrahedron labeled with ferrocene (reprinted with permission from ref. 105. Copyright 2012 American Chemical Society).

to Cu^+ if added to the $(dpp)_2$ duplex. If four-coordinate Ag^+ was added to the tpy:dpp duplex, it could be easily displaced by Cu^{2+}. The preference of Fe^{2+} to the $(tpy)_2$ ligand sphere was so strong that if it was added to the tpy:dpp duplex a rearrangement of the DNA would occur to form dimers linked by $Fe(tpy)_2{}^{2+}$.

The Sleiman group also demonstrated the placement of metal ions into a triangle that had single stranded sides and Cu^+ ions at the corners (Figure 2.28C).[102] The coordination sphere for the Cu^+ was two diphenylphenanthroline ligands templated by double stranded DNA extending from the triangle corners. One notable advantage of this approach was that Cu^+, which is usually not stable in water, could be added to the DNA nanoassembly as a result of the unique coordination environment. More importantly, the single stranded sides imparted controlled, dynamic character to the Cu^+–Cu^+ distances, because reversibly adding and removing an additional strand could modulate the triangle structure and the distance between metal complexes.

This work with DNA duplexes and triangles was taken into three dimensions with the formation of a prismatic DNA cage with pre-programmed placement of six metal ions at the vertices (Figure 2.28D).[103] To generate the self-assembled 3D cage, two triangles with $(dpp)_2$ ligand environments were linked *via* their single stranded sides. Metals could be added either to the completed 3D cage, or to the component triangles prior to cage assembly. Cu^+, Ag^+, Zn^{2+}, Co^{2+}, Cd^{2+}, Au^+, and Eu^{2+} could all be added to the prism. In this case the prism contained six of the same metal complex, but in principle, different metal coordination spheres could be incorporated to allow precise location of different metal ions at each vertex.

In recent work by Yan and Fan, ferrocenes were positioned (post-DNA synthesis) at controlled locations in a self-assembled 3D DNA tetrahedron that was immobilized onto a gold electrode (Figure 2.29, top; this example does not involve polypyridine ligands, however we include it in this section because

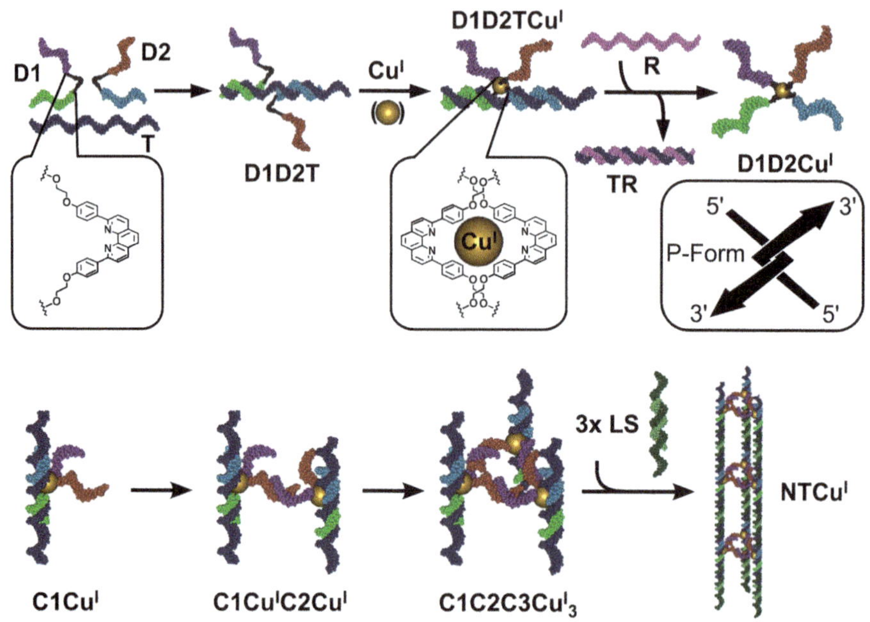

Figure 2.30 Metal-mediated assembly of chiral four-armed DNA junction and its assembly into metal-containing nanotube.[106]

it illustrates the use of a 3D-DNA structure to organize metal centers). The ferrocene complex was attached to the DNA after DNA synthesis. Charge transfer rates between ferrocene and the electrode surface varied depending on the precise location of the ferrocene complex.[104] Furthermore, a structurally switchable tetrahedron, with one vertical edge conjugated to a ferrocene and shortened by a hairpin loop, was also bound to a gold electrode surface (Figure 2.29, bottom).[105] In the presence of a DNA stimulus strand, the hairpin loop was reconfigured, allowing full extension of the shortened edge and increasing the distance between the ferrocene and the electrode. This is an interesting step towards the construction of stimulus-induced, electronically addressable nanoswitching devices based on surface-immobilized 3D DNA assemblies.

The coordination of metal ions was also used by our group to create chiral DNA junctions that could be extended to form DNA nanotubes.[106] The chiral junctions were made by first synthesizing two 20-mer DNA strands with diphenylphenanthroline (dpp) ligands in the middle. A template strand was then used to bring the two dpp–DNA strands together, and Cu^+ coordination locked the two strands as a single unit. Removal of the template strand left a stable chiral four-way DNA junction with four different DNA arms and a Cu^+ ion at the center (Figure 2.30, top). The controlled orientation of the four arms meant that hybridization with proper linking strands could extend the junctions first into triangular "rungs" (2D) and then into extended nanotubular structures (3D) with metals at precise locations (Figure 2.30, bottom).

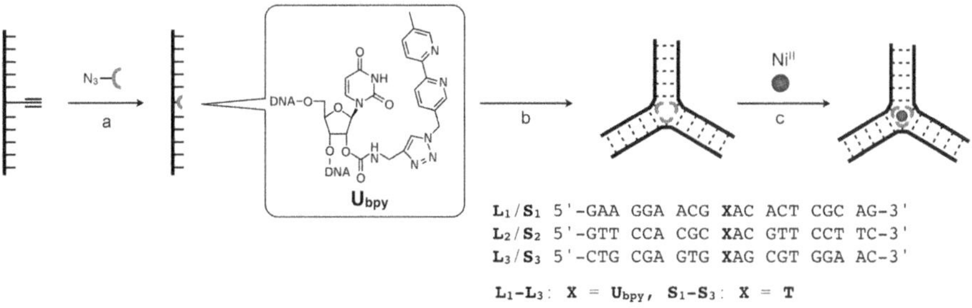

Figure 2.31 Metalation of a three-way junction.[107]

Recent interesting work from Shionoya involved the metallation of a three-way DNA junction.[107] Three DNA strands that usually assemble into a three-way junction were functionalized at the junction bases with bipyridine (bpy) moieties, creating a pre-organized coordination site (Figure 2.31). Metal ions that prefer tris(bpy) octahedral complexes, such as Ni^{2+}, Fe^{2+} and Co^{2+}, increased the stability of the junction. Other metals, such as Zn^{2+}, Mn^{2+} and Cu^{2+}, which bind more weakly into the tris(bpy) pocket provided little or no stabilization.

These examples highlight some interesting ways that a variety of metal ions can be incorporated into DNA nanostructures of various shapes and functionalities.

2.7 Locked Nucleic Acids (LNA) and Peptide Nucleic Acids (PNA)

DNA is the predominant construction material among nucleic acid biopolymers, but peptide nucleic acids (PNA) and locked nucleic acids (LNA) have distinct features that make them useful candidates for metal ion arrangement (Figure 2.32).

LNAs are DNA analogues that have a locked conformation that enhances base stacking and backbone pre-organization, improving hybridization and resulting in increased duplex stability (Figure 2.32, middle). Individual LNA nucleotides can be incorporated into DNA or pure LNA strands can be generated using standard DNA automated synthesis.[108,109]

In DNA nanotechnology, LNAs are potentially useful tools allowing additional control over DNA hybridization properties. Some studies involve attaching metal-chelating tpy or bpy-based moieties to LNA monomers for incorporation into DNA oligomers.[110,111] In one example, terpyridine ligands were attached to LNA bases incorporated into DNA oligomers.[111] Adding 1 equivalent of Ni^{2+}, Cu^{2+}, or Zn^{2+} ions resulted in notable increases in duplex stability, and this was believed to be due to the formation of a metal complex in the minor groove of the duplex (Figure 2.33). Interestingly, additional metal was found to decrease duplex stability, and the stability of the duplexes could also be modulated by the subsequent addition of EDTA.

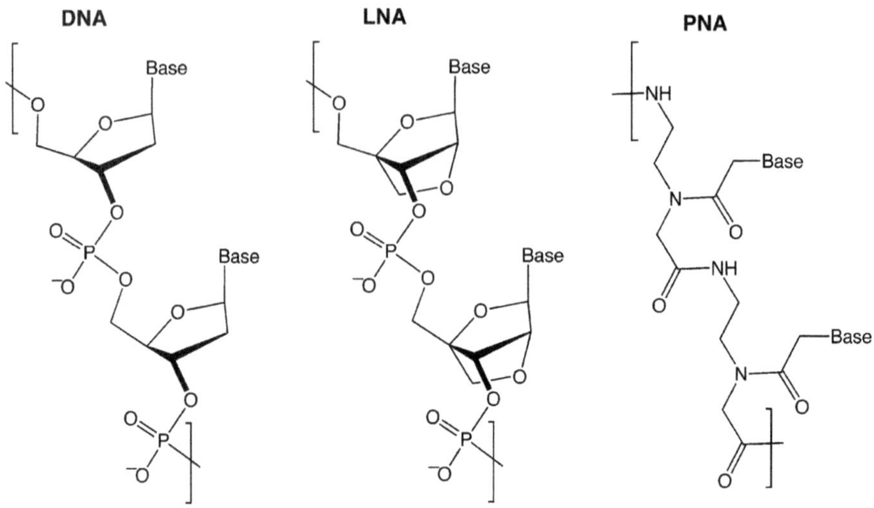

Figure 2.32 DNA (left), locked nucleic acid LNA (middle) and peptide nucleic acid PNA (right) (reprinted with permission from ref. 108. Copyright 2005 American Chemical Society).

PNAs are a different synthetic analogue of DNA. PNAs have the same nucleobases as DNA, but they are connected through a pseudopeptide backbone commonly composed of aminoethyl glycine units (Figure 2.32, right). Unlike the phosphate backbone of DNA, the backbone of PNA is neutral. As a result, DNA–PNA hybrids are not affected by electrostatic repulsion and can exhibit increased thermal stability compared to DNA–DNA hybrids and greater sensitivity to base mismatches than DNA.[108,112] Furthermore, the PNA backbone does not have a chiral center, meaning that PNA duplexes do not have an inherent preferred handedness. They can adopt one, however, if a chiral center is introduced into the PNA, such as a chiral amino acid at the end of the oligomer or through modification to the peptide-like backbone.[113] PNA–PNA double helices have a smaller twist and therefore more base-pairs per turn than B-DNA.

Similarly to DNA, metal complexes can be added to PNA oligomers or duplexes through covalent binding to the backbone and formation of metallo–base pairs. The addition of metal complexes to PNA structures has been thoroughly reviewed.[112,114] PNA–metal conjugates are studied for their applications as probes for DNA detection, metal–PNA artificial proteases or nucleases, fluorescent- or radioactive probes, modulation of cell uptake of PNA, and metal–PNAs as therapeutics, or for their utility in arranging redox-active molecules.

Achim and coworkers have modified PNA monomers with metal-binding ligands for incorporation into PNA oligomers.[113,115–118] Metallo–base pairs based on Cu^{2+}, Co^{2+}, Ni^{2+}, and Zn^{2+} with ligands such as tpy, bpy, and 8-hydroxyquinoline have been incorporated into PNA helices and the effects

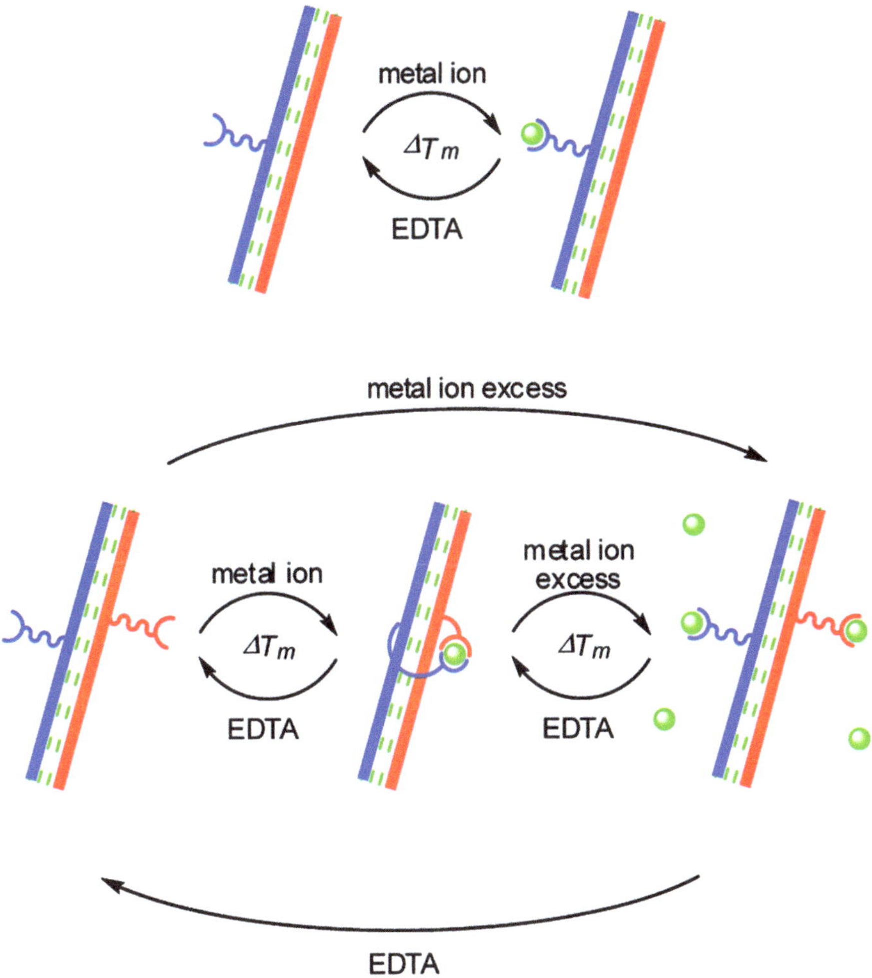

Figure 2.33 Modulation of duplex stability by metal addition to terpyridine-modified LNA bases in DNA oligomers (reprinted with permission from ref. 111. Copyright 2007 American Chemical Society).

on helix handedness, thermal stability, and charge transfer properties have been examined. In one example, ligands such as terpyridine, pyridine or tetrazole were combined to replace a natural base pair in PNA oligomers.[113] In all cases, adding Cu^{2+} to any duplex containing the terpyridine ligand resulted in the formation of $Cu(tpy)_2^{2+}$ complexes, even if the artificial base across from the terpyridine was a pyridine or tetrazole. The result was, in these cases, a complex between two PNA duplexes (Figure 2.34). Furthermore, the preferred handedness of the PNA oligomer could be switched upon metal binding to the coordination site.

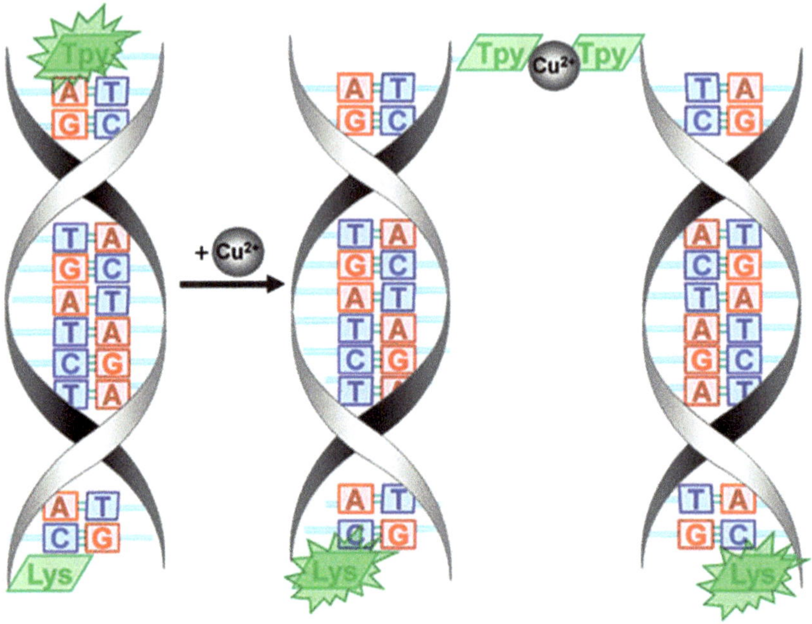

Figure 2.34 Formation of duplex dimers upon coordination of Cu^{2+} with terpyridine bases (reprinted with permission from ref. 113. Copyright 2011 American Chemical Society).

In a recent example of the use of PNA to arrange functional metal complexes, PNA monomers derivatized with $Ru(bpy)_3^{2+}$ and Cu(8-hydroxyquinolate)$_2$ were positioned along a PNA duplex using typical solid-state synthesis procedures. The luminescence quenching of the $Ru(bpy)_3^{2+}$ donor by the Cu(8-hydroxyquinolate)$_2$ acceptor, *via* photoinduced charge transfer, was found to be dependent on the length and sequence of the PNA bridge between them, as well as their positions within the PNA oligomer (terminal *vs.* internal).[119]

2.8 Conclusion

Supramolecular self-assembly and crystal engineering have led to a range of multi-metallic symmetrical structures. On the other hand, the organization of metals *one by one* into arbitrarily designed asymmetric patterns is much more challenging. This approach is what one may need, in order to assemble, *e.g.*, a heterometallic metal–organic framework, an artificial photosynthetic system, or a multimetallic catalytic "factory", among others. As shown in this review, DNA can be used to direct this site-specific and anisotropic organization. A number of robust techniques have been developed, from binding transition metals to the interior of the DNA duplex, to covalently placing a range of metal centers at specific positions on a DNA strand, to non-covalently binding metallosupramolecules to DNA. DNA can thus add programmability and anisotropy to metallosupramolecular structures. It can be used to pattern different transition metals into 2D- and 3D-structures, and these structures are dynamic and stimuli responsive. On the other hand, from the perspective of DNA assembly and DNA biology, transition metals can add a number of desirable properties. Metals can significantly increase the stability of DNA double helices. They introduce orthogonal interactions to the regular A:T, G:C base-pairing code, thus expanding the range of nanostructures that can be formed. Finally, they add much needed functionality to the normally passive DNA structures, such as redox, photophysical, magnetic and catalytic properties. The next step is to start developing the applications of metal–DNA structures. Biological applications include sensing (some of which are starting to emerge), imaging, diagnostics and therapy. Materials applications will stem from the ability of DNA to finely organize transition metals and to switch this organization with external stimuli. They will include multimetallic catalysis, artificial photosynthesis, enzyme mimics, nanoelectronics, nanooptics and spintronics. It is hoped that more young scientists will enter this area, as the future of DNA metallosupramolecular structures is bright indeed.

References

1. S. Park and H. Sugiyama, *Molecules*, 2012, **17**, 12792.
2. A. Kumar and V. Kumar, *Chem. Rev.*, 2014, **114**, 7044.
3. L. H. Tan, H. Xing and Y. Lu, *Acc. Chem. Res.*, 2014, **47**, 1881.
4. O. I. Wilner and I. Willner, *Chem. Rev.*, 2012, **112**, 2528.
5. F. A. Aldaye, A. L. Palmer and H. F. Sleiman, *Science*, 2008, **321**, 1795.
6. E. Winfree, F. Liu, L. A. Wenzler and N. Seeman, *Nature*, 1998, **394**, 539.
7. H. Yan, S. H. Park, G. Finkelstein, J. H. Reif and T. H. LaBean, *Science*, 2003, **301**, 1882.
8. Y. He, Y. Chen, H. Liu, A. E. Ribbe and C. Mao, *J. Am. Chem. Soc.*, 2005, **127**, 12202.
9. S. Hamada and S. Murata, *Angew. Chem. Int. Ed.*, 2009, **48**, 6820.
10. C. Zhang, M. Su, Y. He, X. Zhao, P. Fang, A. E. Ribbe, W. Jiang and C. Mao, *Proc. Natl. Acad. Sci. U. S. A.*, 2008, **105**, 10665.
11. P. W. Rothemund, *Nature*, 2006, **440**, 297.
12. F. A. Aldaye and H. F. Sleiman, *J. Am. Chem. Soc.*, 2007, **129**, 13376.
13. P. K. Lo, P. Karam, F. A. Aldaye, C. K. McLaughlin, G. D. Hamblin, G. Cosa and H. F. Sleiman, *Nat. Chem.*, 2010, **2**, 319.
14. H. Yang, K. L. Metera and H. F. Sleiman, *Coord. Chem. Rev.*, 2010, **254**, 2403.
15. S. Katz, *J. Am. Chem. Soc.*, 1952, **74**, 2238.
16. Z. Kuklenyik and L. G. Marzilli, *Inorg. Chem.*, 1996, **35**, 5654.
17. Y. Miyake, H. Togashi, M. Tashiro, H. Yamaguchi, S. Oda, M. Kudo, Y. Tanaka, Y. Kondo, R. Sawa, T. Fujimoto, T. Machinami and A. Ono, *J. Am. Chem. Soc.*, 2006, **128**, 2172.

18. H. Torigoe, A. Ono and T. Kozasa, *Chem.–Eur. J.*, 2010, **16**, 13218.
19. H. Urata, E. Yamaguchi, T. Funai, Y. Matsumura and S.-i. Wada, *Angew. Chem., Int. Ed.*, 2010, **49**, 6516.
20. S. Johannsen, S. Paulus, N. Düpre, J. Müller and R. K. O. Sigel, *J. Inorg. Biochem.*, 2008, **102**, 1141.
21. J. Kondo, T. Yamada, C. Hirose, I. Okamoto, Y. Tanaka and A. Ono, *Angew. Chem., Int. Ed.*, 2014, **53**, 2385.
22. H. Yamaguchi, J. Sebera, J. Kondo, S. Oda, T. Komuro, T. Kawamura, T. Dairaku, Y. Kondo, I. Okamoto, A. Ono, J. V. Burda, C. Kojima, V. Sychrovsky and Y. Tanaka, *Nucleic Acids Res.*, 2014, **42**, 4094.
23. A. Ono, S. Cao, H. Togashi, M. Tashiro, T. Fujimoto, T. Machinami, S. Oda, Y. Miyake, I. Okamoto and Y. Tanaka, *Chem. Commun.*, 2008, 4825.
24. H. Torigoe, I. Okamoto, T. Dairaku, Y. Tanaka, A. Ono and T. Kozasa, *Biochimie*, 2012, **94**, 2431.
25. H. Urata, E. Yamaguchi, Y. Nakamura and S.-i. Wada, *Chem. Commun.*, 2011, **47**, 941.
26. J. Liu and Y. Lu, *Angew. Chem., Int. Ed.*, 2007, **46**, 7587.
27. G. Mor-Piperberg, R. Tel-Vered, J. Elbaz and I. Willner, *J. Am. Chem. Soc.*, 2010, **132**, 6878.
28. Z. Zhu, L. Xu, X. Zhou, J. Qin and C. Yang, *Chem. Commun.*, 2011, **47**, 8010.
29. C.-W. Liu, Y.-T. Hsieh, C.-C. Huang, Z.-H. Lin and H.-T. Chang, *Chem. Commun.*, 2008, 2242.
30. G. Pelossof, R. Tel-Vered, X.-Q. Liu and I. Willner, *Chem.–Eur. J.*, 2011, **17**, 8904.
31. Y. Takezawa and M. Shionoya, *Acc. Chem. Res.*, 2012, **45**, 2066.
32. K. S. Park, C. Jung and H. G. Park, *Angew. Chem., Int. Ed.*, 2010, **49**, 9757.
33. F. Li, Y. Feng, S. Liu and B. Tang, *Chem. Commun.*, 2011, **47**, 6347.
34. S. Bi, B. Ji, Z. Zhang and J.-J. Zhu, *Chem. Sci.*, 2013, **4**, 1858.
35. F. Xuan, X. Luo and I. M. Hsing, *Anal. Chem.*, 2013, **85**, 4586.
36. R. Yang, J. Jin, L. Long, Y. Wang, H. Wang and W. Tan, *Chem. Commun.*, 2009, 322.
37. J. Ding, W. Qin, Y. Zhang and X. Wang, *Biosens. Bioelectron.*, 2013, **45**, 148.
38. J. M. Thomas, H.-Z. Yu and D. Sen, *J. Am. Chem. Soc.*, 2012, **134**, 13738.
39. P. J. Bond, R. Langridge, K. W. Jennette and S. J. Lippard, *Proc. Natl. Acad. Sci. U. S. A.*, 1975, **72**, 4825.
40. J. K. Barton, E. D. Olmon and P. A. Sontz, *Coord. Chem. Rev.*, 2011, **255**, 619.
41. K. E. Erkkila, D. T. Odom and J. K. Barton, *Chem. Rev.*, 1999, **99**, 2777.
42. A. C. Komor and J. K. Barton, *Chem. Commun.*, 2013, **49**, 3617.
43. B. M. Zeglis, V. C. Pierre and J. K. Barton, *Chem. Commun.*, 2007, 4565.
44. B. A. Jackson, V. Y. Alekseyev and J. K. Barton, *Biochemistry*, 1999, **38**, 4655.
45. R. Kieltyka, P. Englebienne, J. Fakhoury, C. Autexier, N. Moitessier and H. F. Sleiman, *J. Am. Chem. Soc.*, 2008, **130**, 10040.

46. A. J. Boersma, B. de Bruin, B. L. Feringa and G. Roelfes, *Chem. Commun.*, 2012, **48**, 2394.
47. A. J. Boersma, R. P. Megens, B. L. Feringa and G. Roelfes, *Chem. Soc. Rev.*, 2010, **39**, 2083.
48. J. D. Slinker, N. B. Muren, S. E. Renfrew and J. K. Barton, *Nat. Chem.*, 2011, **3**, 228.
49. D. T. Odom, E. A. Dill and J. K. Barton, *Nucleic Acids Res.*, 2001, **29**, 2026.
50. J. Malina, M. J. Hannon and V. Brabec, *Chem.–Eur. J.*, 2008, **14**, 10408.
51. A. C. G. Hotze, B. M. Kariuki and M. J. Hannon, *Angew. Chem., Int. Ed.*, 2006, **45**, 4839.
52. L. Cardo and M. J. Hannon, *Inorg. Chim. Acta*, 2009, **362**, 784.
53. Y. Parajo, J. Malina, I. Meistermann, G. J. Clarkson, M. Pascu, A. Rodger, M. J. Hannon and P. Lincoln, *Dalton Trans.*, 2009, 4868.
54. J. Malina, M. J. Hannon and V. Brabec, *Nucleic Acids Res.*, 2008, **36**, 3630.
55. C. Uerpmann, J. Malina, M. Pascu, G. J. Clarkson, V. Moreno, A. Rodger, A. Grandas and M. J. Hannon, *Chem.–Eur. J.*, 2005, **11**, 1750.
56. A. Oleksi, A. G. Blanco, R. Boer, I. Uson, J. Aymami, A. Rodger, M. J. Hannon and M. Coll, *Angew. Chem., Int. Ed.*, 2006, **45**, 1227.
57. J. Malina, M. J. Hannon and V. Brabec, *FEBS J.*, 2014, **281**, 987.
58. I. Vargas-Baca, D. Mitra, H. J. Zulyniak, J. Banerjee and H. F. Sleiman, *Angew. Chem., Int. Ed.*, 2001, **40**, 4629.
59. D. Mitra, N. Di Cesare and H. F. Sleiman, *Angew. Chem., Int. Ed.*, 2004, **43**, 5804.
60. K. M. Stewart and L. W. McLaughlin, *J. Am. Chem. Soc.*, 2004, **126**, 2050.
61. K. M. Stewart, J. Rojo and L. W. McLaughlin, *Angew. Chem., Int. Ed.*, 2004, **43**, 5808.
62. K. Tanaka and M. Shionoya, *J. Org. Chem.*, 1999, **64**, 5002.
63. Y. Takezawa and M. Shionoya, *Acc. Chem. Res.*, 2013, **46**, 604.
64. S. Johannsen, N. Megger, D. Bohme, R. K. O. Sigel and J. Muller, *Nat. Chem.*, 2010, **2**, 229.
65. C. Kaul, M. Muller, M. Wagner, S. Schneider and T. Carell, *Nat. Chem.*, 2011, **3**, 794.
66. S. Atwell, E. Meggers, G. Spraggon and P. G. Schultz, *J. Am. Chem. Soc.*, 2001, **123**, 12364.
67. K. Petrovec, B. J. Ravoo and J. Müller, *Chem. Commun.*, 2012, **48**, 11844.
68. K. Tanaka, A. Tengeiji, T. Kato, N. Toyama, M. Shiro and M. Shionoya, *J. Am. Chem. Soc.*, 2002, **124**, 12494.
69. K. Tanaka, A. Tengeiji, T. Kato, N. Toyama and M. Shionoya, *Science*, 2003, **299**, 1212.
70. S. Liu, G. H. Clever, Y. Takezawa, M. Kaneko, K. Tanaka, X. F. Guo and M. Shionoya, *Angew. Chem., Int. Ed.*, 2011, **50**, 8886.
71. G. H. Clever, K. Polborn and T. Carell, *Angew. Chem., Int. Ed.*, 2005, **44**, 7204.
72. G. H. Clever, S. J. Reitmeier, T. Carell and O. Schiemann, *Angew. Chem., Int. Ed.*, 2010, **49**, 4927.
73. G. H. Clever and T. Carell, *Angew. Chem., Int. Ed.*, 2007, **46**, 250.

74. H. Mei, I. Rohl and F. Seela, *J. Org. Chem.*, 2013, **78**, 9457.
75. H. Mei, H. Yang, I. Röhl and F. Seela, *ChemPlusChem*, 2014, **79**, 914.
76. T. Richters, O. Krug, J. Kosters, A. Hepp and J. Müller, *Chem.–Eur. J.*, 2014, **20**, 7811.
77. C. Switzer and D. Shin, *Chem. Commun.*, 2005, 1342.
78. C. Switzer, S. Sinha, P. H. Kim and B. D. Heuberger, *Angew. Chem., Int. Ed.*, 2005, **44**, 1529.
79. M. Su, M. Tomas-Gamasa, S. Serdjukow, P. Mayer and T. Carell, *Chem. Commun.*, 2014, **50**, 409.
80. Y. Takezawa, K. Tanaka, M. Yori, S. Tashiro, M. Shiro and M. Shionoya, *J. Org. Chem.*, 2008, **73**, 6092.
81. K. Tanaka, G. H. Clever, Y. Takezawa, Y. Yamada, C. Kaul, M. Shionoya and T. Carell, *Nat. Nanotechnol.*, 2006, **1**, 190.
82. J. L. Czlapinski and T. L. Sheppard, *J. Am. Chem. Soc.*, 2001, **123**, 8618.
83. J. L. Czlapinski and T. L. Sheppard, *Bioconjugate Chem.*, 2005, **16**, 169.
84. J. L. Czapinski and T. L. Sheppard, *Chembiochem*, 2004, **5**, 127.
85. J. L. Czlapinski and T. L. Sheppard, *Chem. Commun.*, 2004, 2468.
86. K. V. Gothelf, A. Thomsen, M. Nielsen, E. Clo and R. S. Brown, *J. Am. Chem. Soc.*, 2004, **126**, 1044.
87. R. S. Brown, M. Nielsen and K. V. Gothelf, *Chem. Commun.*, 2004, 1464.
88. T. Nguyen, A. Brewer and E. Stulz, *Angew. Chem., Int. Ed.*, 2009, **48**, 1974.
89. A. Brewer, G. Siligardi, C. Neylon and E. Stulz, *Org. Biomol. Chem.*, 2011, **9**, 777.
90. G. Sargsyan, B. L. MacLeod, U. Tohgha and M. Balaz, *Tetrahedron*, 2012, **68**, 2093.
91. G. Sargsyan and M. Balaz, *Org. Biomol. Chem.*, 2012, **10**, 5533.
92. G. Sargsyan, B. M. Leonard, J. Kubelka and M. Balaz, *Chem.–Eur. J.*, 2014, **20**, 1878.
93. C. Wellner and H. A. Wagenknecht, *Org. Lett.*, 2014, **16**, 1692.
94. M. Endo, T. Shiroyama, M. Fujitsuka and T. Majima, *J. Org. Chem.*, 2005, **70**, 7468.
95. M. Endo, N. C. Seeman and T. Majima, *Angew. Chem., Int. Ed.*, 2005, **44**, 6074.
96. K. Borjesson, J. Wiberg, A. H. El-Sagheer, T. Ljungdahl, J. Martensson, T. Brown, B. Norden and B. Albinsson, *ACS Nano*, 2010, **4**, 5037.
97. J. G. Woller, K. Borjesson, S. Svedhem and B. Albinsson, *Langmuir*, 2012, **28**, 1944.
98. J. S. Choi, C. W. Kang, K. Jung, J. W. Yang, Y. G. Kim and H. Y. Han, *J. Am. Chem. Soc.*, 2004, **126**, 8606.
99. M. Goritz and R. Kramer, *J. Am. Chem. Soc.*, 2005, **127**, 18016.
100. G. Bianke, V. Chaurin, M. Egorov, J. Lebreton, E. C. Constable, C. E. Housecroft and R. Haner, *Bioconjugate Chem.*, 2006, **17**, 1441.
101. H. Yang, A. Z. Rys, C. K. McLaughlin and H. F. Sleiman, *Angew. Chem., Int. Ed.*, 2009, **48**, 9919.
102. H. Yang and H. F. Sleiman, *Angew. Chem., Int. Ed.*, 2008, **47**, 2443.

103. H. Yang, C. K. McLaughlin, F. A. Aldaye, G. D. Hamblin, A. Z. Rys, I. Rouiller and H. F. Sleiman, *Nat. Chem.*, 2009, **1**, 390.
104. N. Lu, H. Pei, Z. L. Ge, C. R. Simmons, H. Yan and C. H. Fan, *J. Am. Chem. Soc.*, 2012, **134**, 13148.
105. A. Abi, M. H. Lin, H. Pei, C. H. Fan, E. E. Ferapontova and X. L. Zuo, *ACS Appl. Mater. Interfaces*, 2014, **6**, 8928.
106. H. Yang, F. Altvater, A. D. de Bruijn, C. K. McLaughlin, P. K. Lo and H. F. Sleiman, *Angew. Chem., Int. Ed.*, 2011, **50**, 4620.
107. J. L. H. A. Duprey, Y. Takezawa and M. Shionoya, *Angew. Chem., Int. Ed.*, 2013, **52**, 1212.
108. P. S. Ng and D. E. Bergstrom, *Nano Lett.*, 2005, **5**, 107.
109. S. Laschi, I. Palchetti, G. Marrazza and M. Mascini, *Bioelectrochemistry*, 2009, **76**, 214.
110. B. R. Babu, P. J. Hrdlicka, C. J. McKenzie and J. Wengel, *Chem. Commun.*, 2005, 1705.
111. M. Kalek, A. S. Madsen and J. Wengel, *J. Am. Chem. Soc.*, 2007, **129**, 9392.
112. G. Gasser, A. M. Sosniak and N. Metzler-Nolte, *Dalton Trans.*, 2011, **40**, 7061.
113. S. Bezer, S. Rapireddy, Y. A. Skorik, D. H. Ly and C. Achim, *Inorg. Chem.*, 2011, **50**, 11929.
114. C. P. Myers and M. E. Williams, *Coord. Chem. Rev.*, 2010, **254**, 2416.
115. R. M. Franzini, R. M. Watson, G. K. Patra, R. M. Breece, D. L. Tierney, M. P. Hendrich and C. Achim, *Inorg. Chem.*, 2006, **45**, 9798.
116. Z. J. Ma, F. Olechnowicz, Y. A. Skorik and C. Achim, *Inorg. Chem.*, 2011, **50**, 6083.
117. T. Z. Markus, A. R. de Leon, D. Reid, C. Achim and R. Naaman, *J. Phys. Chem. Lett.*, 2013, **4**, 3298.
118. E. Wierzbinski, A. de Leon, K. L. Davis, S. Bezer, M. A. Wolak, M. J. Kofke, R. Schlaf, C. Achim and D. H. Waldeck, *Langmuir*, 2012, **28**, 1971.
119. X. Yin, J. Kong, A. De Leon, Y. Li, Z. Ma, E. Wierzbinski, C. Achim and D. H. Waldeck, *J. Phys. Chem. B*, 2014, **118**, 9037.

CHAPTER 3

Constitutionally Dynamic Metallosupramolecular Systems

MIHAIL BARBOIU*[a]

[a]Adaptive Supramolecular Nanosystems Group, Institut Européen des Membranes – UMR CNRS 5635, Place Eugène Bataillon, CC 047, F-34095 Montpellier, France
*E-mail: mihail-dumitru.barboiu@univ-montp2.fr

3.1 Introduction

Scientific interest in complex dynamic systems is continuously growing because of their interesting applications at the boundaries of different scientific disciplines. Research into this field has resulted in important advancements in most disciplines, including biology, mathematics, engineering, physics, economics, computer science, and more recently in chemistry and materials science.

Exciting examples include cognitive models of thinking,[1] the human genome,[2] ecosystems,[3] metabolomics,[4] systems chemistry,[5,6] and materials science.[7,8] Within this context, multivalent interactions, representing multiple copies of a specific recognition event and exhibiting a large range of reversible valencies, can generate adaptive networks capable of varying their composition/structure. This opens the possibility to engineer them in order

RSC Smart Materials No. 15
Functional Metallosupramolecular Materials
Edited by John G. Hardy and Felix H. Schacher

Published by the Royal Society of Chemistry, www.rsc.org

to control the organization of complex matter across various length scales, potentially generating systems in which novel properties not displayed at the molecular level emerge. We can therefore imagine a fundamental transition from molecular design toward up-scaled constitutional selection strategies,[7–10] which might show great potential in a variety of applications. The concepts underpinning this transition are: "*Supramolecular*[6] or *Chemical*[7] *Darwinism*", "*Integrative Self-assembly*"[8] and "*Chemical Collectivism*".[9]

Constitutional dynamic chemistry (CDC)[10] expresses the adaptive dynamic behavior of both molecular and supramolecular systems, and brings into play self-evolution toward the selection of discrete systems from mixtures of exchanging components. This concerns collections of objects reversibly exchanging and continuously reorganizing at the nano- or macroscopic level.[11] Today these new concepts may be connected with the simple definition of dynamic multicomponent self-assembly in chemical and biological systems.[12,13] Within this context, self-organization of metallosupramolecular architectures (*i.e.* double-helices,[14] grids,[15] borromean rings,[16] capsules,[17] polygons,[18] and zippers[19]), based on the utilization of ligands and metal ions mutually reading out structural information through interactional algorithms, can be considered as a dynamic reversible process, with constitutionally dynamic behaviour. Multiple constitutional interactions shed light on strategies used to dynamically drive the coordination (molecular) and the self-assembly (supramolecular) processes of metal-coordinated architectures, under the pressure of interactional algorithms (*i.e.* stacking, H-bonding, *etc.*). This chapter focuses on the implementation of examples of *constitutional dynamic chemistry* principles to manufacture dynamic constitutional metallosupramolecular systems. The basic working principles of emerging metallosupramolecular architectures are provided in order to better understand the requirements during the structural self-selection for the generation of functional systems. Actual and potential applications of such emerging systems presenting the combined features of structural adaptation are discussed.

3.2 Dynamic Metallosupramolecular Libraries (DMLs)—General Considerations

Simple or double level DMLs can be generated *via* metal–ligand and reversible covalent ligand bond exchanges. The metal ion coordination process may help to position ligands in close proximity. Reversible molecular exchanges and supramolecular non-covalent interactions may synergistically occur, leading to rearrangements of the coordinating ligands within a restricted interactional space, defined by the coordination geometry around the metal centers.[20–22] A rather unconventional and as yet relatively uncharted option is multiple coding, offered in the process of simultaneous coordination with mixtures of ligands, producing DMLs of mono- and poly-nuclear components containing distinct metal ions. The design and construction of metallosupramolecular architectures has attracted intense interest not only due to their potential applications as new materials but also for their fascinating structural diversity. It is based on the spatial organization and functional integration within a supramolecular architecture of molecular ligand components presenting dynamic features.[23] Self-organization of metallosupramolecular entities may be directed by design or by constitutional evolution of dynamic combinatorial systems.[24–28] Self-organization by design is based on the utilization of ligands containing specific molecular information stored in the arrangement of suitable binding sites and of metal ions reading out the structural information through the algorithm defined by their coordination geometry. The constitutional selection of one or more components occurs as a function of either internal factors (such as the nature and the geometry of the binding subunits of the ligand, the stoichiometry, *etc.*) or external factors (such as the nature of the medium, the presence of specific molecules or anions, *etc.*). In view of the lability of the coordination bond, numerous coordination processes may present a number of novel features such as cooperativity, diversity, selection, or adaptation. Thus, coordination equilibria may allow the generation of dynamic libraries of metal complexes presenting features such as amplification in solution or in the solid-state of a constituent from an equilibrating mixture of different coordination arrays.[24–31] Starting with seminal publications by Lehn[24] and Sanders,[25] dynamic combinatorial chemistry has emerged as a powerful tool for the development of new receptors, drugs, catalysts and materials. The dynamic features of supramolecular devices serve to bring into play the constitutional evolution of such species, and the groups of Stang,[26] Severin,[27] and Lehn[28] have recently made crucial advancements in this field. One of the key issues in designing complex dynamic combinatorial libraries (DCLs) is the choice of reversible processes that can interconvert the constitutional components. A number of such reactions have been tested including imine exchange, thiol–disulfide exchange, esterification, and ligand exchange in coordination complexes.[24,25] Combination of two such reversible reactions in double-level dynamic combinatorial libraries using “orthogonal” and “communicative” coordination processes have also been reported by Eliseev *et al.*[29]

3.3 Ligand-Driven Selection in Dynamic Metallosupramolecular Libraries

Coordination equilibria may allow amplification from the dynamic mixtures of very specific metallosupramolecular architectures with a unique network topology.[30] Several types of supplementary features, such as coordination behaviour and geometry, stacking interactions, inter- and intra-molecular spatial hindrance, are some of the parameters identified as being responsible for the amplification of topologically favoured architectures in a DCL equilibrium. Indeed, such complicated topologies have attracted tremendous interest because of their simple one-pot synthesis from molecular components.

Examples presenting features such as structural self-selection (self-sorting),[26–28] reversible switching between different coordination arrays,[29,30] or solid-state selection[31] are becoming increasingly feasible. For example, single level ligand- and double level ligand-/aromatic-imine exchanges around a metal center within terpyridine-type architectures leads to the preferential formation of mixtures of heteroduplex complexes instead of the amplification of the pure compounds (Figure 3.1).[19,20] Within these mixtures, the complexes of pairs of two homoligands $[(L_{ii}L_{jj})Zn]^{2+}$ or those formed by the heteroligand L_{ij}: $[(L_{ij}L_{jj})Zn]^{2+}$, $[(L_{ij}L_{ij})Zn]^{2+}$ and $[(L_{ii}L_{ij})Zn]^{2+}$ are preferentially amplified. Within this context we discovered that for each combination of two homoduplex complexes, the fluorescence intensity of the initial compounds systematically decreases for single level ligand exchange mixtures and then for double level imine/ligand heterocomplexes. Such collective reorganization of the system allows the self-sorting of specific heterocomplex architectures presenting an intrinsic property of energy transfer specifically amplified under the pressure of internal structural factors.[19] In addition, the octahedral metal ion complexation is positioning the aromatic chromophore moieties and the pseudo-terpyridine units of the second ligand in close proximity, so that charge transfer in between the choromophore and the pseudoterpyridine can occur. From a conceptual point of view, the constitutional evolution of such dynamic metallosupramolecular systems toward the selection of mixtures of heteroduplex architectures is of special interest for the improvement of specific charge transfer functions.

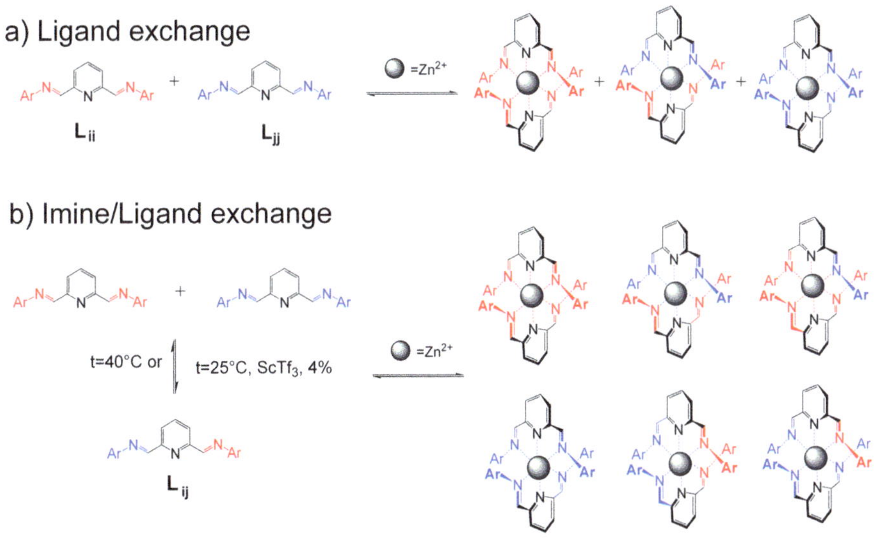

Figure 3.1 (a) Single level ligand-exchange and (b) double level imine/ligand exchange DCLs. (Adapted by permission from the American Chemical Society from ref. 19.)

Similar examples of dynamically instructed mixtures of coordination compounds, where the concept of constitutional self-sorting and a specific function may be in principle associated, have been reported where the crystallization kinetics apparently did not follow the same rules as expected for chemical thermodynamics in solution, and where the dynamic amplification in solution was not simply being transferred to the solid state.[31] A simple inspection of H-bonding distribution on crystal packing reveals that the invariable H-bond network generated by homoligand $\mathbf{L}_{ii}$ is more symmetric and crosslinked than those produced by the two isomers of the heteroligand $\mathbf{L}_{ij}$, is amplified in solution and has reduced crosslinking and asymmetry behaviour in the solid state (Figure 3.2).

The same constitutional approaches can be applied to supramolecular chiral systems. *Supramolecular chirality* obtained from achiral molecular constituents associated through a desymmetrizing interaction mode is by constitution dynamic, and examples of the large scale transcription of such *virtual chirality* have been recently published.[32–34] The *single and double helical systems* represent nice examples of supramolecular chiral systems.[35] The intrinsic features of biological oligosaccharides, nucleic acids and proteins, have made the understanding of the factors governing their *supramolecular chiral self-organization* particularly attractive.[36–39] Usually, such helical structures crystallize in distinct alternative *P* and *M* columns or layers of Δ or Λ mirror enantiomers, *but overall the crystals are racemic*. Intermolecular crystal packing is usually not discriminating; not many previous examples of direct crystallization of enantiopure helical supramolecular single-crystals have been reported.[39,40] Of special interest are terpyridine-type complexes $\mathbf{1}_2Zn^{2+}$–$\mathbf{5}_2Zn^{2+}$, $\mathbf{1}_2Fe^{2+}$ and $\mathbf{1}_2Co^{2+}$ based on achiral ligands **1–5** and metal ions which form double-helical systems presenting *supramolecular helical*

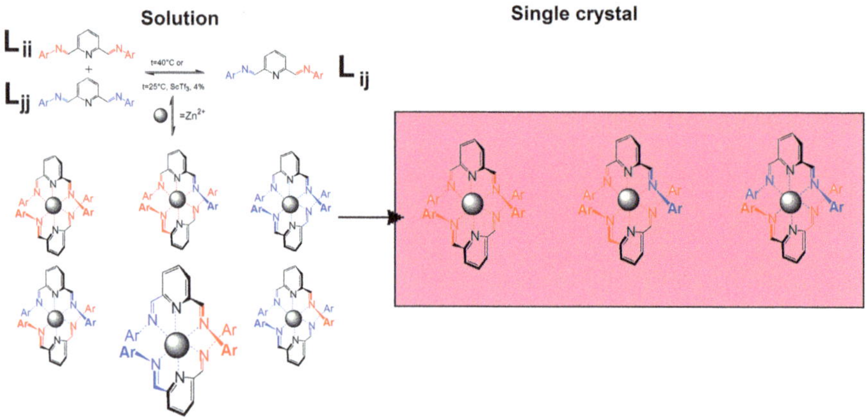

Figure 3.2 Metallosupramolecular dynamic libraries produce in solution the amplification of heteroleptic complexes $[(\mathbf{L}_{ij}\mathbf{L}_{ij})\mathbf{Zn}]^{2+}$, which convert into homoleptic complexes of ligand $\mathbf{L}_{ii}$ *via* constitutional crystallization. (Adapted with permission from ref. 31.)

homochirality (Figure 3.3).[20] The homochirality in the crystalline state is promoted by *constitutional chiral affinity* of supramolecular helices of the same handedness, interacting *via* van der Waals hypersurfaces.[20]

The internal robustness of the columns is mainly responsible for the transmission of the homochiral order at the supramolecular/nanometric level (Figure 3.4). Related sliding biological processes along homochiral 3D hypersurfaces result in the formation of the ion-channel gramicidin A dimers,[41a] the self-aggregation of collagen,[41b] and the self-assembly of the tobacco mosaic virus.[41c]

Along the same lines, simple structural variability could induce significant changes in the composition of the DCLs for a series of halogen substituted pseudoterpyridines: Zn^{2+} homoleptic mononuclear complexes, based on 2,6-pyridinedicarboxaldehydebis(*p*-R-phenylimines) ligands where R = –F, –Cl, –Br, –I.[42] Although isostructurality of halogen analogues is far from being common, –Br, –Cl and –F are found to be isostructural in different degrees, whereas –I is not. Interestingly, although it is closely isostructural to the –Cl and –Br compounds, the F analogue is shown not to form F...O bonds, while the Cl and the Br analogues do form Hal···O bonds. This raises an important question on the role of Hal···O bonds in the structuration of the crystal packing, particularly the stabilization effect. Similarly, while the CH···Hal interaction seems to give one-dimensional cohesion in the –Cl and –Br analogues, this feature is absent in the –F analogue, despite its isostructurality. CH···O interactions appear to dominate the cohesion between the anionic triflate network and the cationic Zn-pyridinedicarboxaldehydebis(*p*-R-phenylimines) network.

For coordination processes, the design of the ligand is crucial. The bipyridine-, terpyridine- or polypyridine-type derivatives are extensively used. Their synthesis is very challenging and requires laborious multistep procedures.[43] The isomorphic correspondence between terpyridine (terpy) **6** and 2,6-(iminoarene)pyridine compound **7** provides a solution

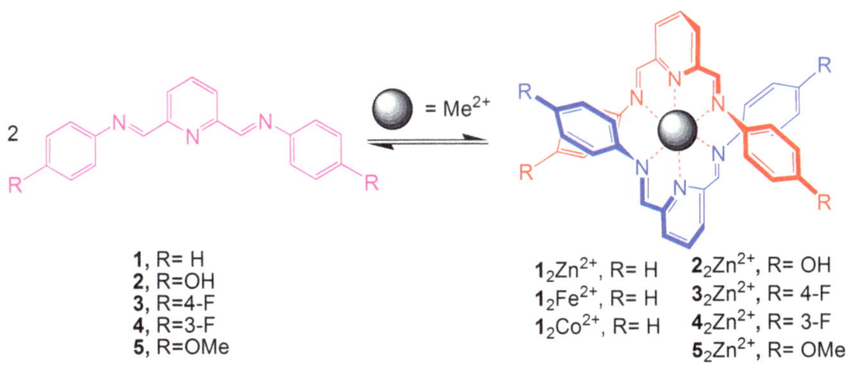

Figure 3.3 Terpyridine-type complexes $\mathbf{1}_2Zn^{2+}$–$\mathbf{5}_2Zn^{2+}$, $\mathbf{1}_2Fe^{2+}$ and $\mathbf{1}_2Co^{2+}$ based on achiral ligands **1–5** and metal ions. (Reproduced with permission from ref. 20.)

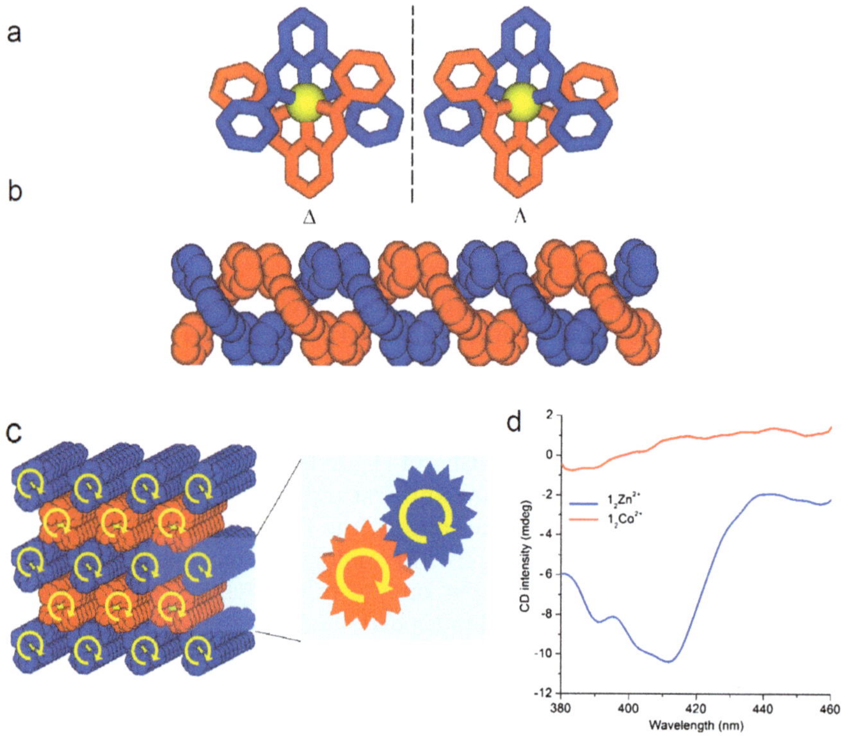

Figure 3.4 (a) Δ and Λ enantiomers and (b) homochiral double helix columns of $\mathbf{1_2Zn^{2+}}$ units; (c) crystal packing of double-helix columns of unique *P* chirality; (d) solid-state CD spectra of $\mathbf{1_2Zn^{2+}}$ (blue) and $\mathbf{1_2Co^{2+}}$ (red)–KBr matrix. (Reproduced with permission from ref. 20.)

for greatly facilitating this by replacing heterocyclic synthesis with simple imino-condensation reactions. Imine generation and exchange has been extensively used for designing dynamic adaptive systems (Figure 3.5).[44] Further developments might consider hydrazine or acylhydrazone functionalities, both providing dynamic character through the reversibility of the imine bond, and may introduce additional coordination, H-bonding, and geometrical effects in the organization of metallosupramolecular architectures through secondary amine and amide groups. One question related to this aim is: how can the coordination to a metallic center and the supramolecular packing of bis-chelate metal complexes $\mathbf{8}_2Zn^{2+}$ and $\mathbf{9}_2Zn^{2+}$ be influenced by introducing secondary amino (–NH–) and acyl (–NH–CO–) spacers in the structure of ligands **8** and **9**?

The complex $\mathbf{8}_2Zn^{2+}$ is quantitatively trapped *via* constitutionally-driven crystallization from binary solutions of **8**–Zn^{2+} 1:1 and 2:1 (mol–mol). The constitutional internal overlap between the phenyl moieties and the central pyridine ring clearly plays an important role for internal location and stabilization of duplex formation by π–π stacking.

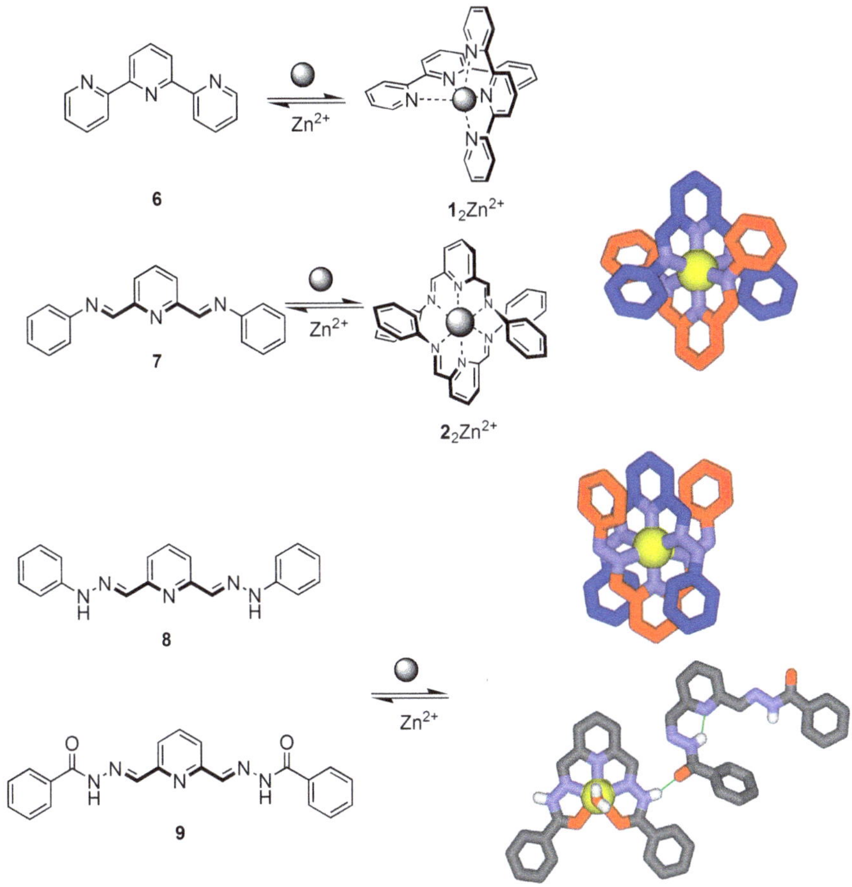

Figure 3.5 Structures of the terpyridine-type ligands **6–9** and of the $\mathbf{6}_2Zn^{2+}$ and $\mathbf{7}_2Zn^{2+}$ metallosupramolecular complexes. Crystal structure as stick representation of the complexes $\mathbf{7}_2Zn^{2+}$, $\mathbf{8}_2Zn^{2+}$ and H-bonding self-assembly of mononuclear complex $\mathbf{9}Zn^{2+}$ and the free ligand **9** in the crystal. The Zn^{2+} ions are shown as yellow spheres. (Adapted by permission from the American Chemical Society from ref. 44.)

Crystallization processes may also be used for the amplification of very specific supramolecular architectures with a unique network topology from dynamic mixtures.[45] Several types of supplementary features, such as coordination behaviour and geometry, stacking interactions, and inter- and intramolecular spatial hindrance are some of the parameters identified as being responsible for the amplification of topologically favoured architectures in DCL equilibria. Indeed, such complicated topologies have attracted tremendous interest because of their simple one-pot synthesis.

Of particular interest are coordination complexes based on terpyridine-type ligands and octahedrally coordinated metal ions, revealing a range of interesting structural and physicochemical properties (Figure 3.6a). Addition of

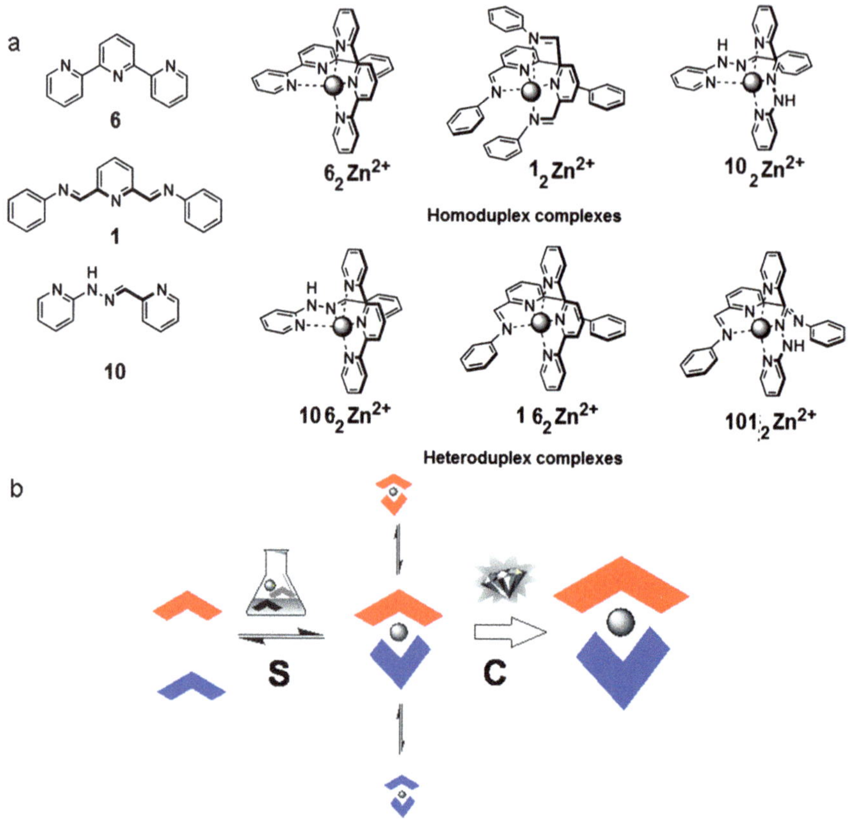

Figure 3.6 (a) Structures of the terpyridine-type ligands **1**, **6** and **10** and the DCL library of their resulting homo- and heteroduplex metallosupramolecular complexes. (b) Statistical pre-amplification in solution (**S**) 25% of each homoduplex and 50% heteroduplex and 100% solid state selection process (**C**) of the heteroduplex complex. (Adapted by permission from Wiley from ref. 46.)

Zn^{2+} to the binary mixture of terpy-type ligands **1**, **6** and **10** should cause the appropriate recruitment of two ligands in a pairwise mode and might be controlled by coordination behaviour and internal interactions between different aromatic moieties of these compounds. Indeed, an investigation of whether the coordination behaviour of these systems on the addition of octahedral Zn^{2+} ions could be selectively expressed in the formation heteroduplex ternary complexes found that in terms of specific self-assembly the homo- and heteroduplex Zn^{2+} complexes of the terpyridine-type ligands self-organize in the solid state in 1D- or 2D-stacked layers, double-helix or dimeric architectures as a function of synergetic intra- and intermolecular π–π stacking interactions between different ligands. The specific self-organization of the heteroduplex architectures represents an attractive example of solid-state structures resulting from synergistic effects of statistically and

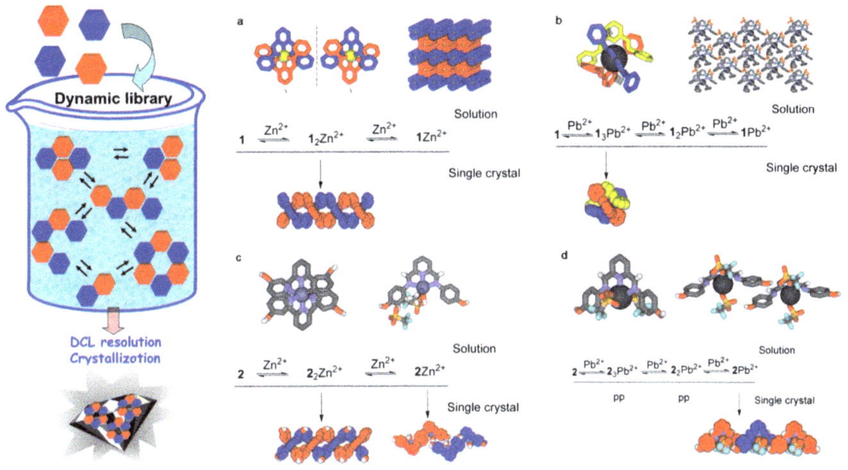

Figure 3.7 Crystallization-induced selection from a supramolecular DCL (left); interconverting metallosupramolecular DCLs self-sorted *via* constitutional crystallization of (a) $\mathbf{1}_2Zn^{2+}$, (b) $\mathbf{1}_3Pb^{2+}$, and sequential crystallization of (c) $\mathbf{2}_2Zn^{2+}$ and $\mathbf{2}Zn^{2+}$ and (d) $\mathbf{2}Pb^{2+}$ (right). (Adapted with permission from ref. 48.)

constitutionally-driven binding of two different ligands by the Zn^{2+} metal ions. Layering the solutions of stoichiometric binary mixtures of homoduplex complexes in acetonitrile with a non-solvent resulted in a unique set of crystals for each experiment. The X-ray structural determinations of the selected single-crystals formed from these stoichiometric binary mixtures revealed that such complexes present unique *heteroduplex superstructures* in the solid state. Re-dissolution of the crystals in CD_3CN results in regeneration of the reaction mixture prior to crystal growth. The heteroduplex Zn^{2+} complexes are quantitatively crystallized in the solid state by statistically and structurally driven selection in solution from a binary mixture of the terpyridine-type complexes and are then trapped by crystallisation in the solid state (Figure 3.6b).[46] These features are similar to other dynamic processes operating in mixtures of dynamic covalent systems[47] or libraries of coordination complexes[48] from which a single member may be trapped by crystallization.

The generation of metallosupramolecular DCLs can be performed over a wide range of ligand–M^{2+} ratios resulting in the formation of an interexchanging equilibrating mixture of subcomponents of different stoichiometries which convert into unique solid state species *via* crystallization.

For example, the use of pseudoterpyridine-type coordination systems obtained from previously presented ligands **1**, **2** and Zn^{2+}, Pb^{2+} metal ions allows the formation of different sets of mononuclear coordination complexes $\mathbf{1}_iZn^{2+}$, $\mathbf{2}_iZn^{2+}$, $i = 1, 2$ and $\mathbf{1}_jPb^{2+}$, $\mathbf{2}_jPb^{2+}$, $j = 1–3$, interconverting in solution (Figure 3.7).[48] From these mixtures, the $\mathbf{1}_2Zn^{2+}$ or $\mathbf{1}_3Pb^{2+}$ mononuclear complexes are amplified *via* solid-state crystallization and promoted by the constitutional π–π stacking affinity of ligands organizing around the metal centres. The internal robustness of the crystal packing and entropically-favored highly packed architectures are mainly responsible for the transmission of the supramolecular order, which is then constitutionally transmitted to the next hierarchical level, producing a crystallization selection process. The introduction of supplementary H-bonding groups in **2** leading to supplementary stabilization factors results in the complete solid state formation of each member of the DCL.

Considering the simplicity of this strategy, the synthesis of more complex heteroarchitectures might allow close to full conversion *via* crystallization processes with no need of further purification.

3.4 Metal-Driven Selection in Dynamic Metallosupramolecular Libraries

All of the examples previously highlighted have demonstrated that dynamic metallosupramolecular libraries can be generated *via* ligand and ligand/reversible covalent bond exchanges. Another, so far uncharted, option for information storage is multiple coding, offered by ligands in the process of simultaneous coordination with *mixtures of metal cations*, producing DMLs of mononuclear and polynuclear components containing distinct metal ions. Competitive constitutional equilibria may also be generated by employing a specific ligand showing different conformations and mixtures of metal ions in order to generate all possible interconverting combinations and geometries in the resulting coordination complexes. It might allow the self-selection of a metal-ion specific constituent from an equilibrating collection of finite mononuclear or/and polymeric metallosupramolecular species.

Also, double-level combinatorial libraries based on metal-ion coordination and *cis–trans* conformational processes can be generated.

The simple building block **11** carrying two 2-iminopyridine functionalities was chosen as a synthetic precursor for the formation of a double level-library (Figure 3.8a).[49] The simultaneous metal-ion coordination/conformational interconversion between the *cis* and *trans* arrangements around the central phenyl moiety of the chelating sites of **11** are expected to act simultaneously, leading to exchanging species in solution. The strategy to employ ligand **10** and different tetrahedral Cu^+ and Ag^+ metal ions in order to facilitate the interconversion of all possible stoichiometries resulted in the generation of a double-level dynamic library of hetero- and homonuclear grid complexes (Figure 3.8c). The present system is able to select the appropriate conformation of the ligand, required by the metal ions to quantitatively form an equimolar mixture of *homonuclear* grid-type complexes $\mathbf{G_1Cu}$ and $\mathbf{G_2Ag}$ (Figure 3.8c). The heteronuclear grid compounds combining different metal ions require the ligands to adapt the conformation to the very different coordination geometries of Cu^+ and Ag^+ (Figure 3.8b). The symmetrical (Cu^+/iminopyridine) and the unsymmetrical (Ag^+/iminopyridine)

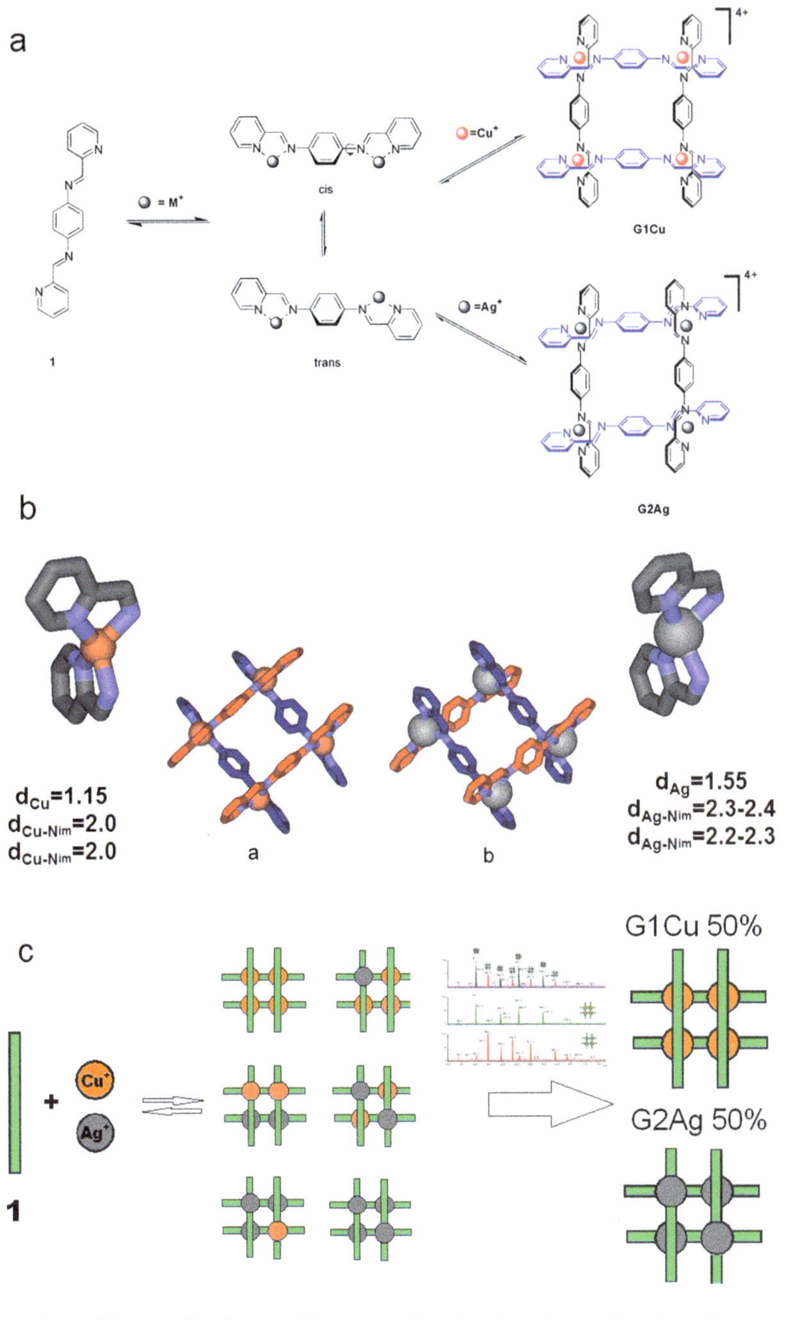

Figure 3.8 (a) Two-level metal-ion coordination/conformational exchange processes of ligand **11** and a mixture of Cu^+ and Ag^+ metal ions. (b) Crystal structure and coordination geometries parameters (stick representation) of $\mathbf{G_1Cu}$ and $\mathbf{G_2Ag}$ grid complexes. Cu^+ and Ag^+ ions are represented as red and grey spheres, respectively, and (c) generation of a dynamic library of the [2 × 2] grids from mixtures of **11**–Cu^+–Ag^+, 2 : 1 : 1 (mol–mol–mol). (Adapted by permission from the American Chemical Society from ref. 49.)

coordination geometries may be unfavourably associated within the same grid entity. The homocomplexation process represents the best compromise of the combination of ligand **11** with Cu^{+} and Ag^{+} metal ions resulting in the kinetic formation of the symmetrical **G$_1$Cu** grid and in the thermodynamic evolution of the interwoven **G$_2$Ag** grid.

These results show that the use of a simple ligand of different conformational geometries, different metal ions and a simple chemical mechanism—*quantitative parallel amplification*—results in the possibility to select the metal-adapted geometries of the ligand in order to amplify the *homonuclear* grid architectures within a double-level DCL by constitutional self-selection.

With ligand **12**, the available pseudo-terpyridine (terpy) and pyridine side-arm (py) coordination sites can be involved in both octahedral and linear coordination binding events of the metal ions, depending on experimental conditions such as metal ion nature, concentration, and ligand–metal molar ratio (Figure 3.9a).

The generation of DMLs of ligand **12** over a range of ligand–metal structures and ratios results in the formation of an inter-exchanging equilibrating mixture of mononuclear, dinuclear, or polymeric species which converts into compact packed mononuclear species in solution and in the solid state

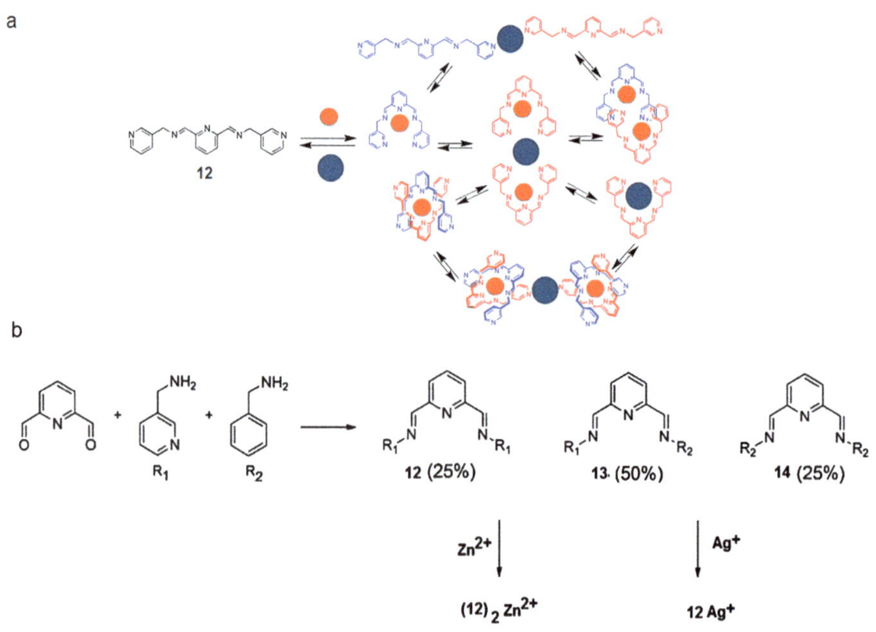

Figure 3.9 Dynamic metallosupramolecular library generated by (a) the simultaneous addition of two metal ions to ligand **12** or (b) *via* ligand selection which could be selectively expressed as an independent (linear) or interfering (cross-over) combination of the above coordination events. (Adapted with permission from ref. 50.)

via crystallization.[50] At this point, the *quantitative "metal-driven" selection* in solution *via* coupled coordination/stacking interactional algorithms is possible: the preferential amplification of $\mathbf{12}_2Zn^{2+}$ and of $\mathbf{12}_2Zn^{2+}/\mathbf{12}Zn^{2+}$ and further crystallization of $\mathbf{12}_2Zn^{2+}$ from a solution of ligand **12** and a mixture of Zn^{2+} and Ag^{+} and Pb^{2+} metal ions confirms the previous observations of the formation of highly packed mononuclear architectures which are entropically-favored[48] against the multinuclear, dimeric $[\mathbf{12}_2Pb^{2+}]_2$ or polymeric $[\mathbf{12}Ag^{+}]_n$ ones (Figure 3.10). The entropic cost of binding the metal ion to the free ligands must be far larger for the polymeric systems than in the case where these ligands are bound in compact and robust monoclusters with restricted conformational entropy.

These mechanisms are also robust enough to induce *ligand selection* from a dynamic pool of different ligands, **12–14**. This is based on the relative constitutional stability of crystalline duplex $\mathbf{12}_2Zn^{2+}$ and/or polymeric $[\mathbf{12}Ag^{+}]_n$ cleft complexes, amplified from the complex mixtures in solution *via* a kinetically-driven crystallization process. The robustness of $\mathbf{12}_2Zn^{2+}$ and $[\mathbf{12}Ag^{+}]_n$ coordination systems results from the compatibility and synergistic effects of the metal ion-binding and of the stacking interactions which are much stronger for the picolyl coordinative side-arms when compared with the competitive benzyl ones. A further step toward processes of increasing complexity could involve the ligands of different dimensionality or library diversification by ligands undergoing structural reversible interconnections by chemical exchange.

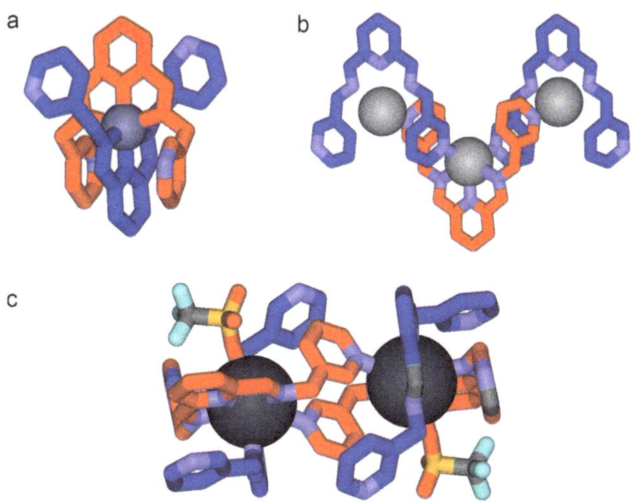

Figure 3.10 Crystal structure (stick representation) coordination of the ligand **12**: (a) $\mathbf{12}_2Zn^{2+}$, (b) $\mathbf{12}Ag^{+}$, and (c) $\mathbf{12}_2Pb^{2+}$. The metal ions are shown as blue (Zn^{2+}), light grey (Ag^{+}) and grey (Pb^{2+}) spheres. The non-coordinating Tf^{-} anions and the H-atoms have been omitted for clarity. (Adapted with permission from ref. 50.)

3.5 Conclusion

The *dynamic constitutional metallosupramolecular systems* described here are nice examples for adaptive devices correlating their morphology under the pressure of internal and external factors. Complex dynamic and positive feedback between molecular and metal ion partners gives rise to metallosupramolecular materials with emergent functional properties with a collective behavior. From a conceptual point of view, these systems express a synergistic adaptive behavior embodying a constitutional self-reorganization (self-adaptation) of their configuration producing the "fittest" architecture between the sum of all possible combinations in response to their medium constraints.

These examples extend the application of *constitutional dynamic chemistry*[11–13] from molecular and supramolecular sciences to *functional constitutional devices*. This feature offers perspectives to coordination chemistry towards self-designed materials that involve modification and control of the intrinsic structural properties of dynamic entities correlated with the dynamic functional building blocks. Prospects for the future include the development of these original methodologies towards *adaptive dynamic constitutional systems* and their application to different processes of interest for the development of a supramolecular approach to nanoscience and nanotechnology and systems of increasing functional complexity.

Operating such evolutive dynamic interactive systems[12] is mainly related to reversible interactions between continually interchanging components within the metallosupramolecular architectures, yielding specific dimensional and temporal behavior, and resulting in the formation of the most efficient functional superstructures in the presence of the applied stimulus, that are selected from a set of diverse less-selective possible architectures which can form *via* self-assembly. This can lead to an adaptive functionality, directed by the evolution of *dynamic encodable* interactions between components self-designing the available space by using dynamic ligands and around informed metallic ion centers.

References

1. E. Koechlin and A. Hyafil, *Science*, 2007, **18**, 594.
2. S. A. Brener and A. M. Sismour, *Nat. Rev.*, 2005, **6**, 533.
3. J. Bascompte, *Science*, 2009, **325**, 416.
4. B. Palsson, *FEBS Lett.*, 2009, **583**, 3900.
5. M. Kindermann, I. Stahl, M. Reimold, W. M. Pankau and G. von Kiedrowski, *Angew. Chem., Int. Ed.*, 2005, **44**, 6750.
6. A. Müller, *Chem. Commun.*, 2003, 803.
7. J.-M. Lehn, *Chem. Soc. Rev.*, 2007, **36**, 151.
8. E. Prouzet, S. Ravaine, C. Sanchez and R. Backov, *New J. Chem.*, 2008, **32**, 1284.
9. F. Menger, *Angew. Chem., Int. Ed.*, 1991, **30**, 1086.

10. (a) J.-M. Lehn, *Chem. Soc. Rev.*, 2007, **36**, 151–160; (b) M. Barboiu and J.-M. Lehn, *Proc. Natl. Acad. Sci. U. S. A.*, 2002, **99**, 5201.
11. J.-M. Lehn, *Proc. Natl. Acad. Sci. U. S. A.*, 2002, **99**, 4763.
12. M. Barboiu, *Chem. Commun.*, 2010, **46**, 7466.
13. (a) Dynamic Combinatorial Chemistry, in *Drug Discovery, Bioorganic Chemistry and Materials Science*, ed. B. L. Miller, John Wiley and Sons, Hoboken, 2010; (b) *Dynamic Combinatorial Chemistry*, ed. J. N. H. Reek and S. Otto. Wiley-VCH, Weinheim, 2010; (c) *Constitutional Dynamic Chemistry*, ed. M. Barboiu, *Top. Curr. Chem.*, Springer Verlag, Berlin, 2012, vol. 322.
14. P. N. W. Baxter, J.-M. Lehn and K. Rissanen, *Chem. Commun.*, 1997, 1323.
15. (a) P. N. W. Baxter, J.-M. Lehn, B. O. Kneisel and D. Fenske, *Chem. Commun.*, 1997, 2231; (b) P. N. W. Baxter, J.-M. Lehn, G. Baum and D. Fenske, *Chem.–Eur. J.*, 2000, **6**, 4510; (c) M. Barboiu, M. Ruben, G. Blasen, N. Kyritsakas, E. Chacko, M. Dutta, O. Radekovich, K. Lenton, D. J. R. Brook and J.-M. Lehn, *Eur. J. Inorg. Chem.*, 2006, 784.
16. C. D. Pentecost, K. S. Chichak, A. J. Peters, G. W. V. Cave, S. J. Cantrill and J. F. Stoddart, *Angew. Chem., Int. Ed.*, 2007, **46**, 218.
17. A. Muller, L. Toma, H. Bogge, C. Scaffer and A. Stammler, *Angew. Chem., Int. Ed.*, 2005, **44**, 7757.
18. B. H. Northrop, H.-B. Yang and P. J. Stang, *Inorg. Chem.*, 2008, **47**, 11257.
19. Y.-M. Legrand, A. van der Lee and M. Barboiu, *Inorg. Chem.*, 2007, **46**, 9540.
20. Y. M. Legrand, F. Dumitru, A. van der Lee and M. Barboiu, *Chem. Commun.*, 2009, 2667.
21. Y.-M. Legrand, F. Dumitru, A. van der Lee and M. Barboiu, *Supramol. Chem.*, 2009, **21**, 229.
22. M. Barboiu, E. Petit and G. Vaughan, *Chem.–Eur. J.*, 2004, **10**, 2263.
23. J.-M. Lehn, *Supramolecular Chemistry-Concepts and Perspectives*, VCH, Weinheim, 1995.
24. J.-M. Lehn, *Chem.–Eur. J.*, 1999, **5**, 2455.
25. P. T. Corbett, J. Leclaire, L. Vial, K. R. West, J.-L. Wietor, J. K. M. Sanders and S. Otto, *Chem. Rev.*, 2006, **106**, 3652.
26. T. Megeyes, H. Jude, T. Grosz, I. Bako, T. Radnai, G. Tarkanyi, G. Palinkas and P. J. Stang, *J. Am. Chem. Soc.*, 2005, **127**, 10731.
27. I. Saur, R. Scopelliti and K. Severin, *Chem.–Eur. J.*, 2006, **12**, 1058.
28. N. Giuseppone, J.-L. Schmitt and J.-M. Lehn, *Angew. Chem., Int. Ed.*, 2004, **43**, 4902.
29. More precisely, if the reversible processes occur independently, they might be addressed as *orthogonal*. When two levels of dynamic exchange cross over, they are considered as *communicative*. See for details: V. Goral, M. I. Nelen, A. V. Eliseev and J.-M. Lehn, *Proc. Natl. Acad. Sci. U. S. A.*, 2001, **98**, 1347.
30. M. Barboiu, G. Vaughan, R. Graff and J.-M. Lehn, *J. Am. Chem. Soc.*, 2003, **125**, 10257.

31. F. Dumitru, Y.-M. Legrand, E. Petit, A. van der Lee and M. Barboiu, *Dalton Trans.*, 2012, **41**, 11860.
32. M. A. Mateos-Timoneda, M. Crego-Calama and D. N. Reinhoudt, *Chem. Soc. Rev.*, 2004, **33**, 363.
33. V. Percec, J. G. Rudick, M. Peterca and P. A. Heiney, *J. Am. Chem. Soc.*, 2008, **130**, 7503.
34. (a) C. Arnal-Hérault, A. Banu, M. Barboiu, M. Michau and A. van der Lee, *Angew. Chem., Int. Ed.*, 2007, **46**, 4268; (b) C. Arnal-Herault, A. Pasc-Banu, M. Michau, D. Cot, E. Petit and M. Barboiu, *Angew. Chem., Int. Ed.*, 2007, **46**, 8409.
35. M. Albrecht, *Chem. Rev.*, 2001, **101**, 3457.
36. P. E. Nielsen, *Acc. Chem. Res.*, 1999, **32**, 624.
37. M. Barboiu, G. Vaughan, N. Kyritsakas and J.-M. Lehn, *Chem.–Eur. J.*, 2003, **9**, 763.
38. M. Ohkita, J.-M. Lehn, G. Baum and D. Fenske, *Chem.–Eur. J.*, 1999, **5**, 3471.
39. R. Kramer, J.-M. Lehn, A. De Cian and J. Fischer, *Angew. Chem., Int. Ed.*, 1993, **32**, 703.
40. N. Asano, T. Harada, T. Sato, N. Tajima and R. Kuroda, *Chem. Commun.*, 2009, 899.
41. (a) D. A. Langs, *Science*, 1988, **241**, 188–191; (b) C. Cohen, *J. Struct. Biol.*, 1998, **122**, 3; (c) A. N. Creager, *The Life of a Virus: Tobacco Mosaic Virus as an Experimental Model*, University of Chicago Press, 2002.
42. F. Dumitru, Y. M. Legrand, M. Barboiu and A. van der Lee, *Acta Crystallogr., Sect. B: Struct. Sci., Cryst. Eng. Mater.*, 2013, **B69**, 43.
43. For reviews and papers on terpyridine-metal complexes, see for example: (a) M. L. Scudder, H. A. Goodwin and I. G. Dance, *New J. Chem.*, 1999, **23**, 695; (b) E. Constable, *Chem. Soc. Rev.*, 2007, **36**, 246.
44. F. Dumitru, Y.-M. Legrand, M. Barboiu, E. Petit and A. van der Lee, *Cryst. Growth Des.*, 2009, **9**, 2917.
45. Y. M. Legrand, F. Dumitru, C. Arnal-Herault, M. Michau and M. Barboiu, *Rev. Roum. Chim.*, 2009, **54**, 421 and references therein.
46. F. Dumitru, E. Petit, A. van der Lee and M. Barboiu, *Eur. J. Inorg. Chem.*, 2005, 4255.
47. M. Angelin, P. Vongvilai, A. Fischer and O. Ramström, *J. Org. Chem.*, 2008, **73**, 3593.
48. M. Barboiu, F. Dumitru, Y.-M. Legrand, E. Petit and A. van der Lee, *Chem. Commun.*, 2009, 2192.
49. M. Barboiu, E. Petit, A. van der Lee and G. Vaughan, *Inorg. Chem.*, 2006, **45**, 484.
50. I. Kocsis, D. Dumitrescu, Y.-M. Legrand, A. van der Lee, I. Grosu and M. Barboiu, *Chem. Commun.*, 2014, **50**, 2621.

CHAPTER 4

Functional Materials Based on Metal-Containing Polymers

ALAA S. ABD-EL-AZIZ*[a], CHRISTIAN AGATEMOR[a] AND NOLA ETKIN[a]

[a]Department of Chemistry, University of Prince Edward Island, 550 University Avenue, Charlottetown, Prince Edward Island, C1A 4P3, Canada
*E-mail: abdelaziz@upei.ca

4.1 Introduction

With an ever-increasing demand for new materials to advance technology and medicine, material scientists have explored several options that include the synthesis of metal-containing polymers. Metal-containing polymers hold a lot of promise as they exhibit properties that complement those of conventional organic-based polymers. With recent advancements in synthetic chemistry, the electronic, optical, and magnetic properties of metals have been integrated into functional materials through the incorporation of metal centers into polymers. The result is a rich library of functional metal-containing polymers that are applicable as catalysts, sensors, magnetic materials, electrical conductors, stimuli-responsive materials, and ceramics precursors.

Although interest in metal-containing polymers started in the 1950s following reports of ferrocene-containing polymers,[1,2] progress in this field was slow as a result of problems with their synthesis, and characterization.[3] The earliest metal-containing polymers were intractable, had low molecular

RSC Smart Materials No. 15
Functional Metallosupramolecular Materials
Edited by John G. Hardy and Felix H. Schacher

Published by the Royal Society of Chemistry, www.rsc.org

weights, were poorly defined, insoluble macromolecules.[4] However, breakthroughs in the synthesis of well-defined, high molecular weight, tractable metal-containing polymers have been accomplished in the last decade of the 20th century, and in the 21st century, as a result of advances in techniques available for their synthesis and characterization. Interest in this class of polymers continues to bourgeon as a result of the potential applications of these polymers in technology and medicine.[1–10] This chapter examines selected functional metallopolymers with a focus on polymers whose intrinsic functions are due to the presence of a metal center.

4.2 Conducting and Redox Active Metal-Containing Polymers

Electrical conductivity in polymers is a desirable property as it offers the possibility of replacing classical electrical conductors with lightweight polymeric materials. Interest in conducting metal-containing polymers stems from the discovery of this property in organic-based polymers, and the potential applicability of these polymers in electronic as well as in sensory devices.[5,6] Although the majority of conductive polymers are organic-based with delocalized electronic states, the incorporation of a redox active metal within the polymer framework can induce or enhance conductivity through the creation of redox sites, and low-lying energetic states.[7] This is facilitated by the overlap of the atomic orbitals of metal centers with the delocalized orbitals of the polymer backbone, if the former is inserted within the conjugated framework of the latter. For instance, in a coordination polymer, $[\{Cu(TANC)\}F_{0.5}]_n$ (**1**) (where TANC is 5,6,11,12-tetraaza-naphthacene), the considerable orbital overlap between the lowest unoccupied molecular orbital (LUMO) of the organic non-semiconductor, TANC, and the copper orbitals as a result of the coordination of the copper atom to the TANC molecules imparts semi-conductivity in **1** (Figure 4.1).[11] A four-probe conductivity measurement along the *a* axis of a single crystal of **1** reveals temperature-dependent conductivity behavior that is characteristic of semi-conductors. At ambient pressure, **1** is highly conductive (50 S cm^{-1} greater than that of most conducting polymers) at 300 K and markedly less so (10^{-6} S cm^{-1}) at 10 K (Figure 4.1).[12]

A key feature of metal-containing conducting polymers is the orbital overlap between the metal centers and the bridging ligands, which usually has delocalized orbitals. For instance, dithiocarbamate ligands with an expanded highest occupied molecular orbital (HOMO) were used to synthesize several semi-conducting coordination polymers that feature halido-bridged mixed valence Cu^I–Cu^{II} or heterometallic Cu^I–Ni^{II} metal centers.[13–19] Tanaka *et al.* reported the synthesis and characterization of a mixed-valence Cu^I–Cu^{II} 1D polymer $[Cu^I_2Cu^{II}X_2(Pip\text{-}dtc)_2(CH_3CN)_2]_n$,[19] where Pip-dtc is piperidine-1-carbodithiate; X is Br (**2a**) or I (**2b**) (Figure 4.2). The conductivities of these polymers were studied with complex impedance spectroscopy. Nyquist plots as a function of temperature revealed the characteristic semi-conductor

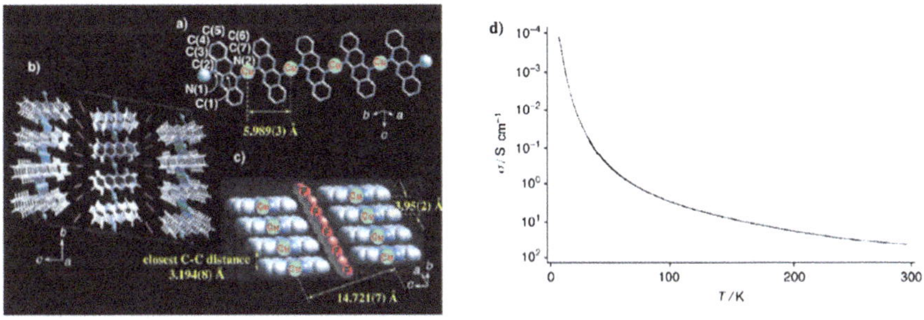

Figure 4.1 1D structure of **1** with Cu^+ alternating linkages with TANC (a), segregated stacking of **1** along the *a* axis (b), perspective view of **1** along the *a* axis (c), and a four-probe dc temperature-dependent conductivity measurement of **1** along the *a* axis at ambient pressure (d). Reproduced with permission from ref. 11. Copyright 2006 WILEY-VCH Verlag GmbH & Co KGaA, Weinheim.

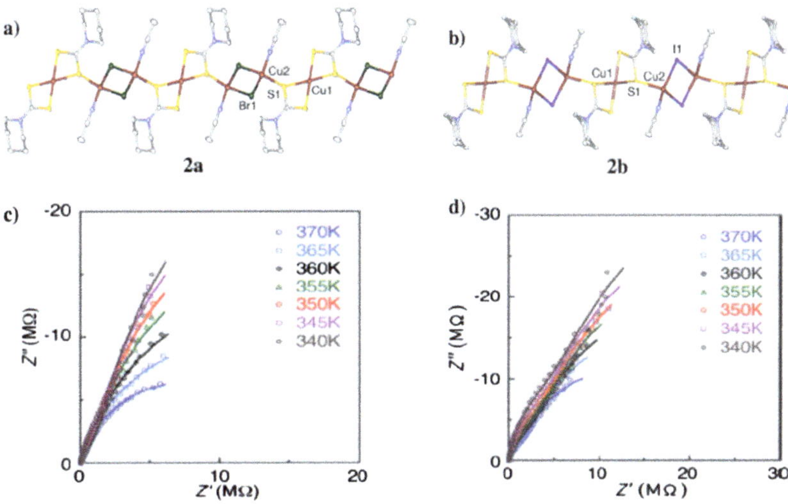

Figure 4.2 Infinite chain structure of **2a** (a), and **2b** (b). Nyquist plot of **2a** (c), and **2b** (d). Reproduced with permission from ref. 19. Copyright 2013 WILEY-VCH Verlag GmbH & Co KGaA, Weinheim.

negative temperature coefficient of resistance for **2a** and **2b** (Figure 4.2). The same group probed the conductivity of isostructural analogues of **2a** and **2b**, $[Cu^{I}_{2}Cu^{II}X_2(Hm\text{-}dtc)_2(CH_3CN)_2]_n$ where Hm-dtc is hexa-methylenedithiocarbamate and X is Br (**2c**) or I (**2d**)[13] and these were found to have higher conductivities (Table 4.1), which is potentially due to diminished overlap of the orbitals of the metal ions with those of the Pip-dtc ligands. As a result of their semi-conductivity, these metal-containing polymers were incorporated into dye-sensitized solar cells (DSSCs) by mixing TiO_2 paste with the polymers in

Table 4.1 Conductive properties of **2a–d**. Reproduced with permission from ref. 19. Copyright 2013 WILEY-VCH Verlag GmbH & Co KGaA, Weinheim.

Polymer	Bulk conductivity[a] (S cm^{-1})	Activation energy[b] (eV)	Photocurrent density[c] (mA cm^{-1})	Power conversion efficiency[d] (%)
2a	8.42×10^{-9}	0.66	0.19	0.051
2b	3.67×10^{-9}	0.50	0.15	0.038
2c	1.07×10^{-7}	0.56	0.41	0.11
2d	2.46×10^{-7}	0.48	0.39	0.11

[a]Bulk conductivity at 340 K.
[b]Activation energy associated with conductivity.
[c]Photocurrent density of DSSCs.
[d]Power conversion efficiency of DSSCs.

an agate mortar.[19] Studies conducted on these DSSCs suggested that **2a** facilitates charge transport more efficiently than **2b**. The photocurrent density of **2a**-based DSSCs is higher than those of **2b** but lower than those of **2c** and **2d** (Table 4.1). This observed difference in photocurrent is attributed to differences in conductivity, or in the energy band gaps between the conduction band of the TiO_2 and the LUMO of the polymers.

Material scientists are not limited to coordination polymers in their quest for conducting metal-containing polymers. Indeed, covalently linked polymers that incorporate various transition metals within their framework have been synthesized and shown to be conductive. Interestingly, incorporation of a metal center pendent to the polymer backbone promotes a charge transfer mechanism even when no direct interaction occurs between the orbitals.[7] Recently, Abd-El-Aziz and coworkers reported the electropolymerization of thiophene and 3,4-ethylenedioxythiophene with tethered η^6-aryl-η^5-cyclopentadienyliron hexafluorophosphates to form conducting polymeric films (**3a–d**) with redox active iron centers (Figure 4.3).[20,21] The electrical properties of these films were modulated by the nature of the tethered sandwich complexes. For instance, the conductivity of **3b** (conductivity = 14.2 mS cm^{-2}) is noticeably reduced when the tethered complex is switched from a bimetallic to monometallic as in **3d** (conductivity = 9.84 mS cm^{-2}); however, this difference is not pronounced between **3a** and **3c** (conductivities = 15.0 mS cm^{-2} and 16.0 mS cm^{-2}, respectively).[20,21] Although no direct interaction between the orbitals of iron centers and those of the thiophene backbone exists in **3a–d**,[20,21] the former still influences the conductivity of the polymer probably through an outer sphere mechanistic pathway that could involve charge hopping between two iron centers as well as between the polymer backbone and iron center. Electronic communication *via* charge hopping was proposed as one of the possible charge transfer mechanistic pathways in tethered ferrocenyl groups as in ferrocenyl-terminated dendrimers where the flexibility of the tether allows easy and fast electronic communication.[22–26]

Redox activity is key to many chemical processes; therefore, functional materials with this behavior are attractive, especially in the design of catalysts, sensors, and molecular batteries where electron transfer plays a pivotal role.[27] As

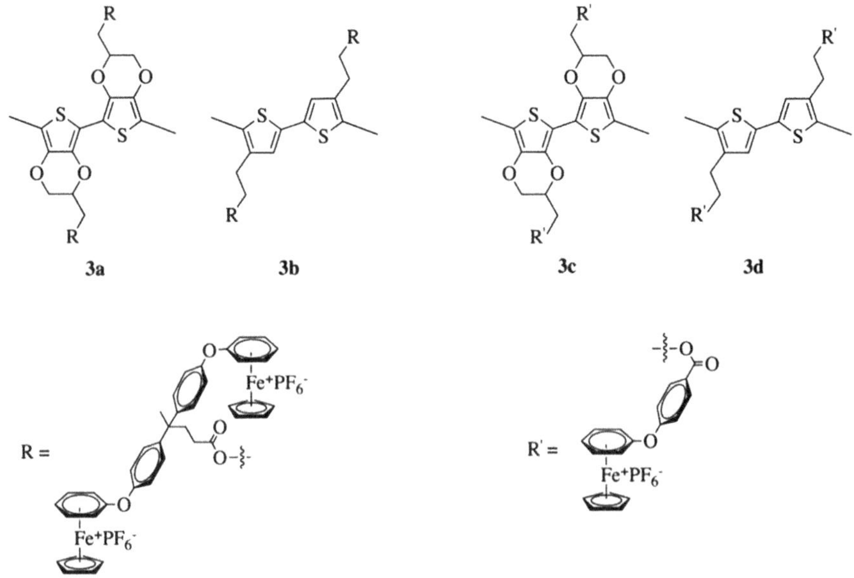

Figure 4.3 Structures of electropolymerized thiophene derivatives **3a–d**.

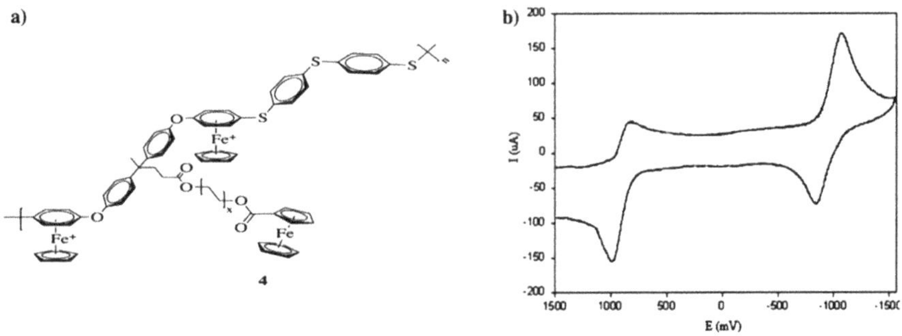

Figure 4.4 Structure of iron-containing polymer **4** (a). Cyclic voltammogram of **4** in propylene carbonate containing 0.1 M tetra-*n*-butylammonium perruthenate (TBAP) at −20 °C (b). Reproduced with permission from ref. 29. Copyright 2005 American Chemical Society.

sandwich complexes are redox active, exhibiting fully reversible redox behavior, incorporation of these complexes into a polymeric framework extends this redox behavior to the macromolecular scale.[28–32] For instance, Abd-El-Aziz and coworkers incorporated iron-based sandwich complexes into polymeric frameworks to form a series of neutral ferrocene, cationic η^6-aryl-η^5-cyclopentadienyliron-containing polymers that exhibit reversible redox waves at −30 °C, although the reduction process is irreversible at 30 °C.[29] Polymer **4** exhibits a reversible redox wave at −20 °C that consists of a reversible reduction at −1.1 V and a reversible oxidation at 0.7 V, which were assigned to the cationic iron and neutral iron centers, respectively (Figure 4.4).[29]

These redox active metal centers have been shown to be useful in the design of sensors. Ferrocene-terminated dendrimers exhibit sensory capability as the single reversible redox wave of ferrocenyl moieties in the dendrimers is sensitive to the presence of oxo anions and transition metal cations (Figure 4.5).[30] In the presence of $H_2PO_4^-$ or ATP^{2-}, the dendrimers are easily oxidized as the anion donates electrons to the iron centers; consequently, the initial redox wave shifts to a less positive potential (Figure 4.5c). In contrast, the presence of a transition metal, such as Cu^+, Cu^{2+}, Pd^{2+}, or Pt^{2+}, renders the iron centers difficult to oxidize as electrons are withdrawn from the iron centers. The result is a new redox wave that appears at a more positive potential

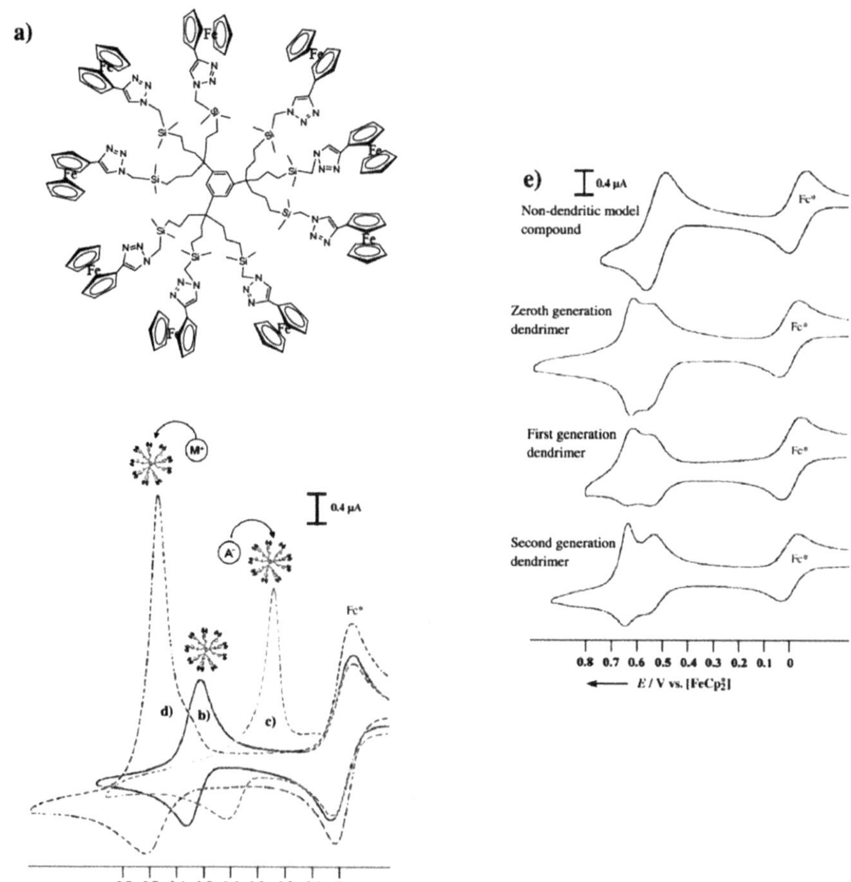

Figure 4.5 Structure of G_0 ferrocenyl dendrimers (5) (a). Redox sensitivity of 5 ($c = 8.37 \times 10^{-5}$ M) without (b), in the presence of $H_2PO_4^-$ (1 equiv. per branch) (c), and in the presence of Pd^{2+} (0.25 equiv. per branch) (d). Positive dendritic effect in the presence of Cu^+ (e). Fc*: [$FeCp^*_2$], Cp* = C_5M_5. Reproduced with permission from ref. 30. Copyright 2007, WILEY-VCH Verlag GmbH & Co KGaA, Weinheim.

(Figure 4.5d). The redox sensitivity is specific to these dendrimers, as a non-dendritic monoferrocenyl model did not respond to the presence of oxo anions or transition metal cations except Pd^{2+}. Except for Cu^{+} (Figure 4.5e), the redox sensitivity to the presence of oxo-anions and transition metal cations decreases slightly (negative dendritic effect) from a zeroth generation (G_0) (Figure 4.5a) through a first generation (G_1) to a second generation (G_2) dendrimer due to steric effects. Ion recognition is not limited to ferrocenyl dendrimers; for instance, a G_2 cobaltocenium dendrimer (**6**) is sensitive to $H_2PO_4^-$.[32] The interaction of the $H_2PO_4^-$ with the dendrimer not only renders the cobaltocenium center less susceptible to reduction but also splits the initial $Co^{III/IV}$ redox wave, and eventually leads to a new wave at a more negative potential (Figure 4.6). The absence of redox sensitivity in G_0 and G_1 cobaltocenium dendrimers is probably due to insufficient electrostatic and supramolecular interactions, and suggests a positive dendritic effect in redox sensitivity.[32]

Sandwich complex-containing dendrimers have been well studied. The reversibility of the redox chemistry of sandwich complexes suggests their potential as molecular batteries, which are molecules that store charges.[25] Giant ferrocenyl and pentamethylferrocenyl dendrimers exhibit electrochromic battery behavior, as suggested from their reversible redox cycle and UV-vis spectra.[33] For example, oxidation of the ferrocenyl dendrimer by a stoichiometric amount of acetylferrocenium tetrafluoroborate affords stable blue solid ferrocenium dendrimers as distinct from an orange neutral ferrocenyl dendrimer.[33] Further, $TiCl_3$ reduction of the blue ferrocenium dendrimer results in a stable orange ferrocenyl dendrimer.[33] Although the

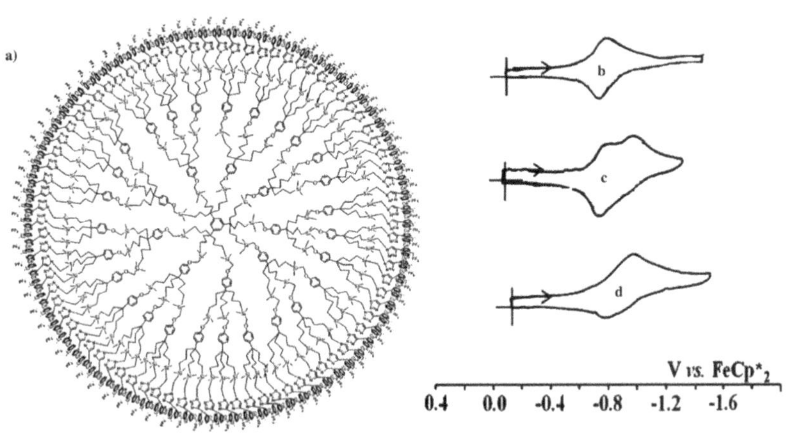

Figure 4.6 Structure of G_2 cobaltocenium dendrimer (**6**) (a). Cyclic voltammogram of **6** without anion (in DMF, [n-Bu_4N][PF_6] 0.1 M) (b), splitting of the $Co^{III/IV}$ redox wave in the presence of $H_2PO_4^-$ as $Bu_4NH_2PO_4$ (c), and appearance of a new redox wave and disappearance of the initial wave in the presence of $H_2PO_4^-$ as $Bu_4NH_2PO_4$ (d). Reproduced with permission from ref. 32. Copyright 2013, American Chemical Society.

reversibility of the ferrocenyl and ferrocenium dendrimers' redox chemistry suggests their applicability as electrochromic batteries, the limited stability of the ferrocenium analogue under ambient conditions at the negative redox potential compromises their potential use as charge storage systems.[25] Sandwich complex-based dendrimers with more negative oxidation potential such as the η^6-aryl-η^5-cyclopentadienyliron series are more robust systems for the design of charge storage systems. The orange, 18-electron Fe^{II} dendrimer [dendr-64-NHCOCpFe^{II}(η^6-C_6Me_6)] (**7**) was successfully reduced to the deep blue, 19-electron form [dendr-64-NHCOCpFe^{I}(η^6-C_6Me_6)] (**9**) by the parent 19-electron [CpFeI(η^6-C_6Me_6)] (**8**) (Scheme 4.1).[34,35] Although **9** decomposes at 0 °C, it is stable at −30 °C and reduces fullerene (C_{60}) at −30 °C to an insoluble black precipitate (**10**) that consists of the 18-electron Fe^{II} complex as well as $C_{60}{}^{\bullet -}$.

The multi-arm architecture of globular-shaped polymers such as dendrimers and hyperbranched polymers makes them promising platforms for incorporating peripheral transition metals as catalytic sites in multi-electron redox catalysis. Although the regular monodispersity of dendrimers ensures excellent control over the number of catalytic sites, the high cost associated with the synthesis of dendrimers makes hyperbranched polymers attractive for catalysis where the exact number of electrons transferred is not important. Thus, hyperbranched polycarbosilane (**11** and **12**) that incorporate redox active and interacting ferrocenyl moieties were employed in the multi-electron catalytic reduction of oxygen as well as the reduction and oxidation of hydrogen peroxide.[36] Hyperbranched polymers **11** and **12** (which exhibit the characteristic redox property of interacting ferrocene moieties in solution as well as on the surface of platinum electrodes) were synthesized *via* the hydrosilylation of diferrocenylmethylsilane with polydiallyl-methylsilane and polymethyldiundecenylsilane, respectively. A platinum electrode modified with polymeric films of **11** or **12** catalyzes the reduction of molecular oxygen *via* a four-electron transfer process by lowering the activation energy of the reaction as revealed from the anodical shift of the reduction potential

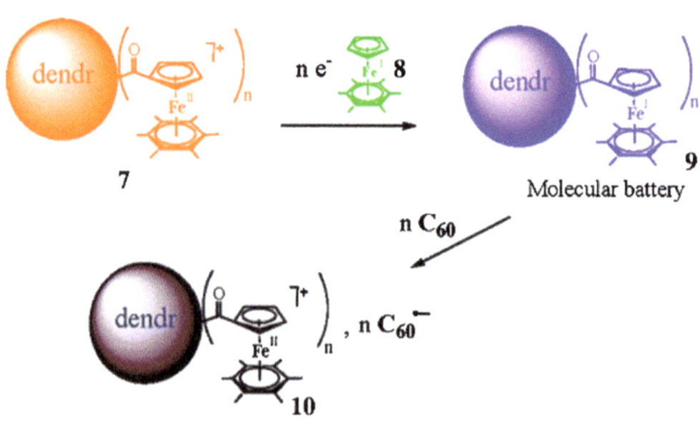

Scheme 4.1 Redox reaction of a molecular battery. Reproduced with permission from ref. 35. Copyright 2010 American Chemical Society.

by 0.1 V (Figure 4.7). As the reduction of oxygen is insignificant at potentials less negative than −0.2 V when an unmodified platinum electrode is used, and the formal potential for the catalyzed reduction of oxygen is less positive than that of a ferrocenium/ferrocene couple, the authors inferred that ferrocene moieties in the polymeric films catalyze the reduction. Further, platinum electrodes modified with polymeric films of **11** or **12** catalyze the reduction and oxidation of hydrogen peroxides (Figure 4.7). With the modified electrodes, reduction of H_2O_2 occurs at a less negative potential relative

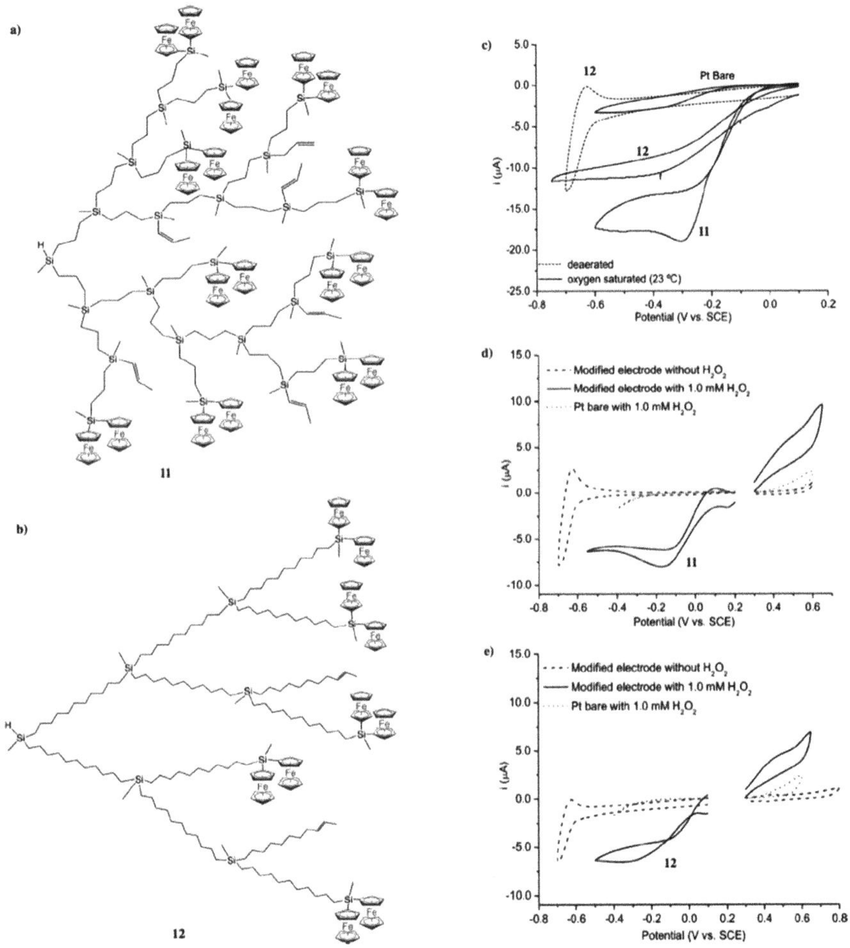

Figure 4.7 Structures of ferrocene-containing hyperbranched polymers **11** (a) and **12** (b). Cyclic voltammograms for the reduction of oxygen with a bare (unmodified) electrode and electrodes modified with polymeric films **11** and **12** (c); and for the reduction and oxidation of H_2O_2 with a bare electrode and electrodes modified with polymeric films **11** (d) or **12** (e). Reproduced with permission from ref. 36. Copyright 2013 WILEY-VCH Verlag GmbH & Co KGaA, Weinheim.

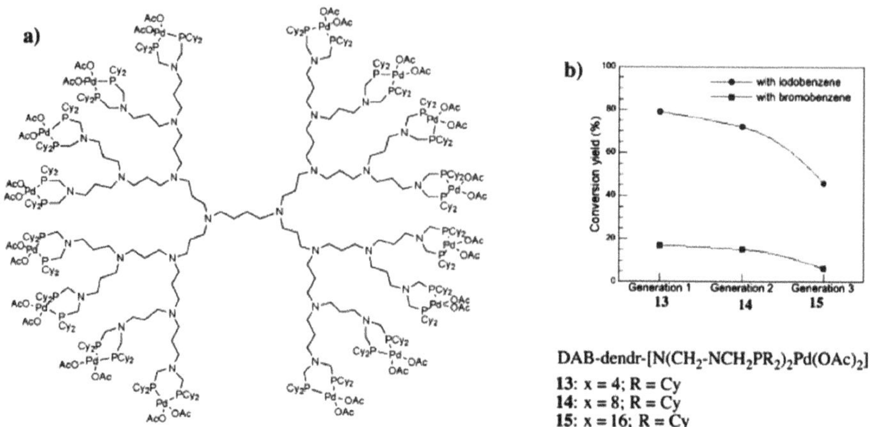

14

DAB-dendr-[N(CH$_2$-NCH$_2$PR$_2$)$_2$Pd(OAc)$_2$]
13: x = 4; R = Cy
14: x = 8; R = Cy
15: x = 16; R = Cy
16: x = 4; R = *t*Bu
17: x = 8; R = *t*Bu
18: x = 16; R = *t*Bu

Figure 4.8 Structure of dendritic palladium catalyst (**14**) (a). Percent conversion as a function of dendrimer generation in the Sonogashira coupling of aryl halide with phenylacetylene catalyzed by **12–14** at 80 °C (b). Reproduced with permission from ref. 37.

to an unmodified electrode. The oxidation wave shifts cathodically and the peak currents are significantly higher when a modified electrode is used compared to an unmodified electrode.

Notwithstanding the high synthetic cost of dendrimers, functionalization of dendrimers with transition metal catalysts remains promising and continues to be explored. For instance, peripheral functionalization of dendrimers with palladium,[37,38] rhodium,[39] and tungsten[40] for catalysis has been reported. The Sonogashira coupling between an aryl halide and an alkyne, which is an important synthetic route to arylalkynes or conjugated enynes as well as several biologically important compounds, is catalyzed by palladium-containing dendrimers.[37,38] In a coupling reaction of phenylacetylene and iodobenzene or bromobenzene in the presence of dendritic palladium catalysts (**13–18**), good-to-excellent conversion is obtained under some reaction conditions.[37] At room temperature, catalyst **16** affords a quantitative conversion within a few hours (Table 4.2); however, a negative dendritic effect limits the activity of these catalysts as the generation of the dendrimers increases (Figure 4.8). Interestingly, catalysts **14** and **15** are recoverable without loss of activity; in contrast, the excellent solubility of **15–17** precludes their recovery from the reaction system.

4.3 Magnetic Metal-Containing Polymers

The design of new magnetic materials is intensively investigated, as these materials are relevant in many areas such as in high tech applications, where they are needed in information storage devices, as well as in medicine where

Table 4.2 Sonogashira coupling of aryl halide with phenylacetylene under different reaction conditions.[a] Reproduced with permission from ref. 37.

Entry	Aryl halide	Solvent	Catalyst (mol%)	Temperature (°C)	Reaction time[b] (h)	Conversion[c] (%)
1	Iodobenzene	Et_3N	**13**	80	24	79
2	Iodobenzene	Et_3N	**14**	80	24	72
3	Iodobenzene	Et_3N	**15**	80	24	46
4	Iodobenzene	Et_3N	**16**	25	15	97
5	Iodobenzene	Et_3N	**17**	25	40	100
6	Iodobenzene	Et_3N	**18**	25	48	100
7	Bromobenzene	Et_3N	**13**	80	48	17
8	Bromobenzene	Et_3N	**14**	80	48	15
9	Bromobenzene	Bu_2NH	**14**	120	20	20
10	Bromobenzene	Et_3N	**15**	80	48	6
11	Bromobenzene	Et_3N	**16**	25	17	100
12	Bromobenzene	Et_3N	**17**	25	48	93
13	Bromobenzene	Et_3N	**18**	25	48	96

[a]Reaction conducted under nitrogen; aryl halide: 2 mmol; phenylacetylene: 3 mmol; solvent: 6 mL; catalyst: 1 mol%.
[b]TLC was used to monitor the reaction.
[c]Product was isolated by column chromatography when reaction was completed.

they are potentially useful in targeted drug delivery. Thus, efforts have been directed towards incorporating magnetic metals such as cobalt (Co) into polymers to generate precursors to magnetic materials.[41–51] Recently, Tew and coworkers reported a Co-containing diblock copolymer, **19a**, which, on thermal treatment, exhibits room temperature ferromagnetism due to its nanoscopic ordering, and cylindrical morphology.[47] The homopolymer, **19b**, which lacks these nanostructural morphologies, is only superparamagnetic despite its higher Co content. Investigation of the magnetic properties of these polymers revealed that **19a** with a coercivity (H_c) of 200 Oe is strongly attracted to a permanent magnet, whereas **19b** with a H_c of zero is unresponsive (Figure 4.9). In subsequent investigations of the magnetic properties of a random copolymer containing Co carbonyl complex and ferrocene moieties (**20**), the room temperature ferromagnetism is modulated by control of the Co content.[48] The report showed that as the Co content is decreased, the H_c decreased until the room temperature magnetism switched from ferromagnetism to superparamagnetism (Figure 4.10). However, the ferromagnetic property was unresponsive to changes in the domain size of the diblock copolymers **19a**.[50] Manners and coworkers also demonstrated the feasibility of switching between ferromagnetism and superparamagnetism in a magnetic ceramic obtained from the pyrolysis of a cross-linked perceramic ferrocene-containing polymer.[41] The polymer (**22**) was obtained from the thermal polymerization of spirocyclic [1]silaferrocenophane (**21**), as shown in Scheme 4.2. The magnetism in the ceramic is tunable by control of the temperature and time of pyrolysis of **22**. For instance, pyrolysis of **22** at 650 °C and 850 °C affords superparamagnetic ceramics that lack hysteresis

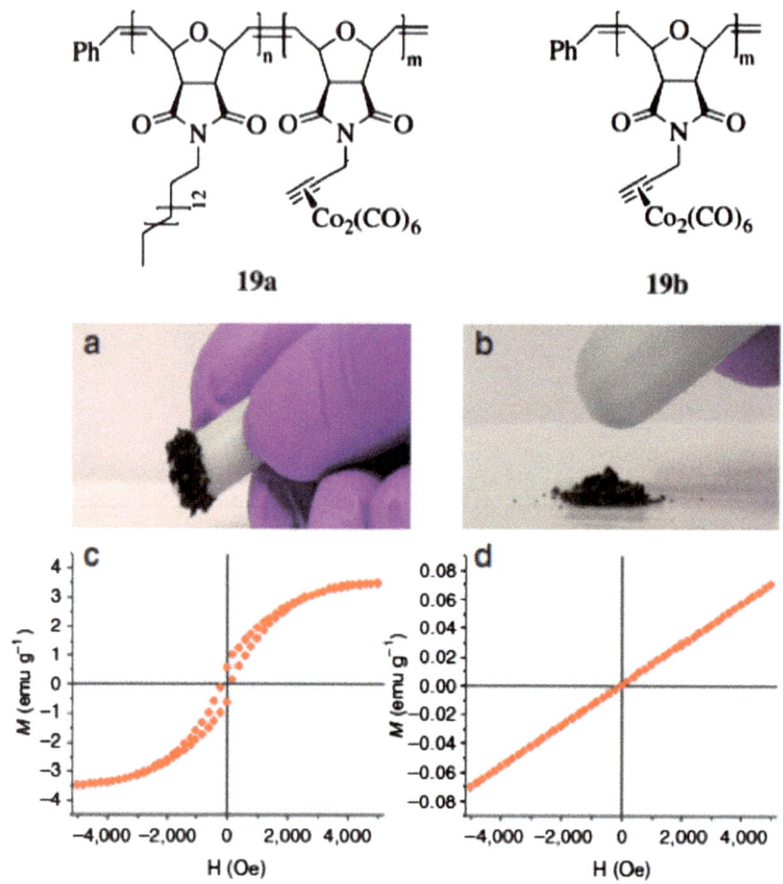

Figure 4.9 Structures of **19a** and **19b** (top). Room temperature response of thermally annealed **19a** (a) and **19b** (b) to a permanent magnet. Superconducting quantum interference device magnetometer measurements that suggest that **19a** is ferromagnetic (c) and that **19b** is paramagnetic (d). Reproduced with permission from ref. 47. Copyright 2011 Nature Publishing Group.

at 100 K or 300 K, whereas pyrolysis at 1000 °C yields ferromagnetic ceramics with room temperature hysteresis (Figure 4.11).

Further, patterned magnetic nanostructures that are promising functional materials in spintronics and in logic circuits have been obtained from metal-containing polymers *via* lithography.[44,52–56] Patterned nanostructures were obtained from cobalt-clusterized polyferrocenylsilane (Co-PFS) (**23**), molybdenum-clusterized polyferrocenylsilane (Mo-PFS) (**24**), and nickel-clusterized polyferrocenylsilane (Ni-PFS) (**25**), which function as resistors in electron-beam lithography (EBL).[54,55] Specifically, Co-PFS (**23a**) was shown to be a promising precursor to nanobars, dots and curve lines *via* EBL (Figure 4.12).[54] The EBL process involved the spin-coating of a solution of

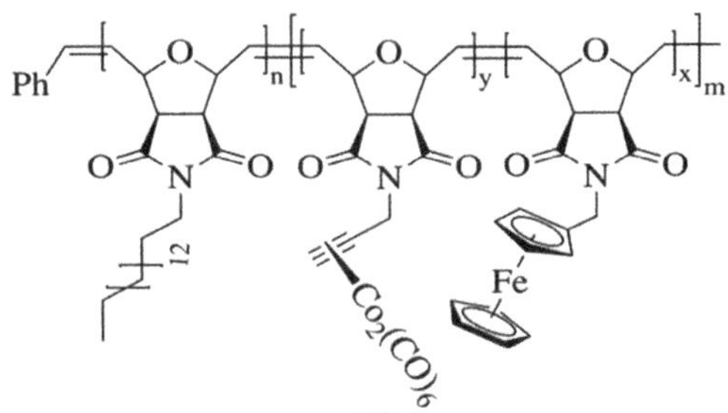

20

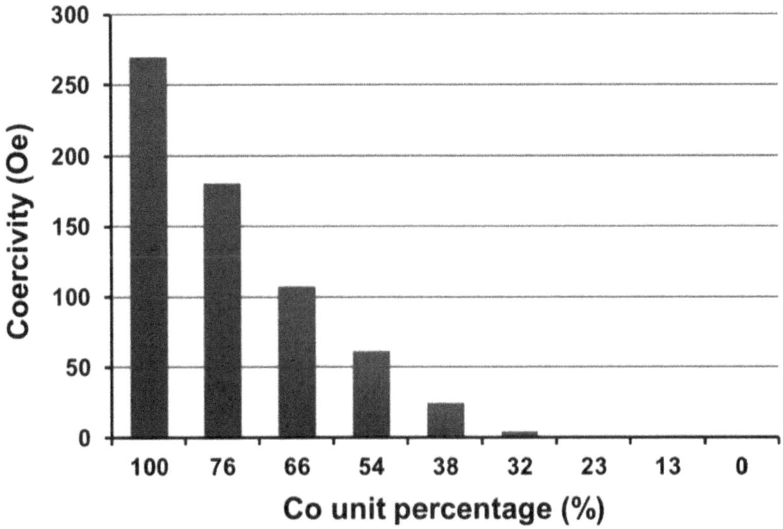

Figure 4.10 Structure of **20** (above) and effect of the percentage of the Co moiety in the cylindrical domains on the coercivity (H_c) of the nanostructured **20** (below). Reproduced with permission from ref. 48. Copyright 2012 American Chemical Society.

the clusterized polyferrocenylsilane (PFS) onto a silicon substrate with subsequent exposure of the substrate to an electron beam in a converted thermonic DSM 940 (Zeiss) scanning electron microscope. These clusterized PFSs function in the negative-tone mode in which the exposed polymeric film is firmly attached to the silicon substrate, and the unexposed film is removed (Figure 4.13). The characteristic negative-tone is attributed to electron beam-induced homolytic sigma bond cleavage, and a radical dimerization process that ultimately results in the crosslinking of polymer chains.[54] Pyrolysis of these nanostructures could afford patterned magnetic metal-containing ceramics that are useful in catalysis, spintronics, and in the formation of logic circuits. The clusterized PFSs were obtained *via* reaction of acetylide-substituted PFSs with a metal source, usually a carbonyl,[54,55] an approach employed by Abd-El-Aziz and coworkers to incorporate cobalt into iron-containing polynorbornenes and polymethacrylates.[45,49]

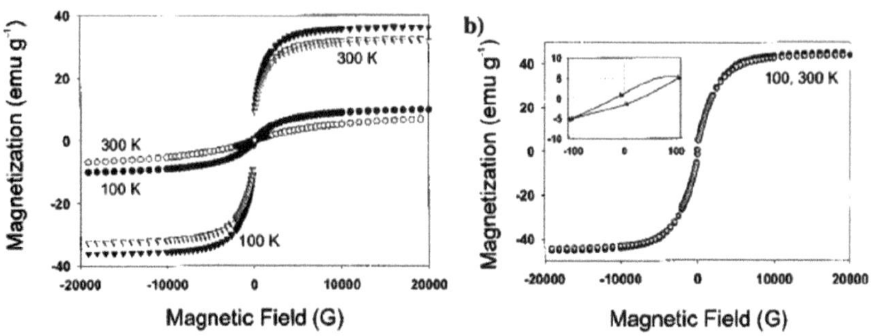

Figure 4.11 Magnetization *versus* field measurements of pyrolyzed **22** at 650 °C [at 100 K (block circle) and 300 K (circle)] and 850 °C [at 100 K (block inverted triangle) and 300 K (inverted triangle)] (a), and at 1000 °C [at 100 K (block inverted triangle) and 300 K (inverted triangle)] (b). The inset in (b) shows that pyrolysis of **22** at 1000 °C yields a ceramic that exhibits room temperature hysteresis with a small remanent magnetization. Reproduced with permission from ref. 41. Copyright 2000 American Association for the Advancement of Science.

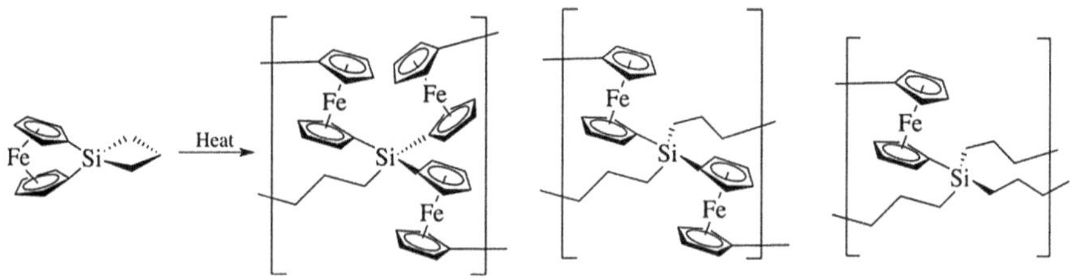

21 **22**

Scheme 4.2 Polymerization of spirocyclic [1]silaferrocenophane (**21**); the three possible microstructures of the cross-linked perceramic metal-containing polymeric product (**22**).

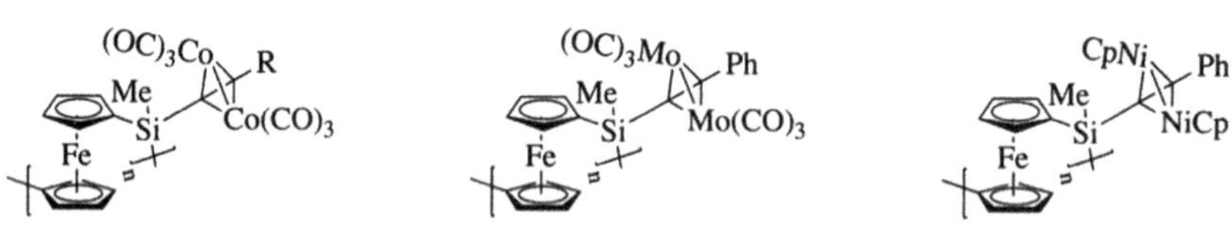

23: **a**, R = Ph
b, R = *n*-Bu

24

25

Figure 4.12 Structures of clusterized polyferrocenylsilane.

With increasing amounts of information generated in this century, there is a concomitant demand for storage devices. Bit patterned media based on magnetic iron–platinum (FePt) nanoparticles (NPs) show a lot of promise as the next generation of ultrahigh-density magnetic data storage systems. The remarkably large bulk magnetocrystalline anisotropy (K_u) ($K_u \approx 7 \times 10^6$ J m^{-3})

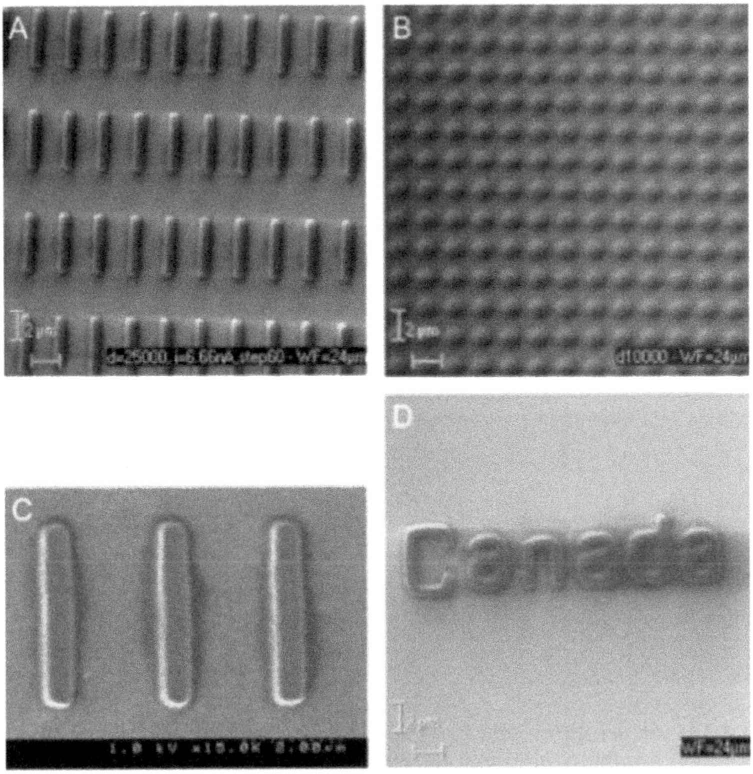

Figure 4.13 Scanning electron micrographs of bars (A and C), dots (B), and curve lines (D) fabricated by EBL using Co-PFS (**23a**) as a resistor. Reproduced with permission from ref. 54. Copyright 2004 WILEY-VCH Verlag GmbH & Co KGaA, Weinheim.

of face-centered tetragonal ($L1_0$) FePt NPs makes these NPs ideal materials for the design of ultrahigh-density data recording systems.[57] Recently, a bimetallic polymer, polyferroplatinyne (**26**) that incorporated iron and platinum was reported to be an excellent precursor for the design of $L1_0$-ordered FePt NPs bit patterned media *via* a novel single step nanoimprint lithography (NIL) that took advantage of the excellent solubility of **26** in organic solvents.[57] The NIL procedure involves drop-casting a saturated chloroform solution of **26** onto a silicon substrate, and subsequent pattern imprinting onto the substrate using a polydimethylsiloxane (PDMS) stamp. Exposure of the substrate to UV irradiation evaporates the chloroform, and transforms the low molecular weight **26** to a high molecular weight cross-linked polymer. A negative copy of nanoscale line and dot array patterns is successfully reproduced onto the FePt-containing polymer when the PDMS stamp is removed (Figure 4.14). The FePt-containing polymer is transformed into FePt NPs upon annealing at 800 °C under argon for 1 h after removal of residue from the trenches. A room temperature coercivity of 14 kOe exhibited by the

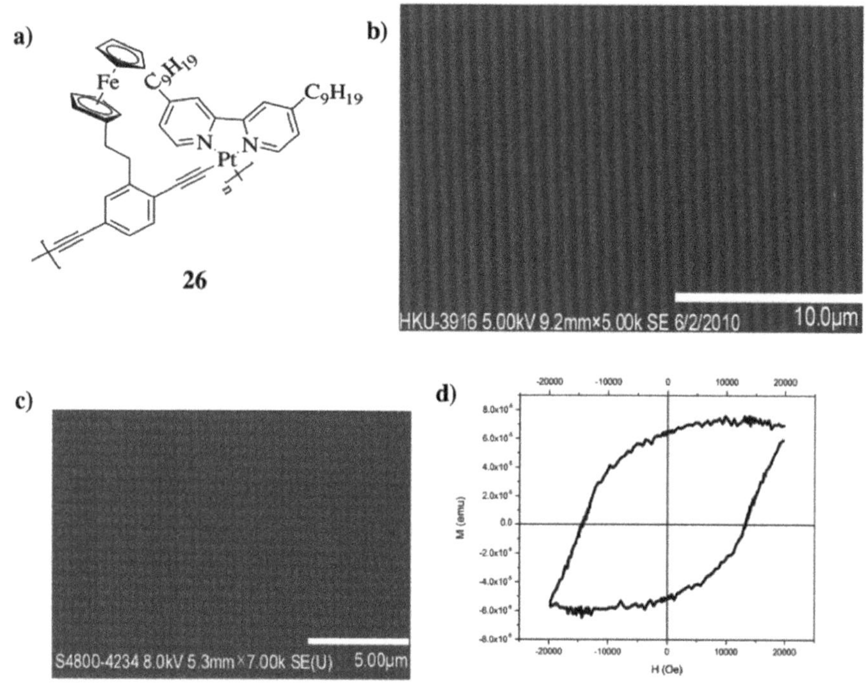

Figure 4.14 Structure of FePt-based polymer (**26**) (a). Scanning electron micrographs of the nanoimprinted line array pattern (b), and dot array pattern (c) of **26**. Room temperature hysteresis loop measurement of **26** using a vibrating sample magnetometer (d). Reproduced with permission from ref. 57. Copyright 2012 WILEY-VCH Verlag GmbH & Co KGaA, Weinheim.

FePt NPs-based nanopatterned dot arrays matches those currently exhibited by magnetic data storage systems currently in use.[57]

4.4 Optical Metal-Containing Polymers

Increasing awareness of the danger of sudden exposure of sensitive optical components such as the human eyes and optical sensors to intense laser pulses has led to the search for optical power limiters that possess high linear transmissions and fast response speed, as well as high solubility.[58–63] Materials with optical power limiting feature high transmittance for low intensity inputs such as images, while limiting the transmittance of high intensity inputs such as laser pulses. In addition to protecting sensitive optical components, these materials offer a means of optical signal enhancement *via* suppression of noise in the laser beam[64,65] as well as optical signal regeneration.[64,66] Although metallophthalocyanines, and metalloporphyrins are effective optical power limiters, their poor solubility hinders processibility during the fabrication of optical power limiting devices.[58,59] Materials such

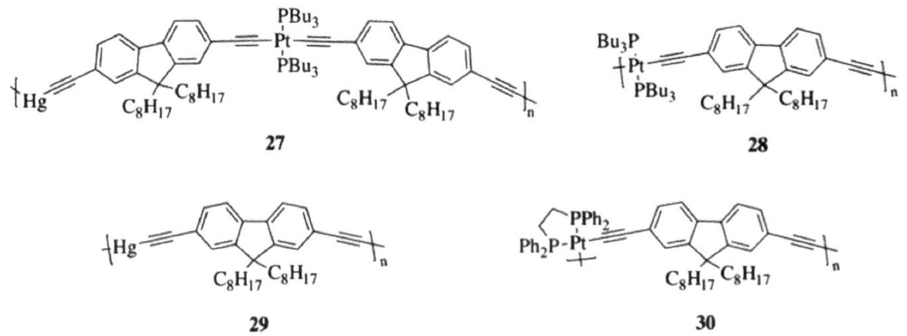

Figure 4.15 Structures of homo and heterometallic polyynes **27–30**.

as organic polyynes[58,59,67] and small molecule platinum acetylides[58,59,68] feature good optical power limiting; however, their absorption beyond 400 nm limits their commercial use. Thus, Zhou *et al.* designed a series of homo and heterometallic polyynes that feature solution-processibility, as well as optical power limiting with good transparency tradeoff (Figure 4.15).[58,59] They compared the optical power limiting of **27–30** to those of classic optical power limiters such as C_{60}, Zn^{II}-tetrakis(4-phenyl)-porphyrin (Zn-TPP), and Cu^{II}-tetrakis(*tert*-butyl)phthalocyanine (CuPc-*t*Bu_4).[59] Overall, this series of homo and heterometallic polyynes features excellent optical power limiting that is superior to those of C_{60} and Zn-TPP as well as CuPc-*t*Bu_4 (Figure 4.16). The absorbed energy associated with the transitions of the first triplet state ($T_1 \rightarrow T_n$) is key to the nonlinear optical power limiting process, as the lifetime of the first singlet is short. The crucial role of the triplet state in the optical power limiting behavior is supported by photoluminescence (PL) spectroscopic measurements of the singlet and the triplet states, which show an intense triplet emission at 77 K (Figure 4.16)[56] that is diagnostic of a strong spin–orbit coupling between the ligand π and the metal d orbitals in the polyynes.[69,70] Molecular orbital (MO) calculations suggest that the triplet state is not pivotal to the optical power limiting behavior of **30** but a $S_0 \rightarrow S_1$ transition that is due to π(C≡CR) → π*(C≡CR) orbitals is responsible for the optical power limiting property.[59,71] The MO calculations are corroborated by PL results that show the absence of room temperature triplet emission in **30** (Figure 4.16). Typically, improving the transparent window of the optical power limiter by reducing π-conjugation compromises the optical nonlinearity, which eventually reduces the optical power limiting behavior. However, in this study, the transparency window is improved without compromising the optical power limited behavior through reduction of π-conjugation by incorporation of Hg as in the heterometallic **27** or through a switch to the *cis*-isomer as in **30**. In contrast to classic optical power limiters, C_{60}, CuPc-*t*Bu_4, Zn-TPP, these polyynes (**27–30**) are white to off-white powders with excellent transparencies in the visible region as shown by UV-vis spectra (Figure 4.16), and are therefore potentially useful as optical power limiters.

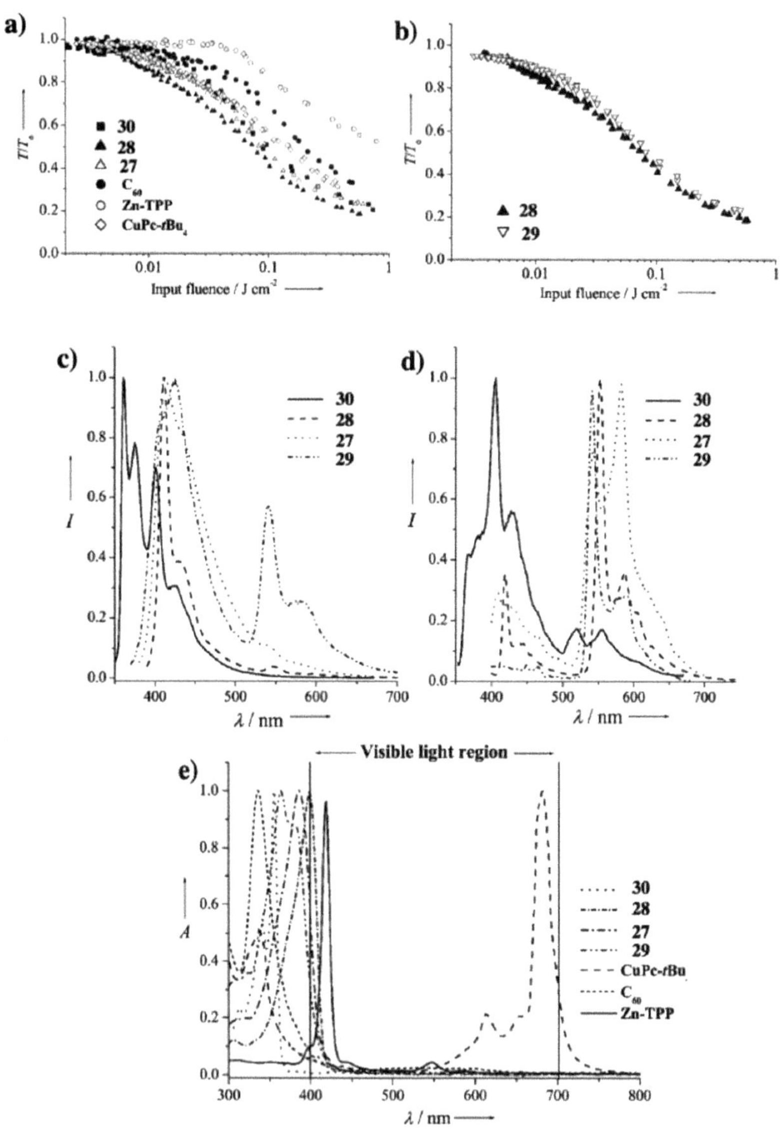

Figure 4.16 Normalized transmittance (T/T_0) of homo and heterometallic polyynes (T_0 = 92%) and some classic optical power limiters (T_0 = 86%) *versus* incident fluence (a) and (b). Photoluminescence spectra of homo and heterometallic polyynes in $CHCl_2$ at 298 K (c) and 77 K (d). Optical transparency of homo and heterometallic polyynes and some classic optical power limiters (e). Reproduced with permission from ref. 59. Copyright 2006 WILEY-VCH Verlag GmbH & Co KGaA, Weinheim.

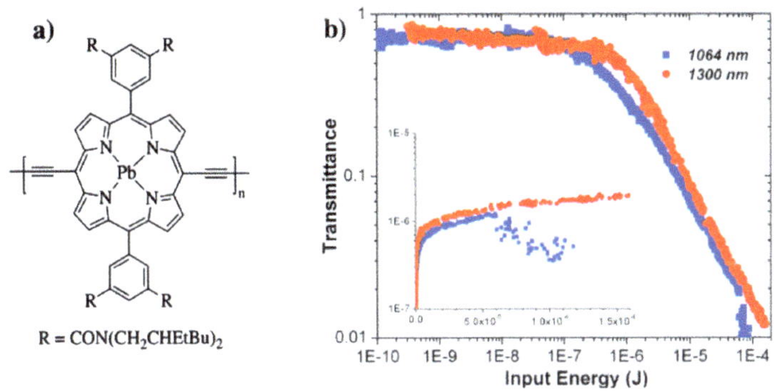

Figure 4.17 Structure of Pb-containing metalloporphyrin polymer (**31**) (a). Nanosecond pulsed optical power limiting of chlorobenzene solution of **31** in fused silica microcapillary waveguides and inset shows output energy *versus* input energy (b). Reproduced with permission from ref. 64. Copyright 2009 Optical Society of America.

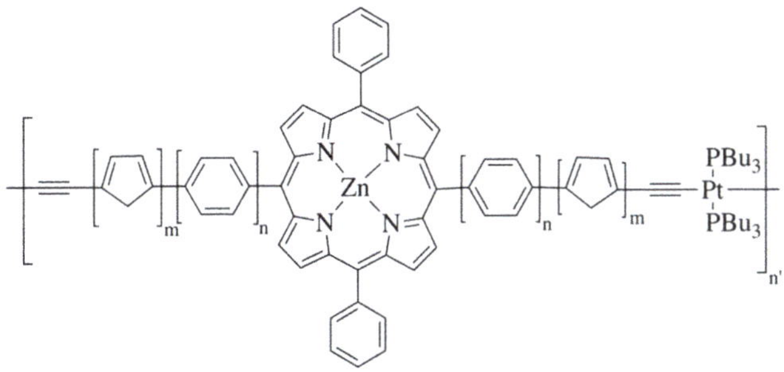

32: m = 0, n = 1
33: m = 1, n = 1
34: m = 1, n = 0

Figure 4.18 Structures of metalloporphyrin-containing polyplatinyne polymers.

Notwithstanding the limited solubility of metalloporphyrins,[58,59] considerable efforts are still being devoted to investigating their optical power limiting behavior.[64] For example, a Pb-containing porphyrin polymer (**31**) (Figure 4.17) features an optical power limiting behavior over a range of 500 nm (*ca.* 1050–1600 nm) in the near IR region, and over five orders-of-magnitude (75 fs to 40 ns) for laser pulsewidths.[64] Nonlinear multimode waveguides featuring optical power limiting with ns pulses and high linear transmittance ($T_L \approx 80\%$) were formed by the incorporation of a solution of **31** in chlorobenzene in a fused silica microcapillary waveguide (Figure 4.17). Conjugated metalloporphyrin polymers are also of interest in the design of solar cells due to the low band gap associated with their excited state.[72,73] For example, a series of metalloporphyrin-containing polyplatinyne polymers **32–34** was designed and their photovoltaic properties were investigated (Figure 4.18).[73] The solar cells were

Table 4.3 Photovoltaic properties of metalloporphyrin-containing polymers. Reproduced with permission from ref. 73. Copyright 2011 American Chemical Society.

Polymer–PCBM	V_{oc}[a] (V)	J_{sc}[b] (mA cm^{-2})	FF[c]	PCE[d] (%)
32 (1:4)	0.72	2.74	0.34	0.68
33 (1:4)	0.78	3.02	0.30	0.71
34 (1:4)	0.77	3.42	0.39	1.04

[a]Open-circuit voltage.
[b]Short-circuit current density.
[c]Fill factor.
[d]Photocurrent efficiency.

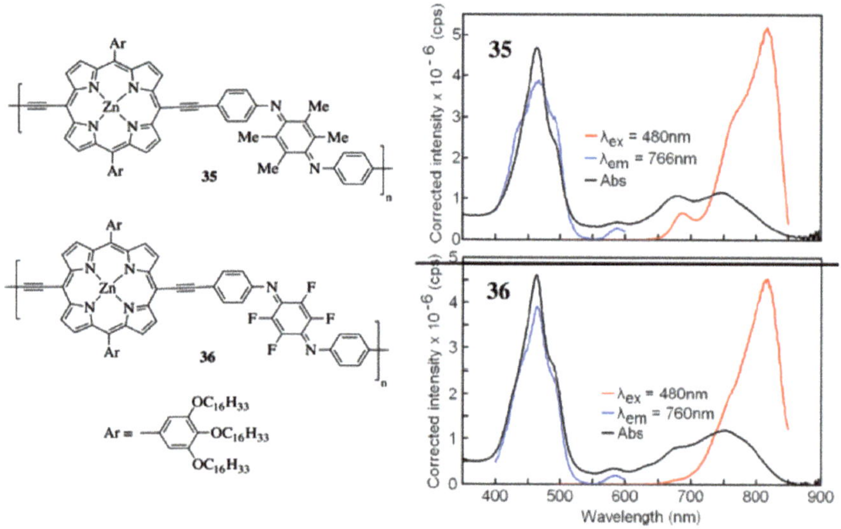

Figure 4.19 Structures of polymers **35** (top left) and **36** (bottom left). Absorption, emission and excitation spectra of **35** (top right) and **36** (bottom right). Reproduced with permission from ref. 72.

fabricated with **32**, **33** or **34** as electron donors and [6,6]-phenyl C_{61}-butyric acid methyl ester as an electron acceptor. Interestingly, **34** exhibits a photocurrent efficiency of 1.04% (Table 4.3), which is among the highest reported for conjugated porphyrin polymer-based bulk heterojunction solar cells.

The inclusion of zinc(II) porphyrin into a polyaniline backbone yielded metal-containing polymers (**35** and **36**) with a low energy band gap of 1.59 eV[72] as shown by the zinc(II) porphyrin to quinone diimine charge transfer[72,74,75] absorption band at 780 nm (Figure 4.19). Electronic communication between the porphyrin rings in these polymers is apparent as Soret bands with a shoulder are observed.[72,75,76] The polymers are also emissive as the luminescence spectra feature bands at 680 and 820 nm that superimpose on the charge transfer absorption bands (Figure 4.19). These emissions, which are mostly phosphorescence with a total quantum yield of ~0.23, are attributed to $S_2 \rightarrow S_0$ and $T_2 \rightarrow S_0$ transitions, and suggest that the polymers are applicable for light emitting diodes in near IR applications.

Interest in light emitting metal-containing polymers is growing, as they are potential materials for the design of organic light emitting devices (OLEDs). This interest is not limited to metalloporphyrin-based polymers; indeed, incorporation of phosphorescent transition metal complexes such as iridium(III) complexes into a polymeric framework also offers a route to OLED precursors.[77,78] The emission color of the iridium(III) complex-based light emitting polymers are tunable as changing the nature of the attached ligands tunes the color from red to blue.[77,79,80] Further, in contrast with conjugated metalloporphyrin-based polymer light emitters, the iridium(III) complex systems feature high triplet energy levels that not only offer the opportunity to deal with high band gap emitters such as blue emitters[77,81] but also prevent luminescence quenching through back energy transfer from the emitter to the polymer.[77] Recently, the phosphorescent iridium(III) complex was incorporated into a polymeric host by the free radical copolymerization of methyl methacrylate functionalized with an iridium(III) complex monomer with either methyl methacrylate or methyl methacrylate functionalized with carbazole to afford photoactive copolymers (**37** and **38**), which are potentially OLED precursor materials.[77] Photoexcitation of **38** in dichloromethane reveals efficient intrachain energy transfer from the polymeric host to the phosphorescent emitter that suggests its applicability in the design of OLEDs. The phosphorescence of **38** results exclusively from the iridium(III) complex as excitation of the complex (λ_{ex} = 475 nm) in addition to excitation in the carbazole absorption band (λ_{ex} = 569 nm) results exclusively in the complex phosphorescent (λ_{ex} = 569 nm) without an emission from the carbazole unit (Figure 4.20). Intrachain energy transfer from the polymeric host to the complex is demonstrated by the predominance of the carbazole absorption band in the excitation spectra at the different emission wavelengths (425, 569 and 612 nm) (Figure 4.20).

Material scientists have also explored the possibility of designing white light emitting metal-containing polymers. Generally, white light emitting polymers are potentially useful in the design of light emitting diodes, backlights in flat panel displays, full color flat panel displays, as well as solid-state lighting devices.[82–85] An approach toward this class of luminescence polymers involves the incorporation of a blue and an orange or blue, green and red chromophore in the right proportion within a polymeric framework.[85] For instance, the coordination of a red emitting Eu(III) ion onto cyan (mixture of blue and green) emitting alternating copolymers of 9,9-dioctyl-9*H*-fluorene-2,7-diyl and 4-ethynyl-2,6-bis(pyrazolyl)pyridine (BPP) (**39**) or octylated BPP (**40**) results in an emitted light with Commission Internationale de l'Eclairage (CIE) coordinates that match closely with those of white light.[85] The metal chelating ability of the BPP moieties in the polymer facilitates the incorporation of the Eu(III) ions into the polymeric framework. This was achieved by dipping the copolymer films into a methanolic solution of Eu(III) ions for one hour to afford the Eu-complexed polymers, which on exposure to UV light emitted a brilliant white color due to the blending of the 430, 546 and 625 emissions

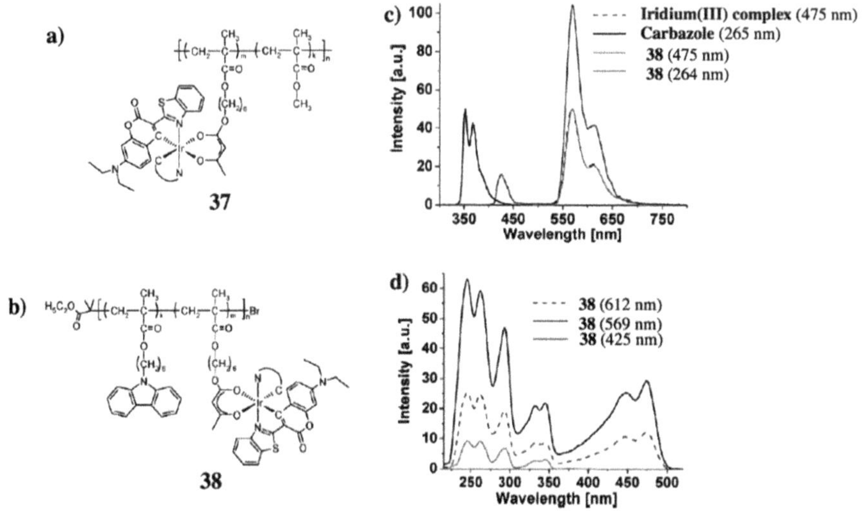

Figure 4.20 Structures of **37** (a) and **38** (b). Normalized emission spectra of methyl methacrylate functionalized iridium(III) complex, methyl methacrylate functionalized carbazole and the copolymer **38** with the excitation wavelengths in brackets (c). The emissions of **38** after excitation at 475 and 264 nm, respectively are given in the found proportion. Excitation spectra of **38** at the different emission maxima in brackets (d). Reproduced with permission from ref. 77. Copyright 2008 WILEY-VCH Verlag GmbH & Co KGaA, Weinheim.

(Figures 4.21 and 4.22). Based on a chromaticity diagram (Figure 4.22), CIE coordinates of (0.31, 0.35) and (0.31, 0.32), which are close to those of perfect white light (0.33, 0.33), were observed for the Eu-complexed copolymers of **39** and **40**. The phosphorescence of the Eu(III) metal center was attributed to intrachain energy transfer from the polymer backbone to the Eu(III) center.

The metal to ligand charge transfer in ruthenium complexes of polypyridine, which usually results in luminescence, is well-exploited in photonic[86,87] as well as in biological applications such as *in vivo* imaging of cell compartments[87,88] and *in vivo* monitoring of active molecules.[87,89] A biocompatible, water soluble, luminescent Ru-containing polymer (**41**), obtained from the reaction of a 1,10-phenanthroline-functionalized poly-(amidoamine) with $[Ru(phen)_2(CF_3SO_3)_2]$ complex, self-aggregates in an aqueous medium to form nanoparticles with a hydrodynamic radius of 18 ± 2 nm.[87] These characteristics are ideal for potential biological applications such as *in vivo* imaging. The UV-vis absorption spectrum of **41** revealed an absorption band at 450 nm (Figure 4.23) characteristic of the metal to ligand charge transfer of $[Ru(phen)_3]$.[87,90] Further, an emission band at 614 nm (Figure 4.23) with a lifetime of 580 ns and photoluminescence yield of 4.1% is revealed by photoluminescence spectroscopy. Polymer **41**, which is stable under physiological conditions and in the presence of excess cysteine, did not exhibit any toxic effect on human embryonic kidney (HEK-293) cells. This biocompatible Ru-containing polymer, **41**, was also investigated for its cellular

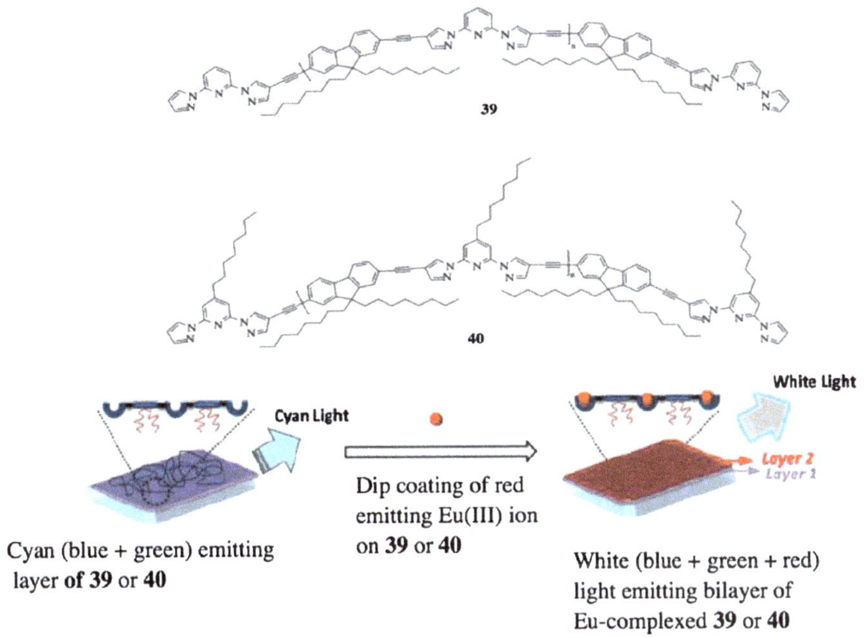

Figure 4.21 Structures of cyan-emitting copolymers **39** (top) and **40** (center). Schematics of the dip-coating of Eu(III) ions onto **39** or **40** (bottom). Reproduced with permission from ref. 85. Copyright 2013 American Chemical Society.

internalization and intracellular distribution by using fluorescence confocal microscopy (FCM). The FCM images suggest that **41** is internalized within the vesicular compartments of the cytoplasm in 6 hours, and is homogeneously distributed within the cytoplasm in 12 hours (Figure 4.23).

Luminescent metal-containing polymers are also used as probes for the detection of small quantities of biomolecules. Polymeric probes with multiple luminescent centers are desired in this regard as the standard fluorescence probes require labor-intensive concentration of the analyte, whereas the former save time and labor by simply enhancing the signal.[91,92] Unlike organic fluorophores, luminescent iridium(III) complexes feature large Stokes shifts that render them less susceptible to self-quenching,[92,93] and are also resistant to photobleaching.[92,94] Further, the tunability of the emission of these iridium-containing luminescent polymers over a wide range of wavelengths, in addition to their long luminescence lifetimes,[92,94] makes these polymers good probes for the detection of a small amount of biomolecules. A series of iridium(III) complex-containing amphiphilic polymers (**42–44**), synthesized *via* ring-opening metathesis polymerization (ROMP), was investigated for their potential as fluorescence probes in solid phase immunoassays with an antibody-conjugated biotin.[92] Micellization of the polymers in a 9:1 water–acetonitrile mixture resulted in an increase in the

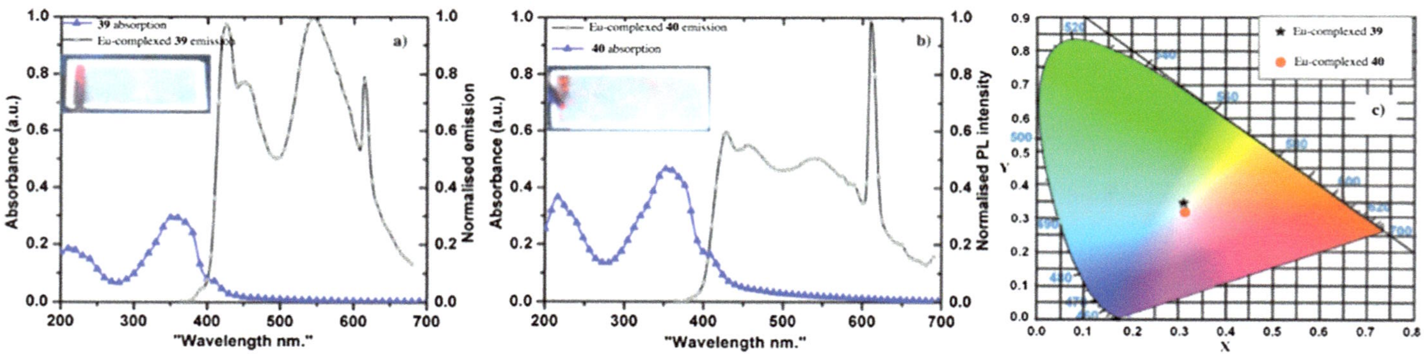

Figure 4.22 Absorption and emission spectra of white light emitting Eu-complexed **39** (a) and **40** (b). Chromaticity diagram showing the emission color of Eu-complexed **39** and **40** films in a CIE 1931 2° (c). Reproduced with permission from ref. 85. Copyright 2013 American Chemical Society.

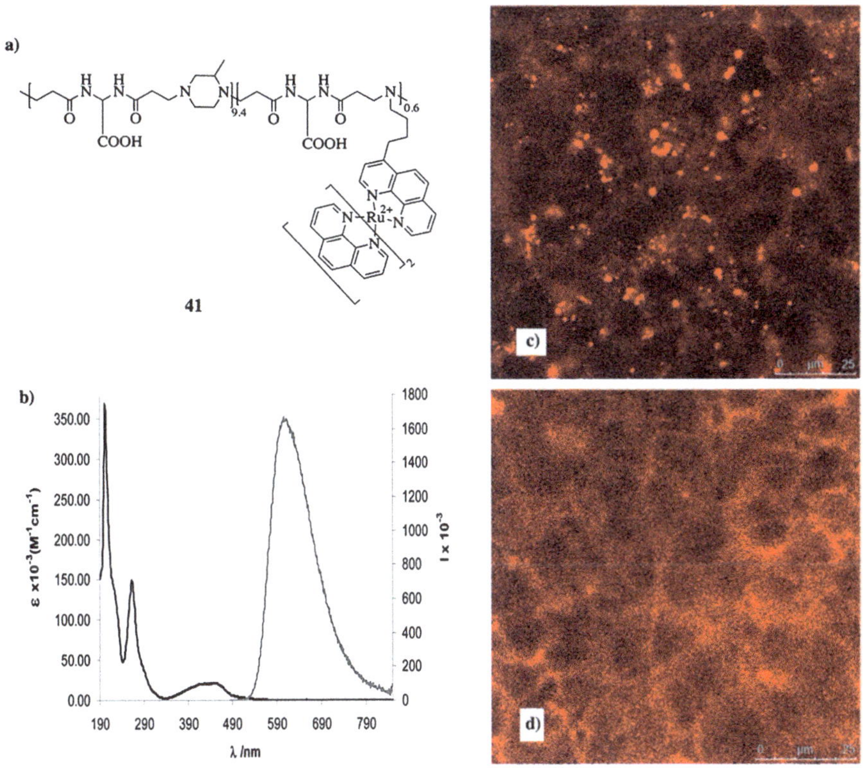

Figure 4.23 Structure of ruthenium-coordinated 1,10-phenanthroline-functionalized poly(amidoamine) (**41**) (a). UV-vis absorption spectrum (black) and photoluminescence spectrum (gray) of **41** (b). Confocal microscopy images of human embryonic kidney (HEK-293) cells incubated with **41** (*ca.* 5 μM) for 6 hours (c) and 12 hours (d). Reproduced with permission from ref. 87. Copyright 2012 American Chemical Society.

quantum yield and a blue shift in the emission spectra relative to an acetonitrile solution of the polymers. These micellization-induced changes in optical properties were attributed to the increased rigidity and desolvation of the iridium(III) complex-containing hydrophobic core.[92,95–98] In contrast with polymers **42** and **43**, polymer **44** specifically binds to streptavidin-coated magnetic beads through the biotin–streptavidin interactions, as shown by the noticeable decrease in luminescence after incubation with streptavidin-coated magnetic beads (Figure 4.24). The absence of biotin in **42** and the presence of a long polyethylene chain adjacent to the biotin in **43** precludes interactions of these micelles with the streptavidin-coated magnetic beads. Specific biotin–streptavidin interactions in **44**, which are further substantiated by fluorescence microscopy, suggest the potential application of this polymer as a fluorescent probe for the detection of macromolecules *in vivo* (Figure 4.24).

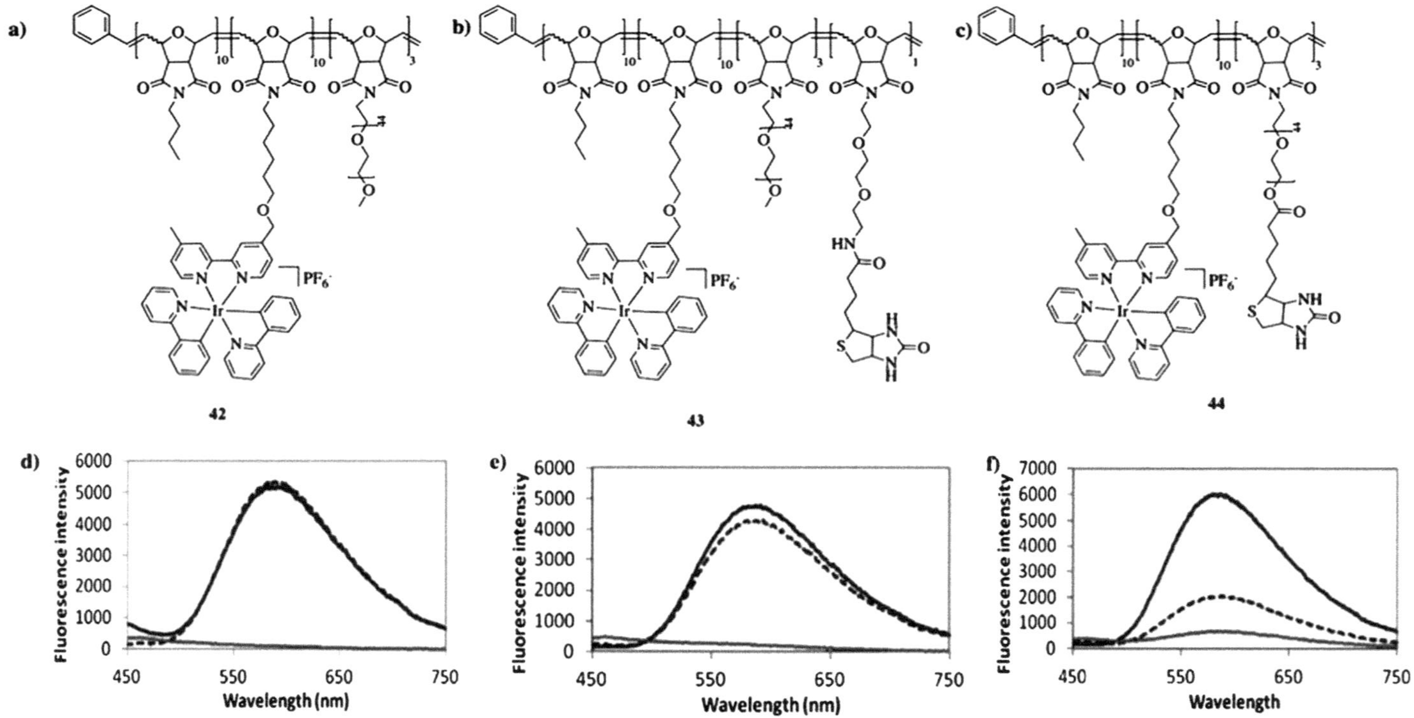

Figure 4.24 Structures of iridium(III)-containing polymers **42** (a), **43** (b), and **44** (c). Fluorescence spectra of biotin–streptavidin reaction solution for **42** (d), **43** (e), and **44** (f). Solid black curve: control solution without streptavidin-coated magnetic beads; dotted black curve: **42**, **43** or **44** with streptavidin-coated magnetic beads; gray curve: 0.1% Tween 20 solution used for washing beads after reaction. Reproduced with permission from ref. 92. Copyright 2012 American Chemical Society.

The use of fluorescent probes is not limited to biological applications; indeed, these probes are increasingly used in environmental science, especially in oxygen sensing. As transition metal complexes have highly stable phosphorescence, good quantum yield, tunable emission, as well as resistance to self-quenching and photobleaching, their use in the environmental monitoring of oxygen is advantageous, since the ground state of oxygen quenches the triplet phosphorescence.[99] An efficient approach of utilizing phosphorescent transition metal complexes is through copolymerization of the complex monomer with an organic fluorescent monomer to obtain a dual fluorescent/phosphorescent emission for the ratiometric sensing of oxygen. Indeed, a series of poly([5,15-bis(pentafluorophenyl)-10,12-bis(phenyl)porphyrinato]platinum(II)-9,9-dioctylfluorene) copolymers (**45–49**) based on the fluorescent fluorene monomer and phosphorescent platinum(II) porphyrin monomer was synthesized and screened for the ratiometric sensing of oxygen.[99] The photoexcitation of these copolymers at 375 nm in degassed CH_2Cl_2 at room temperature results in the expected blue fluorescence of the fluorene, as well as the red phosphorescence of the Pt(II) porphyrin moieties. The relative intensities of these emissions are modulated by percentage composition of the monomers in the copolymer. Clearly, the photoluminescence spectrum of **46** is fluorescence-dominated whereas that of **48** is phosphorescence-dominated (Figure 4.25). The photoluminescence spectra of **48** in the presence of different oxygen concentrations clearly show that as the level of oxygen decreases, the intensity of the phosphorescence increases; although the intensity of the fluorescence remains unchanged (Figure 4.25). This observation, as well as the good linearity of the plot of I_F/I_P (I_F is blue fluorescence intensity and I_P is phosphorescence intensity) as a function of oxygen concentration, suggests that **48** is potentially useful in the design of probes for the ratiometric sensing of oxygen.

Increasing awareness of public security has also driven research towards the design of sensors for the detection of nitro compounds, which are basic components of some explosives.[100,101] Fluorescent metal-containing coordination polymers have been exploited especially in the detection of nitro aromatic compounds as the charge transfer from the excited state of the coordination polymer to the electron deficient aromatic ring results in selective quenching of the luminescent polymer.[101,102] For instance, coordination polymer $[Zn_2(L)(bipy)(H_2O)]\cdot(H_2O)_3(DMF)$ (**50**) (H_4L is bis-(3,5-dicarboxyphenyl) terephthalamide, bipy is 4,4′-bipyridine, and DMF is *N*,*N*′-dimethylformamide) (Figure 4.26a), exhibits fluorescence, and could be used in the selective detection of nitrobenzene *via* the redox induced fluorescence quenching process.[101] Photoluminescence spectroscopy of **50** in the solid state and in acetonitrile revealed strong emissions at 476 nm and 450 nm, respectively, which were attributed by the authors to coordination metal-perturbed ligand-centered electronic transitions.[101] While the addition of nitrobenzene to the acetonitrile solution of **50** remarkably decreases the fluorescence intensity of **50**, the addition of other organic solvents has a negligible effect on the fluorescence intensity (Figure 4.26).[101] Further, the quenching process is selective

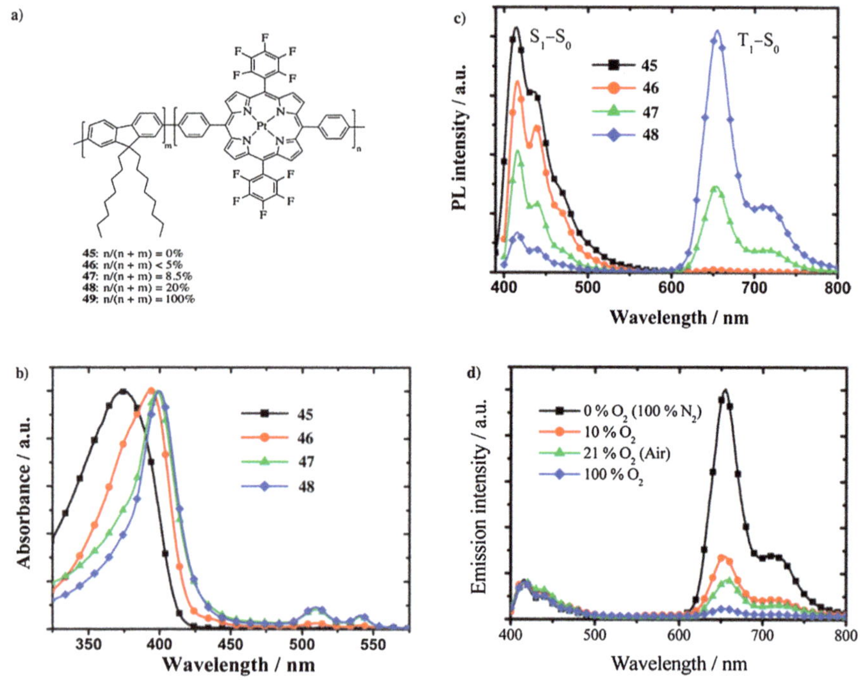

Figure 4.25 Structures of copolymers **45–49** (a). Absorption spectra of **45–48** in degassed CH_2Cl_2 (b). Photoluminescence spectra of **45–48** (c). Photoluminescence spectra of **48** at various oxygen concentrations (d). Reproduced with permission from ref. 99. Copyright 2012 American Chemical Society.

as the addition of other organic solvents to a mixture of nitrobenzene and acetonitrile solution of the polymer did not affect the fluorescence intensity of the polymer (Figure 4.26). Polymer **50** also demonstrates good sensitivity towards the presence of nitrobenzene as 90% of the fluorescence is quenched with 2000 ppm of nitrobenzene (Figure 4.26). Moreover, the good linear correlation between quenching efficiency and nitrobenzene concentration suggests the applicability of the polymer as a probe for the quantitative analysis of nitrobenzene.

4.5 Conclusion and Future Outlook

The incorporation of a metal into polymeric frameworks has not only expanded the frontiers of material science, but has also advanced technology and medicine. As presented in this chapter, the unique properties of metal-containing functional polymers are well explored in the design of catalysts, sensors, semi-conductors, magnetic storage devices, and optical power limiters. These breakthroughs notwithstanding, the field of metallopolymers is far from mature, and we believe that the synthesis of new functional

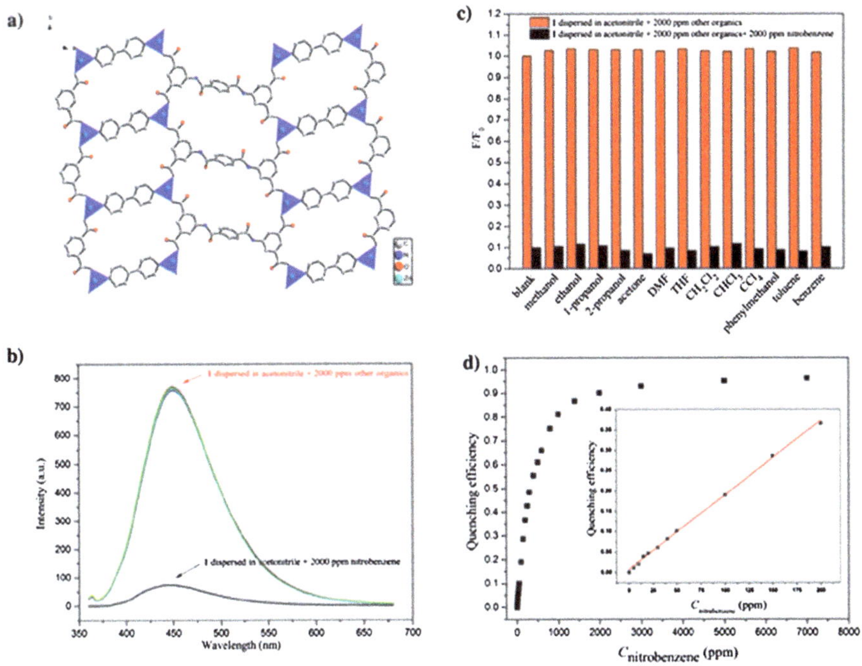

Figure 4.26 Layer structure of **50** (a). Photoluminescence spectra of acetonitrile dispersion of **50** in the presence of nitrobenzene or other organic solvents (b). Histogram of ratio of fluorescence intensity (at 450 nm) of acetonitrile dispersion of **50** in the presence of other organic solvents (red) and nitrobenzene (black) (c). Plot of quenching efficiency (at 450 nm) of acetonitrile dispersion of **50** as a function of the concentration of nitrobenzene (d), and linear relationship of quenching efficiency with concentrations of nitrobenzene in the range of 5–200 ppm (d inset). Reproduced with permission from ref. 101.

metallopolymers that are affordable yet feature intriguing properties will advance human civilization. For instance, a number of the metallopolymers discussed in the chapter incorporate expensive and rare metals such as platinum. Incorporation of such metals limits the competitive advantage of these metallopolymers during commercialization. Thus, it should be worthwhile to explore the synthesis of affordable metallopolymers based on cheap and readily available metals.

References

1. A. S. Abd-El-Aziz and E. A. Strohm, *Polymer*, 2012, **53**, 4879.
2. A. S. Abd-El-Aziz and I. Manners, *J. Inorg. Organomet. Polym.*, 2005, **15**, 157.
3. G. R. Whittell, M. D. Hager, U. S. Schubert and I. Manners, *Nat. Mater.*, 2011, **10**, 176.

4. P. Nguyen, P. Gomez-Elipe and I. Manners, *Chem. Rev.*, 1999, **99**, 1515.
5. J. Janata and M. Josowicz, *Nat. Mater.*, 2003, **2**, 19.
6. I. Manners, *Science*, 2001, **294**, 1664.
7. B. J. Holliday and T. M. Swager, *Chem. Commun.*, 2005, 23.
8. C. Friebe, M. D. Hager, A. Winter and U. S. Schubert, *Adv. Mater.*, 2012, **24**, 332.
9. M. Ramanathan, Y.-C. Tseng, K. Ariga and S. B. Darling, *J. Mater. Chem. C*, 2013, **1**, 2080.
10. J. M. Stanley and B. J. Holliday, *Coord. Chem. Rev.*, 2012, **256**, 1520.
11. M. Tadokoro, S. Yasuzuka, M. Nakamura, T. Shinoda, T. Tatenuma, M. Mitsumi, Y. Ozawa, S. Toriumi, T. Tatenuma, M. Mitsumi, Y. Ozawa, K. Toriumi, H. Yoshino, D. Shiomi, K. Sato, T. Takui, T. Mori and K. Murata, *Angew. Chem., Int. Ed.*, 2006, **45**, 5144.
12. A. K. Cheetham and C. N. R. Rao, *Science*, 2007, **318**, 58.
13. T. Okubo, N. Tanaka, K. H. Kim, H. Yone, M. Maekawa and T. Kuroda-Sowa, *Inorg. Chem.*, 2010, **49**, 3700.
14. K. H. Kim, T. Okubo, N. Tanaka, N. Mimura, M. Maekawa and T. Kuroda-Sowa, *Chem. Lett.*, 2010, **39**, 792.
15. T. Okubo, N. Tanaka, K. H. Kim, H. Anma, S. Seki, A. Saeki, M. Maekawa and T. Kuroda-Sowa, *Dalton Trans.*, 2011, **40**, 2218.
16. K. H. Kim, T. Ueta, T. Okubo, S. Hayami, H. Anma, K. Kato, T. Shimizu, J. Fujimori, M. Maekawa and T. Kuroda-Sowa, *Chem. Lett.*, 2011, **40**, 1184.
17. T. Okubo, N. Tanaka, H. Anma, K. H. Kim, M. Maekawa and T. Kuroda-Sowa, *Polymers*, 2012, **4**, 1613.
18. T. Okubo, H. Anma, N. Tanaka, K. Himoto, S. Seki, A. Saeki, M. Maekawa and T. Kuroda-Sowa, *Chem. Commun.*, 2013, **49**, 4316.
19. N. Tanaka, T. Okubo, H. Anma, K. H. Kim, Y. Inuzuka, M. Maekawa and T. Kuroda-Sowa, *Eur. J. Inorg. Chem.*, 2013, **2013**, 3384.
20. A. S. Abd-El-Aziz and S. S. Dalgakiran, *J. Inorg. Organomet. Polym.*, 2013, **23**, 126.
21. A. S. Abd-El-Aziz, S. Sezgin Dalgakiran and L. Bichler, *Eur. Polym. J.*, 2012, **48**, 1901.
22. C. Amatore, Y. Bouret, E. Maisonhaute, J. I. Goldsmith and H. D. Abruna, *Chem.–Eur. J.*, 2001, **7**, 2206.
23. F. Hauquier, J. Ghilane, B. Fabre and P. Hapiot, *J. Am. Chem. Soc.*, 2008, **130**, 2748.
24. A. Wang, C. Ornelas, D. Astruc and P. Hapiot, *J. Am. Chem. Soc.*, 2009, **131**, 6652.
25. D. Astruc, C. Ornelas and J. Ruiz, *Chem.–Eur. J.*, 2009, **15**, 8936.
26. D. Astruc, *Nat. Chem.*, 2012, **4**, 255.
27. D. Astruc, P. Zhao, L. Liang, A. Rapakousiou, R. Djeda, A. Diallo, T. Kusamoto, J. Ruiz and C. Ornelas, *J. Inorg. Organomet. Polym.*, 2013, **23**, 41.
28. D. Astruc, M. Daniel and J. Ruiz, *Chem. Commun.*, 2004, 2637.
29. A. S. Abd-El-Aziz, E. K. Todd, R. M. Okasha, P. O. Shipman and T. E. Wood, *Macromolecules*, 2005, **38**, 9411.

30. C. Ornelas, J. Ruiz Aranzaes, E. Cloutet, S. Alves and D. Astruc, *Angew. Chem., Int. Ed.*, 2007, **46**, 872.
31. C. Ornelas, J. Ruiz and D. Astruc, *Organometallics*, 2009, **28**, 4431.
32. A. Rapakousiou, Y. Wang, C. Belin, N. Pinaud, J. Ruiz and D. Astruc, *Inorg. Chem.*, 2013, **52**, 6685.
33. C. Ornelas, J. Ruiz, C. Belin and D. Astruc, *J. Am. Chem. Soc.*, 2009, **131**, 590.
34. J. Ruiz, C. Pradet, F. Varret and D. Astruc, *Chem. Commun.*, 2002, 1108.
35. R. Djeda, C. Ornelas, J. Ruiz and D. Astruc, *Inorg. Chem.*, 2010, **49**, 6085.
36. G. de la Cruz, H. Schuele, J. Losada, M. P. Garcia-Armada, H. Frey, B. Alonso and C. M. Casado, *Eur. J. Inorg. Chem.*, 2013, **2013**, 44.
37. K. Heuze, D. Mery, D. Gauss and D. Astruc, *Chem. Commun.*, 2003, 2274.
38. K. Heuze, D. Mery, D. Gauss, J. Blais and D. Astruc, *Chem.–Eur. J.*, 2004, **10**, 3936.
39. S. Lu and H. Alper, *J. Am. Chem. Soc.*, 2003, **125**, 13126.
40. S. Nlate, L. Plault and D. Astruc, *New J. Chem.*, 2007, **31**, 1264.
41. M. J. MacLachlan, M. Ginzburg, N. Coombs, T. W. Coyle, N. P. Raju, J. E. Greedan, G. A. Ozin and L. Manners, *Science*, 2000, **287**, 1460.
42. X. Wang, A. Arsenault, G. A. Ozin, M. A. Winnik and I. Manners, *J. Am. Chem. Soc.*, 2003, **125**, 12686.
43. S. B. Clendenning, S. Fournier-Bidoz, A. Pietrangelo, G. Yang, S. Han, P. M. Brodersen, C. M. Yip, Z. Lu, G. A. Ozin and I. Manners, *J. Mater. Chem.*, 2004, **14**, 1686.
44. K. Liu, S. Fournier-Bidoz, G. A. Ozin and I. Manners, *Chem. Mater.*, 2009, **21**, 1781.
45. A. S. Abd-El-Aziz, D. J. Winram, P. O. Shipman and L. Bichler, *Macromol. Rapid Commun.*, 2010, **31**, 1992.
46. X. Hu, Y. Li and C. Xu, *Solid State Sci.*, 2011, **13**, 1478.
47. Z. M. Al-Badri, R. R. Maddikeri, Y. Zha, H. D. Thaker, P. Dobriyal, R. Shunmugam, T. P. Russell and G. N. Tew, *Nat. Commun.*, 2011, **2**, 482.
48. Y. Zha, H. D. Thaker, R. R. Maddikeri, S. P. Gido, M. T. Tuominen and G. N. Tew, *J. Am. Chem. Soc.*, 2012, **134**, 14534.
49. A. S. Abd-El-Aziz, D. J. Winram, P. O. Shipman, C. L. Rock, M. S. Vandel and B. O. Patrick, *Macromol. Chem. Phys.*, 2012, **213**, 2136.
50. Y. Zha, R. R. Maddikeri, S. P. Gido and G. N. Tew, *J. Inorg. Organomet. Polym.*, 2013, **23**, 89.
51. J. Kong, T. Schmalz, G. Motz and A. H. E. Mueller, *J. Mater. Chem. C*, 2013, **1**, 1507.
52. S. B. Clendenning, S. Han, N. Coombs, C. Paquet, M. S. Rayat, D. Grozea, P. M. Brodersen, R. N. S. Sodhi, C. M. Yip, Z. Lu and I. Manners, *Adv. Mater.*, 2004, **16**, 291.
53. A. Y. Cheng, S. B. Clendenning, G. Yang, Z. Lu, C. M. Yip and I. Manners, *Chem. Commun.*, 2004, 780.
54. S. B. Clendenning, S. Aouba, M. S. Rayat, D. Grozea, J. B. Sorge, P. M. Brodersen, R. N. S. Sodhi, Z. Lu, C. M. Yip, M. R. Freeman, H. E. Ruda and I. Manners, *Adv. Mater.*, 2004, **16**, 215.

55. W. Y. Chan, S. B. Clendenning, A. Berenbaum, A. J. Lough, S. Aouba, H. E. Ruda and I. Manners, *J. Am. Chem. Soc.*, 2005, **127**, 1765.
56. S. Greenberg, S. B. Clendenning, K. Liu, I. Manners, S. Aouba and H. E. Ruda, *Macromolecules*, 2005, **38**, 2023.
57. Q. Dong, G. Li, C. Ho, M. Faisal, C. Leung, P. W. Pong, K. Liu, B. Tang, I. Manners and W. Wong, *Adv. Mater.*, 2012, **24**, 1034.
58. G. Zhou, W. Wong, C. Ye and Z. Lin, *Adv. Funct. Mater.*, 2007, **17**, 963–975.
59. G. Zhou, W. Wong, Z. Lin and C. Ye, *Angew. Chem., Int. Ed.*, 2006, **45**, 6189.
60. G. Zhou, W. Wong, D. Cui and C. Ye, *Chem. Mater.*, 2005, **17**, 5209.
61. D. G. McLean, R. L. Sutherland, M. C. Brant, D. M. Brandelik, P. A. Fleitz and T. Pottenger, *Opt. Lett.*, 1993, **18**, 858.
62. L. W. Tutt and T. F. Boggess, *Prog. Quantum Electron.*, 1993, **17**, 299.
63. C. W. Spangler, *J. Mater. Chem.*, 1999, **9**, 2013.
64. J. M. Hales, M. Cozzuol, T. E. O. Screen, H. L. Anderson and J. W. Perry, *Opt. Express*, 2009, **17**, 18478.
65. G. S. He, L. Yuan, J. D. Bhawalkar and P. N. Prasad, *Appl. Opt.*, 1997, **36**, 3387.
66. M. R. E. Lamont, M. Rochette, D. J. Moss and B. J. Eggleton, *IEEE Photonics Technol. Lett.*, 2006, **18**, 1185.
67. G. Zhou, S. Zhang, P. Wu and C. Ye, *Chem. Phys. Lett.*, 2002, **363**, 610.
68. P. Nguyen, G. Lesley, T. B. Marder, I. Ledoux and J. Zyss, *Chem. Mater.*, 1997, **9**, 406.
69. T. J. McKay, J. A. Bolger, J. Staromlynska and J. R. Davy, *J. Chem. Phys.*, 1998, **108**, 5537.
70. J. Staromlynska, T. McKay, J. Bolger and J. Davy, *J. Opt. Soc. Am. B*, 1998, **15**, 1731.
71. J. E. Rogers, T. M. Cooper, P. A. Fleitz, D. J. Glass and D. G. McLean, *J. Phys. Chem. A*, 2002, **106**, 10108.
72. S. Lamare, S. M. Aly, D. Fortin and P. D. Harvey, *Chem. Commun.*, 2011, **47**, 10942.
73. H. Zhan, S. Lamare, A. Ng, T. Kenny, H. Guernon, W. Chan, A. B. Djurišić, P. D. Harvey and W. Wong, *Macromolecules*, 2011, **44**, 5155.
74. K. Gagnon, S. Mohammed Aly, A. Brisach-Wittmeyer, D. Bellows, J. Berube, L. Caron, A. S. Abd-El-Aziz, D. Fortin and P. D. Harvey, *Organometallics*, 2008, **27**, 2201.
75. D. Fortin, S. Clement, K. Gagnon, J. Berube, M. P. Stewart, W. E. Geiger and P. D. Harvey, *Inorg. Chem.*, 2009, **48**, 446.
76. G. Langlois, S. M. Aly, C. P. Gros, J. Barbe and P. D. Harvey, *New J. Chem.*, 2011, **35**, 1302.
77. C. Ulbricht, C. R. Becer, A. Winter, D. Veldman and U. S. Schubert, *Macromol. Rapid Commun.*, 2008, **29**, 1919.
78. E. Holder, B. M. W. Langeveld and U. S. Schubert, *Adv. Mater.*, 2005, **17**, 1109.
79. S. Lamansky, P. Djurovich, D. Murphy, F. Abdel-Razzaq, H. Lee, C. Adachi, P. E. Burrows, S. R. Forrest and M. E. Thompson, *J. Am. Chem. Soc.*, 2001, **123**, 4304.

80. Y. You and S. Y. Park, *J. Am. Chem. Soc.*, 2005, **127**, 12438.
81. M. Sudhakar, P. I. Djurovich, T. E. Hogen-Esch and M. E. Thompson, *J. Am. Chem. Soc.*, 2003, **125**, 7796.
82. B. W. D'Andrade and S. R. Forrest, *Adv. Mater.*, 2004, **16**, 1585.
83. Y. Sun, N. C. Giebink, H. Kanno, B. Ma, M. E. Thompson and S. R. Forrest, *Nature*, 2006, **440**, 908.
84. G. Schwartz, M. Pfeiffer, S. Reineke, K. Walzer and K. Leo, *Adv. Mater.*, 2007, **19**, 3672.
85. S. Basak, Y. S. L. V. Narayana, M. Baumgarten, K. Mullen and R. Chandrasekar, *Macromolecules*, 2013, **46**, 362.
86. M. Gratzel, *Acc. Chem. Res.*, 2009, **42**, 1788.
87. D. Maggioni, F. Fenili, L. D'Alfonso, D. Donghi, M. Panigati, I. Zanoni, R. Marzi, A. Manfredi, P. Ferruti, G. D'Alfonso and E. Ranucci, *Inorg. Chem.*, 2012, **51**, 12776.
88. L. Cosgrave, M. Devocelle, R. J. Forster and T. E. Keyes, *Chem. Commun.*, 2010, **46**, 103.
89. R. Zhang, Z. Ye, Y. Yin, G. Wang, D. Jin, J. Yuan and J. A. Piper, *Bioconjugate Chem.*, 2012, **23**, 725.
90. A. Juris, V. Balzani, F. Barigelletti, S. Campagna, P. Belser and A. Von Zelewsky, *Coord. Chem. Rev.*, 1988, **84**, 85.
91. N. L. Rosi and C. A. Mirkin, *Chem. Rev.*, 2005, **105**, 1547.
92. K. L. Metera, K. D. Hanni, G. Zhou, M. K. Nayak, H. S. Bazzi, D. Juncker and H. F. Sleiman, *ACS Macro Lett.*, 2012, **1**, 954.
93. N. B. Sankaran, A. Z. Rys, R. Nassif, M. K. Nayak, K. Metera, B. Chen, H. S. Bazzi and H. F. Sleiman, *Macromolecules*, 2010, **43**, 5530.
94. W. Jiang, Y. Gao, Y. Sun, F. Ding, Y. Xu, Z. Bian, F. Li, J. Bian and C. Huang, *Inorg. Chem.*, 2010, **49**, 3252.
95. Y. You, K. S. Kim, T. K. Ahn, D. Kim and S. Y. Park, *J. Phys. Chem. C*, 2007, **111**, 4052.
96. Q. Zhao, L. Li, F. Li, M. Yu, Z. Liu, T. Yi and C. Huang, *Chem. Commun.*, 2008, 685.
97. Y. You, H. S. Huh, K. S. Kim, S. W. Lee, D. Kim and S. Y. Park, *Chem. Commun.*, 2008, 3998.
98. K. Huang, H. Wu, M. Shi, F. Li, T. Yi and C. Huang, *Chem. Commun.*, 2009, 1243.
99. H. Xiang, L. Zhou, Y. Feng, J. Cheng, D. Wu and X. Zhou, *Inorg. Chem.*, 2012, **51**, 5208.
100. A. Rose, Z. Zhu, C. F. Madigan, T. M. Swager and V. Bulovic, *Nature*, 2005, **434**, 876.
101. G. Wang, L. Yang, Y. Li, H. Song, W. Ruan, Z. Chang and X. Bu, *Dalton Trans.*, 2013, **42**, 12865.
102. B. Gole, S. Shanmugaraju, A. K. Bar and P. S. Mukherjee, *Chem. Commun.*, 2011, **47**, 10046.

CHAPTER 5

Recent Advances in Immobilized Ferrocene-Containing Polymers

MARKUS GALLEI*[a] AND JOHANNES ELBERT[a]

[a]Technische Universität Darmstadt, Darmstadt, Germany
*E-mail: m.gallei@mc.tu-darmstadt.de

5.1 Introduction

By functionalizing substrates with stimuli-responsive polymers, *i.e.*, polymers that feature the intrinsic capability to change their conformation, solubility or even break or form covalent bonds caused by an external trigger, so-called *smart surfaces* can be produced.[1–3] The resulting materials can be stimulated in terms of polarity or swelling behaviour by the change of solvent, temperature, ionic strength, light or mechanical stress, and it is possible to prepare materials that respond to a combination of different stimuli which is indeed favourable for some applications. Although the vast majority of reports dealing with external triggers for smart surfaces focus on such classical stimuli, other stimuli such as the influence of an electric or magnetic field are of growing interest.[3,4] Compared to reports dealing with classical external triggers, relatively few examples are known and well investigated for redox-responsive polymers. This chapter will focus on recent advances in immobilized metallopolymers bearing the ferrocene motif as part of the polymer main chain or side chain which are attached or confined on (non-) functional flat substrates, nanoparticles, in pores, in gels or at the surface of electrodes.

RSC Smart Materials No. 15
Functional Metallosupramolecular Materials
Edited by John G. Hardy and Felix H. Schacher

Published by the Royal Society of Chemistry, www.rsc.org

In the last decade, ferrocene-containing polymers have been the focus of scientists' interest due to their excellent combination of redox, mechanical, semi-conductive, photo-physical and optoelectronic properties.[5–8] While oxidizing the ferrocene moieties (electro)chemically, polymer chains undergo a remarkable transition from hydrophobic to hydrophilic accompanied with a swelling effect due to the generation of additional charges.[9–11] The oxidation/reduction cycles of such redox-active polymers have been utilized for various applications. This book chapter will not exhaustively discuss all the investigated applications, but it will highlight some feasible and novel applications that polymer scientists are currently interested in. For this purpose, the chapter is divided into the following sections: first, an introduction to the surface-immobilization of ferrocene-containing polymers is given with a view to the use of the redox-responsive nature of the materials. The second section addresses the crucial role of changes in terms of surface polarity, while the third section is attributed to redox stimulus-induced release of different payloads from the interior of metallopolymer multilayers, capsules and gels. Approaches highlighted in the fourth section are based on the ability of ferrocene to form host–guest complexes with cyclodextrins (CD). Here, ferrocene moieties in the reduced state fit perfectly into the CD cavities, while in the oxidized state decomplexation occurs. This redox-responsive host–guest complexation between ferrocene and CD has led to manifold interesting applications. The subsequent section takes the modification of electrodes into account. Although these systems are already widely used for glucose sensing and for enzyme-mediated biosensors, electrode modification with ferrocene-containing oligomers and polymers will only be briefly addressed. In the penultimate section, immobilized or confined ferrocene-containing polymers are discussed regarding their applications as ceramic precursors, which are finally followed by a brief description of some fascinating optical applications based on metallopolymer colloidal crystals with high periodicities.

5.2 Redox Switchable Polarity

In general, two basic synthetic approaches are widely used for functionalizing substrates with stimuli-responsive polymers: first, the *grafting from* methodology, by which polymer chains "grow" from an initiator-functionalized substrate and second, the *grafting to* method, which necessitates end functionalized polymers for substrate modification. The first method allows the preparation of dense polymer brushes while the latter leads to comparably low grafting densities. A high grafting density allows access to smart polymer brushes for many important applications: mechanical actuation, sensing, drug release, cell growth and separation techniques are only a few examples.[12] Controlled radical polymerization strategies—especially atom transfer radical polymerization (ATRP)—seem to be the most promising and adjustable route for decorating surfaces with dense polymer brushes. The control during this polymerization technique leads to—in terms of constitution and molar masses—precisely defined polymers.

Moreover, in some cases the reaction can be carried out in the presence of water featuring a remarkable tolerance towards many functional groups. This chapter focusses on ferrocene-containing polymers, immobilized by the methodologies mentioned, in order to change the polarity of the obtained metallopolymer surfaces.

It is well known that ferrocene is intrinsically capable of switching reversibly between two oxidation states, with both the reduced and the oxidized forms being chemically stable.[13,14] In the case of surface-attached ferrocene compounds, the change in oxidation state accompanied with its change of polarity was demonstrated by Whitesides *et al.* in 1994.[15] The authors studied the influence of oxidation on the wettability of a self-assembled monolayer of a ferrocene-containing thiol derivative attached to a gold substrate. In this particular case, a repeated potential-dependent cycling reaction in the presence of water leads to a progressively decreasing response of the ferrocene moieties, which was attributed to ferrocene degradation. On the other hand, ferrocene-containing polymers for the manipulation of surface polarity gained interest as ferrocene degradation does not significantly take place for some of the polymer/redox systems reported in the literature.

The influence of oxidation on the surface properties of immobilized ferrocene-containing polymers was thoroughly studied by Vancso *et al.* for polyferrocenylsilane (PFS)-based layers on gold electrodes.[16] By grafting thiol end-functionalized PFSs onto gold substrates, redox responsive monolayers were prepared and their adherence and friction behaviour was studied using atomic force microscopy (AFM). Upon oxidation, the adhesion and friction of a Si_3N_4 or a carboxylic acid-modified AFM probe tip increases. The origin of these observations was attributed to changes in the electrostatic interactions between the AFM tip and PFS film upon an applied potential. This effect was also influenced by electrolyte anions and ferrocenium association.

Recently, Elbert and co-workers took advantage of ferrocene-containing polymers bearing the ferrocene motif in the side-chain in order to investigate the influence of oxidation and polymeric structure on surface wettability.[17] There, the wettability was studied for polyvinylferrocene (PVFc) and poly(2-(methacryloyloxy)ethyl ferrocenecarboxylate) (PFcMA) grafted onto silica wafers. For this purpose, alkoxy-silane-functionalized PVFc[18]—accessible *via* sequential anionic polymerization protocols—was immobilized in a *grafting to* approach, while PFcMA brushes were accessible by surface-initiated ATRP (SI-ATRP) protocols. The water contact angle (CA) decreases by 25° due to oxidation in the case of low molar masses of PVFc (Figure 5.1).

For PFcMA brushes, a dependency of CA with molecular weight was reported whereby the CA for the highest molecular weight in the reduced form was increased up to 104°. Remarkably, oxidation of PFcMA brushes led to a dramatic drop in the CA of approximately 70° for the highest molecular weight evaluated in this study (Figure 5.1). Also, the effect different oxidation reagents had on CA change was studied, resulting in a more pronounced decrease in CA for $FeCl_3$ (70°) than $(BrC_6H_4)_3NSbCl_6$ (40°). This difference was attributed to counterion effects as reported for other polyelectrolyte brushes.[19]

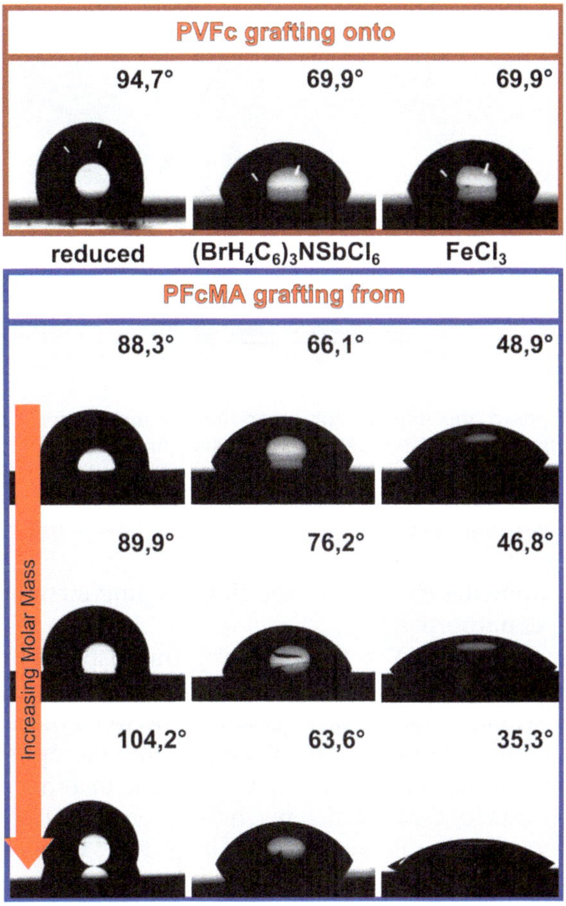

Figure 5.1 Water contact angles obtained for PVFc- and PFcMA-grafted silica wafers before and after oxidation with iron(III) chloride or $(BrC_6H_4)_3NSbCl_6$. For comparison, the bare purified silicon wafer revealed a contact angle of 64.4°.[17] Reproduced with permission from ref. 17. Copyright 2013 American Chemical Society.

The change in surface polarity and conformational change of surface-anchored ferrocene-containing polymer chains can also be used for the modulation of surface-immobilized catalysts.[18] For this purpose, alkoxy silane-functionalized PVFc and *N*-heterocyclic carbene (NHC) ligands were co-deposited onto silica nanoparticles prior to the conversion of the NHC into the active Grubbs 2nd generation catalyst. In this case, the metallopolymer acts as a modulator for the accessibility of norbornene for intended ring opening metathesis polymerization (Figure 5.2). Oxidizing PVFc leads to a tremendous swelling of the PVFc layer. The authors claim that the modulation is both attributed to swelling and increase in polarity while oxidizing the ferrocene moieties. Notably, *in situ* reduction of the PVFc layer using

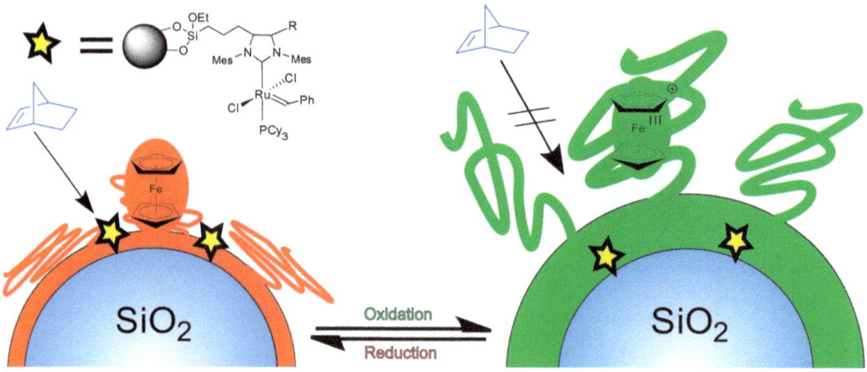

Figure 5.2 Proposed mechanism for reversible catalyst activity modulation of immobilized ROMP catalysts by using immobilized redox-responsive ferrocene-containing polymers (R = $-(CH_2)_3-Si(OEt)_3$).[18] Reproduced with permission from ref. 18. Copyright 2013 American Chemical Society.

L-ascorbic acid immediately initiates the ring-opening metathesis polymerization (ROMP) of norbornene.

Mao *et al.* investigated PVFc for switching the dispersibility of carbon nanotubes (CNTs) in non-aqueous media.[20] PVFc was immobilized at the CNT surface by strong π–π stacking interactions between the π-electron-rich cyclopentadiene rings of PVFc and the CNTs, without disruption of the long range π conjugation of the nanotube. In contrast to other non-covalent attached dispersants for CNTs, which reach CNT concentrations of less than 0.1 mg mL^{-1}, PVFcs with a molar mass of 50 kg mol^{-1} were capable of dispersing CNTs in concentrations up to 2 mg mL^{-1}. Furthermore, this protocol has also been applied to disperse CNTs in a polystyrene matrix.[20] When the ferrocene moieties are oxidized (electro)chemically, CNTs precipitate immediately and can be redispersed by reduction (Figure 5.3). This phenomenon can be exploited for the preparation of CNT functionalized surfaces by convenient electrochemical deposition.

The same group reported the application of PVFc/CNT hybrid materials as supercapacitors.[21] By changing the ratio of PVFc to CNTs, the nanostructure and the electrochemical properties of the material were systematically manipulated, ranging from CNTs in a PVFc matrix to CNTs being only partially coated with PVFc. A morphology with a PVFc-covered CNT network revealed an unusually high capacitance of 1452 F g^{-1} and an energy density of 79.5 W h kg^{-1}, outperforming other pseudo-capacitive materials. Moreover, this novel PVFc/CNT hybrid showed a high cycling stability up to 3000 charge–discharge cycles. The remarkable electrochemical properties were attributed to the interconnected, porous network of PVFc-coated nanotubes facilitating electron transport and ion diffusion to the ferrocene redox centres. This led to a high ferrocene utilization efficiency of 63% while maintaining the mechanical integrity.

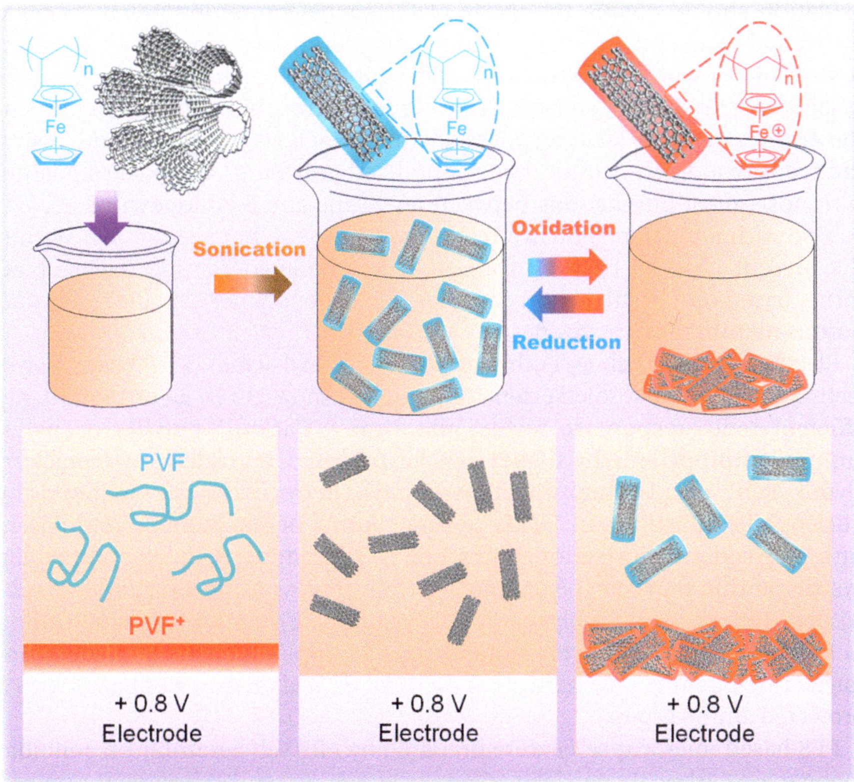

Figure 5.3 Redox-controlled dispersion and precipitation of carbon nanotubes (CNTs) using polyvinylferrocene (PVFc).[20] Top: conceptual illustration of the preparation and redox-controlled dispersion and precipitation of PVFc/CNT complexes. Bottom: illustration of the redox behaviour of PVFc (left), CNTs (middle) and the PVFc/CNT complex (right). Reproduced with permission from ref. 20. Copyright 2013 American Chemical Society.

Very recently, ferrocene moieties in random copolymers of vinylferrocene and styrene or methyl methacrylate have been used to graft obtained polymers onto graphene oxide (GO) by substituting one cyclopentadiene ligand by one of the aromatic rings of the GO surface.[22] The degree of grafting was reported to be 60% as determined by thermogravimetric analysis (TGA). The grafted polymers enhanced the dispersibility of GO in common organic solvents.

5.3 Immobilized Ferrocene Polymers for Controlled Release and Permeability

As shown in the previous section, the polarity of surfaces functionalized with ferrocene-containing polymers is tremendously influenced by the oxidation state of the ferrocene moieties. In general, the huge change in polarity upon oxidation should enable the usage of ferrocene-containing polymers for a controlled release of payloads by two considerations: first, either by disassembly of the whole structure embedding the payload (*e.g.* in layer-by-layer assemblies, LBL) or second, by changing the permeability in certain parts of the entire structure, such as pores, channels, gels as a result of addressing the redox-responsiveness of the immobilized polymers. Approaches dealing with host–guest interactions between ferrocene and β-cyclodextrins accompanied with a redox-responsive release are addressed in section 5.4. Within the present chapter, recent advances of ferrocene-containing release platforms based on LBL films and capsules, block copolymer capsules, gels and porous membranes are given.

Electrostatic LBL self-assembly of poly(ferrocenylsilane) (PFS)-based polyelectrolytes has been studied extensively and was shown to be an extraordinarily efficient redox-responsive release platform.[23] Cationic poly(ferrocenyl(3-ammoniumpropyl)-methylsilane) can be prepared by transition metal catalysed ROP of (3-chloropropyl)methylsilyl[1]ferrocenophane followed by nucleophilic substitution.[24] This polymer forms organometallic multilayers with poly(sodium vinylsulfonate) by LBL self-assembly. Fully organometallic multilayer thin films can be prepared by this process using a carboxylate substituted PFS instead of poly(sodium vinylsulfonate).[25] Another route leading to PFS-based polyelectrolytes suitable for LBL applications is the living photolytic ring-opening polymerization (ROP) of *ansa*-silaferrocenophanes with a protected amino group.[26]

PFS-based microcapsules were prepared by LBL self-assembly on colloidal manganese carbonate particles (diameter ~ 8.5 μm) and subsequent removal of the $MnCO_3$ template with ethylenediaminetetraacetic acid.[27] The thickness of the capsule wall depends on the number of PFS bilayers. A single bilayer shows an average thickness between 5 and 8 nm. The permeability of the capsule wall was probed with tetramethylrhodamine isothiocyanate (TRITC)-labelled dextran (M_w ~ 4400 g mol^{-1}). The microcapsules are impermeable for the probe molecule when the bilayer number is higher than four, suggesting an effective mesh size of the capsule wall of less than 2 nm. The permeability of these microcapsules can be switched by oxidation of the ferrocene moieties.[11] Upon chemical oxidation, these capsules expand, accompanied by an increase in permeability of the capsule wall (Figure 5.4).

The expansion follows a first order kinetic with a rate constant of up to $k = 0.010\ s^{-1}$. In addition to expansion of the capsules, disintegration—as indicated by a decrease of the wall thickness—takes place to maintain charge neutrality of the capsule wall.

The disassembly of PFS-based multilayers was further studied for LBL thin films on flat substrates.[28] Electrochemical oxidation of the ferrocene moieties creates positive charges in the PFS main chain (Figure 5.5), resulting in a charge imbalance and thus disassembly by electrostatic repulsion within the multilayer film. The disassembly can be used for the electrochemically-controlled release of molecular payloads (dye-labeled dextran) incorporated in the multilayer.[28]

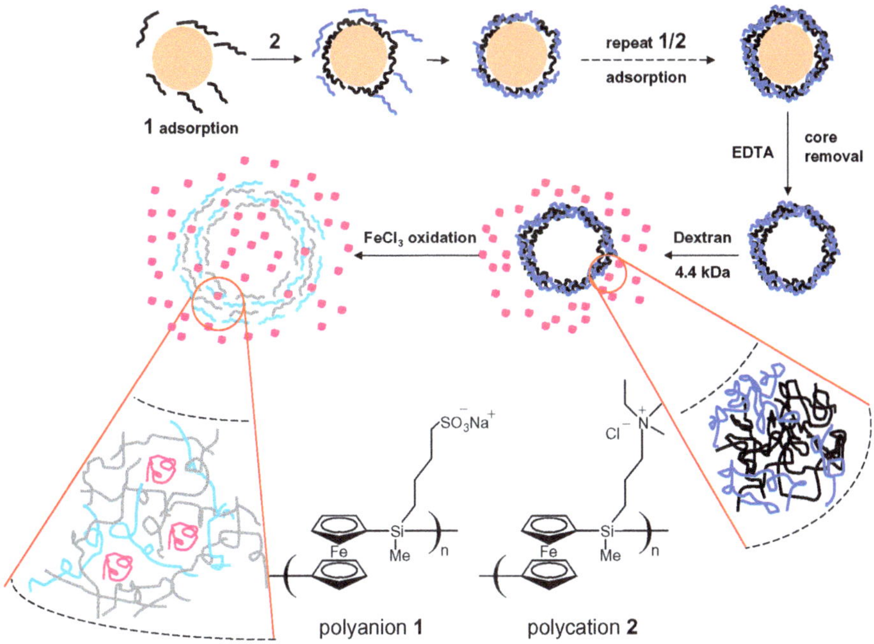

Figure 5.4 Schematic diagram of the formation and permeability control of organometallic multilayer capsules. PFS-based polyelectrolytes (black and blue) were used in electrostatic self-assembly followed by template removal. By oxidation with $FeCl_3$, dextran molecules (red) could permeate the capsule wall.[11] Reproduced with permission from ref. 11. Copyright 2006 Nature Publishing Group.

The release kinetics are influenced by the applied potential and the number of PFS layers on top of the payload. A potential of 0.4 V corresponding to complete oxidation of PFS leads to a faster release than 0.1 V, which only oxidizes every second ferrocene moiety in the polymer. Another way to slow down the release is by increasing the number of bilayers on top of the payload. Additionally, the release can be influenced by the incorporation of blocking layers, which consist of redox-inert polyelectrolytes such as poly(acrylic acid) and poly(allylamine hydrochloride), into the multilayer structure.[29] If the blocking layers are located on top of the redox-active bilayers, there is no considerable influence on the disassembly process, but if placed between electrode and PFS bilayers, the non-conducting blocking layers influence the disassembly by preventing oxidation of the PFS. Only two blocking layers result in a 97% decrease of addressable ferrocene moieties.

More complex release profiles can also be created by using PFS-based LBL thin films.[30] A payload incorporated near the top of a multilayer thin film is released faster than another located near the bottom of the structure. Spatial control of the release can be achieved by using microelectrode arrays. This approach is not only limited to thin films but can also be used for the preparation of redox-responsive nanotubes.[31] Porous membranes such as

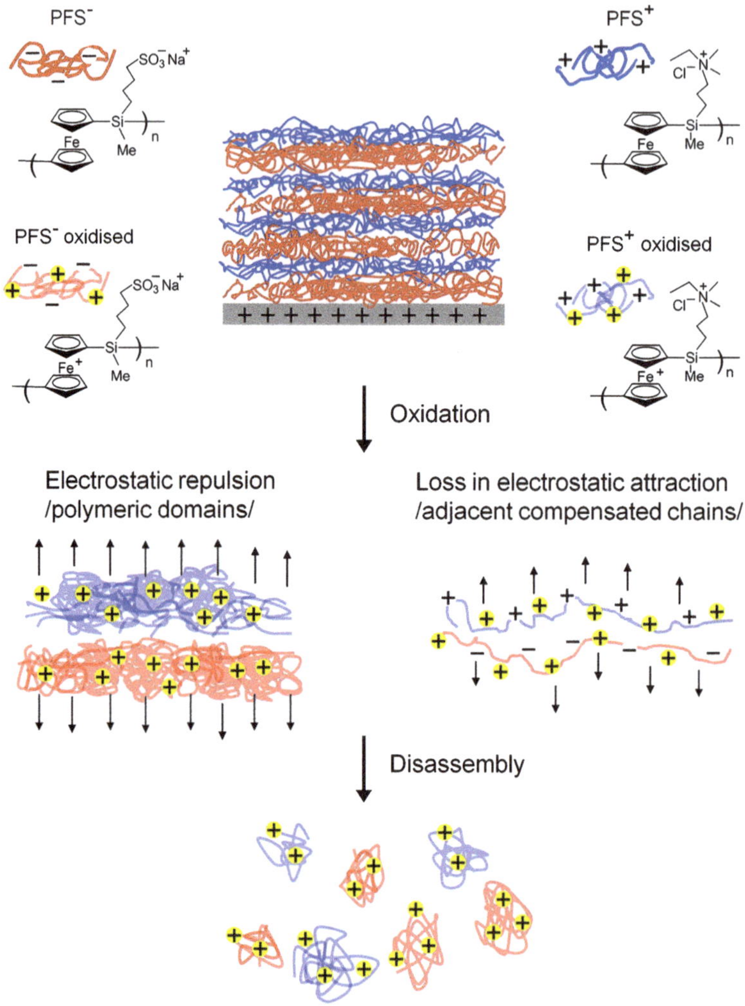

Figure 5.5 PFS-based multilayer disassembly mechanism by oxidizing ferrocene moieties.[28] Reproduced with permission from ref. 28.

anodized aluminium oxide or track-etched polycarbonate were used as templates for the LBL self-assembly of PFS-based polyelectrolytes. This route results in a coating of the inner walls of the membranes and subsequent template removal yields redox-responsive nanotubes. These nanotubes can be loaded with large quantities of guest molecules due to their high surface area if compared to flat thin films. The encapsulated payloads can readily be released electrochemically.

In addition to LBL thin films, ferrocene-containing gels have also been used for the catch and release of organic molecules.[32] Therefore, polyvinylferrocene (PVFc)-based redox-active gels were prepared by free radical copolymerization of up to 22% vinylferrocene with hydroxybutylmethacrylate and ethyleneglycol

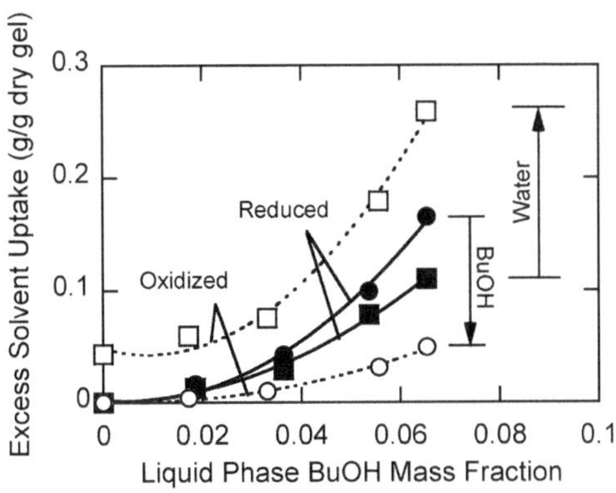

Figure 5.6 Catch and release of butanol (BuOH) by PVFc-based gels: excess uptake of water and butanol relative to a reduced, water-saturated, butanol-free gel. Solid and broken lines (and filled and open symbols) indicate the reduced and oxidized states, respectively, while circles and squares denote butanol and water. The changes in solute uptake upon oxidation under butanol-saturated conditions are indicated by the arrows.[32] Reproduced with permission from ref. 32. Copyright 2011 American Chemical Society.

dimethacrylate. These gels selectively extract organic substances from aqueous solutions. The equilibrium distribution coefficient for butanol between the gel and water is up to nine times higher than the distribution coefficient for water (Figure 5.6, solid lines and symbols). That means that butanol has a higher affinity for the polymer gel than water, leading to an enrichment of butanol in the polymer gel. The reduced gel can be used to selectively extract butanol from aqueous solutions. Upon oxidation, the selectivity for butanol is altered tremendously. Due to the presence of charged species within the gel, the need of hydration increases, the hydrophobicity of the gel decreases and butanol is released (Figure 5.6). This offers the possibility to construct novel separation systems for the removal of organic substances from water.

Another structure that can be used for the controlled release of payloads is patchy nanocapsules based on polyvinylferrocene block copolymers.[33] Polyvinylferrocene-*b*-poly(methyl methacrylate) (PVFc-*b*-PMMA) was synthesized by sequential anionic polymerization involving dimethylsilacyclobutane-mediated reactivation of inactive PVFc-macro anions.[34] Nanocapsules with a PVFc-*b*-PMMA shell were prepared from droplet templates of aqueous emulsions. As a result of microphase separation between both polymer blocks, PVFc creates equally distributed nanopatches within the PMMA capsule wall. The patches with an average size of 25 nm can be addressed selectively by various oxidation reagents. Upon oxidation, the patches swell in water due to the polyelectrolyte effect in the polyvinylferrocenium domains and an encapsulated payload (pyrene) can be released (Figure 5.7). While no release was detected for reduced polyvinylferrocene, oxidation with H_2O_2 leads to fast release of encapsulated pyrene. Using $KMnO_4$ the release was less pronounced, which

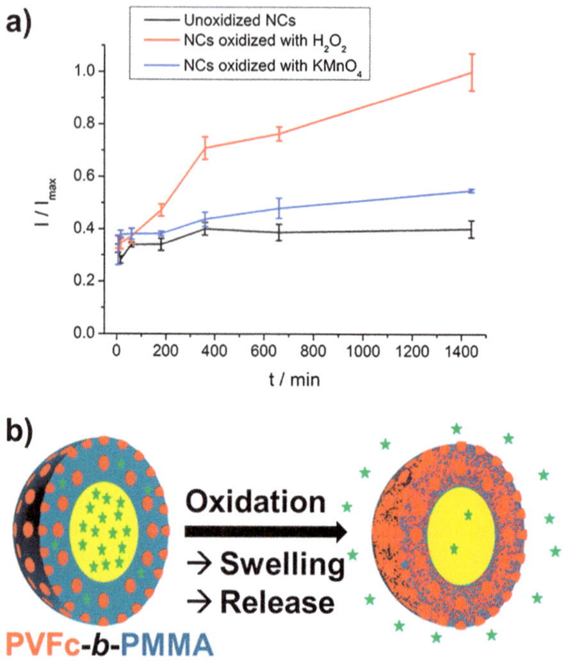

Figure 5.7 Redox-responsive release from patchy nanocapsules: release kinetics of pyrene from PVFc-*b*-PMMA patchy nanocapsules: (a) the fluorescence intensity of the oxidized nanocapsules increases with time, indicating a release of pyrene, whereas the intensity of the non-oxidized nanocapsules remains almost constant. (b) Schematics showing the oxidation responsive release of the hydrophobic payload from the PVFc-*b*-PMMA patchy nanocapsules.[33] Reproduced with permission from ref. 33. Copyright 2012 American Chemical Society.

can be explained by less effective diffusion of $KMnO_4$ into the PVFc domains, thus creating less ferrocenium ions and resulting in lower permeability.

Ferrocene-containing polymers are not only able to release organic molecules, but are also capable of redox-responsive switching of ion permeability.[35] Mesoporous silica thin films were functionalized with PVFc and poly(2-(methacryloyloxy)ethylferrocenecarboxylate) (PFcMA), by using either a *grafting to*, or a *grafting from* approach. For the first method, immobilized polymers were achieved using PVFc with a triethoxysilane end group,[18] while established SI-ATRP protocols were used to graft FcMA from the silica membranes.[35] The latter approach leads to modification across the entire film thickness, while the *grafting to* approach leads to a polymer layer predominantly located at the mesoporous film surface. This PVFc layer resulted in a diffusion barrier, which was attributed to a hydrophobic outer surface. Moreover, that thin hydrophobic layer was not capable of completely shielding electrostatic interactions between cations in solution and the negatively charged silica membrane beneath, resulting in transport of the cations across the hydrophobic diffusion layer. Compared to this, dense PFcMA brushes led to a tremendous decrease in ion flux across the membrane due to increased hydrophobicity and pore filling. Upon oxidation, positively charged ferrocenium ions are created within the pores resulting in complete exclusion of positively charged ions. On the other hand, the amount of anions diffusing through the pores increases if no charged silanol groups are present. At basic pH the silanol groups are deprotonated creating negative charges at the pore walls. The relatively low amount of oxidized ferrocene moieties cannot completely compensate the silica wall charges, resulting in a less effective transport of anions under basic conditions.

5.4 Host–Guest Complexation of Ferrocene Polymers with Cyclodextrin

The inclusion of the hydrophobic part of a monomer into the hydrophobic interior of cyclodextrin (CD) cavities has led to many interesting novel materials.[36,37] This host–guest complexation yields water soluble complexes so that even unpolar ferrocene monomers (*e.g.* vinylferrocene) can be polymerized by free radical polymerization protocols in water.[38] Some fascinating macromolecular properties based on ferrocene/CD complexation and redox-responsive ferrocenium/CD decomplexation—which occurs due to changes in polarity of the ferrocene moieties after oxidation—will be addressed in this section. Reversible assembly/disassembly can be carried out either by using redox reagents or an electrical stimulus as an external trigger.

Dubacheva *et al.* prepared ferrocene-functionalized chitosan-based polymers for electrochemically controlled adsorption of self-assembled β-CD derivative monolayers on gold substrates (Figure 5.8).[39] The electrochemical trigger applied to these functionalized substrates led to an efficient detachment of the ferrocene-containing polymer chains proving the stimuli-responsiveness and suitability for an electrically driven release platform.

Polystyrene featuring a β-CD as a functional end group and ferrocene-terminated poly(ethyleneoxide) (PEO) form a supramolecular diblock copolymer which self-assembles into vesicles in water.[40] These vesicles can be loaded with rhodamine B and by applying an external potential, the vesicles disassemble accompanied with a release of the embedded payload.

Possibly the most promising example for β-CD complexation with ferrocene-containing polymers was reported by Harada and co-workers for self-healing applications (Figure 5.9).[41–43] The authors fabricated several self-healing hydrogels based on β-CD- and ferrocene-modified poly(acrylic acid) (PAA).[41] Upon mixing both polymers in water, gelation occurs as a consequence of the complexation of ferrocene and β-CD. The obtained material is reversibly stimulus-responsive with regard to the switch between the

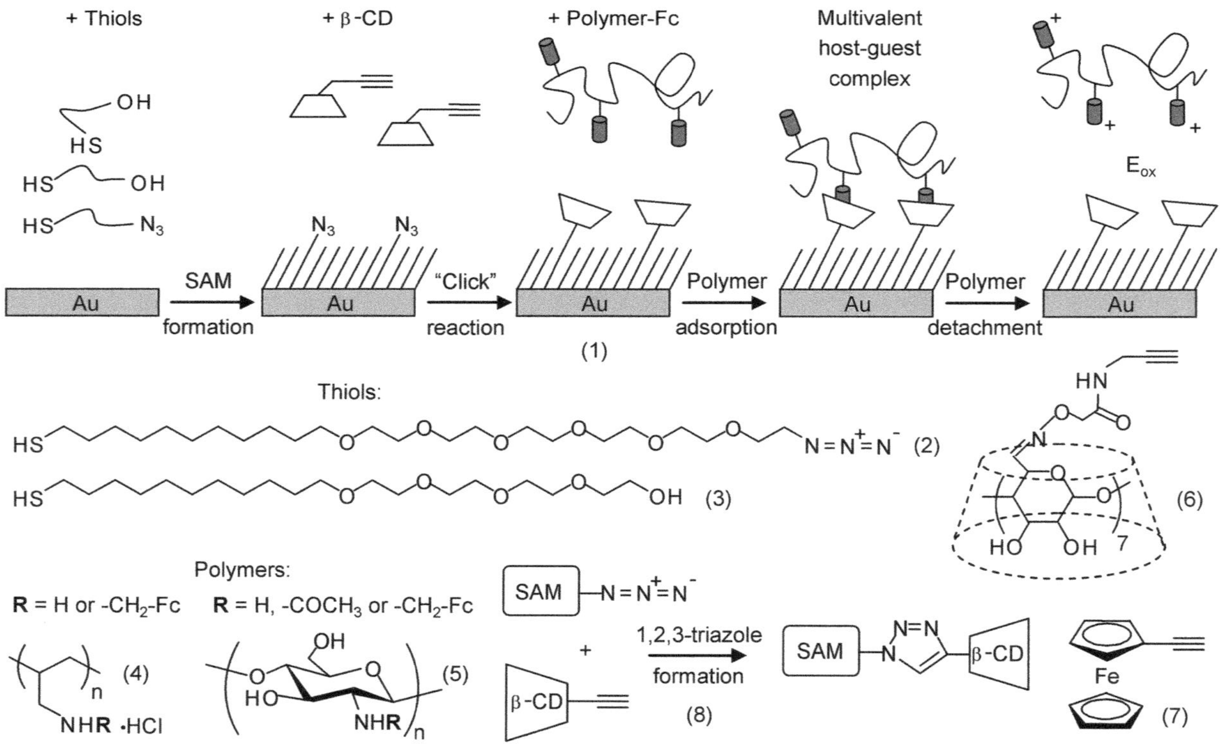

Figure 5.8 Schematic representation of self-assembled monolayer formation and β-CD attachment *via* click reaction followed by the adsorption/desorption of ferrocene-modified polymers.[39] Reproduced with permission from ref. 39. Copyright 2010 American Chemical Society.

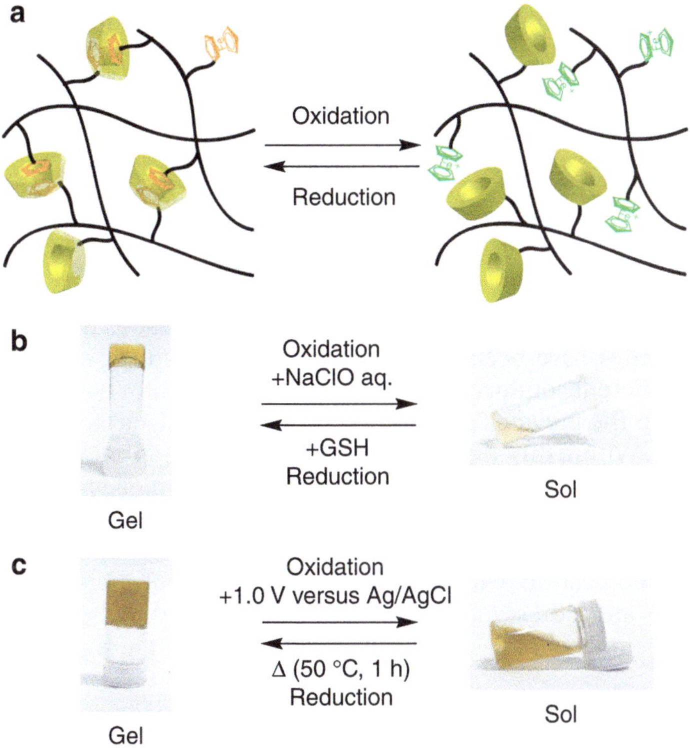

Figure 5.9 (a) Schematic illustration of sol–gel transitions in β-CD/ferrocene crosslinked hydrogels. (b) Sol–gel transition experiment using chemical reagents. Adding NaClO to the PAA-6 β-CD/PAA-Fc hydrogel induced a phase transition into the sol state, and continuous addition of glutathione to the sol recovered the elasticity to yield the hydrogel again. (c) Sol–gel transition experiment using electrochemical reactions. Electrochemical oxidation (+1.0 V *versus* Ag/AgCl) transformed the hydrogel into the sol, whereas reduction recovered the viscosity, reverting it back to the hydrogel.[41] Reproduced with permission from ref. 41. Copyright 2011 Nature Publishing Group.

sol state and gel state by adding oxidation or reducing agents such as glutathione. Remarkably, after cutting the hydrogel with a razor blade into two pieces, it can be rejoined and the crack sufficiently self-healed to re-form the gel with its original properties.

Recently, a similar concept was used for fabricating a redox-responsive shape memory polymer based on the inclusion of ferrocene into β-CD cavities.[44] In this study, the authors deformed ferrocene-modified branched poly(ethylene imine) (Fc-PEI) crosslinked with β-CD-modified chitosan into different temporary shapes and received the original material's shape after adding different concentrations of an oxidation reagent. Once more, these examples show the potential for ferrocene-containing polymers to carry out a redox-responsive release or even act as a structure-guiding platform.

Moreover, the host–guest chemistry of CD and ferrocene can be used to modify electrodes and various conductive surfaces. Applications focussing on the electrochemistry of electrode-attached ferrocene-containing polymers will be part of the next section.

5.5 Electrodes Modified with Ferrocene-Containing Polymers

The modification of electrodes with ferrocene-containing polymers has led to manifold sensing applications in the last decade. In general, various approaches have been carried out in order to attach metallopolymer chains to different conductive substrates. Synthetic routes have already been addressed in the previous sections for non-functional surfaces (*cf.* Sections 5.2 and 5.4). Within this section, these methods will be extended to electrode modifications starting with layer-by-layer (LBL) techniques of polyelectrolytes, followed by *grafting to* and *grafting from* methodologies and electro-polymerization of ferrocene monomers. Here, only more recent advances in electrode-modification with ferrocene-containing polymers for different applications are addressed, although it is noteworthy that the first ferrocene polymer electrode coatings appeared in literature by Daum and co-workers in 1979.[45,46]

5.5.1 Layer-by-Layer (LBL) Technique Examples for Electrode Modifications

The applications of PFS-based polyelectrolyte multilayer thin films are not only limited to non-functional surfaces *e.g.* for controlled release by redox reagents, but they can also be used for sensing applications on modified electrodes as reported by Feng *et al.*[47] For this purpose, the multilayer films have to be crosslinked *e.g.*, with poly(ethylene imine), for stability reasons as depicted in Figure 5.10. In this way, modified electrodes are capable of hydrogen peroxide sensing applications.

The LBL technique was successfully applied for covering an enzyme electrode with poly(2-methacryloyloxyethyl phosphorylcholine-*co*-(*p*-vinylphenylboronic acid)-*co*-vinylferrocene) (PMVF) as a polymeric electron transfer mediator.[48] Glucose oxidase (GOx) is one of the most examined enzymes for electrodes modified with ferrocene-containing polymers. Immobilization of the metallopolymer at the surface of gold electrodes in a multilayer system with GOx-modified poly(vinyl alcohol) (PVA) results in a feasible platform for bioelectronic devices with high diffusion rates for glucose molecules.[48]

Recently, DeLuca *et al.* reported a LBL approach of ferrocene-modified poly(ethylene imine), poly(acrylic acid), and poly(glutamic acid) or GOx as a negatively charged counterpart to produce redox-responsive films on gold electrodes.[49] The authors expect applications such as nano-scaled biosensors for glucose monitoring.

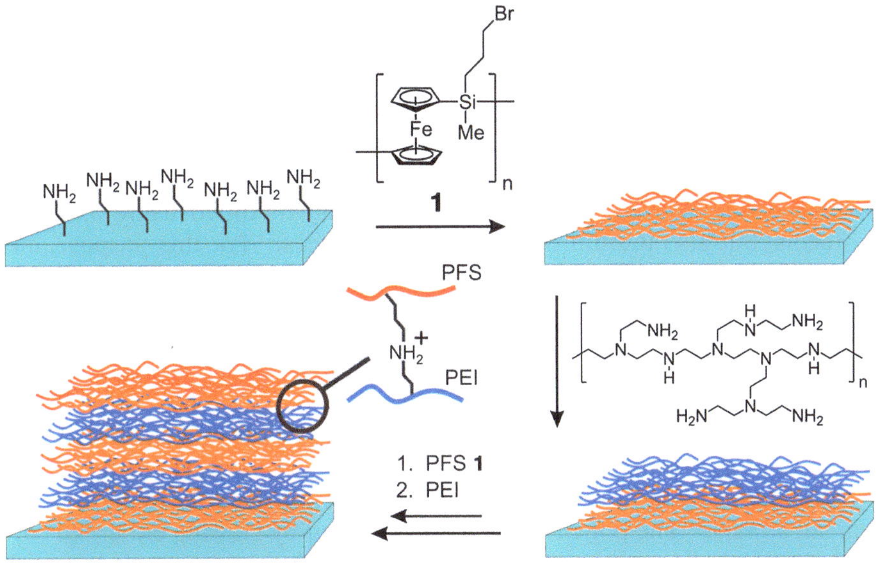

Figure 5.10 Schematic representation of a crosslinked PFS poly(ethylene imine) multilayer system on an amine-functionalized substrate, which can be used in hydrogen peroxide sensing.[47] Reproduced with permission from ref. 47. Copyright 2013 American Chemical Society.

5.5.2 Grafting To and Grafting From Approaches for Electrode Modifications

In particular, PFS-based metallopolymers have been used for electrode modifications by using the *grafting to* technique. As PFS is accessible by living anionic ring-opening polymerization of *ansa*-silaferrocenophanes, end functionalization of the well-defined polymer chains can be carried out with various functionalities for surface attachment. For example, functionalization of PFS with thiol groups allows an efficient *grafting to* strategy for polymer immobilization onto gold substrates as already described in Section 5.2.[16] This strategy is not only useful for switching the surface polarity by electrochemical means, but also for the production of thin film devices. Vancso and co-workers furthermore succeeded in building a redox-driven sensing platform for ascorbic acid by functionalizing amine-terminated monolayers on gold substrates with poly(ferrocenyl(3-iodopropyl)methylsilane). The PFS-based sensor provided high sensitivity and stability.[50] The same group thoroughly studied the mechanical properties of PFS polymer chains attached to an atomic force microscopy (AFM) tip on conductive substrates prior to and after oxidation of the metallopolymer. These interesting findings were also referred to as a "redox-driven macromolecular motor".[51,52]

As mentioned earlier, surface-initiated (SI) polymerization strategies, especially SI-ATRP, are feasible methods for the formation of dense immobilized polymer layers, also known as polymer brushes. An electrochemical immunosensor was obtained *via* SI-ATRP of ferrocenyl methyl methacrylate (FMMA) and glycidyl methacrylate (GMA) carried out from gold substrates.[53] After polymerization, tumor necrosis factor-alpha antigen (TNF-α) was introduced to the PGMA-backbone by carbodiimide coupling. The sensitivity could be adjusted by varying the monomer contents in the polymer structure.

It is not only gold electrodes that have been modified using SI-ATRP protocols, but also indium tin oxide (ITO) surfaces.[54] For this purpose, an ATRP-initiator was immobilized on the surface of ITO by silanization chemistry, which is suitable for the ATRP of FMMA and GMA. Glucose oxidase (GOx) was subsequently anchored to the epoxide moieties of GMA for efficient redox-mediated glucose detection.

The electron transport in redox polymer electrodes is crucial for all redox-mediated applications. Recently, Hatton and co-workers studied the diffusion of electrons through films of random copolymers consisting of hydroxybutyl methacrylate and vinylferrocene (poly(HBMA-*co*-VF)) on carbon substrates.[55] The polymers were immobilized by grafting either "to" or "from" the substrate surfaces. The authors found that transport of electrons predominantly occurs *via* two mechanisms of electron diffusion, bounded diffusion and electron hopping. An important result of these findings was that bounded diffusion mechanisms play a major role in non-crosslinked polymers, while in crosslinked materials, electron hopping was dominant due to hindered polymer chain mobility.[55]

Bunte and Rühe reported the synthesis and immobilization of redox terpolymers with pendant ferrocene moieties by a photochemical crosslinking approach for the modification of electrodes bearing enzymes in the redox-responsive hydrogels. Importantly, the polymeric system revealed good stability under physiologically relevant conditions.[56]

Sung *et al.* developed a protocol for ITO modification with a multifunctional random copolymer.[57] Four different monomers were incorporated into the polymer structure featuring different functionalities, namely 3-(trimethoxysilyl)propyl methacrylate for polymer immobilization on ITO, poly(ethylene glycol) methyl ether methacrylate to introduce antifouling properties, glycidyl methacrylate for biomolecule immobilization through epoxide ring-opening reactions, and FMMA for redox-responsiveness. After electrode dip coating, a sensitive immune sensor using ferrocene-containing copolymers as an effective electron transfer mediator was achieved.[57]

Redox-responsive polymeric nanoparticles and nanocomposites were investigated regarding their usability as glucose biosensors.[58] Therefore, branched poly(ethylene imine) was functionalized with ferrocenecarboxyaldehyde, followed by reduction. After functionalization with GOx, carbon electrodes were prepared as depicted in Figure 5.11. It was shown that poly(3,4-ethylenedioxythiophene):poly(styrenesulfonate) (PEDOT:PSS) can advantageously be used for receiving high glucose oxidation currents. Glucose-sensitivity could be increased by a factor of 2.5 compared to common materials.

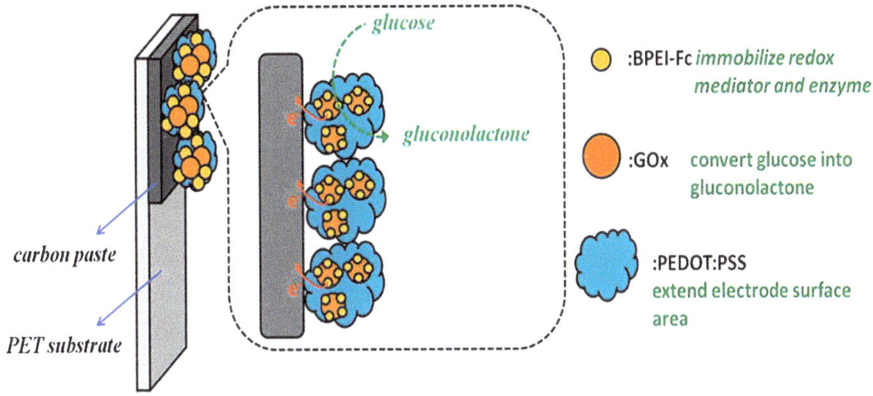

Figure 5.11 Schematic diagram of the immobilization of a redox mediator and enzyme at carbon electrodes.[58] Reproduced with permission from ref. 58. Copyright 2013 American Chemical Society.

5.5.3 Electropolymerization Strategies for Electrode Modification

Bis-diiodo-alkyl ferrocene derivatives have been cathodically reduced at the surface of glassy carbon electrodes forming a redox-responsive polymer film. Three possible reaction mechanisms are given in Figure 5.12.[59] The electrode coverage process was shown to be dominated by the length of the alkyl chains, yielding oligomeric redox-responsive materials.

Within the study reported by Dursun *et al.*, pyrrole was electrochemically polymerized in the presence of poly(glycidyl methacrylate-*co*-vinylferrocene). Again, enzyme immobilization was carried out by postfunctionalization using the pendant epoxy groups of GMA.[60] Thus, obtained electrodes featured facilitated electron transfer by means of the conductive polymer mediator and could be used for the amperometric determination of phenolic compounds.

Vinylferrocene was electrochemically polymerized at gold electrodes with different ionic liquids and the resulting materials were investigated regarding their gas sensing properties in different polymer oxidation states. Both the sensing capability and possibilities for distinguishing between different classes of analytes were studied. The authors found that oxidation with 0.1 M $NaClO_4$ led to the highest sensitivity for most of the analytes tested—benzene, natural gas, methane, formaldehyde and butylmethylimidazolium camphorsulfonate.[61]

The combination of a thermo- and redox-responsive copolymer attached at electrodes *via* a cysteamine linkage for regulating enzyme activity was reported by Nagel *et al.*[62] The electrode surface accessibility for the enzyme was modulated either by thermo-responsiveness of poly(*N*-isopropylacrylamide) or redox-responsiveness of PVFc. With these modified electrodes it was possible to detect glucose dehydrogenase in the nanomolar range.

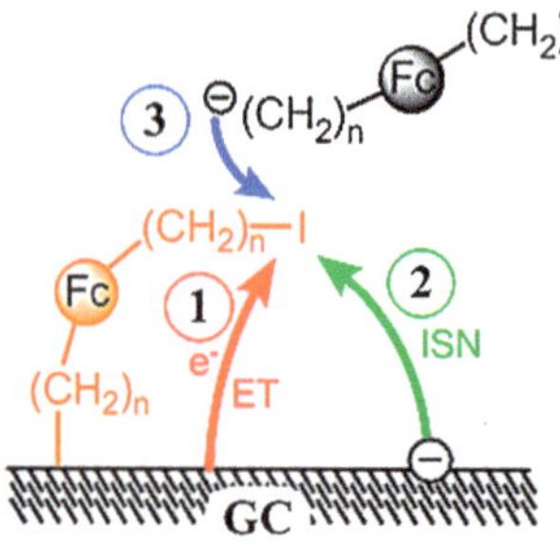

Figure 5.12 Schematic diagram of the possible immobilization strategies of bis-diiodo-alkyl ferrocene derivatives at the surface of glassy carbon (GC) electrodes. Path 1 represents classical heterogeneous electron transfer, path 2 represents nucleophilic substitution by charges induced at the carbon electrodes and path 3 represents homogeneous nucleophilic substitution *via* Wurtz-type reactions.[59] Reproduced with permission from ref. 59. Copyright 2012 Springer Science + Business Media.

5.6 Immobilized or Confined Ferrocene Polymers as Ceramic Precursors

The fabrication of nano-structured materials, *e.g.* hollow spheres, nanowires, nanorods, nanotubes, or fibres with adjustable dimensions, has gained considerable attention due to their tremendous potential for various applications. In general, ferrocene was found to be a feasible precursor for the preparation of magnetic iron oxide nanoparticles.[63–65] However, ceramic yields were even higher when ferrocene-containing polymers were used as iron precursors. Moreover, the self-assembly of organic/ferrocene-containing block copolymers can be used in order to generate ceramic structures at the nano-scale, converting the inorganic block into a ceramic by pyrolysis of the block copolymer. Within this section, the formation of well-defined ferrocene-polymer-templated ceramic materials featuring nano- or microstructures are briefly addressed.

In particular, polymers accessible by ring-opening polymerization (ROP) of ring-strained *ansa*-silaferrocenophanes—leading to the PFS-based polymers found by Manners and co-workers as described earlier—have attracted enormous attention for the preparation of well-defined nanostructured ceramics.[66–69] For example, poly(ferrocenylethylmethylsilane) (PFEMS) doped with palladium nanoparticles or by mixing with $Pd(acac)_2$ yielded ferromagnetic FePd alloys after pyrolysis of the PFS hybrid material.[70]

Aluminium oxide templates have been used to confine PFS as a ferrocene-containing polymeric precursor. Rather large areas with highly ordered magnetic nano arrays could be obtained by aluminium oxide template removal prior to the pyrolysis of the PFS nanorods.[71]

Ceramic nanolines have been prepared after reactive ion etching of PFS-based diblock copolymer micelles.[66] In this case, the ferrocene-containing

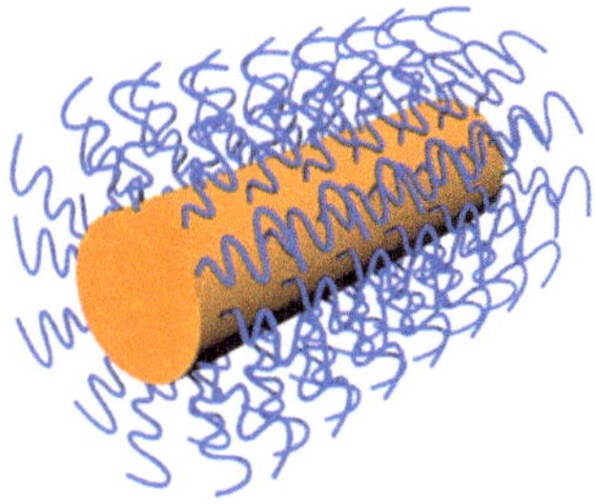

Figure 5.13 Schematic diagram illustrating the structure of the cylindrical PFS-based block copolymer micelle (orange cylinder: PFS; blue corona chains: polyisoprene).[66] Adapted with permission from ref. 66. Copyright 2003 Wiley VCH.

block forms the crystallizable inner cylinder, while polyisoprene is used as a corona-forming block (Figure 5.13). It has been shown that crosslinking of the polyisoprene corona chains led to remarkably high ceramic yields during the pyrolysis step.[72] Furthermore, the obtained cylinders could be used in microfluidic nanopatterning.

Nanolines with a remarkably high aspect ratio could be fabricated, which might be useful as etching resists for 2D quantum wires. Also polystyrene (PS) was used as second block segment for PFS. In the bulk, PFS forms cylinders in a PS matrix independent of the substrates used.[68] These self-assembled block copolymer microphases were capable of producing magnetic ceramic nanodomains after UV-mediated crosslinking of the PS matrix, followed by pyrolysis. Inorganic SiC/C-containing iron crystallites could be obtained by this approach. The scope of this strategy has been widened by the same group to fabricate ferromagnetic ceramic films featuring nanoline domains after reactive ion etching. The authors took advantage of a cobalt-cluster-doped PFS-based metallopolymer as a suitable precursor material.[67]

Recently, the self-assembly and the template synthesis of triblock copolymers featuring a poly(2-(methacryloyloxy)ethyl ferrocenecarboxylate) (PFcMA) block was investigated by Tang and co-workers to yield ordered iron oxides.[73] Compared to these studies, which focus on the self-assembly of ferrocene-containing block copolymers with the intrinsic capability of one block segment to form iron-based ceramics, other different strategies have been used to produce advanced magnetic materials with high ceramic yields.

As another approach, hyperbranched poly(ferrocenylene)s were studied with regard to pyrolytic transformation into magnetic ceramics.[74] Here, the ferrocene moieties are fused into a 3D macromolecular structure, which is feasible for obtaining high ceramic yields for magnetic iron clusters. Kong *et al.* reported the synthesis and pyrolysis of hyperbranched polyferrocenyl(boro)carbosilanes.[75] The additional ferrocene moieties were introduced into the hyperbranched structure of polycarbosilane bearing

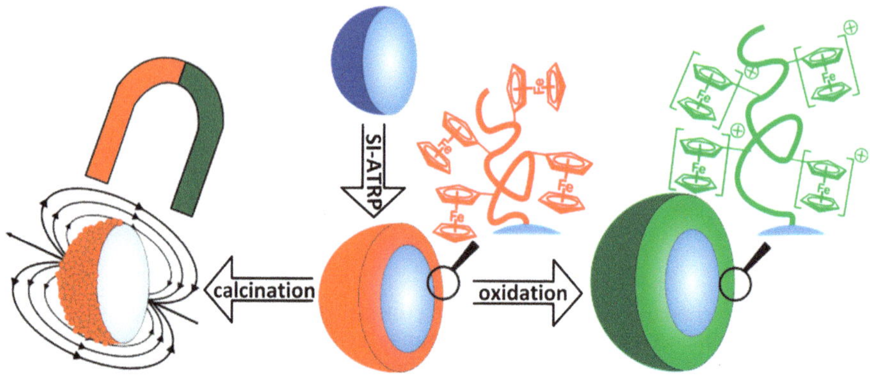

Figure 5.14 Scheme for the functionalization of polystyrene particles featuring an ATRP initiator in the particle shell for the formation of ferrocene-containing polymer (PFcMA) brushes.[77] Besides the redox-responsiveness of the polymers by oxidation (right), the core–shell particles can be converted into spherical magnetic spheres by thermal treatment (left). Reproduced with permission from ref. 77. Copyright 2012 American Chemical Society.

terminal vinylic groups by reaction with 1,1′-bis(dimethylsilyl)ferrocene. At the end, Fe_3Si-nanostructured magnetoceramics were directly obtained *via* pyrolysis.

Hollow tubular nanostructures can be obtained by using a nanoporous alumina template filled with PFS.[76] The organometallic nanorods were converted into γ-Fe_2O_3 ceramic materials while maintaining the shape of the polymeric precursor.

Immobilized ferrocene-containing polymers at the surface of well-defined organic nanoparticles have been found to be feasible precursors for spherical ceramic materials.[77] The thermal conversion of dense poly(2-(methacryloyloxy)ethyl ferrocenecarboxylate) (PFcMA) brushes, *i.e.* PFcMA featuring a high polymer grafting density at the particle surface, attached to organic particles led to spherical magnetic iron oxides (Figure 5.14). It seems that PFcMA-based polymers in particular are promising candidates not only for reversible redox-mediated control of particle polarity and swelling behavior as described earlier, but also as single-source precursors for iron oxides.

As another benefit of the 2-(methacryloyloxy)ethyl ferrocenecarboxylate (FcMA), this ferrocene monomer was shown to react in convenient emulsion polymerization protocols leading to uniform ferrocene-containing particles.[78] Introducing as-prepared nanoparticles with synthetically adjustable diameters into a Stöber process has been shown to produce various types of core–shell particles. Furthermore, after thermal treatment, nanorattles featuring a magnetic iron oxide core could be prepared (Figure 5.15).

Furthermore, immobilized ferrocene polymers can be part of highly ordered structures for optical applications. Some recent advances dealing with such systems as well as their ceramized ordered structures will be described in the following section.

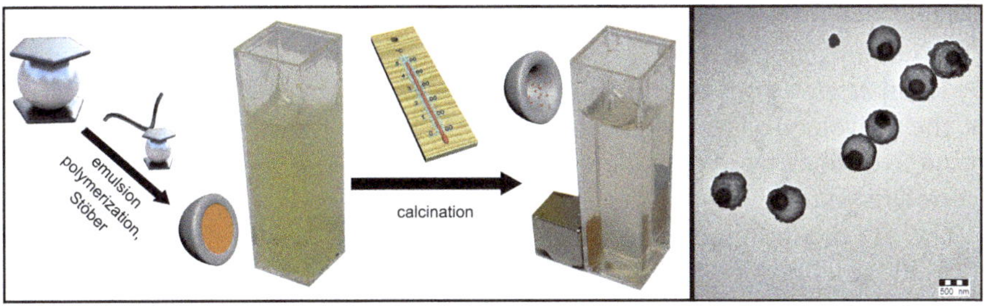

Figure 5.15 Simplified scheme for the controlled build-up of PFcMA core–shell particles by using emulsion polymerization strategies followed by a Stöber process.[78] Thermal treatment led to the formation of magnetic nanorattles out of a single-source precursor material. Reproduced with permission from ref. 78. Copyright 2014 American Chemical Society.

5.7 Confined and Immobilized Ferrocene Polymers for Optical Applications

Photonic band gap materials based on colloidal crystals have attracted enormous attention as promising candidates in optoelectronic sensing devices or as actuation systems.[79] A great benefit of colloidal crystals is their inexpensive and convenient bottom-up preparation giving a good optical performance with iridescent reflection colours caused by Bragg diffraction of visible light.[80–82] Recent efforts in the field of tunable colloidal crystals have been based on particles either featuring a stimuli-responsive core particle or a stimuli-responsive polymer matrix.[83–86]

The most prominent example of a ferrocene-containing polymer embedded in a colloidal crystal structure was reported by Arsenault *et al.* The material properties investigated by this group were based on the redox-activity of PFS-derivatives as matrix materials, which has proven to be reversibly electrically addressable.[87,88] The Bragg diffraction of highly ordered silica spheres embedded in a PFS-based matrix material can be chemo-mechanically varied by oxidizing the ferrocene moieties. The ferrocene-based photonic crystal device is capable of increasing the lattice distances of the close-packed silica spheres accompanied with a fast optical response.[89]

An electro-active inverse opal has been prepared by a silica-templated infiltration/etching route as depicted in Figure 5.16. The position of the Bragg peak can be varied over a wide range by electrical actuation of the ferrocene-containing inverse opal film.[90]

A PFS-based photonic crystal gel has been shown to be tunable regarding its optical properties by using different solvents and redox chemistry.[10] The electrochemical stimulus was used to oxidize or reduce the ferrocene moieties accompanied with a fast and reversible swelling/deswelling behaviour of the system. PFS features the intrinsic capability to be addressed by rather low applied voltage (<0.5 V). The authors also expect potential for the investigated ferrocene-containing materials not only as stimuli-responsive materials but also as self-healing materials.[10]

The optical properties of a PFS-based polyelectrolyte multilayer film obtained by using the previously described LBL technique were investigated by Kooij *et al.*[91] Film thickness was controlled by varying the ionic strength of the polyelectrolytes used. Furthermore, it was found that values for the refractive index were considerably high compared to more common polyelectrolyte systems.

The 3D confinement and morphologies of PFS-*block*-polystyrene, which formed lamellae or cylinders in the bulk state, were investigated in a colloidal crystal structure. It was found that the colloidal silica structure with its

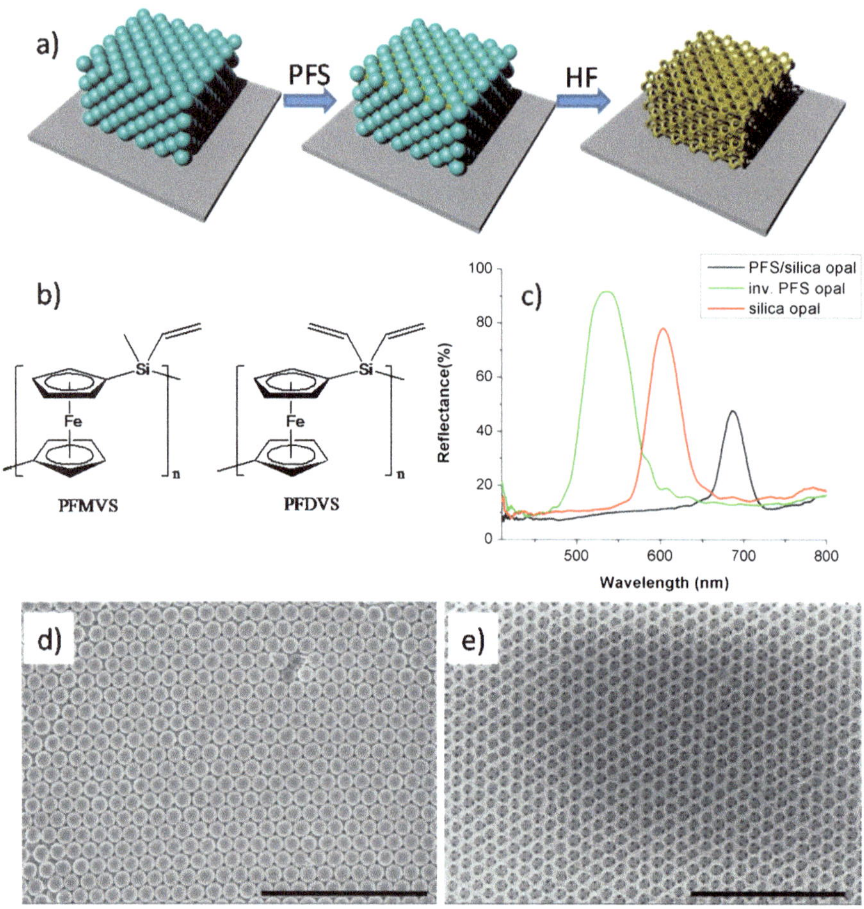

Figure 5.16 (a) General scheme for the preparation of PFS-based inverse opals; (b) molecular structure of the electroactive polymers employed in active opals; (c) UV/vis spectra showing the Bragg peaks after every preparation step; (d) exemplary SEM image of hexagonally packed silica spheres; (e) SEM image of the PFS-based inverse polymer-gel opal templated by silica spheres (scale bars correspond to 3 μm).[90] Reproduced with permission from ref. 90. Copyright 2008 Wiley VCH.

polar surface and spatial confinement tremendously influenced the diblock copolymer microphase separation leading to unprecedented 3D morphology transitions.[92,93]

Very recently, metallopolymers bearing the ferrocene motif in the polymer side-chain were also successfully used for producing photonic materials.[94] Compared to previous studies, hard core–soft shell particles were used. Here, the soft material consisted of FcMA and ethyl acrylate copolymers which could be subjected to the so-called melt shear organization technique yielding elastomeric opal disc films with diameters of up to 15 cm.[86,95] These elastomeric opal films showed distinct iridescent and angle-dependent reflection colours according to Bragg's law (Figure 5.17).[94]

Due to the elastomeric properties of the matrix material, the opal films showed fully reversible mechanochromic behaviour by applying strains of

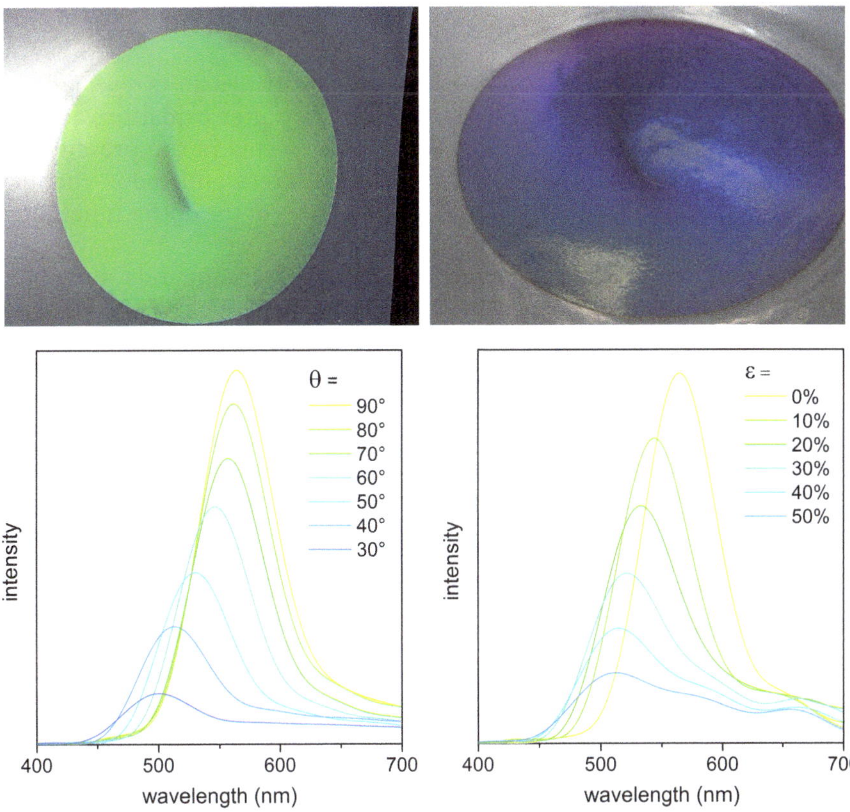

Figure 5.17 Top: photographs of ferrocene-containing elastomeric opal films at different angles of light incidence, the respective angles of view at normal light incidence (top left), and at an angle of view of 45° (top right).[94] Bottom: corresponding UV/vis reflection measurements: angle-dependent (bottom left) and at different strains ε (bottom right). Reproduced with permission from ref. 94.

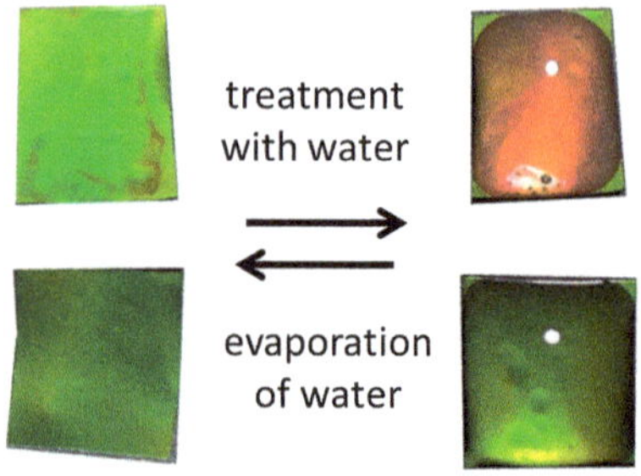

Figure 5.18 Redox-responsive behaviour of a PFcMA-containing elastomeric opal film after treatment with oxidizing reagents and water accompanied with a change of reflection colour from green to red (top), while treatment of the opal film with water in the absence of oxidation reagents maintained the original reflection colour (bottom).[94] Reproduced with permission from ref. 94.

up to 50%. Moreover, the ferrocene-containing repetitive units revealed the intrinsic capability to change the particle polarity after oxidation, accompanied with a tremendous particle swelling especially in the presence of water (Figure 5.18). Both stimuli—mechanical stress and redox chemistry—can individually address the optical properties of the entire functional material.[94]

As described in the previous section, ferrocene compounds and their polymeric structures have been proven to be suitable ceramic precursors. Self-assembly of monodisperse inorganic or organic particles can lead to materials with high periodicity with suitability for photonic applications. Thermal treatment and conversion into the corresponding inorganic material has been shown by Ozin and Manners *et al.* The authors prepared colloidal PFS-based inverse opals and, after calcination, magnetic ceramic opals with well-defined periodicities for photonic applications.[96]

5.8 Conclusion and Outlook

Within this chapter it has been shown that the incorporation of ferrocene moieties as redox-responsive switches or precursor materials into polymeric structures, which furthermore can be attached to various surfaces, in gels or different confinements, have led to manifold fascinating advanced materials over the last few years. As a result of the huge field of potential approaches dealing with immobilized ferrocene-containing polymers, it is expected that such materials will play a significant part in creating novel next generation materials over the next few years. In particular, the intrinsic capability of ferrocene to switch quickly, with high efficiency and in a reversible manner between a polar and rather unpolar state, is furthermore expected to be a powerful tool for sensing and release platforms.

References

1. D. Roy, J. N. Cambre and B. S. Sumerlin, *Prog. Polym. Sci.*, 2010, **35**, 278.
2. M. Ballauff and Y. Lu, *Polymer*, 2007, **48**, 1815.
3. H. Lee, J. Pietrasik, S. S. Sheiko and K. Matyjaszewski, *Prog. Polym. Sci.*, 2010, **35**, 24.
4. S. Minko, *J. Macromol. Sci. Polym. Rev.*, 2006, **46**, 397.
5. G. R. Whittell and I. Manners, *Adv. Mater.*, 2007, **19**, 3439.
6. V. Bellas and M. Rehahn, *Angew. Chem., Int. Ed.*, 2007, **46**, 5082.
7. A. S. Abd-El-Aziz, C. Agatemor and N. Etkin, *Macromol. Rapid Commun.*, 2014, **35**, 513.
8. M. Gallei, *Macromol. Chem. Phys.*, 2014, **215**, 699.
9. M. I. Giannotti, H. Lv, Y. Ma, M. P. Steenvoorden, A. R. Overweg, M. Roerdink, M. A. Hempenius and G. J. Vancso, *J. Inorg. Organomet. Polym. Mater.*, 2005, **15**, 527.
10. M. A. Hempenius, C. Cirmi, F. L. Savio, J. Song and G. J. Vancso, *Macromol. Rapid Commun.*, 2010, **31**, 772.
11. Y. Ma, W.-F. Dong, M. A. Hempenius, H. Möhwald and G. J. Vancso, *Nat. Mater.*, 2006, **5**, 724.
12. R. Barbey, L. Lavanant, D. Paripovic, N. Schüwer, C. Sugnaux, S. Tugulu and H.-A. Klok, *Chem. Rev.*, 2009, **109**, 5437.
13. D. H. Evans, *Chem. Rev.*, 2008, **108**, 2113.
14. I. Manners, *Synthetic Metal-Containing Polymers*, Wiley VCH, Weinheim, Germany, 2004.
15. N. L. Abbott and G. M. Whitesides, *Langmuir*, 1994, **10**, 1493.
16. J. Song and G. J. Vancso, *Langmuir*, 2011, **27**, 6822.
17. J. Elbert, M. Gallei, C. Rüttiger, A. Brunsen, H. Didzoleit, B. Stühn and M. Rehahn, *Organometallics*, 2013, **32**, 5873.
18. J. Elbert, J. Mersini, N. Vilbrandt, C. Lederle, M. Kraska, M. Gallei, B. Stühn, H. Plenio and M. Rehahn, *Macromolecules*, 2013, **46**, 4255.
19. O. Azzaroni, A. A. Brown and W. T. S. Huck, *Adv. Mater.*, 2007, **19**, 151.
20. X. Mao, G. C. Rutledge and T. A. Hatton, *Langmuir*, 2013, **29**, 9626.
21. X. Mao, F. Simeon, D. S. Achilleos, G. C. Rutledge and T. A. Hatton, *J. Mater. Chem. A*, 2013, **1**, 13120.
22. I. Cha, Y. Yagi, T. Kawahara, K. Hashimoto, K. Fujiki, S. Tamesue, T. Yamauchi and N. Tsubokawa, *Colloids Surf., A*, 2014, **441**, 474.
23. Y. Ma, M. A. Hempenius and G. J. Vancso, *J. Inorg. Organomet. Polym. Mater.*, 2007, **17**, 3.
24. M. A. Hempenius, N. S. Robins, R. G. H. Lammertink and G. J. Vancso, *Macromol. Rapid Commun.*, 2001, **22**, 30.
25. M. A. Hempenius, M. Peter, N. S. Robins, E. S. Kooij and G. J. Vancso, *Langmuir*, 2002, **18**, 762.
26. Z. Wang, G. Masson, F. C. Peiris, G. A. Ozin and I. Manners, *Chem.–Eur. J.*, 2007, **13**, 9372.
27. Y. Ma, W.-F. Dong, E. S. Kooij, M. A. Hempenius, H. Möhwald and G. J. Vancso, *Soft Matter*, 2007, **3**, 889.

28. J. Song, D. Jańczewski, Y. Ma, M. Hempenius, J. Xu and G. J. Vancso, *J. Mater. Chem. B*, 2013, **1**, 828.
29. J. Song, D. Janczewski, Y. Ma, M. Hempenius, J. Xu and G. J. Vancso, *J. Colloid Interface Sci.*, 2013, **405**, 256.
30. J. Song, D. Jańczewski, Y. Ma, L. van Ingen, C. Ee Sim, Q. Goh, J. Xu and G. J. Vancso, *Eur. Polym. J.*, 2013, **49**, 2477.
31. J. Song, D. Janczewski, Y. Guo, J. Xu and G. J. Vancso, *Nanoscale*, 2013, **5**, 11692.
32. A. Akhoury, L. Bromberg and T. A. Hatton, *Appl. Mater. Interfaces*, 2011, **3**, 1167.
33. R. H. Staff, M. Gallei, M. Mazurowski, M. Rehahn, R. Berger, K. Landfester and D. Crespy, *ACS Nano*, 2012, **6**, 9042.
34. M. Gallei, R. Klein and M. Rehahn, *Macromolecules*, 2010, **43**, 1844.
35. J. Elbert, F. Krohm, C. Rüttiger, S. Kienle, H. Didzoleit, B. N. Balzer, T. Hugel, B. Stühn, M. Gallei and A. Brunsen, *Adv. Funct. Mater.*, 2014, **24**, 1591.
36. S. Tan, K. Ladewig, Q. Fu, A. Blencowe and G. G. Qiao, *Macromol. Rapid Commun.*, 2014, **35**, 1166.
37. B. V. K. J. Schmidt, M. Hetzer, H. Ritter and C. Barner-Kowollik, *Progr. Polym. Sci.*, 2014, **39**, 23.
38. H. Ritter, B. E. Mondrzik, M. Rehahn and M. Gallei, *BeilsteinJ. Org. Chem.*, 2010, **6**, 60.
39. G. V. Dubacheva, A. Van der Heyden, P. Dumy, O. Kaftan, R. Auzely-Velty, L. Coche-Guerente and P. Labbe, *Langmuir*, 2010, **26**, 13976.
40. Q. Yan, J. Yuan, Z. Cai, Y. Xin, Y. Kang and Y. Yin, *J. Am. Chem. Soc.*, 2010, **132**, 9268.
41. M. Nakahata, Y. Takashima, H. Yamaguchi and A. Harada, *Nat. Commun.*, 2011, **2**, 511.
42. I. Tomatsu, A. Hashidzume and A. Harada, *Macromol. Rapid Commun.*, 2006, **27**, 238.
43. A. Harada and S. Takahashi, *J. Inclusion Phenom.*, 1984, **2**, 791.
44. Z. Q. Dong, Y. Cao, Q. J. Yuan, Y. F. Wang, J. H. Li, B. J. Li and S. Zhang, *Macromol. Rapid Commun.*, 2013, **34**, 867.
45. P. Daum and R. W. Murray, *J. Electroanal. Chem. Interfacial Electrochem.*, 1979, **103**, 289.
46. P. Daum and R. W. Murray, *J. Phys. Chem.*, 1981, **85**, 389.
47. X. Feng, A. Cumurcu, X. Sui, J. Song, M. A. Hempenius and G. J. Vancso, *Langmuir*, 2013, **29**, 7257.
48. Z. Li, T. Konno, M. Takai and K. Ishihara, *Biosens. Bioelectron.*, 2012, **34**, 191.
49. J. L. DeLuca, D. P. Hickey, D. A. Bamper, D. T. Glatzhofer, M. B. Johnson and D. W. Schmidtke, *ChemPhysChem*, 2013, **14**, 2149.
50. X. Sui, X. Feng, J. Song, M. A. Hempenius and G. J. Vancso, *J. Mater. Chem.*, 2012, **22**, 11261.
51. W. Shi, M. I. Giannotti, X. Zhang, M. A. Hempenius, H. Schonherr and G. J. Vancso, *Angew. Chem., Int. Ed.*, 2007, **46**, 8400.

52. S. Zou, M. A. Hempenius, H. Schönherr and G. J. Vancso, *Macromol. Rapid Commun.*, 2006, **27**, 103.
53. L. Yuan, W. Wei and S. Liu, *Biosens. Bioelectron.*, 2012, **38**, 79.
54. Z. B. Zhang, S. J. Yuan, X. L. Zhu, K. G. Neoh and E. T. Kang, *Biosens. Bioelectron.*, 2010, **25**, 1102.
55. A. Akhoury, L. Bromberg and T. A. Hatton, *J. Phys. Chem. B*, 2013, **117**, 333.
56. C. Bunte and J. Rühe, *Macromol. Rapid Commun.*, 2009, **30**, 1817.
57. D. Sung and S. Yang, *Electrochim. Acta*, 2014, **133**, 40.
58. J. Y. Wang, L. C. Chen and K. C. Ho, *ACS Appl. Mater. Interfaces*, 2013, **5**, 7852.
59. V. Jouikov and J. Simonet, *J. Appl. Electrochem.*, 2012, **42**, 527.
60. F. Dursun, S. K. Ozoner, A. Demirci, M. Gorur, F. Yilmaz and E. Erhan, *J. Chem. Technol. Biotechnol.*, 2012, **87**, 95.
61. K. Y. Hou, A. Rehman and X. Zeng, *Langmuir*, 2011, **27**, 5136.
62. B. Nagel, A. Warsinke and M. Katterle, *Langmuir*, 2007, **23**, 6807.
63. G. M. Bhalerao, A. K. Sinha, H. Srivastava and A. K. Srivastava, *Appl. Phys. A: Mater. Sci. Process.*, 2009, **95**, 373.
64. A. Bhattacharjee, A. Rooj, M. Roy, J. Kusz and P. Gütlich, *J. Mater. Sci.*, 2012, **48**, 2961.
65. D. Amara and S. Margel, *J. Mater. Chem.*, 2011, **21**, 15764.
66. L. Cao, J. A. Massey, M. A. Winnik, I. Manners, S. Riethmüller, F. Banhart, J. P. Spatz and M. Möller, *Adv. Funct. Mater.*, 2003, **13**, 271.
67. S. B. Clendenning, S. Han, N. Coombs, C. Paquet, M. S. Rayat, D. Grozea, P. M. Brodersen, R. N. S. Sodhi, C. M. Yip, Z.-H. Lu and I. Manners, *Adv. Mater.*, 2004, **16**, 291.
68. K. Temple, K. Kulbaba, K. N. Power-Billard, I. Manners, K. A. Leach, T. Xu, T. P. Russell and C. J. Hawker, *Adv. Mater.*, 2003, **15**, 297.
69. K. Kulbaba and I. Manners, *Macromol. Rapid Commun.*, 2001, **22**, 711.
70. K. R. Thomas, A. Ionescu, J. Gwyther, I. Manners, C. H. W. Barnes, U. Steiner and E. Sivaniah, *J. Appl. Phys.*, 2011, **109**, 073904.
71. K. Liu, S. Fournier-Bidoz, G. A. Ozin and I. Manners, *Chem. Mater.*, 2009, **21**, 1781.
72. X.-S. Wang, A. Arsenault, G. A. Ozin, M. A. Winnik and I. Manners, *J. Am. Chem. Soc.*, 2003, **125**, 12686.
73. C. G. Hardy, L. Ren, S. Ma and C. Tang, *Chem. Commun.*, 2013, **49**, 4373.
74. M. Häußler, Q. Sun, K. Xu, J. W. Y. Lam, H. Dong and B. Z. Tang, *J. Inorg. Organomet. Polym. Mater.*, 2005, **15**, 67.
75. J. Kong, T. Schmalz, G. Motz and A. H. E. Müller, *J. Mater. Chem. C*, 2013, **1**, 1507.
76. M. Ginzburg-Margau, S. b. Fournier-Bidoz, N. Coombs, G. A. Ozin and I. Manners, *Chem. Commun.*, 2002, 3022.
77. M. Mazurowski, M. Gallei, J. Li, H. Didzoleit, B. Stühn and M. Rehahn, *Macromolecules*, 2012, **45**, 8970.
78. D. Scheid, G. Cherkashinin, E. Ionescu and M. Gallei, *Langmuir*, 2014, **30**, 1204.

79. J. F. Galisteo-López, M. Ibisate, R. Sapienza, L. S. Froufe-Pérez, Á. Blanco and C. López, *Adv. Mater.*, 2011, **23**, 30.
80. J. Zhang, Y. Li, X. Zhang and B. Yang, *Adv. Mater.*, 2010, **22**, 4249.
81. A. Vlad, A. Frölich, T. Zebrowski, C. A. Dutu, K. Busch, S. Melinte, M. Wegener and I. Huynen, *Adv. Funct. Mater.*, 2013, **23**, 1164.
82. G. von Freymann, V. Kitaev, B. V. Lotsch and G. A. Ozin, *Chem. Soc. Rev.*, 2013, **42**, 2528.
83. A. Monge-Marcet, R. Pleixats, X. Cattoën and M. W. C. Man, *Catal. Sci. Technol.*, 2011, **1**, 1544.
84. C. I. Aguirre, E. Reguera and A. Stein, *Adv. Funct. Mater.*, 2010, **20**, 2565.
85. C. G. Schäfer, D. A. Smolin, G. P. Hellmann and M. Gallei, *Langmuir*, 2013, **29**, 11275.
86. C. G. Schäfer, M. Gallei, J. T. Zahn, J. Engelhardt, G. P. Hellmann and M. Rehahn, *Chem. Mater.*, 2013, **25**, 2309.
87. A. C. Arsenault, D. P. Puzzo, I. Manners and G. A. Ozin, *Nat. Photonics*, 2007, **1**, 468.
88. G. A. Ozin and A. C. Arsenault, *Mater. Today*, 2008, **11**, 44.
89. A. C. Arsenault, H. Miguez, V. Kitaev, G. A. Ozin and I. Manners, *Adv. Mater.*, 2003, **15**, 503.
90. D. P. Puzzo, A. C. Arsenault, I. Manners and G. A. Ozin, *Angew. Chem., Int. Ed.*, 2009, **48**, 943.
91. E. S. Kooij, Y. Ma, M. A. Hempenius, G. J. Vancso and B. Poelsema, *Langmuir*, 2010, **26**, 14177.
92. D. A. Rider, J. I. L. Chen, J.-C. Eloi, A. C. Arsenault, T. P. Russell, G. A. Ozin and I. Manners, *Macromolecules*, 2008, **41**, 2250.
93. A. C. Arsenault, D. A. Rider, N. Tetreault, J. I.-L. Chen, N. Coombs, G. A. Ozin and I. Manners, *J. Am. Chem. Soc.*, 2005, **127**, 9954.
94. D. Scheid, C. Lederle, S. Vowinkel, C. G. Schäfer, B. Stühn and M. Gallei, *J. Mater. Chem. C*, 2014, **2**, 2583.
95. C. G. Schäfer, B. Viel, G. P. Hellmann, M. Rehahn and M. Gallei, *ACS Appl. Mater. Interfaces*, 2013, **5**, 10623.
96. J. Galloro, M. Ginzburg, H. Miguez, S. M. Yang, N. Coombs, A. Safa-Sefat, J. E. Greedan, I. Manners and G. A. Ozin, *Adv. Funct. Mater.*, 2002, **12**, 382.

CHAPTER 6

Metallosupramolecular Soft Materials: Metallogels

CHARLOTTE PO[a] AND VIVIAN WING-WAH YAM*[a]

[a]Institute of Molecular Functional Materials (Areas of Excellence Scheme, University Grants Committee, Hong Kong) and Department of Chemistry, The University of Hong Kong, Pokfulam Road, Hong Kong, P. R. China
*E-mail: wwyam@hku.hk

6.1 Introduction

In general, a gel is a soft solid-like material that can immobilize a significant amount of liquid by the adhesion and encapsulation of a solvent in a three-dimensional network.[1,2] Traditionally, gels are made from organic polymers and have been used for centuries in a variety of areas such as food, cosmetics, medicine and materials. It was only in the late 20th century, with the growth of interest to conduct investigations into the propensity of gelation by small molecules, known as "low molecular weight gelators" (LMWGs) or "low molecular mass gelators" (LMMGs),[3] and the emergence of a new branch of chemistry "supramolecular chemistry" as realized by the award of the Nobel prize to Cram, Lehn and Pedersen in 1987, that research in utilizing metal complexes for gelation to give supramolecular metallogels evolved.

In contrast to traditional polymer gels which use covalent cross-linkages to form the three-dimensional networks for gelation, supramolecular gels utilize non-covalent interactions, such as π–π stacking interactions,

RSC Smart Materials No. 15
Functional Metallosupramolecular Materials
Edited by John G. Hardy and Felix H. Schacher

Published by the Royal Society of Chemistry, www.rsc.org

hydrophobic–hydrophobic interactions, hydrogen bonding and others, as the driving force for gelation. A supramolecular gel is formed by the supramolecular assembly of gelator molecules *via* specific interactions to give nanostructures, most commonly observed as elongated fibres, strands or tapes. These one-dimensional fibrous structures join themselves together by branching or by entangling to give three-dimensional networks that immobilize a solvent to give a gel. Supramolecular gels could be prepared by heating a solution of the gelators in an appropriate solvent at a concentration higher than the critical gelation concentration (CGC) to give a clear solution (sol form), and cooling the supersaturated solution to the temperature below the gelation temperature or the sol–gel transition temperature (T_{gel}) to give the gel. A stable gel could be identified by the inversion test in which the gel can withstand its own weight without falling when put upside down. As the three-dimensional networks are formed by supramolecular assemblies *via* weak non-covalent interactions, the gelation processes are usually thermoreversible and could readily be interrupted by the alteration of the microenvironment, such as pH and addition of additives, making supramolecular gels important candidates for stimuli-responsive materials, which have various possible applications in materials science, and also in biomedical sciences.

Early discoveries of supramolecular gels were not by design, but rather by serendipity.[3] Gelators were usually discovered during recrystallization processes and structural modifications of these accidentally discovered gelators yielded new examples. It was only during the 1990s that attempts were made to relate the molecular structure to the gelating properties to enable the rational design of new gelators.[1–4]

Over years of research by different research groups on organic low molecular weight gelators,[1–4] some structural features have been identified as important building blocks for gelators and have been commonly incorporated into different functional motifs to derive new gelators with specific functions. For example, non-polar hydrocarbons such as steroids and long alkyl chains are commonly incorporated to enhance hydrophobic–hydrophobic interactions; moieties such as sugars, peptides, ureas and nucleobases are used to enhance hydrogen bonding and aromatic rings are incorporated to induce π–π stacking interactions.

In addition to the control of the intermolecular forces between gelators for supramolecular assembly, it is also important to strike a balance between the tendency for the gelators to dissolve or to aggregate in a particular solvent.[4] Overly strong self-aggregation in a particular solvent may render the solubility of the gelators to be low, while a low degree of aggregation would give a completely soluble solution. Thus, depending on the solvent, different intermolecular interactions are utilized for gelation of different systems. For example, the hydrogen bonding interaction is a common driving force in gels formed in organic solvents (organogel), while it loses its strength in water, and hydrophobic interactions become important to induce significant aggregation. However, due to the limited solubility of hydrophobic moieties in water, an amphiphilic architecture is normally employed in the construction of gels in aqueous solution (hydrogel).[4]

Metallogels share many structural features with low molecular weight gelators and the diversity of metal–ligand coordination has provided further means to tune the supramolecular assembly of the gelators, which influences the gels' properties. Ligands with more than one coordination site may form coordination polymers with metal ions *via* metal–ligand coordination, resulting in polymeric metallogels. On the other hand, metal ions may also bind to ligands to enhance their supramolecular assembly, which results in supramolecular gels. Due to the preferential coordination geometries (linear, trigonal, square planar, tetrahedral, octahedral *etc.*) of different metal centres, a large variety of metallogel architectures can be achieved by incorporation of different metal centres. In some cases, the inter-conversion between different oxidation states, addition of anions or other external stimuli might interrupt the metal–ligand coordination and cause large structural changes in the supramolecular gels to give stimuli-responsive gels.[5–11,17–19,22–30,38–47,49–68]

Transition metals, which are capable of forming multivalent ions and exhibiting various coordination geometries, are ideal candidates for constructing stimuli-responsive materials. In addition to conventional non-covalent interactions used by organic systems, such as π–π stacking interactions, hydrophobic–hydrophobic interactions, and hydrogen bonding, some transition metal complexes, especially for d^8 and d^{10} metal complexes, might also involve metal–metal interactions in the gelation processes, giving rise to interesting structural as well as spectroscopic properties.[6,7,41,42,56–58,64,65,68–72] Together with the intrinsic magnetic, optical and catalytic properties of transition metals, supramolecular gels of transition metals have opened up new possibilities to construct functional materials. In the past decade, a number of transition metal-based metallogels have been reported to show interesting properties.[6] In this chapter, the recent development of metallogels involving the use of transition metals will be reviewed and the metallogels are grouped according to the metal centres for discussion and the special structural, spectroscopic and stimuli-responsive behaviour of the metallogels will also be highlighted.

6.2 Silver

Silver(I) centres normally adopt a linear geometry and have a high affinity towards nitrogen donor atoms. In the literature, several groups have reported the use of pyridine-, imidazole- and melamine-containing ligands for Ag(I) coordination to give metallogels.[5] Pioneering work by Shinkai and co-workers showed the enhancement of gelating properties *via* the addition of Ag(I) ions.[8] A pyridine-containing cholesterol compound (Figure 6.1) in diphenyl ether was found to form a gel instantly upon the addition of AgOTf.[8] The metallogel formed was found to possess a higher sol–gel phase transition temperature (T_{gel}) than its organic counterpart. From the scanning electron microscopy (SEM) images, it was noted that the rod-like clusters arising from the organic compound alone had changed to a well-defined three-dimensional fibrous network upon the addition of AgOTf. It was also found that the enhancement in gelation properties by the addition of Ag^+ is unique to the compound with the pyridyl nitrogen at the *meta* position and it was

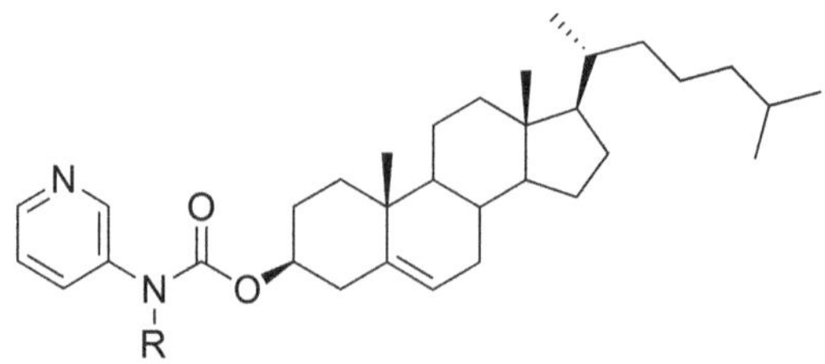

Figure 6.1 A pyridine-containing cholesterol compound showing enhanced gelating property on adding Ag(I) ions. From ref. 8.

suggested that since the pyridyl nitrogen at the *meta* position is perpendicularly oriented to the radius of the columnar aggregates formed by the cholesterol units, the two pyridyl nitrogen atoms in the neighbouring molecules can face each other at approximately 180°. This favours silver–pyridine interactions and leads to the formation of three-dimensional fibrous networks and, hence, spontaneous metallogel formation and an enhancement in gel stability.[8]

An interesting work has been reported by Lee and co-workers for the use of a 1,3-diethylenebenzene bridged bipyridine ligand appended with a triethylene glycol (TEG) dendron for Ag(I) coordination.[9,10] It has been shown that by increasing the size of the counter-anion, the secondary structure of the Ag(I)-coordinated polymer chain would change from helical chains (with NO_3^- and BF_4^-) to dimeric macrocycles (with OTf^-), and further to zigzag chains (with $C_3F_7COO^-$). Only the helical chain and the dimeric macrocyclic structures that could result in columnar stacking would show gelation properties.[9] Such columnar structures were believed to be stabilized by the encapsulation of an anion of critical size inside the cavity of the helix or ring (Figure 6.2).[9] Further investigation was conducted to utilize the system for stimuli-responsive gels.[10] By addition of $C_2F_5COO^-$ or F^- to the BF_4–Ag-gel in aqueous solution, a gel-to-sol transition occurred. It was suggested that the addition of F^- would cause depolymerization by abstracting the Ag(I) ion and the addition of a relatively large amount of $C_2F_5COO^-$ would cause counter-anion exchange, leading to a conformational change from the helical to the zigzag chain.[10]

By using the pyridine group for Ag(I) coordination, Wu and coworkers reported the synthesis of a pyridine-containing ligand with three alkyl chains and an amide group for enhancing hydrophobic–hydrophobic interactions and intermolecular hydrogen bonding interactions, respectively.[11] When AgOTf in ethanol was added to a toluene solution of the ligand, spontaneous metallogel formation was observed and from UV-vis absorption studies it was confirmed that the molar ratio of metal–ligand coordination between the ligand and Ag^+ is 2 : 1, suggesting a linear coordination geometry (Figure 6.3a). Interestingly, the metallogel was found to be stimuli-responsive and would undergo a gel-to-sol phase transition upon addition of a chemical, while on further addition of Ag^+ ions, the gel could be restored. It was suggested that the Ag^+ ions in the gel would be abstracted by halide (X^- = Cl^-, Br^-, I^-), S^{2-} and

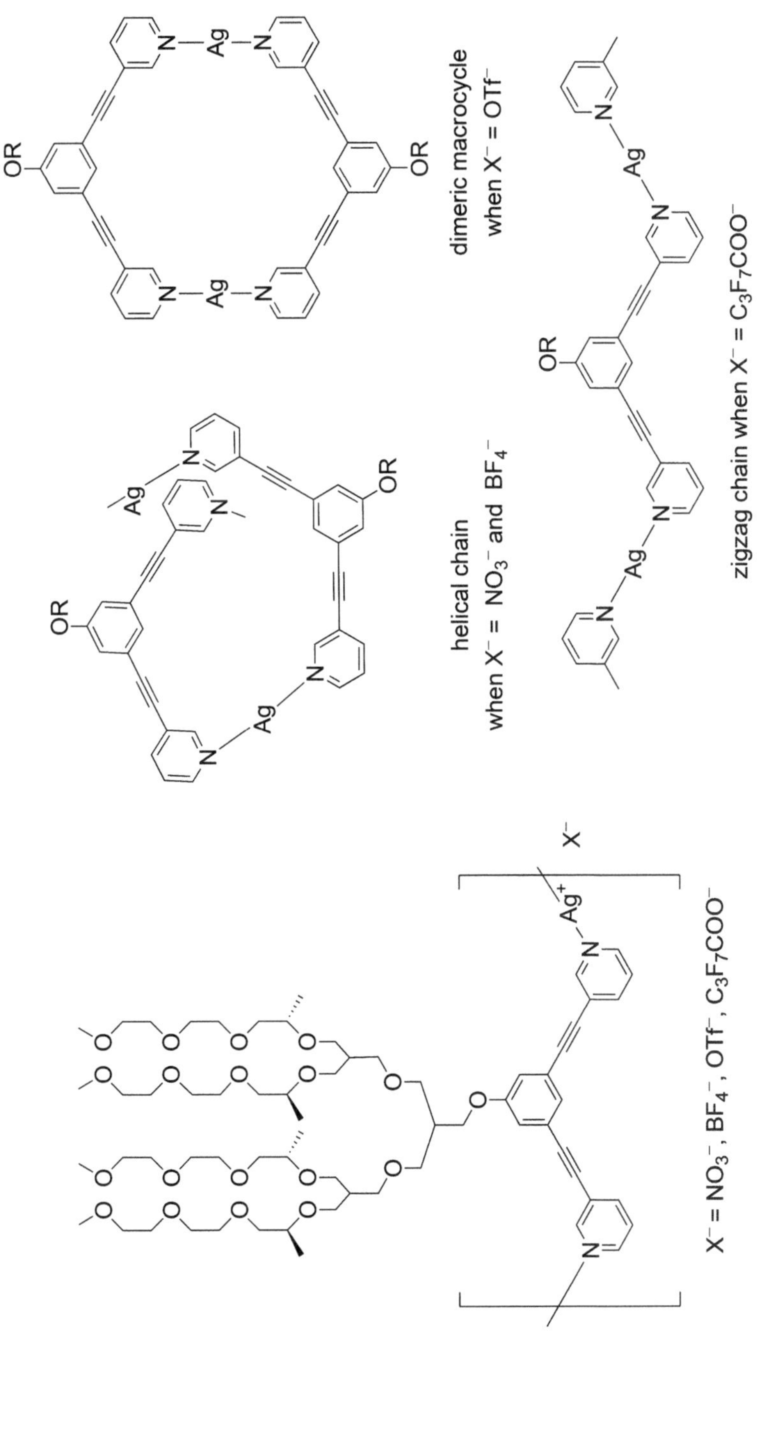

Figure 6.2 Stimuli-responsive Ag(I) polymer gels reported by Lee and co-workers. From ref. 10.

NH_3 to give AgX, Ag_2S and $[Ag(NH_3)_2]^+$ respectively, leading to the disruption of the gel structure. From the transmission electron microscopy (TEM) images, it could be clearly seen that the addition of H_2S and I^- would lead to the disruption of the fibrous network of the gel, giving only small discrete structures or short belt-like fibres that were not able to immobilize the solvent to give the organogelation properties (Figure 6.3b). In addition, the gel was also found to be responsive towards changes in pH. On decreasing the pH by adding an acid, the pyridine group of the ligand would be protonated, while on increasing the pH by adding NaOH, Ag^+ would react to give Ag_2O. In both cases, the Ag–pyridine coordination would be disrupted, losing its gel property.[11]

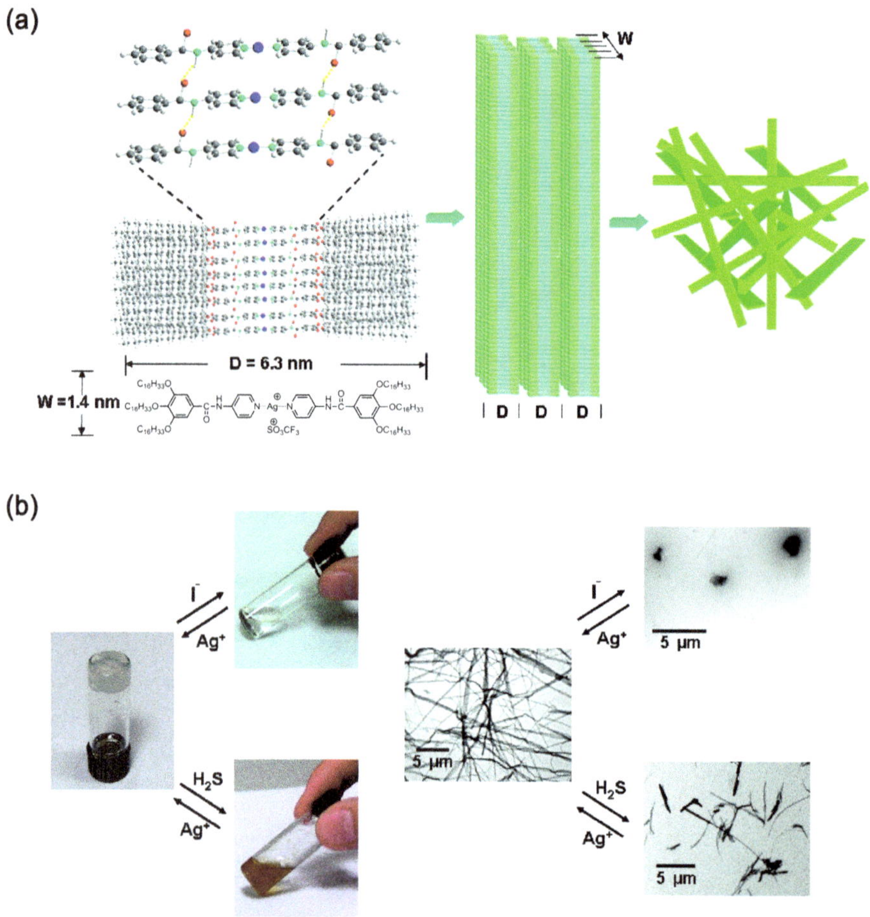

Figure 6.3 (a) Hierarchical structural model of the Ag(I)-coordinated metallogel. (b) Photographs and TEM images of the responsive properties of the Ag(I)-coordinated gel before and after the addition of potassium iodide and H_2S gas. Adapted with permission from ref. 11. Copyright 2007 American Chemical Society.

In addition to using pyridine as the binding site, other N-donor ligands have also been reported for the formation of Ag metallogels.[8–11] You and coworkers reported the utilization of imidazole-containing ligands for the formation of a transparent Ag(I) gel by the addition of a 1 : 1 molar ratio of $AgNO_3$ to the aqueous solution of the ligands at room temperature (Figure 6.4).[12] Unlike the pyridine-containing Ag gel, which is pH-responsive, the imidazole-containing Ag gel was found to be stable under acidic conditions at pH 1 to 7, although the 3-N position of the imidazolyl unit has been known to be readily protonated. It was also interesting to note that the achiral molecules could induce a helical chiral nanotubular structure in the gel state as shown from the atomic force microscopy (AFM) height image and the strong Cotton effect in the circular dichroism (CD) spectra. This has been further supported by theoretical density functional theory (DFT) calculations. It was suggested that the helical architecture would be a result of the strong directional interactions between the rigid bent bridging ligands and the Ag^+ ion, which adopts a linear coordination geometry. It was suggested that once an accidental occurrence of one excess metal–ligand coordination in one helical direction took place due to statistical fluctuation, new aggregates would follow to form secondary structures with the same helical direction, resulting in an overall helical chiral nanotubular structure.[12]

Nandi and co-workers reported the formation of a thermoreversible hydrogel by mixing $AgNO_3$ with melamine.[13] It was found that a stable gel that consisted of a continuous fibrous network would be formed with a 1 : 2 molar composition of Ag^+ to melamine, while a 1 : 1 molar composition would result in a metastable gel that would age gradually to give crystals. With reference to their crystal structures, it was suggested that the stability of the gel resulted from the linear coordination of Ag^+ to the nitrogen atoms of melamine. This linear coordination would assist the formation of extensive hydrogen bonding between the melamine molecules and would enable the bilayer structure to grow longitudinally and laterally to give fibrous structures that were kinetically and thermodynamically stable. In contrast, with the 1 : 2 molar ratio of Ag^+ to melamine, the coordination of silver would no longer be linear and the bent N–Ag–N bond would cause the breaking of the hydrogen bond and result in a less stable structure, and hence over time, it would transform into a more stable structure through aging to give crystals. In addition to demonstrating the chemically-responsive properties of the gelators towards pH,

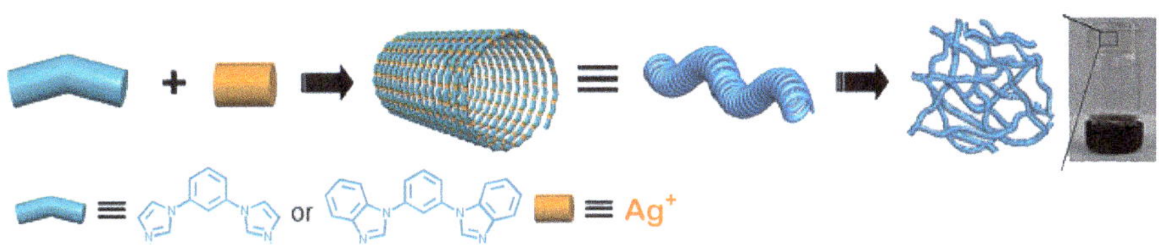

Figure 6.4 A schematic representation of the self-assembly processes of the Ag-coordination polymer gels. Adapted with permission from ref. 12.

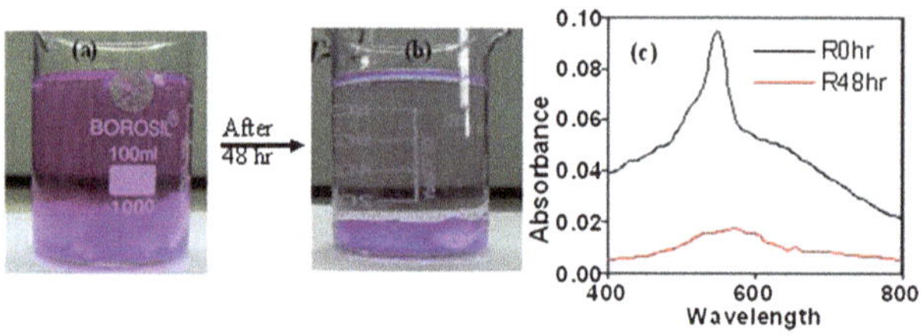

Figure 6.5 (a) Ag-gel in rose Bengal dye and (b) after 48 h of absorption. (c) UV-vis spectra of dye at 0 h and after 48 h. Adapted with permission from ref. 13.

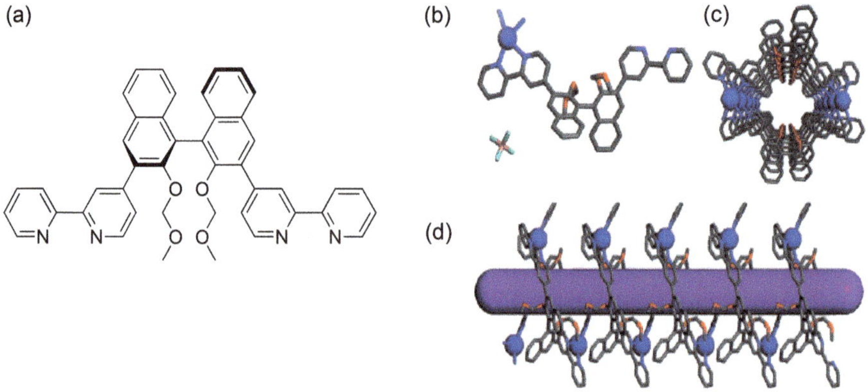

Figure 6.6 (a) Structure of the (S)-form of the ligand. (b) The repeating unit of the Ag(I)-coordination polymer, and (c) perspective and (d) side views of the nanotubular *P*-helical coordination polymer. Hydrogen atoms are omitted for clarity. Adapted with permission from ref. 14.

halides and counter-anions as in other Ag gels, the stable Ag–melamine gel was also found to show effective absorption of anionic dyes (Figure 6.5) while it was found to be unable to absorb dyes that are cationic or neutral in nature. Although the specific affinity to anionic dyes has remained unclear, it was suggested that host–guest interactions are the reason.[13]

On top of the formation of a stable gel through the linear coordination of Ag(I), Bian and Gao reported the formation of a stable Ag(I) polymer gel using a chiral binaphthyl-bearing bis-bipyridine ligand through the coordination of the Ag(I) to two bipyridyl units, each from one ligand, in a distorted tetrahedral geometry.[14] Adjacent Ag(I) centres have been found to be bridged by the ligand and due to the geometrical constraint of the chiral binaphthyl moiety, an extended nanotubular single-stranded *P*-helical coordination polymer was formed (Figure 6.6). The *P*-helix then, with the help of π–π stacking interactions, was found to grow and give multi-layered structures and finally nanofibres with

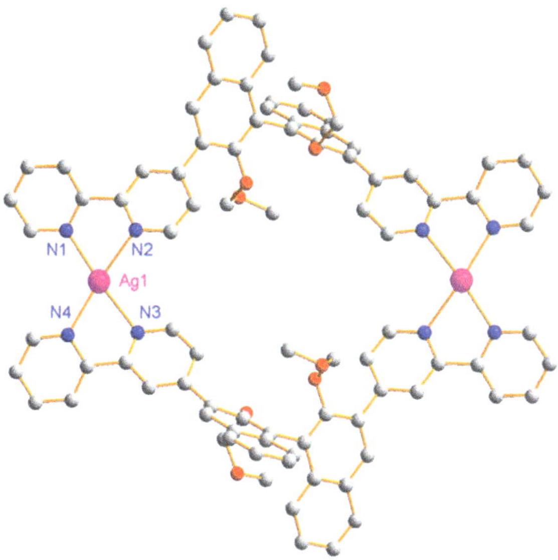

Figure 6.7 Perspective drawing of a dinuclear heterochiral metallocycle from the racemic ligand. Hydrogen atoms and counter-ions are omitted for clarity. Adapted with permission from ref. 15.

lamellar structures to give a gel. As an extension, the group has also investigated the gelation properties of the racemic ligand[15] and it was found that the use of the racemic ligand suppresses gelation. It was reported that more time and a higher concentration would be required for the racemic mixture to undergo gelation and it was also found that the metallogel formed by the racemic ligand would also be less stable than the enantiopure gel as shown by its lower gel-to-sol transition temperature and the occurrence of a gel-to-single crystal transition at room temperature.[15] From the X-ray crystal structure (Figure 6.7), it has been shown that the two enantiomers (each of the (*S*)- and (*R*)-form) are linked by two Ag(I) ions to give a dinuclear metallocycle (Figure 6.7), suggestive of assembly processes involving chiral self-discrimination. It was suggested that the dinuclear metallocycle would undergo self-assembly through π–π stacking interactions to give lamellar structures which led to the gel formation.[14]

6.3 Copper

Copper(I) and copper(II) have high affinities towards nitrogen and oxygen donors, and could coordinate to various N- and O-donor ligands such as pyridine and carboxylate to give copper-coordinated complexes or polymers;[7–9] some of which undergo supramolecular assembly to give suitable nanostructures and lead to the formation of gels. On the other hand, the copper ions may also modify the gel properties of organogels *via* copper coordination.

Owing to the existence of various oxidation states of copper, the inter-conversion between copper(I) and copper(II) may also lead to redox-responsive gels.

As an early example, Hanabusa and co-workers reported the enhancement in gelation properties of organogels through the addition of Cu(II) ions.[16] It was found that upon the addition of $CuCl_2$ in triethylamine, the T_{gel} of a methanol-based gel increased dramatically and the gel was restricted to return to its sol-form even at the boiling point of methanol and, on further heating, the solvent was found to be gradually released from the gel. It was also noted that the copper gel gave very high contrast TEM images even without additional staining, which suggested that heavy metals might have been involved in the gel structure and the fibres were reinforced by cross-linking through copper(II)-coordination.[16]

Kulkarni and co-workers reported the formation of hydrogels by a trinuclear copper(II) complex with inositol and 2,2′-bipyridine (bpy) as the ligands triggered by pH (Figure 6.8).[17] A copper(II) precursor was prepared *in situ* by reacting an equal molar concentration of $Cu(CH_3COO)_2$ with bpy and the gelator was formed by reacting 3 equivalents of the copper(II) precursor with inositol in aqueous solution. A subsequent increase in pH to 12.4 by sodium hydroxide yielded a transparent green gel. It was found that the reaction with only 1 or 2 equivalents of bpy ligand did not lead to the formation of a gel at all pH values and that the addition of sodium hydroxide has also been crucial for gel formation. A xerogel was prepared by drying the gel at 120 °C for 3 h until a constant mass was obtained. On washing the xerogel with ethanol to remove loosely held sodium hydroxide, the residue was found to be a single compound with a formula of $[Cu_3(inositol)(bpy)_3(H_2O)_9]$ as confirmed by thermogravimetric and electrospray ionization mass spectrometry (ESI-MS) analysis, and on addition of an aqueous solution at pH 12.4, a reversible hydrogel was formed. Since hydroxide was found to be crucial in the gel formation, it was suggested that the hydroxide ion assisted the formation of fibrous structures for gelation through bridging the trinuclear copper(II) complex molecules *via* axial coordination and hydrogen bonding between the inositol moieties.[17]

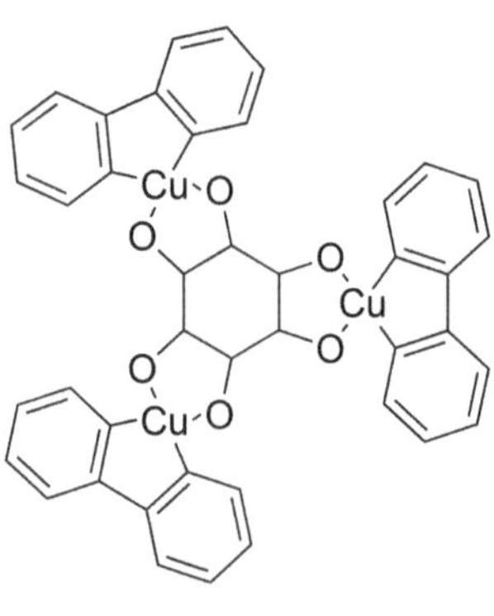

Figure 6.8 A trinuclear copper(II) complex for pH-triggered hydrogel formation. Reproduced with permission from ref. 17.

Hardie and co-workers reported a series of copper(II) metallogels with cyclotriveratrylene-type (CTV) ligands, which exhibited chemo- and mechano-responsive properties (Figure 6.9).[18] Similar to what was commonly observed in silver(I) gels,[11,12] the addition of HCl to protonate the N-donor in pyridine and bipyridine led to a loss in gel properties due to the disruption of metal–ligand interactions. The complex with a thiourea group was found to undergo a gel-to-sol transition in dichloromethane upon a mechanical disturbance such as sonication or shaking and reformation of gel without disruption was observed upon standing. It was suggested that sonication and shaking would destroy the fibrous network, and the hydrogen bonding interactions between the molecules could assist the re-establishment of the fibrous structure.[18]

In addition to chemo- and mechano-responsive gels, a redox-responsive copper(I) metallogel was reported by Shinkai and co-workers.[19] Cu(I) was coordinated to cholesterol-modified bipyridine (Figure 6.10) and the gel was found to show an interesting thermochromic change from reddish-brown to bluish-green *via* a sol–gel transition. It was suggested that the temperature-mediated colour change was not a result of oxidation of Cu(I) to Cu(II), but rather the distortion of the Cu(I) coordination from a tetrahedral geometry in the gel state due to a confined environment in the gel fibrils. On addition of a reducing agent (ascorbic acid) to the Cu(II) complex, the bluish-green gel would be formed, while addition of an oxidizing agent ($NOBF_4$) to the Cu(I) complex resulted in a pale blue precipitate, suggesting that the gel formation was selective towards Cu(I).[19]

Bian and Gao reported the use of a series of redox-responsive copper(I) coordinated polymer gels with binaphthylbis-bipyridine-based ligands (Figure 6.11).[20] On addition of $NOBF_4$ as the oxidant, the dark red gel collapsed to give a pale green solution, typically of Cu(II) and on addition of ascorbic acid, the gel could be regenerated.[20] Such findings suggested that the oxidation state of Cu is crucial for metallogel formation. The stable Cu(I)

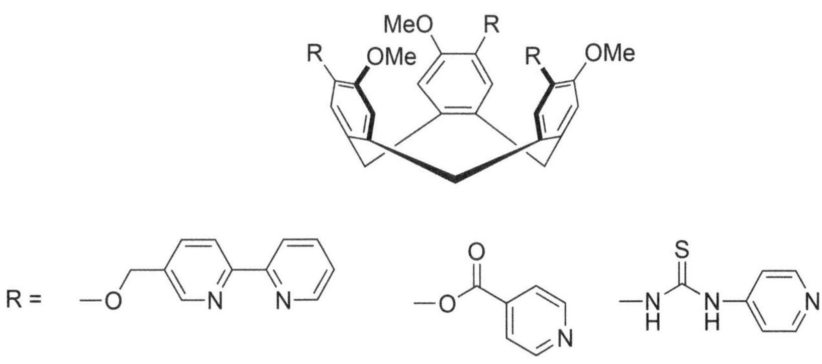

Figure 6.9 Cyclotriveratrylene-type (CTV) ligands for the formation of copper(II) metallogels. Reproduced from ref. 18 with permission from the Centre National de la Recherche Scientifique (CNRS) and the Royal Society of Chemistry.

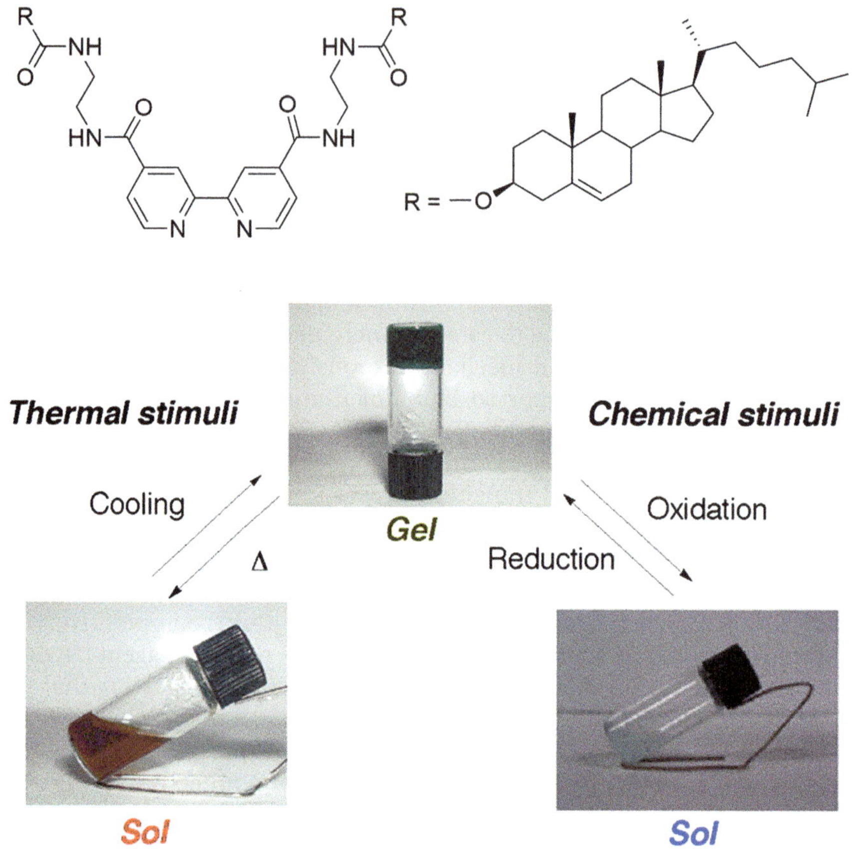

Figure 6.10 Cholesterol-modified bipyridine for a redox-responsive copper(I) gel reported by Shinkai and co-workers and the sol–gel phase transition of the metallogel in response to thermal and chemical stimuli. Adapted with permission from ref. 19. Copyright 2004 American Chemical Society.

xerogel has also been shown to catalyze copper-mediated 1,3-dipolar cycloaddition reactions in water, in which the catalytic activity could be maintained even after one year.[20] It was found that the gel that contained a long alkyl chain gave a better catalytic performance and it has been suggested that the long alkyl chain provided a more hydrophobic microenvironment to facilitate the diffusion of reactants.[20]

Liu has reported a copper(II) hydrogel that showed catalytic activity towards asymmetric Diels–Alder cycloaddition *via* the use of a bolaamphiphilic glutamic acid, *N*,*N*′-hexadecanedioyl-di-L-glutamic acid (L-HDGA), as the ligand.[21] It was noted that in the presence of the Cu(II) ion, the monolayer helical nanotubes formed by the pure ligand were transformed to multilayer nanotubes, as revealed by the increase in the tubular wall thickness shown by the electron microscopy images as well as the appearance of new XRD

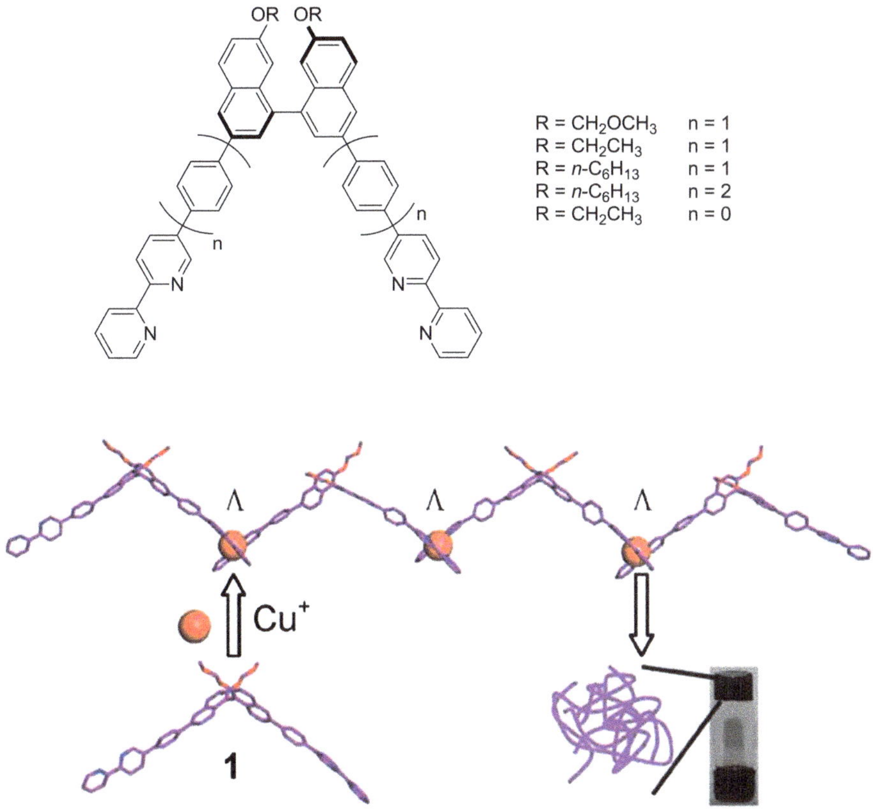

Figure 6.11 Binaphthylbis-bipyridine-based ligands for copper(I) gels and the schematic representation of the possible self-assembly processes of the copper(I) coordination polymer. Adapted with permission from ref. 20.

diffraction peaks and the enhancement of diffraction signals in the presence of Cu(II) ions. For other bolaamphiphiles studied, no gelation was observed on addition of Cu(II) ions and they showed a lower enantioselectivity and reaction rate towards asymmetric Diels–Alder reactions.[21] It was suggested that as the Cu(II) centres were aligned on the surface of the chiral nanotubes in the Cu(II) gel formed by L-HDGA (Figure 6.12), it would provide a high density of catalytic sites as well as a stereochemically favoured environment for the asymmetric Diels–Alder reaction to take place.[21]

As discussed earlier, supramolecular gels usually undergo a gel-to-sol transition at high temperature due to the breaking of intermolecular forces at high temperature. In sharp contrast, Nitschke and co-workers reported an interesting gelation process of a copper(I) polymer at high temperatures.[22] For this polymer, it was suggested that the reversible sol–gel transition was a consequence of the reversible crosslinking by the Cu–N_4 bond between the oligomer chains ($2[CuN_2P_2]^+ \rightleftharpoons [CuN_4]^+ + [CuP_n]^+ + (4-n)P$). It was proposed that at a

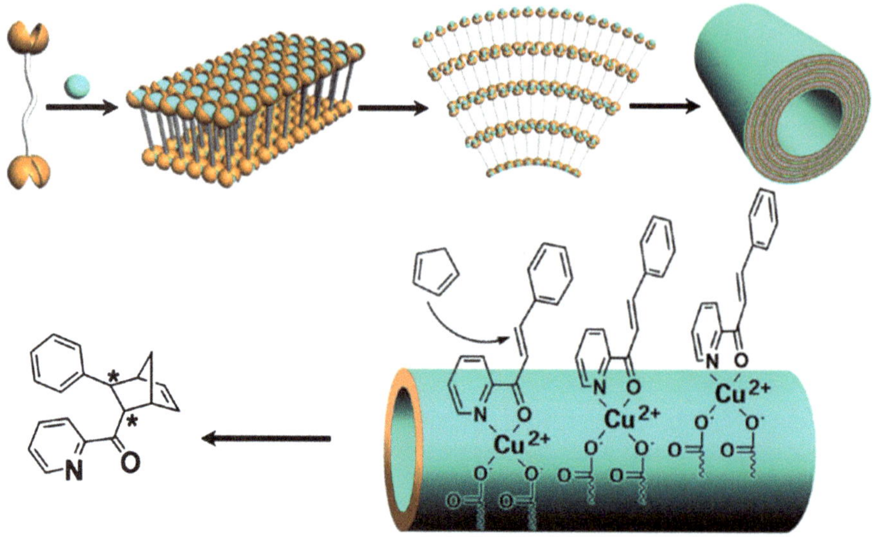

Figure 6.12 Illustration of the assembly mechanism of the copper(II)(L-HDGA) nanotubular structure and its asymmetric catalysis for the Diels–Alder reaction of aza-chalcone with cyclopentadiene. Adapted with permission from ref. 21. Copyright 2011 American Chemical Society.

high temperature, the equilibrium would favour the formation of [CuN_4^+] due to an increase in entropy and crosslinks would be formed. At a lower temperature, due to the labile Cu-coordination, ligand rearrangement would occur to give linear polymer chains and the gel would collapse (Figure 6.13).[22]

6.4 Iron

Iron(II) and iron(III) coordinated polymer gels have been reported.[23–26] Weiss and co-workers reported the gelation of Fe(III) by complexing with a series of bis-phosphonic acid monoesters, bis-phosphonate esters and bis-phosphonic acids.[23] Tong and co-workers reported a Fe(III) gel formed by coordination of Fe(III) with the carboxylate group on the poly(acrylic acid) chain to give a redox-responsive hydrogel.[24] When Fe(III) was reduced to Fe(II) upon photo-irradiation, a gel-to-sol transition occurred due to the reduction in binding affinity of Fe(II) towards carboxylates, and the gel could be restored *via* oxidation of Fe(II) to Fe(III).[24] Fe(II) gels were found to exhibit spin-crossover behavior as reported by Clérac and co-workers.[25,26] At a low temperature, a purple diamagnetic gel was formed, while on increasing the temperature, a gel-to-sol transition occurred and a high spin paramagnetic colourless solution was found with increases in magnetization. However, later it was reported that the gelation and spin crossover were independent properties and the change in spin might not be a direct consequence of the sol–gel transition.[25]

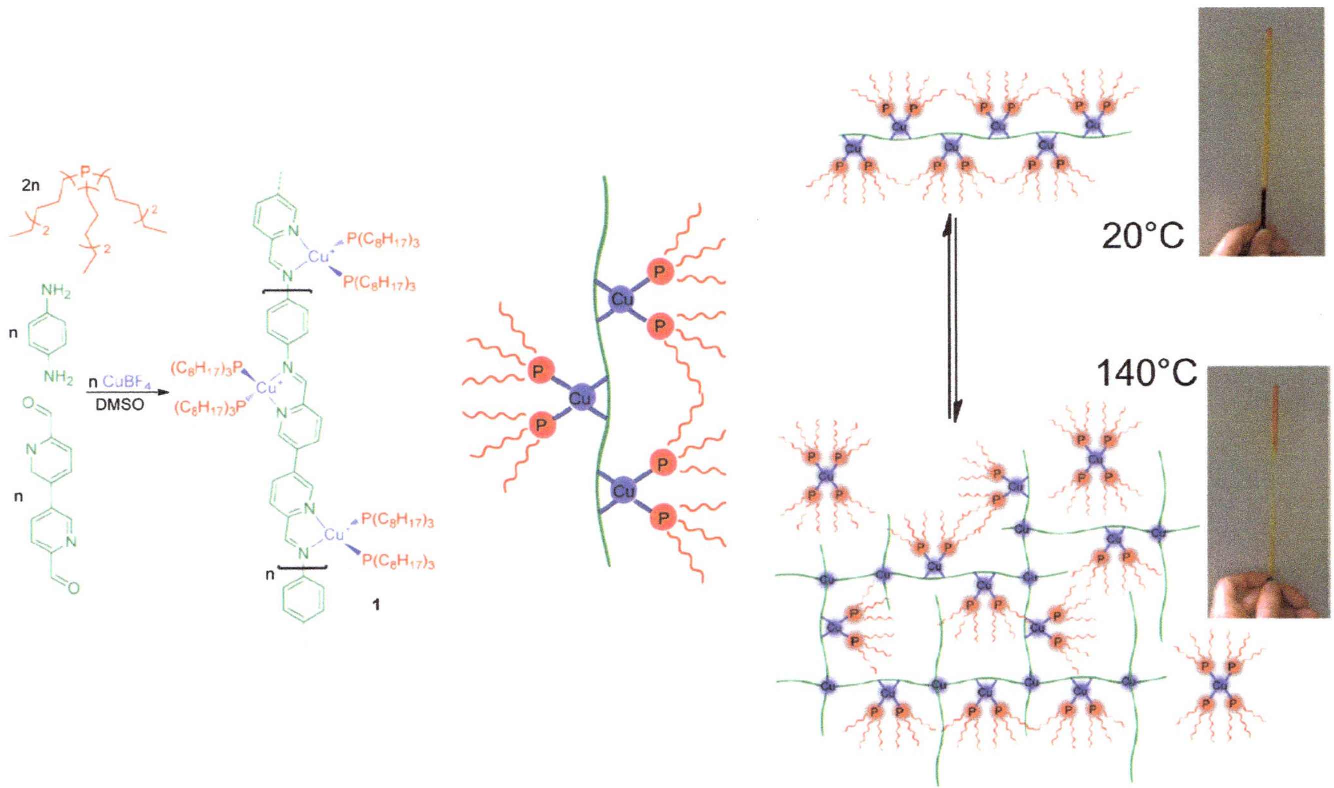

Figure 6.13 Schematic diagram of the unusual gelation of a Cu(I) polymer at a high temperature reported by Nitschke and co-workers. Adapted with permission from ref. 22. Copyright 2011 American Chemical Society.

6.5 Cobalt

Kimizuka and co-workers have reported an unusual gel-to-sol transition of a cobalt(II) gel upon cooling.[26] At 25–27 °C, a blue-coloured gel was observed and on lowering the temperature to 0 °C, a pale pink solution was obtained. As indicated by the colour change, the phase transition was suggested to be due to the coordination geometry change of Co(II) from tetrahedral (in blue gel) to octahedral (in pink solution). This observation suggested that polymer chains formed by octahedral coordination were difficult to aggregate.[27] Shinkai and co-workers reported the use of Co(II) polymer gels for chemosensing.[28] The polymer gels were formed by adding Co(II) ions to a 1,2,4,5-tetra(2*H*-tetrazole-5-yl)benzene solution to give red metallogels in various salts except for the addition of $CoCl_2$, which gave a blue metallogel. The colour of the red gel was suggested to be due to the presence of octahedral complexes while the colour of the blue gel was due to the presence of tetrahedrally coordinated Co(II). This spectroscopic change was utilized for the detection of chloro-containing gases, in which exposure of the red gels to HCl, $SOCl_2$, $COCl_2$ or $(COCl)_2$, generated blue gels.[28]

6.6 Rhenium

Yam and co-workers reported a series of luminescent organogels of alkynylrhenium(I) tricarbonyl diimine complexes,[29] in which the gelation was assisted by π–π stacking interactions between aromatic rings and the hydrophobic–hydrophobic interactions between long alkyl chains on the alkynyl ligands. By modifying the molecular design and attaching the long alkyl chains to the diimine ligands of the Re(I) complexes, a series of chemoresponsive gels was obtained (Figure 6.14).[29] It was found that on addition of Cu(I) or Ag(I) salts, spontaneous gelation with luminescence enhancement in the visible region

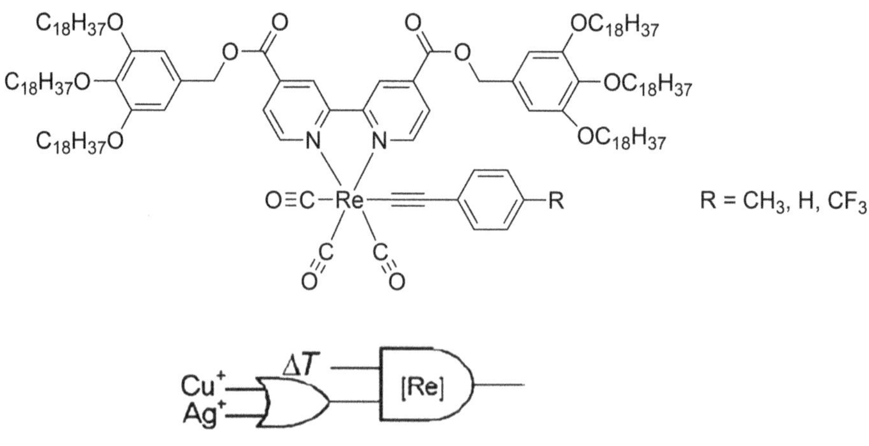

Figure 6.14 A series of chemoresponsive Re(I) metallogelators and molecular logic representing the logic flow of the alkynylrhenium metallogelator. Adapted with permission from ref. 30.

could be observed. It was suggested that the π-coordination of the Cu(I) or Ag(I) to the alkynyl unit in an η^2-fashion acted as an additional driving force for molecular assembly and gelation. It was also found that the sol–gel phase transition could be repeated many times by controlling the temperatures, and this multi-responsive feature has allowed the gel to be used as a molecular AND logic gate which may find application in molecular computation (see Chapter 9).[30]

6.7 Palladium

Palladium(II) complexes are well-known catalysts in organic synthesis, and supramolecular gels may serve as supports for the palladium centres, which may help to enhance the catalytic activity. Thus, much research on palladium(II) metallogels has focused on the catalytic properties of such materials.[33–36]

Xu and co-workers made use of the well-known capability of $[Pd(en)(H_2O)_2]^{2+}$ (en = ethylenediamine) to form extended molecular structures with pyridine to construct coordination polymeric gels. It was reported that by reacting 3-pyridineazocalix[4]arene with $[Pd(en)(H_2O)_2]^{2+}$, a metallogel could be formed in dimethyl sulfoxide (DMSO) (Figure 6.15).[31] At a high concentration of metallogelators, the gel stability was found to be significantly enhanced and it remained stable even upon the addition of hydrophilic and hydrophobic solvents and at elevated temperatures. Taking advantage of the enhanced stability and the hydrophobicity of the 3-pyridineazocalix[4]arene, the gel was tested in the uptake of non-ionic organic molecules at low concentrations (ppm). It was found that toluene could be absorbed into the metallogel from the freshly prepared aqueous solution of toluene and the absorption process could be followed *via* UV-vis absorption studies with time. As this finding has provided an important means to monitor low concentrations of hazardous organics in water, the research group extended their studies towards other organic molecules and the Pd(II) gel was found to be capable of extracting chlorobenzene and Malachite green at a very low

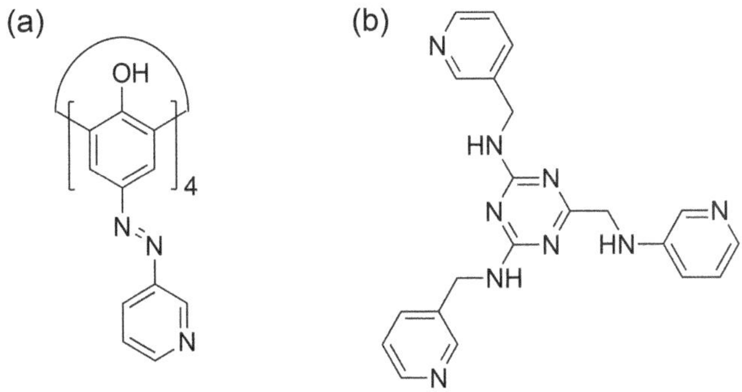

Figure 6.15 (a) 3-Pyridineazocalix[4]arene for forming a palladium(II) gel to extract organic molecules and (b) tridentate N-donor ligand for forming a palladium(II) gel with catalytic properties. From ref. 31–33.

concentration from aqueous solutions.[32] Furthermore, it was found that the extraction could also occur in the gas phase and the gel was found to extract naphthalene vapour two times more efficiently than activated carbon. Using the same ligand and another tridentate N-donor ligand with $Pd(OAc)_2$, the Pd(II) gels were found to be catalytically active towards oxidation of benzyl alcohol to benzaldehyde.[33] When compared to pure $Pd(OAc)_2$, the Pd(II) gels were found to show enhanced catalytic activity and stability. The Pd(II) gels were found to give a higher turnover rate and less decomposition after catalytic reactions. The stability was suggested to be due to the immobilization of the Pd(II) within the gel network. Furthermore, it was also noted that a higher turnover rate could be achieved in the wet gel as compared to the dry gel. The suppression of catalytic activity in the dry gel was rationalized by the decrease in efficiency to deliver reaction substrates to the Pd(II) catalytic centres in the dry state.[33]

Later, Zhang and co-workers modified the tridentate N-donor ligand reported by Xu and studied the catalytic activity of the Pd gel towards Suzuki–Miyaura coupling reactions.[34] A morphological transformation from spheres to fibres was observed by decreasing the Pd to ligand ratio from 1:4 to 1:1 and the gels with fibrous structures were found to give a higher catalytic activity.[34] The research group further extended the work by incorporating superparamagnetic magnetite Fe_3O_4 nanoparticles to the gel structure to give magnetic metallogels (Figure 6.16).[35] From TEM images, it was observed that the Fe_3O_4 nanoparticles were encapsulated in the gel fibrous network and the loading of Fe_3O_4 nanoparticles did not change the catalytic activity of the gel. The xerogel has been shown to catalyze Suzuki and Heck reactions and it has been shown that the xerogel catalyst could be recycled and separated from the reaction mixture of the Suzuki reaction very simply with a magnet.[35]

In addition to N-donor ligands, Dötz and co-workers reported the use of a Pd(II) gel with a pincer ligand to catalyze a double Michael addition

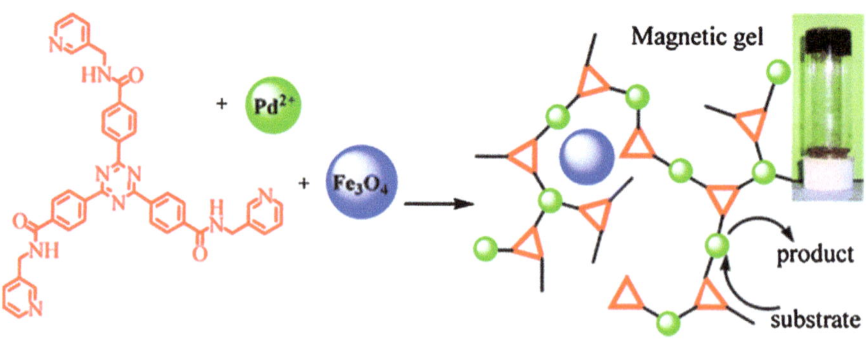

Figure 6.16 Schematic diagram showing the formation of a magnetic metallogel. Adapted with permission from ref. 35. Copyright 2011 American Chemical Society.

(Figure 6.17).[36] Uozumi and coworkers reported the use of a C_3-symmetric triphosphine (Figure 6.18) to form 3D coordination metal–network complexes.[37] The flexible hydrophobic alkyl chains were suggested to be crucial to form a pliable network such that organic substrates could be captured to promote catalytic reactions. When $PdCl_2(CH_3CN)_2$ was allowed to react with the triphosphine ligand in toluene, a yellow gel was formed. The Pd gel was found to catalyze the Suzuki–Miyaura cross-coupling reaction between aryl halides and arylboronic acids in water under atmospheric conditions to give nearly quantitative yields.[37] The catalyst could be used at least four times and could be recycled by just filtering with boiling water to give the insoluble Pd complex as the residue. The Suzuki–Miyaura cross-coupling product could be readily crystallized from the aqueous solution upon cooling to room temperature.[37] The minimal use of organic solvent in the course of the reaction and in the purification process, as well as the good yield obtained under normal atmospheric conditions, are attractive in terms of their implications for green chemistry.

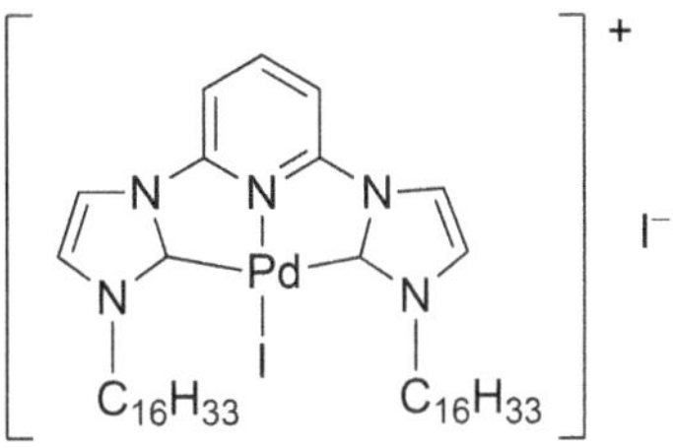

Figure 6.17 A Pd(II) metallogelator catalyst for a double Michael addition. From ref. 36.

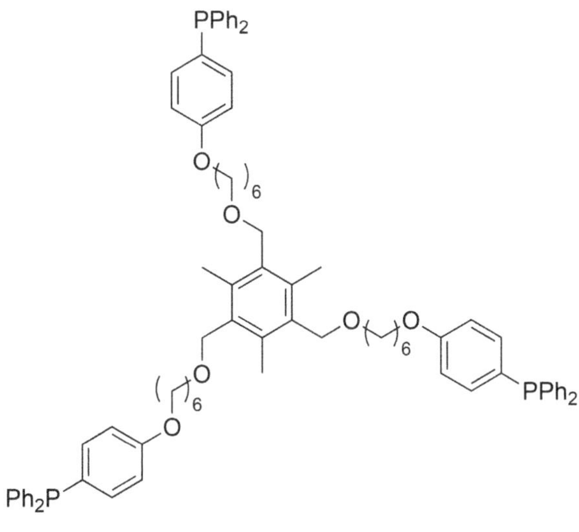

Figure 6.18 C_3-Symmetric triphosphine ligand for a palladium(II) gel. From ref. 37.

While mechanical disturbances were usually reported to disrupt the gel structure, Naota and co-workers reported an interesting sonication-induced gel formation by making use of a palladium(II) complex.[38] It has been found that only a dinuclear palladium(II) complex bridged by alkyl chains with five carbons and with an anti-configuration (Figure 6.19) formed a gel. Upon a short (3–10 s) exposure to ultrasound, the palladium(II) complex was found to undergo gelation in various organic solvents such as acetone, CCl_4, 1,4-dioxane and ethyl acetate. It was found that gelation could only be achieved by sonication, and not by other mechanical or thermal means such as vigorous shaking or microwave irradiation, and heating the gel was found to induce the gel-to-sol phase transition to give a solution. The optically pure (−)-anti complex did not show any gelation. By investigating the optical activities of the partial gel and the remaining solution formed by the 42% ee of the anti-palladium complex in benzene, it was found that the enantiomeric excess in the gel was almost 0%, indicating alternating stacking of the (+) and (−) form has occurred in the gelation process. It was suggested that sonication would cause a self-lock-interlock conversion, and the complex would undergo conformational changes from its clothespin-like, bent structure to the heterochiral interpenetrative dimer which would initiate the assembly process to give supramolecular polymers and lead to the formation of the gel *via* heterochiral chain association.[38]

Later, a similar ultrasound-induced gelation of a palladium(II)-porphyrin complex was reported by Jung and co-workers.[39] It is well known that porphyrins can align into one-dimensional supramolecular structures through H- or J- type aggregation and the sol–gel transition observed was found to be closely related to interconversion between the two aggregate forms. Upon sonication of the porphyrin with one equivalent of palladium(II) ions, a green gel was formed. On heating the green gel, a red solution was obtained and the solution would not return to its gel form upon cooling (Figure 6.20). The gel-to-sol phase transition was monitored by UV-vis absorption studies, in which the green metallogel was found to show a Soret band at 402 nm, characteristic of H-aggregates, while upon heating, the Soret band was red-shifted to 490 nm, indicating a change in aggregation mode to a J-type. It was suggested

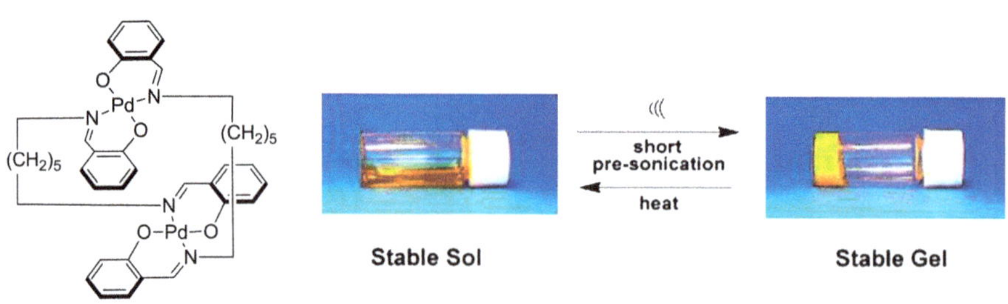

Figure 6.19 Structure of a dinuclear palladium(II) complex for sonication induced gelation and the photographs of the sol and gel form after heating and sonication respectively. Adapted with permission from ref. 38. Copyright 2005 American Chemical Society.

that ultrasound has enhanced the translational motion of the complex molecules and led to the change in aggregation type. In addition, the two different types of aggregation modes have led to two distinct morphologies as revealed by TEM images. The gel formed by ultrasound treatment showed interconnected sheet-like structures while after heat treatment, discrete cubic structures were observed and the loss in gel properties could be rationalized by the decrease in solvent encapsulation property of the cubic structure.

Recently, Huang and co-workers reported a moldable shape-persistent Pd(II) polymeric gel which showed good mechanical properties.[40] The monomer consists of a triazole unit, linking a benzo-21-crown-7 ring host and its complementary guest dialkylammonium salt. At a high monomer concentration, the host–guest interactions between the benzo-21-crown-7 ring and the dialkylammonium salt were found to lead to the formation of linear polymer chains. Addition of Pd(II) resulted in coordination of Pd to the triazole unit and the formation of crosslinks (Figure 6.21). The gel was found to be multi-responsive and would undergo a gel-to-sol transition by an increase in the temperature, in pH, the addition of K^+ ions or competitive ligands. The gel was also found to exhibit an excellent elastic property, in which the gel could bounce to a height of 20 cm when dropped from a height of 25 cm, and its good elastic properties were suggested to be due to the dynamically reversible assembly of the benzo-21-crown-7 ring and the dialkylammonium salt.[40]

Very recently, Fernaández and co-workers reported a Pd(II) gel which showed drastic colour changes from colourless to yellow upon sol–gel transition *via* cooling.[41] The colour change was suggested to be attributed to intermolecular Pd···Pd interactions. This report represented a rare example of utilizing Pd···Pd interactions as a driving force for gelation (Figure 6.22).[41]

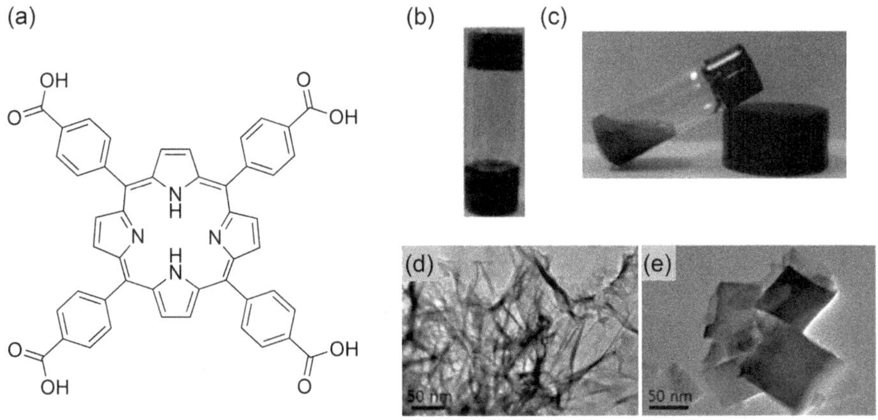

Figure 6.20 (a) Structure of the porphyrin ligand and the photographs of (b) the gel and (c) the solution state of the palladium(II)-porphyrin complex obtained *via* ultrasonication and heating in acetonitrile respectively. TEM images of the sample after (d) sonication and (e) heating. Adapted with permission from ref. 39.

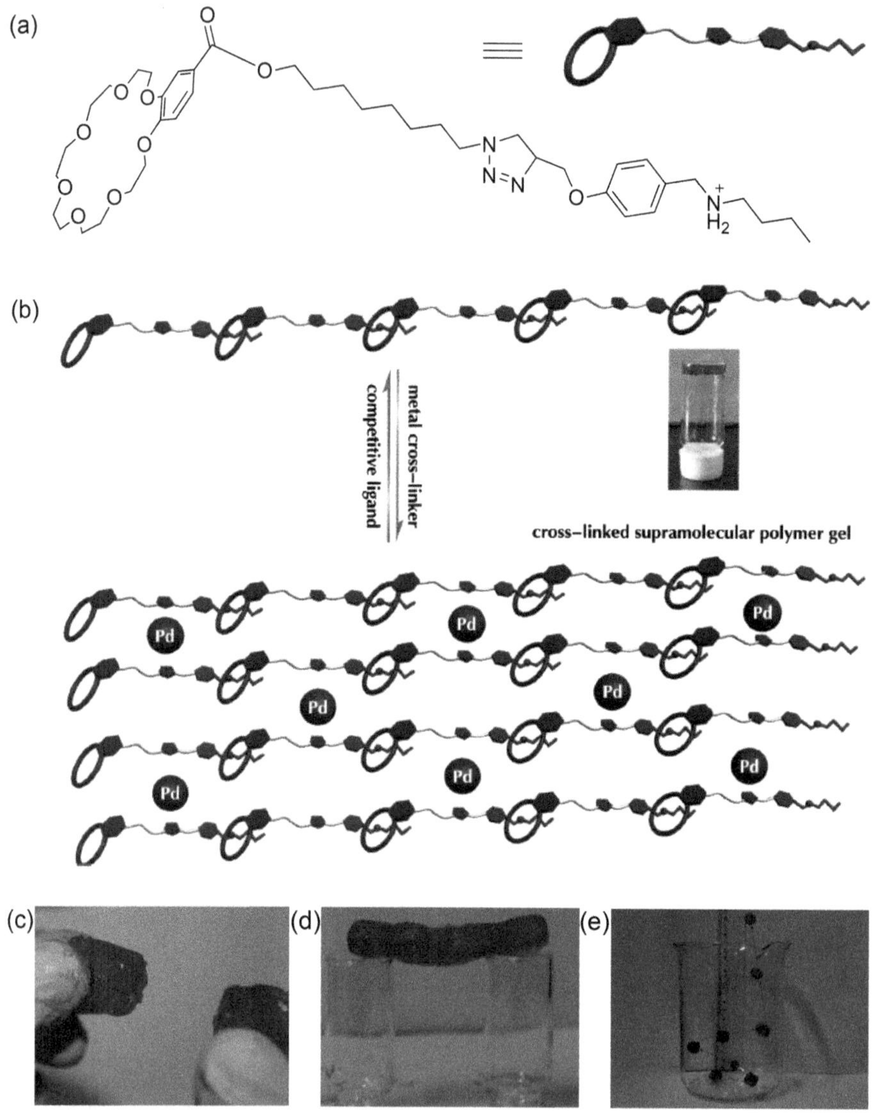

Figure 6.21 (a) The benzo-21-crown-7 ring and the dialkylammonium salt-containing ligand for Pd(II) gel formation. (b) Schematic representation of the self-assembly process to form crosslinks upon addition of palladium. Photographs showing (c) two freshly prepared metallogel blocks, (d) adhesion of four gel blocks to give a free-standing object and (e) the overlaid pictures showing the trajectory of the gel block after bouncing from the bottom of a beaker. Adapted with permission from ref. 40.

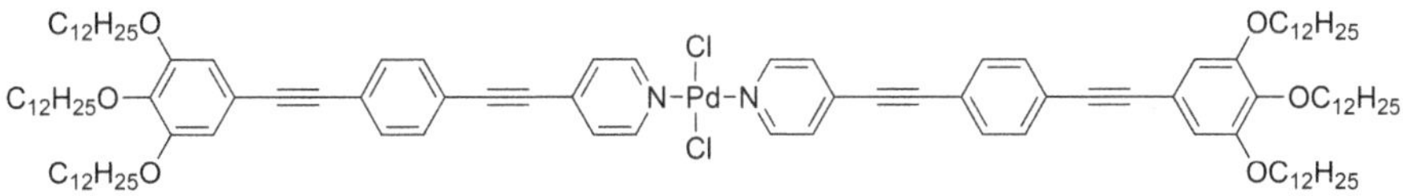

Figure 6.22 A Pd(II) metallogelator reported by Fernaández and co-workers showing Pd···Pd interactions upon gelation. From ref. 41.

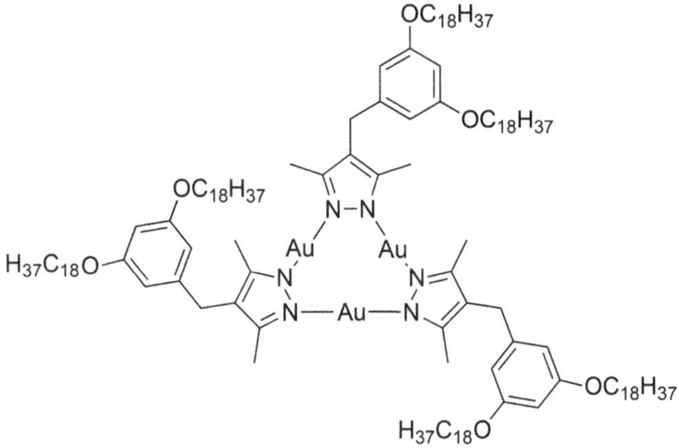

Figure 6.23 A trinuclear Au(I) pyrazolate metallogelator reported by Aida and co-workers showing a RGB emission change upon Ag(I) doping and de-doping. From ref. 42.

6.8 Gold

During research to manipulate metal–metal interactions through gelation, a remarkable example was illustrated by Aida and co-workers in 2005.[42] A trinuclear Au(I) pyrazolate complex with long alkyl chains was found to give drastic emission changes between red, green and blue (RGB) upon reversible sol–gel transitions, Ag(I) doping and de-doping (Figure 6.23).[42] The hexane gel of the Au(I) complex gave red phosphorescence attributed to the presence of Au(I)···Au(I) interactions and on heating, a colourless solution was formed and the emission was found to be further red-shifted to 708 nm, which was hardly detectable by the naked eye. On addition of 0.01 equivalents of AgOTf to the hot solution, a green-emissive solution was observed. On cooling, it gave a blue-emissive gel (Figure 6.24). This blue-shifted emission was suggested to be derived from an emission resulting from heterometallic Au(I)–Ag(I) interactions. As it was known that self-assembly *via* metal–metal interactions would result in a red shift in emission, the unusual blue shift in emission upon sol-to-gel transitions could not be due to enhancement in metal–metal interactions, and was suggested to be a result of distortion of geometry adopted by the complexes in the gel state for fibre growth.[42]

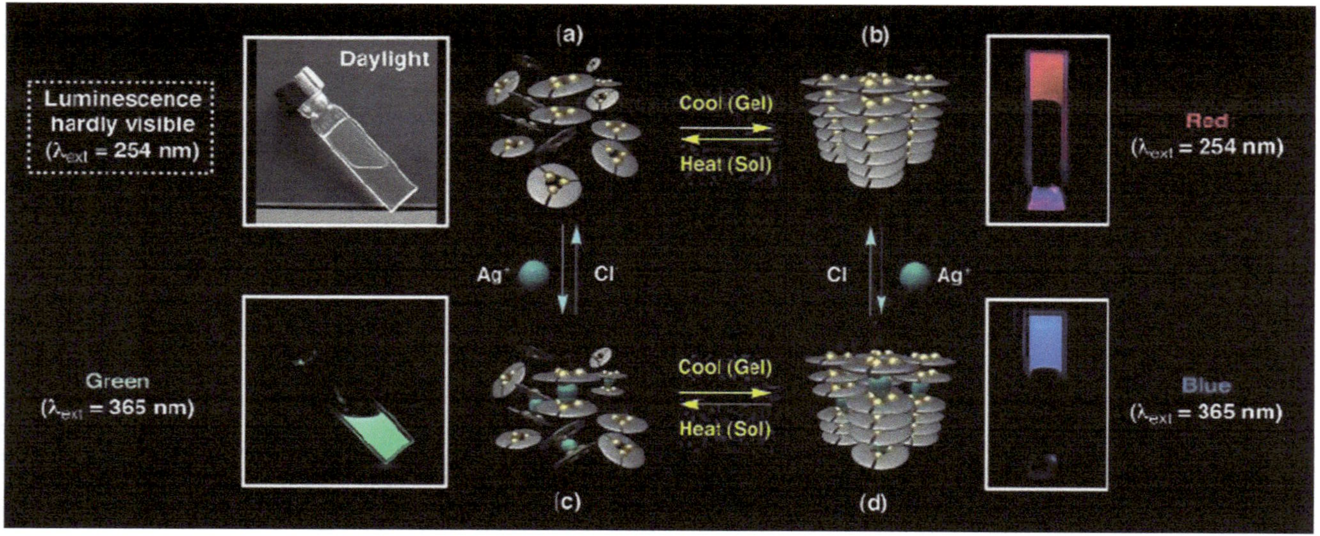

Figure 6.24 Luminescence profiles of the Au(I) organogel in hexane. Pictures and schematic self-assembling structures of (a) sol, (b) gel, (c) sol containing AgOTf (0.01 equiv.), and (d) gel containing AgOTf (0.01 equiv.). Adapted with permission from ref. 42. Copyright 2005 American Chemical Society.

Odriozola and co-workers reported a pH-responsive gold(I) polymeric gel by using a naturally-occurring tripeptide as the ligand.[43] The gel was obtained by reacting a gold(III) salt with the thiol groups of the peptide to give polymeric two-dimensional sheets of gold(I) thiolates (Figure 6.25). The 2D sheet was suggested to be stabilized by Au–Au interactions, commonly known as aurophilic interactions. It was suggested that by careful control of the non-covalent interactions between the side-chains R, the 2D sheets could be used to immobilize water to give hydrogels. It was found that the gold(I) thiolates formed by glutathione (GSH) could form transparent gels, which could immobilize water up to 100 times of the weight of the gelator. The gel has also been shown to be thermally stable with no noticeable decomposition observed even when heated up to 100 °C. In addition, the gel property was only found to be maintained in the pH range of 1.8–2.4; a higher or lower pH would result in either the formation of a suspension or a colorless solution. It was rationalized that the gel property was only maintained when the glutathione gelator was overall neutral near its isoelectric point and any pH change would result in coulombic repulsion and polymer disassembly (Figure 6.26).[44]

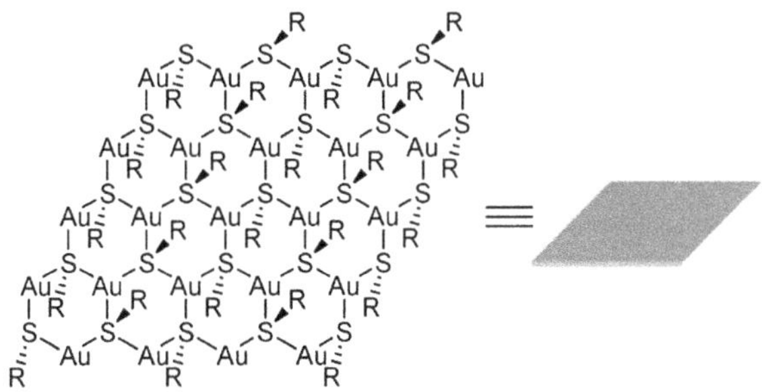

Figure 6.25 Polymeric two-dimensional sheet of gold(I) thiolates. Adapted with permission from ref. 43.

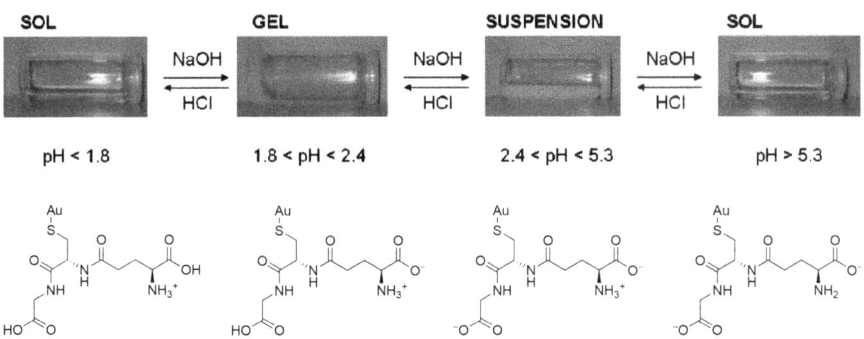

Figure 6.26 Reversible pH-responsive sol–gel transition of gold(I) polymeric gel *via* the protonation and deprotonation of the glutathione ligand. Adapted with permission from ref. 43.

Omary and co-workers have incorporated a water-soluble gold(I) complex (Figure 6.27) into the well-known stimuli-responsive poly(*N*-isopropylacrylamide) (PNIPAM) polymer to give a highly phosphorescent stimuli-responsive hydrogel.[44] It was found that the loading of the gold(I) complex into the hydrogel could be significantly affected by pH and, hence, control of the emission properties of the complex. In acidic and neutral conditions, the amine group in PNIPAM would be protonated and the coulombic attraction between the positively charged polymer and the negatively charged gold(I) complex would cause an efficient loading of the gold(I) complex and a bright green emission originating from the gold(I) complex was observed, while at pH > 9, the loading of the gold(I) complex to the microgel would be inefficient and a suppression in emission intensity was observed. In addition, it was observed that the emission intensity would be enhanced by increasing the temperature to 37 °C and this luminescence enhancement was suggested to be due to the change of microenvironment from hydrophilic to hydrophobic at an elevated temperature above the lower critical solution temperature (LCST) of the PNIPAM (Figure 6.27). The high luminescence intensity of the metallogel at physiological temperature and pH was suggested to be useful for bioimaging and drug delivery.[44]

Lin and co-workers reported a gold(I) N-heterocyclic carbene complex which would undergo assembly to give a metallogel in DMSO (Figure 6.28).[45] Unlike other gels that commonly show a network of fibres, the fibres in this metallogel were found to bundle to give various nanostructures such as helical belts and hollow lantern-shaped structures (Figure 6.29). The involvement of hydrogen bonds, hydrophobic–hydrophobic interactions and coulombic interactions was suggested in the self-assembly.[45]

Apart from gold(I) complexes, gold(III) complexes have also been utilized for the formation of metallogels. Yam and co-workers have demonstrated the use of a cyclometalated C^N^C ligand to form luminescent gold(III) complexes, and the presence of aromatic π–π stacking interactions has indicated

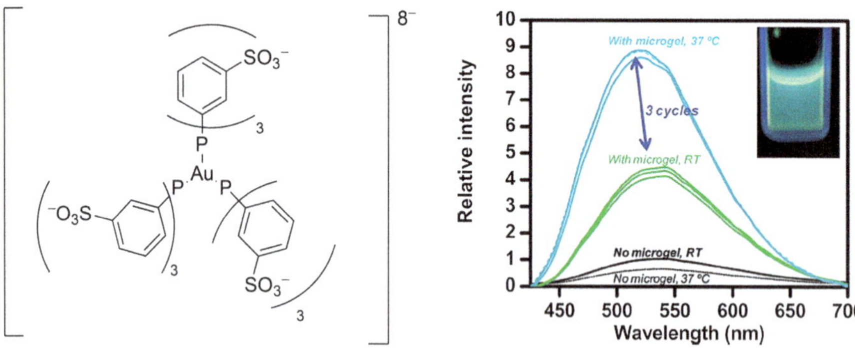

Figure 6.27 Water-soluble gold(I) complex and the temperature-dependent photoluminescence enhancement of the Au(I) gel upon heating from room temperature to 37 °C. Adapted with permission from ref. 44. Copyright 2010 American Chemical Society.

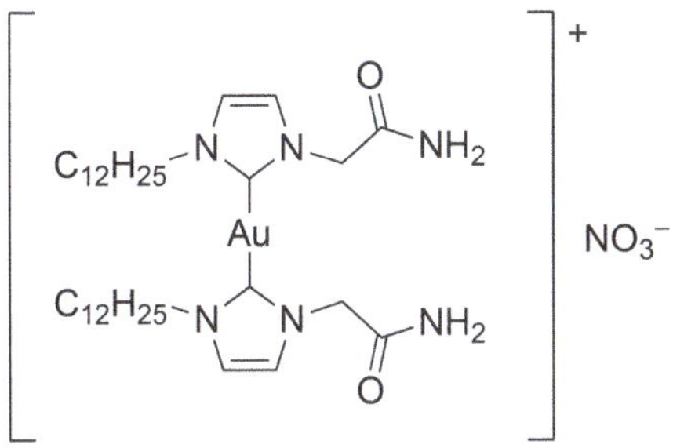

Figure 6.28 Structure of gold(I) N-heterocyclic carbene complex. From ref. 45.

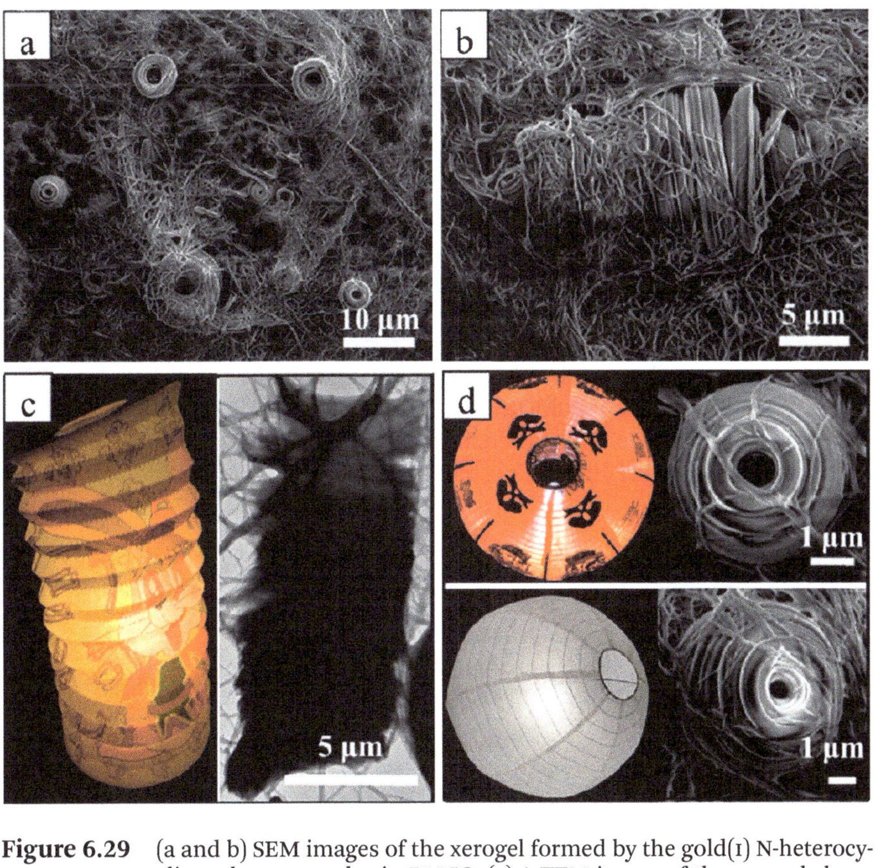

Figure 6.29 (a and b) SEM images of the xerogel formed by the gold(I) N-heterocyclic carbene complex in DMSO. (c) A TEM image of the xerogel showing a lantern-shaped morphology and a picture of a lantern. (d) SEM images of lantern-shaped fibrous bundles of xerogel and a picture of a lantern. Adapted with permission from ref. 45. Copyright 2011 American Chemical Society.

the possibility of utilizing the system for supramolecular assembly.[46,47] Later, using the same C^N^C ligand reported by Yam and co-workers, Che and co-workers reported the gelation of a cyclometalated Au(III) complex with an anti-angiogenic ligand, 4-diamino-6-(4-pyridyl)-1,3,5-triazine (4-dpt) (Figure 6.30).[48] It has been known that the 4-dpt ligand in the Au(III) complex could be substituted by chloride ions in the buffer or by GSH in cancer cells, causing the release of the anti-angiogenic 4-dpt. It was found that through the gelation of the Au(III) complex in acetonitrile *via* supramolecular assembly, the release of 4-dpt in buffer could be significantly slowed down and 4-dpt could be more selectively released in the cancer cells due to the high concentration of GSH, giving a sustained cytotoxicity and selectivity towards cancer cells for cancer treatment. In addition, the gel could also be used as a drug carrier for other cytotoxic agents such as $[Au^{III}(TPP)]Cl$ (H_2TPP = *meso*-tetraphenylporphyrin) to achieve sustained-release behaviour.

At the same time, Yam and co-workers independently investigated the propensity of a series of luminescent bis-cyclometalated alkynylgold(III) complexes to gel in non-polar organic solvents.[49] When compared to the charged complex reported by Che and co-workers, these complexes, with the incorporation of alkynyl ligands, are charge-neutral and possess a better solubility in non-polar solvents. Among all complexes studied, it was found that complexes that possess three $C_{18}H_{37}$ chains (Figure 6.31) showed good gelation properties in non-polar solvents such as *n*-hexane and cyclohexane, while other complexes precipitated. Variable-temperature UV-vis absorption studies showed a clear isosbestic point upon the gel-to-sol phase transition upon heating the gel, suggesting a clean conversion from supramolecular assemblies to monomeric species.[49] The lack of gelation behaviour in the other complexes was attributed to the imbalance of intermolecular interactions. Yam and co-workers have further extended the work of gold(III) metallogels to a new class of gold(III) complexes with a dianionic 2,6-bis(benzimidazol-2′-yl)pyridine (bzimpy) ligand.[50] The complex with three long alkyl chains (Figure 6.32) was found to form gels

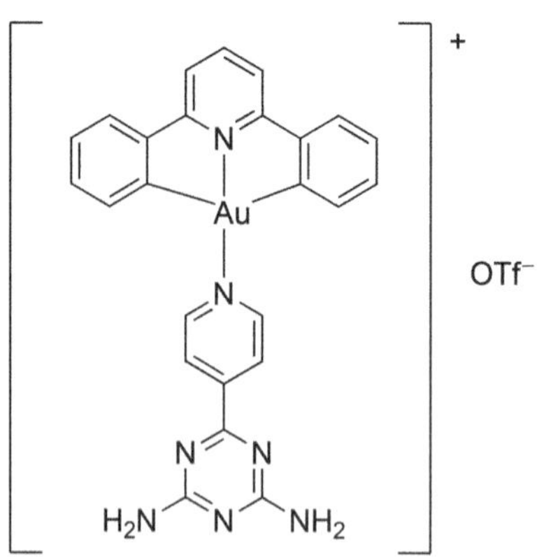

Figure 6.30 A cyclometalated Au(III) metallogelator for anticancer treatment. From ref. 48.

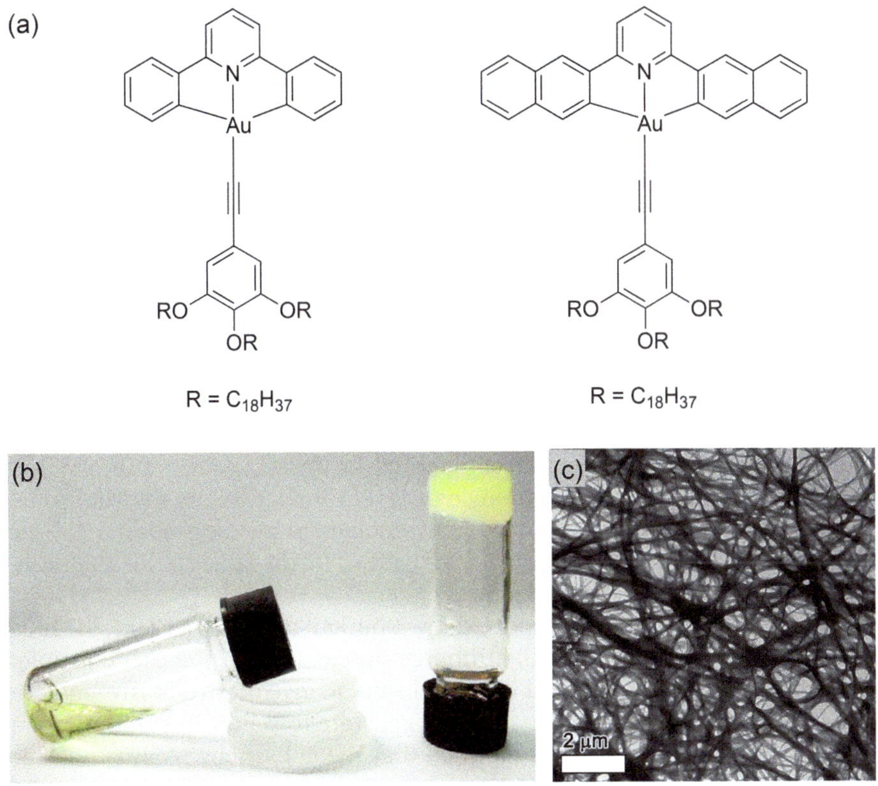

Figure 6.31 (a) Structures of luminescent bis-cyclometalated alkynylgold(III) complexes that form gel in non-polar solvents. (b) Photograph of the cyclohexane gel and the corresponding sol forms at elevated temperature. (c) TEM image of the xerogel prepared from the cyclohexane gel. From ref. 49.

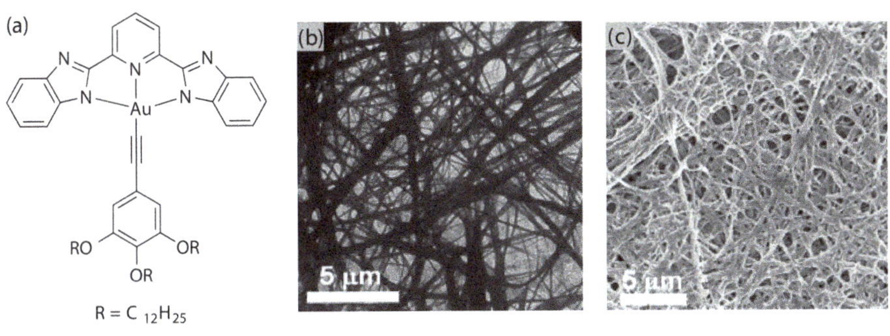

Figure 6.32 (a) Structure of gold(III) bzimpy complex for gelation in chloroform at 0 °C. (b) TEM and (c) SEM images of the xerogel. From ref. 50.

when cooled to about 0 °C in chloroform. It was observed from concentration- and temperature-dependent ^{1}H NMR studies that both the aromatic and the alkyl proton signals showed upfield shifts upon an increase in concentration or a decrease in temperature, and thus, π–π stacking and hydrophobic–hydrophobic interactions were suggested to be involved in the supramolecular assembly of the complexes. The broadening of proton NMR signals at low temperatures was also in line with the possible increase in π–π stacking interactions upon gelation.

6.9 Platinum

Square planar platinum(II) complexes have been extensively studied in the past two decades owing to their interesting spectroscopic and luminescence properties.[51–54] It has been shown that the change in the extent of Pt···Pt interactions has always been accompanied by drastic colour and/or luminescence changes. Thus, by employing Pt···Pt interactions as the additional driving force for gelation, the change in molecular packing upon a sol–gel transition could possibly be probed, providing further insight for the rational design of gelators.

The first Pt(II) gel was reported by Shinkai and co-workers in 2005 (Figure 6.33).[55] It was shown that dioxygen quenching of the excited triplet state could be largely inhibited in the gel state to give more emissive materials. A colour change from orange in the gel state to yellow in solution upon heating was observed. It was suggested that the colour change was not due to the inference of Pt···Pt interactions, but rather a result of the formation of J-aggregates upon gelation. This assignment was further supported by work from the same group, in which the gels of Cu(II) and Pd(II) of the same ligand showed similar red shifts in absorption bands in the gel state as compared to the solution state.[56]

In 2007, Yam[57] and Ziessel[58] independently reported the utilization of alkynylplatinum(II) terpyridine complexes as metallogelators and have successfully demonstrated the contribution from Pt···Pt interactions as additional driving forces for gelation. Ziessel and co-workers reported the use of a diacylamido framework to link six dodecyl (C_{12}) chains to tune the solubility

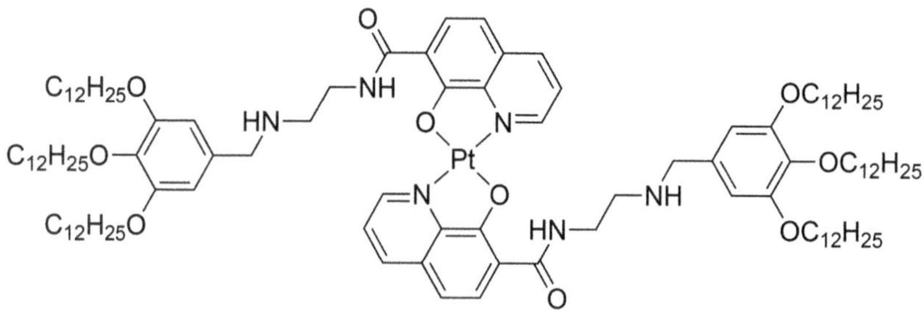

Figure 6.33 A Pt(II) metallogelator reported by Shinkai and co-workers showing J-aggregation upon gelation. From ref. 56.

and enhance the van der Waals' interactions (Figure 6.34). The Pt(II) complex was found to form gel in dodecane to give a deep green gel which showed a thermochromic change from a deep green gel to a deep red solution upon a sol–gel transition. By AFM, powder X-ray diffraction (PXRD) and molecular dynamic studies, it was suggested that extended columnar structure formation was facilitated by Pt···Pt and Pt–alkynyl interactions.[58]

Yam and co-workers reported the gelation of Pt(II) terpyridine complexes with 3,4,5-tris(dodecyloxy)phenylacetylene as the alkynyl ligand.[57] It was found that in the presence of unsubstituted terpyridine, the gelating ability was enhanced. For instance, the complex was found to form a more stable gel than the ligand and was found to have a lower CGC and a higher T_{gel} as compared to its *tert*-butyl substituted terpyridine (^{t}Bu$_3$tpy) analogue. The purple DMSO gel of the unsubstituted Pt(II) terpyridine complex gave metal–metal-to-ligand charge transfer (MMLCT) absorption and emission bands attributed to Pt···Pt and π–π stacking interactions. Upon increasing the temperature, the purple gel collapsed to give an orange solution. The drastic colour change indicated the involvement of Pt···Pt and π–π stacking interactions as an additional driving force for gelation (Figure 6.35).[57] For the *tert*-butyl substituted analogue, a yellow gel was formed in DMSO, indicating the absence of Pt···Pt and π–π stacking interactions in the presence of bulky substituents and it was suggested that the gel formation was supported by hydrophobic–hydrophobic interactions exerted by the *tert*-butyl groups. The influence of a counter-anion on gelation was also studied and it has been shown that by changing the counter-anion of the unsubstituted terpyridine complex from OTf^- to BF_4^-, ClO_4^- and PF_6^-, a colour change and a blue shift in the MMLCT absorptions and emissions were observed, which indicated a decrease in Pt···Pt interactions. This decrease in the extent of Pt···Pt interactions was found to influence the gel stability as correlated by

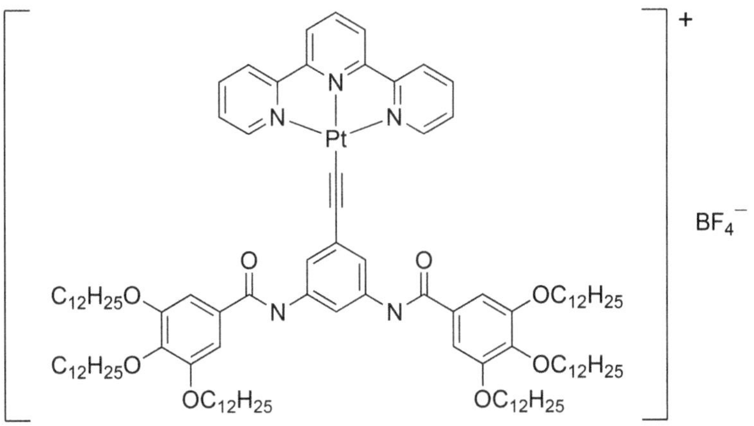

Figure 6.34 Structure of an alkynylplatinum(II) terpyridine complex using a diacylamido framework to link six dodecyl (C_{12}) chains for gelation. From ref. 58.

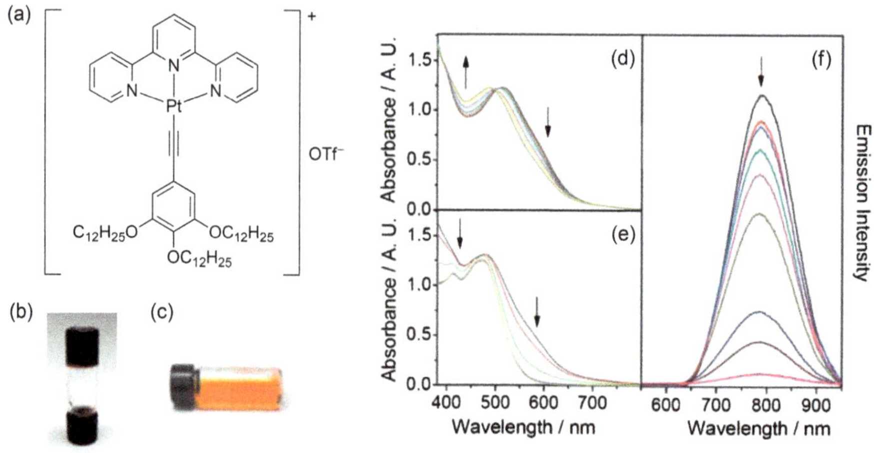

Figure 6.35 (a) Structure of the Pt(II) complex with unsubstituted terpyridine for gelation. Photographs of (b) its DMSO gel at room temperature and (c) sol form at elevated temperature. UV-vis absorption spectra on increasing the temperature (d) from 10 to 41 °C and (e) from 44 to 80 °C. (f) Emission spectra of the DMSO gel on increasing the temperature from 20 to 56 °C. Adapted with permission from ref. 57.

the increase of CGC and decrease of T_{gel}.[59] This represents the first direct piece of evidence on the correlation of the extent of Pt···Pt interactions with the stability of the metallogel, indicating the importance of metal–metal interactions in the stabilization of the supramolecular assembly. Later, Yam and co-workers extended the work to study the influence of counterions for the stabilization of helical structures in a Pt(II) terpyridine alkynyl gel.[60] It has been shown that only in the presence of OTf⁻ ions, the helical superstructure formed by the Pt···Pt and π–π stacking interactions was stabilized as revealed by the large circular dichroism (CD) signals and ordered helical structure as observed from electron microscopy (Figure 6.36).[60]

Che and co-workers reported the gelation of alkynylplatinum(II) terpyridine complexes without the aid of long alkyl chains. A dinuclear (^{t}Bu$_3$tpy) Pt(II) complex bridged by an oxadiazole[61] and a mononuclear (^{t}Bu$_3$tpy)Pt(II) biphenylacetylide complex[62] were reported to give an orange gel in acetonitrile and in acetonitrile–alcohol mixtures. Due to the presence of the bulky *tert*-butyl groups, Pt···Pt interactions were not involved in gelation and π–π stacking interactions were suggested to be the major contributor for gelation. The unsubstituted terpyridine analogue of the biphenylacetylide complex was found to possess limited solubility, with the exception of its Cl⁻ salt, which has been shown to display gelation with water after standing for more than 12 hours to give a brown hydrogel as assisted by Pt···Pt interactions (Figure 6.37).[62]

In 2008, Schanze and co-workers reported a blue shift in absorption and emission attributed to the formation of H-aggregates in the gel state in a

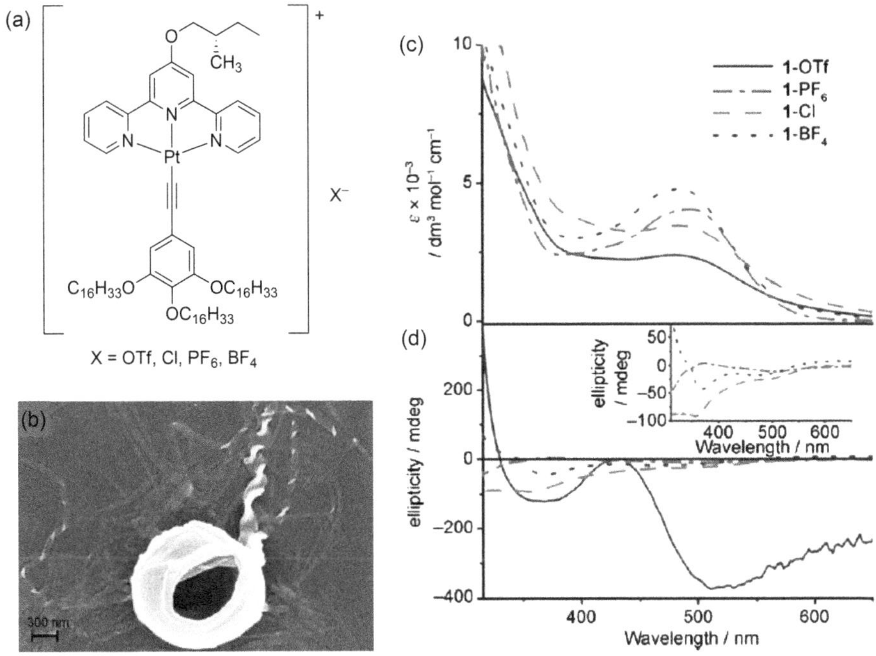

Figure 6.36 (a) Structure of the chiral Pt(II) metallogelator. (b) SEM image of the OTf gel showing a coil-shaped superstructure. (c) UV-vis absorption spectra and (d) CD spectra of the DMSO gel with different counterions. Adapted with permission from ref. 60.

series of phosphine Pt(II) alkynyl oligomers (Figure 6.38).[63] No Pt···Pt interactions were found to be involved in the gelation and it was suggested that the formation of H-aggregates might be due to the presence of the relatively bulky PMe_3 on the Pt(II) centre which prevented the oligomers from packing in an all-planar conformation for J-aggregates to form.[63] Interestingly, through a series of phosphorescence spectroscopic studies, triplet exciton diffusion and trapping have been observed, representing the first examples of such processes in the molecular aggregate system. It was found that mixed aggregates and gels consisting of complexes with 1,4-phenylene as a triplet donor (host) and 2,5-thienylene as an acceptor (trap) were able to show triplet energy transfer. The triplet energy transfer was suggested to involve exciton diffusion among host chromophores followed by a Dexter exchange energy transfer to the trap chromophore. The energy transfer was found to be observable in the aggregate state and the process has been shown to be efficient even with a low acceptor concentration (1–2 mol%).[63]

As seen from these examples, the rigid medium of the gel was found to enhance the luminescence by inhibiting dioxygen quenching of the excited triplet state[55] or by slowing down the non-radiative decay due to the restriction of motion in the gel state. In contrast to this phenomenon, Yam and co-workers reported the gelation of a series of alkynylplatinum(II) bzimpy

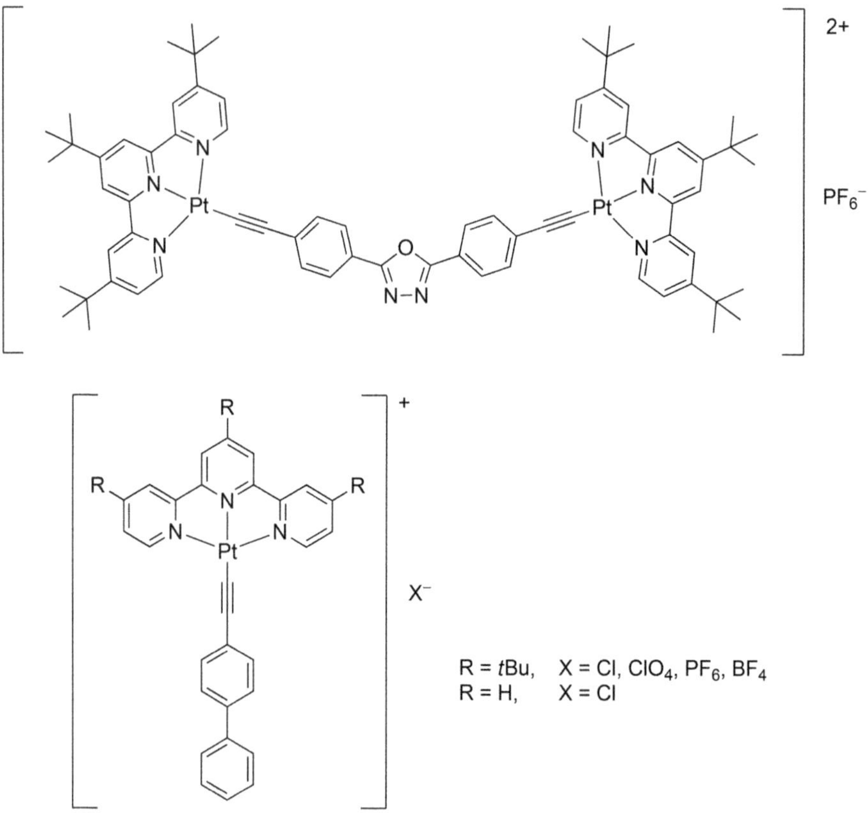

Figure 6.37 Pt(II) metallogelators reported by Che and co-workers. From ref. 61 and 62.

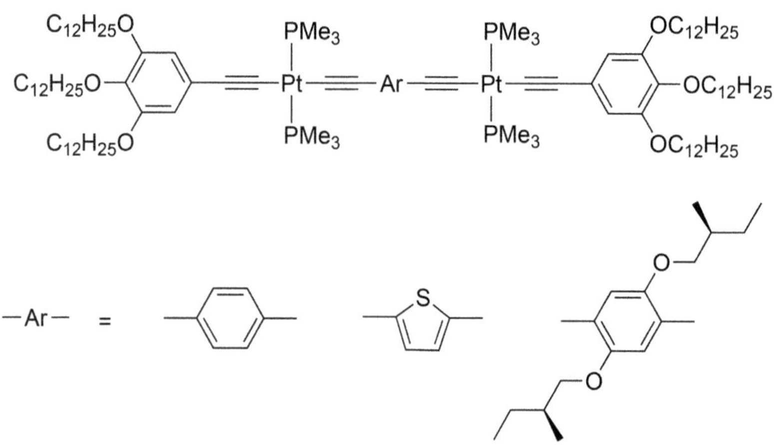

Figure 6.38 Pt(II) metallogelators showing H-aggregation upon gelation. From ref. 63.

complexes (Figure 6.39), which had shown unusual luminescence enhancement upon gel-to-sol transition.[64] Yellow gels were formed at room temperature in benzene and were found to give yellow intraligand (IL) emission typical of monomeric species, indicating that Pt···Pt and π–π stacking interactions were minimal in the gel state. On heating the gel, the complex without the incorporation of a bulky methyl group melted to give a red solution with enhancement of red emission, which was assigned to an excimeric emission due to enhanced Pt···Pt and π–π stacking interactions in the excited state. The unusual behavior was suggested to be due to the misalignment of molecules under the restricted geometry in the gel state; and on increasing the temperature, the degrees of freedom of motion would increase and result in a higher chance for the molecules to assemble through Pt···Pt and π–π stacking interactions. By the incorporation of a sterically bulky methyl group at the *ortho* position of the phenylalkynyl ligand, Pt···Pt and π–π stacking interactions were minimized and no luminescence enhancement was observed.[64]

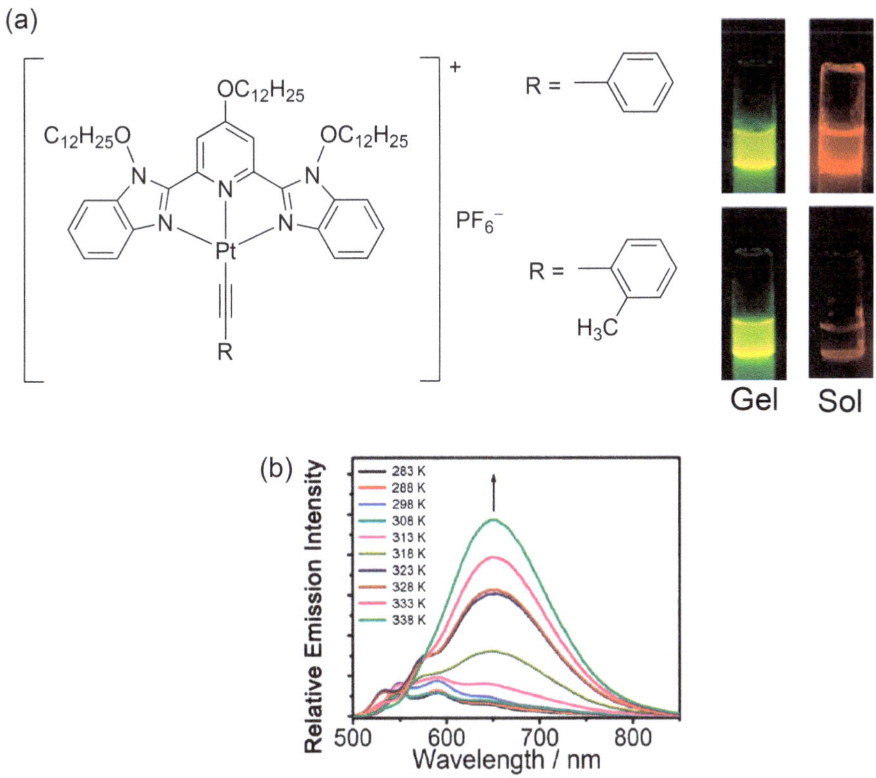

Figure 6.39 (a) Structures of a series of Pt(II) gels and corresponding photographs of the sol and gel forms under UV illumination. (b) Emission spectra showing unusual luminescence enhancement of the gel on increasing temperature. Adapted with permission from ref. 64. Copyright 2009 American Chemical Society.

Later, a similar luminescence enhancement at an elevated temperature upon gel-to-sol transition was also observed by Sun and co-workers in a bis-alkynylplatinum(II) bipyridine complex and the observation was explained similarly by excimeric emission due to relaxation in molecular geometry at elevated temperatures.[65]

In addition to the use of long alkyl chains to facilitate gelation, Yam and co-workers have also incorporated cholesteryl moieties in Pt(II) complexes for gelation. They reported a spironaphthoxazine-containing bis-alkynylplatinum(II) bipyridine complex with the incorporation of cholesteryl moieties on the alkynyl ligands which formed a highly stable gel in dodecane (Figure 6.40).[66] The gel was found to be thermally stable even at 117 °C, near its decomposition temperature and resistant to deformation upon 10 min of sonication. In addition, the incorporation of photoactive spironaphthoxazine on the bipyridine has imparted the gel with a photochromic property and on prolonged light irradiation, the gel turned from yellow to blue due to the ring-opening reaction of spironaphthoxazine to give its merocyanine form with an increase in the extent of π-conjugation.[66]

Yi and co-workers reported an example of ultrasound-assisted gelation of cholesterol-based alkynylplatinum(II) terpyridine complexes.[67] It was found that gelation could only be achieved *via* sonication and precipitation occurred upon thermal treatment. In the (^{t}Bu$_3$tpy)Pt(II) analogue observed by electron microscopy, long fibres were formed upon sonication which contributed to the gelation, whereas formation of vesicles by heat treatment gave precipitation. The unsubstituted terpyridine was found to be difficult to gel and only gave a weak gel after prolonged sonication. It was suggested that the poor gelation properties were a consequence of its poor solubility.[67]

Recently, Yam and co-workers reported another series of cholesterol-appended alkynylplatinum(II) complexes with 2,6-bis(1-alkylpyrazol-3-yl) pyridine as the tridentate N-donor ligand.[68] Better gelating properties, such

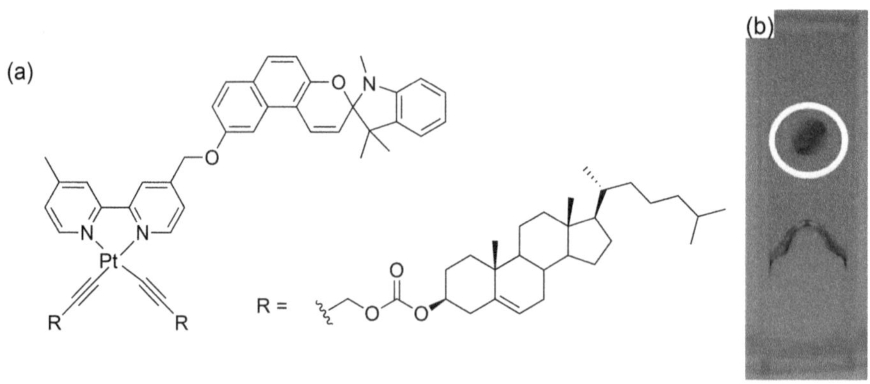

Figure 6.40 (a) Structure of a spironaphthoxazine-containing bis-alkynylplatinum(II) bipyridine complex. (b) Photograph of its dodecane gel showing a colour change from yellow to blue (indicated by circle). Adapted with permission from ref. 66.

as shorter gelation time and more stable gels, were achieved by complexes with a longer alkyl chain. In addition to being thermoresponsive, the gel was found to be mechanoresponsive, in which shaking the gel in a vibrator would lead to disruption of the gel while standing for about half an hour would lead to its re-formation (Figure 6.41). This represents a self-healing process of the gel through supramolecular assembly.

De Cola and co-workers reported the coordination of 2,6-bis(tetrazolyl) pyridine to Pt(II) to give emissive gels with a high luminescence quantum yield of up to 90% and the system was suggested to be useful for fabricating solution-processable OLED devices (Figure 6.42).[69] Due to the fact that the complexes were non-emissive in dilute solutions at room temperature and yellow emission could only be observed at a concentration larger than 10^{-4} M, the origin of the yellow emission was suggested to originate from a MMLCT state.[69,70]

Recently, Yam and co-workers have also investigated the influence of hydrogen bonding on gelation through comparing the gelation properties of a series of platinum(II) terpyridine complexes which have incorporated valine-modified alkynyls with an increasing number of hydrogen bonding sites.[71] It was found that only the complex with an unsubstituted terpyridine and one valine unit on the alkynyl could form a gel in acetonitrile. A drastic colour change from red to yellow was observed when heating the gel to give a sol. The red shift in absorption in the gel-form as compared to the sol-form together with the complete suppression of gelation upon the incorporation of bulky *tert*-butyl groups on the terpyridine ring, strongly suggest

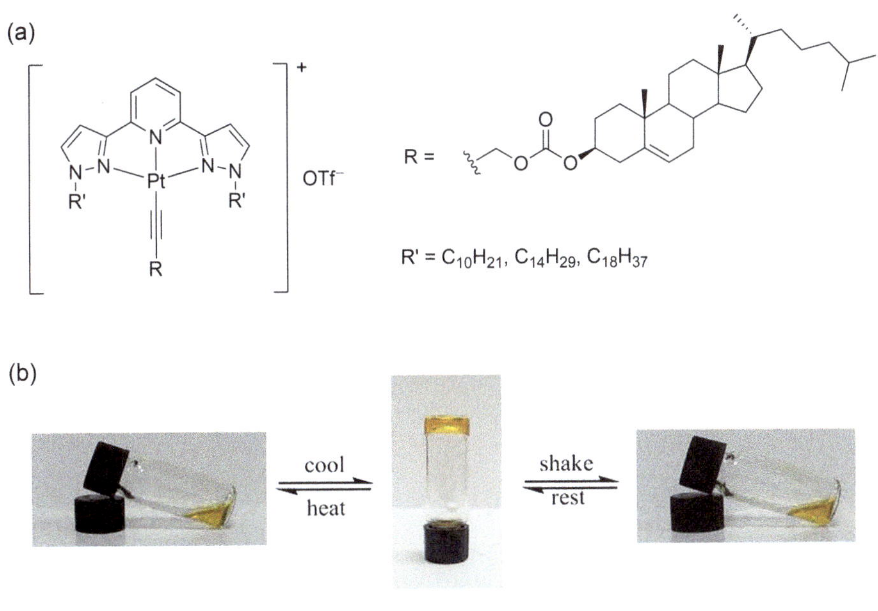

Figure 6.41 (a) Structures of a Pt(II) gel showing thermo- and mechanoresponsive properties. (b) Black and white photographs showing the reversible formation of the gel after heating and shaking. Adapted with permission from ref. 68.

the involvement of Pt···Pt interactions as a driving force for gelation (Figure 6.43). Moreover, the gelation properties of the complex were found to be in contrast to the corresponding alkynyl ligands. For instance, the alkynes were shown to possess an increasing tendency for gelation with an increasing number of valine units; however, incorporation of alkynyls with an increasing number of valine units to the platinum(II) complexes led to precipitation instead of gelation. This has been suggested to be due to the presence of the

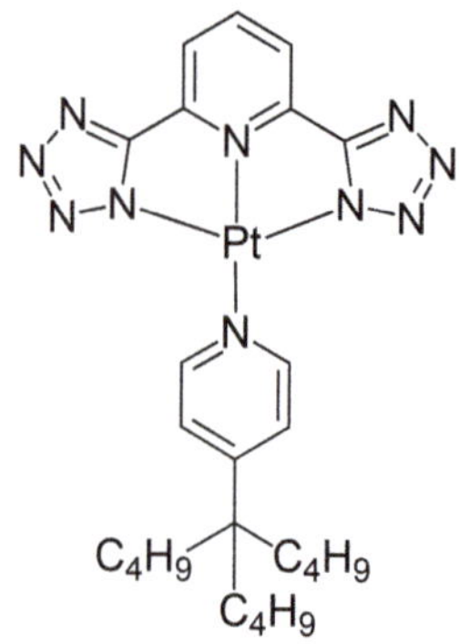

Figure 6.42 A Pt(II) metallogelator reported by De Cola and co-workers showing a high luminescence quantum yield in the gel state. From ref. 69.

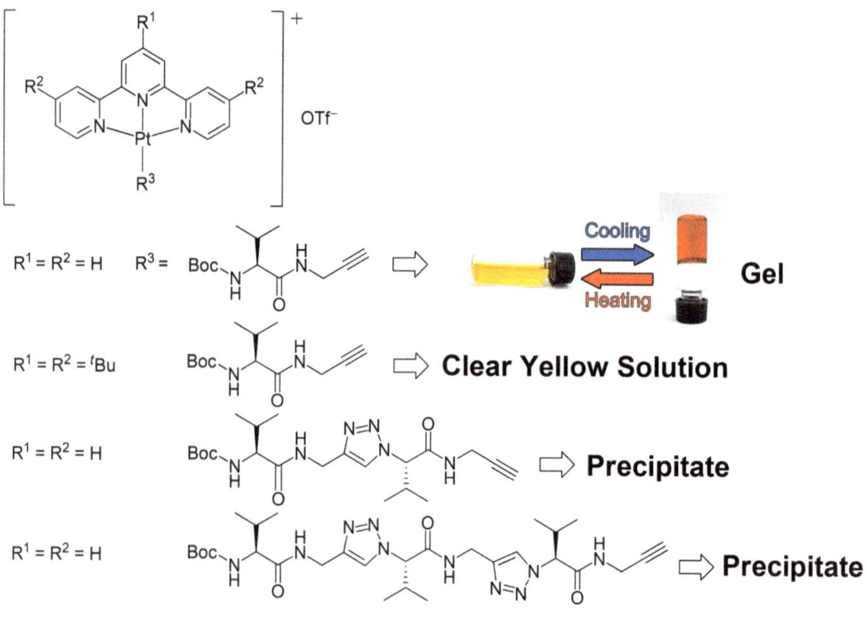

Figure 6.43 Structures of alkynylplatinum(II) complexes with ligands of different numbers of valine units and photographs showing the drastic colour change from yellow to red upon a sol–gel transition. Adapted with permission from ref. 71.

strong Pt···Pt interactions, such that the incorporation of alkynyls with more valine units which would give rise to a more extensive formation of hydrogen bonding would cause an overly strong interaction, and in the absence of other solubilizing groups, precipitation would occur instead of gelation. This work has highlighted the importance of maintaining the delicate balance between different intermolecular forces in the design of gelators.

Very recently, Che and co-workers utilized oligo(oxyethylene)-bridged dinuclear cyclometalated platinum(II) complexes as the cross-linkers for the formation of hydrogels (Figure 6.44).[72] Mononuclear cyclometalated

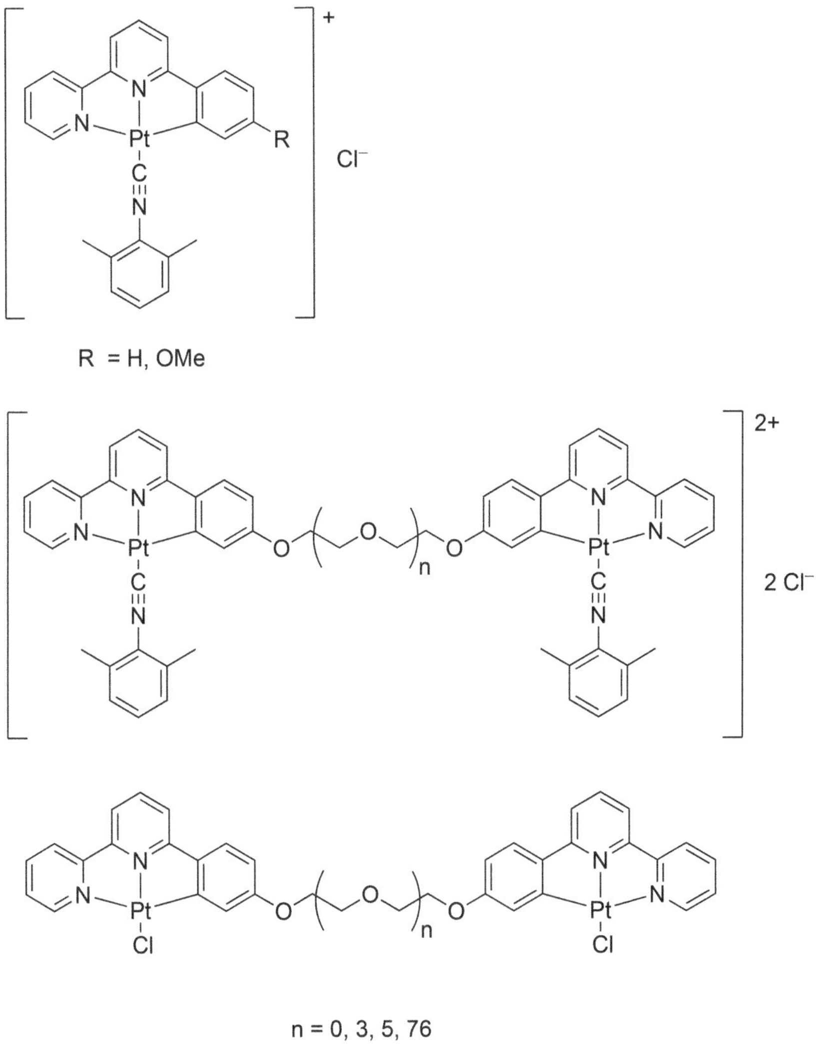

Figure 6.44 Structures of mononuclear and dinuclear Pt(II) complexes investigated for gelation *via* co-assembly. Reprinted with permission from ref. 72.

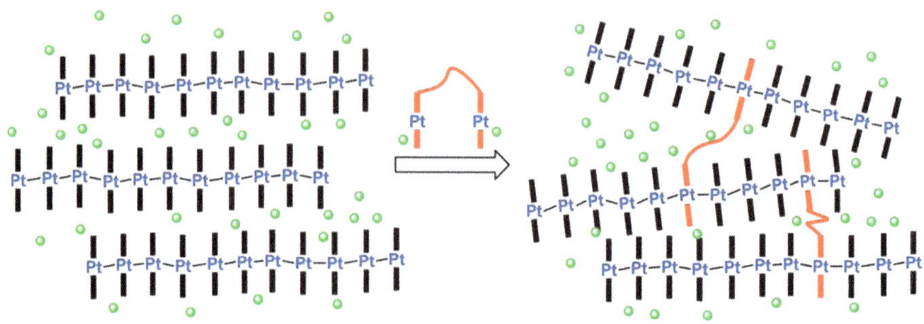

Figure 6.45 Schematic diagram showing formation of a 3D network *via* the co-assembly of dinuclear platinum(II) complexes. Adapted with permission from ref. 72.

platinum(II) complexes were known to undergo supramolecular assembly to give 1D chains *via* Pt···Pt and π–π stacking interactions and these 1D chains alone did not lead to the formation of a gel; but when a small amount of dinuclear complexes was added, spontaneous gelation was induced. It was suggested that the dinuclear Pt(II) complexes could undergo co-assembly with the mononuclear complexes and could interconnect the discrete 1D chains to give 3D networks which led to gelation (Figure 6.45). It was noted that if the oligo(oxyethylene) chain was too short (n = 3, 5), intramolecular Pt···Pt and π–π stacking interactions would dominate and no gelation would occur. From spectroscopy and emission studies, it was found that if cationic dinuclear complexes were added as cross-linkers, the absorption and emission bands remained more or less similar to that of the 1D chain formed by the monomer. If neutral dinuclear complexes were added, a blue hydrogel was obtained, with a significant red shift observed in absorption and the emission band was found to shift to the near-IR region. It was rationalized that through co-assembly with neutral Pt(II) complexes, a decrease in electrostatic repulsion between two cationic Pt(II) centres would result, leading to an enhancement in intermolecular interactions.[72]

6.10 Conclusion

With the involvement of transition metal centres in the gel structures, metallogels have demonstrated unique and interesting properties, which could hardly be achieved by organic compounds alone. In the past two decades, through continuous investigations by different research groups over the world, highly phosphorescent, catalytic, magnetic and multi-stimuli-responsive metallogels have been reported. In addition, with the incorporation of d^{10} and d^8 metal complexes, as exemplified by Au(I), Au(III) and Pt(II), metal–metal interactions have also been utilized to tune the gel properties. The metal–metal interactions do not only act as additional driving forces to stabilize the gels, but may also help to report the microenvironment and the packing of the gelator molecules. This would be a result of the drastic colour and emission changes due to the formation or disruption of metal–metal interactions during sol–gel transitions upon the introduction of external stimuli, giving valuable information for the more rational design of gelators. With the fascinating results reported and the increase in the understanding of the interplay of different non-covalent interactions and supramolecular assembly, there is no doubt that the research in metallogels will continue to grow and more interesting functional materials can be anticipated.

Acknowledgements

This work was supported by the University Grants Committee Areas of Excellence Scheme (AoE/P-03/08), the General Research Fund (GRF) Grant (HKU 7060/12P) from the Research Grants Council of Hong Kong Special Administrative Region, China.

References

1. *Molecular Gels. Materials with Self-Assembled Fibrillar Networks*, ed. R. G. Weiss and P. Terech, Springer, Dordrecht, The Netherlands, 2005.
2. P. Terech and R. G. Weiss, *Chem. Rev.*, 1997, **97**, 3133.
3. N. M. Sangeetha and U. Maitra, *Chem. Soc. Rev.*, 2005, **34**, 821.
4. L. A. Estroff and A. D. Hamilton, *Chem. Rev.*, 2004, **104**, 1201.
5. A. Y. Y. Yam and V. W. W. Yam, *Chem. Soc. Rev.*, 2013, **42**, 1540.
6. M. O. M. Piepenbrock, G. O. Lloyd, N. Clarke and J. W. Steed, *Chem. Rev.*, 2010, **110**, 1960.
7. J. Zhang and C. Y. Su, *Coord. Chem. Rev.*, 2013, **257**, 1373.
8. S. Kawano, N. Fujita, K. J. C. van Bommel and S. Shinkai, *Chem. Lett.*, 2003, **32**, 12.
9. H. J. Kim, W. C. Zin and M. Lee, *J. Am. Chem. Soc.*, 2004, **126**, 7009.
10. H. J. Kim, J. H. Lee and M. Lee, *Angew. Chem., Int. Ed.*, 2005, **44**, 5810.
11. Q. Liu, Y. Wang, W. Li and L. Wu, *Langmuir*, 2007, **23**, 8217.
12. S. Zhang, S. Yang, J. Lan, S. Yang and J. You, *Chem. Commun.*, 2008, 6170.
13. P. Bairi, B. Roy and A. K. Nandi, *J. Mater. Chem.*, 2011, **21**, 11747.
14. Y. He, Z. Bian, C. Kang and L. Gao, *Chem. Commun.*, 2010, **46**, 5695.
15. Y. He, Z. Bian, C. Kang and L. Gao, *Chem. Commun.*, 2011, **47**, 1589.
16. K. Hanabusa, Y. Maesaka, M. Suzuki, M. Kimura and H. Shirai, *Chem. Lett.*, 2000, 1168.
17. S. A. Joshi and N. D. Kulkarni, *Chem. Commun.*, 2009, 2341–2343.
18. A. Westcott, C. J. Sumby, R. D. Walshaw and M. J. Hardie, *New J. Chem.*, 2009, **33**, 902.
19. S. I. Kawano, N. Fujita and S. Shinkai, *J. Am. Chem. Soc.*, 2004, **126**, 8592.
20. Y. He, Z. Bian, C. Kang, Y. Cheng and L. Gao, *Chem. Commun.*, 2010, **46**, 3532.

21. Q. Jin, L. Zhang, H. Cao, T. Wang, X. Zhu, J. Jiang and M. Liu, *Langmuir*, 2011, **27**, 13847.
22. X. de Hatten, N. Bell, N. Yufa, G. Christmann and J. R. Nitschke, *J. Am. Chem. Soc.*, 2011, **133**, 3158.
23. M. George, G. P. Funkhouser, P. Terech and R. G. Weiss, *Langmuir*, 2006, **22**, 7885.
24. F. Peng, G. Li, X. Liu, S. Wu and Z. Tong, *J. Am. Chem. Soc.*, 2008, **130**, 16166.
25. O. Roubeau, A. Colin, V. Schmitt and R. Clérac, *Angew. Chem., Int. Ed.*, 2004, **43**, 3283.
26. P. Grondin, O. Roubeau, M. Castro, H. Saadaoui, A. Colin and R. Clérac, *Langmuir*, 2010, **26**, 5184.
27. K. Kuroiwa, T. Shibata, A. Takada, N. Nemoto and N. Kimizuka, *J. Am. Chem. Soc.*, 2004, **126**, 2016.
28. H. Lee, S. H. Jung, W. S. Han, J. H. Moon, S. Kang, J. Y. Lee, J. H. Jang and S. Shinkai, *Chem.–Eur. J.*, 2011, **17**, 2823.
29. S. T. Lam, G. Wang and V. W. W. Yam, *Organometallics*, 2008, **27**, 4545.
30. S. T. Lam and V. W. W. Yam, *Chem.–Eur. J.*, 2010, **16**, 11588.
31. B. Xing, M. F. Choi and B. Xu, *Chem. Commun.*, 2002, 362.
32. B. Xing, M. F. Choi, Z. Zhou and B. Xu, *Langmuir*, 2002, **18**, 9654.
33. B. Xing, M. F. Choi and B. Xu, *Chem.–Eur. J.*, 2002, **8**, 5028.
34. Y. Liu, L. He, J. Zhang, X. Wang and C. Y. Su, *Chem. Mater.*, 2009, **21**, 557.
35. Y. Liao, L. He, J. Huang, J. Zhang, L. Zhuang, H. Shen and C. Y. Su, *ACS Appl. Mater. Interfaces*, 2010, **2**, 2333.
36. T. Tu, W. Assenmacher, H. Peterlik, R. Weisbarth, M. Nieger and K. H. Dötz, *Angew. Chem., Int. Ed.*, 2007, **46**, 6368.
37. Y. M. A. Yamada, Y. Maeda and Y. Uozumi, *Org. Lett.*, 2006, **8**, 4259.
38. T. Naota and H. Koori, *J. Am. Chem. Soc.*, 2005, **127**, 9324.
39. Y. Cho, J. H. Lee, J. Jaworski, S. Park, S. S. Lee and J. H. Jung, *New J. Chem.*, 2012, **36**, 32.
40. X. Yan, D. Xu, X. Chi, J. Chen, S. Dong, X. Ding, Y. Yu and F. Huang, *Adv. Mater.*, 2012, **24**, 362.
41. M. J. Mayoral, C. Rest, V. Stepanenko, J. Schellheimer, R. Q. Albuquerque and G. Fernaández, *J. Am. Chem. Soc.*, 2013, **135**, 2148.
42. A. Kishimura, T. Yamashita and T. Aida, *J. Am. Chem. Soc.*, 2005, **127**, 179.
43. I. Odriozola, I. Loinaz, J. A. Pomposo and H. J. Grande, *J. Mater. Chem.*, 2007, **17**, 4843.
44. S. Marpu, Z. Hu and M. A. Omary, *Langmuir*, 2010, **26**, 15523.
45. T. H. T. Hsu, J. J. Naidu, B. J. Ysang, M. Y. Jang and I. J. B. Lin, *Inorg. Chem.*, 2011, **51**, 98.
46. V. W. W. Yam, K. M. C. Wong, L. L. Hung and N. Zhu, *Angew. Chem., Int. Ed.*, 2005, **44**, 3107.
47. K. M. C. Wong, L. L. Hung, W. H. Lam, N. Zhu and V. W. W. Yam, *J. Am. Chem. Soc.*, 2007, **129**, 4350.
48. J. J. Zhang, W. Lu, R. W. Y. Sun and C. M. Che, *Angew. Chem., Int. Ed.*, 2012, **51**, 4882.

49. V. K. M. Au, N. Zhu and V. W. W. Yam, *Inorg. Chem.*, 2013, **52**, 558.
50. K. C. Yim, E. S. H. Lam, K. M. C. Wong, V. K. M. Au, C. C. Ko, W. H. Lam and V. W. W. Yam, *Chem.–Eur. J.*, 2014, **20**, 9930.
51. D. R. McMillin and J. J. Moore, *Coord. Chem. Rev.*, 2002, **229**, 113.
52. F. N. Castellano, I. E. Pomestchenko, E. Shikhova, F. Hua, M. L. Muro and N. Rajapakse, *Coord. Chem. Rev.*, 2006, **250**, 1819.
53. K. M. C. Wong and V. W. W. Yam, *Coord. Chem. Rev.*, 2007, **251**, 2477.
54. K. M. C. Wong and V. W. W. Yam, *Acc. Chem. Res.*, 2011, **44**, 424.
55. M. Shirakawa, N. Fujita, T. Tani, K. Kanekob and S. Shinkai, *Chem. Commun.*, 2005, 4149.
56. M. Shirakawa, N. Fujita, T. Tani, K. Kaneko, M. Ojima, A. Fujii, M. Ozaki and S. Shinkai, *Chem.–Eur. J.*, 2007, **13**, 4155.
57. A. Y. Y. Tam, K. M. C. Wong, G. Wang and V. W. W. Yam, *Chem. Commun.*, 2007, 2028.
58. F. Camerel, R. Ziessel, B. Donnio, C. Bourgogne, D. Guillon, M. Schmutz, C. Iacovita and J. P. Bucher, *Angew. Chem., Int. Ed.*, 2007, **46**, 2659.
59. A. Y. Y. Tam, K. M. C. Wong, N. Zhu, G. Wang and V. W. W. Yam, *Langmuir*, 2009, **25**, 8685.
60. A. Y. Y. Tam, K. M. C. Wong and V. W. W. Yam, *Chem.–Eur. J.*, 2009, **15**, 4775.
61. W. Lu, Y. C. Law, J. Han, S. S. Y. Chui, D. L. Ma, N. Zhu and C. M. Che, *Chem.–Asian J.*, 2008, **3**, 59.
62. J. Wang, Y. Chen, Y. C. Law, M. Li, M. X. Zhu, W. Lu, S. S. Y. Chui, N. Zhu and C. M. Che, *Chem.–Asian J.*, 2011, **6**, 3011.
63. T. Cardolaccia, Y. Li and K. S. Schanze, *J. Am. Chem. Soc.*, 2008, **130**, 2535.
64. A. Y. Y. Tam, K. M. C. Wong and V. W. W. Yam, *J. Am. Chem. Soc.*, 2009, **131**, 6253.
65. K. C. Chang, J. L. Lin, Y. T. Shen, C. Y. Hung, C. Y. Chen and S. S. Sun, *Chem.–Eur. J.*, 2012, **18**, 1312.
66. Y. Li, A. Y. Y. Tam, K. M. C. Wong, W. Li, L. Wu and V. W. W. Yam, *Chem.–Eur. J.*, 2011, **17**, 8048.
67. K. Liu, L. Meng, S. Mo, M. Zhang, Y. Mao, X. Cao, C. Huang and T. Yi, *J. Mater. Chem. C*, 2013, **1**, 1753.
68. Y. Li, E. S. H. Lam, A. Y. Y. Tam, K. M. C. Wong, W. H. Lam, L. Wu and V. W. W. Yam, *Chem.–Eur. J.*, 2013, **19**, 9987.
69. C. A. Strassert, C. H. Chien, M. D. G. Lopez, D. Kourkoulos, D. Hertel, K. Meerholz and L. De Cola, *Angew. Chem., Int. Ed.*, 2011, **50**, 946.
70. N. K. Allampally, C. A. Strassert and L. de Cola, *Dalton Trans.*, 2012, **41**, 13132.
71. C. Po, Z. Ke, A. Y. Y. Yam, H. F. Chow and V. W. W. Yam, *Chem.–Eur. J.*, 2013, **19**, 15735.
72. X. S. Xiao, W. Lu and C. M. Che, *Chem. Sci.*, 2014, **5**, 2482.

CHAPTER 7

Metal-Organic Frameworks as Chemical Sensors

NOLAN W. WAGGONER[a], ALISHA M. BOHNSACK[a] AND SIMON M. HUMPHREY*[a]

[a]Department of Chemistry, The University of Texas at Austin, 105 E. 24th Street A5300, Austin, TX 78712-1224, USA
*E-mail: smh@cm.utexas.edu

7.1 Introduction

Metal–organic frameworks (MOFs) are a popular class of emerging materials that show great promise in a wide variety of advanced solid-state applications. MOFs are crystalline materials with infinite, periodic structures composed of organic ligands or 'linkers' and inorganic 'nodes'.[1] Commonly, the linkers are aromatic molecules with two or more anionic donors (*e.g.*, carboxylic acids or pyrazolates) that bridge between adjacent metal cations or metal cluster nodes to form open, polymeric arrays. The rapid and continuing expansion of MOF research can be attributed (at least in part) to their exceptional chemical tunability and structural diversity, which result in access to a wide range of chemical and physical properties that can be advantageous in material design.[2] MOFs based on open, 3-D network structures tend to incorporate uniform pores or channels with dimensions in the approximate range of 0.5–2 nm. MOFs are classified as microporous solids and commonly exhibit large bulk surface areas (*ca.* 500–5000 $m^2\ g^{-1}$);[3,4] the

RSC Smart Materials No. 15
Functional Metallosupramolecular Materials
Edited by John G. Hardy and Felix H. Schacher

Published by the Royal Society of Chemistry, www.rsc.org

void spaces can account for up to 90% of the total volume of such materials.[1] MOFs that display permanent porosity (*i.e.*, MOFs that can resist collapse of the pore structure when devoid of matter) are useful for the adsorption of a broad range of 'guest' molecules that are small enough to enter the pore windows—akin to the action of a molecular sieve.[5] For the past two decades, MOFs have attracted a great deal of attention for the high-capacity storage of light gases such as H_2 for fuel cell and automotive applications,[6–9] for selective CO_2 sequestration in air and flue gases,[10–12] and for complex molecular separations.[13–15]

MOFs are most commonly constructed using harder (acidic) metal ions, such as 3d transition metal ions or lanthanide Ln(III) ions, and hard organic anions. The mixture of covalent and ionic/electrostatic metal–ligand bonding results in substitutionally-inert materials that possess sufficient chemical and thermal stability for subsequent utilisation. Metal ions and/or clusters of the 3d or 4f elements also facilitate orthogonal bonding between adjacent organic linkers, which allows for the formation of polymers with a range of 3-D network topologies. These materials display superior thermal and (to a lesser extent) chemical stability compared to classical, amorphous organic polymers.[16–19] The geometries of the organic linkers and the preferred coordination modes of the metal species tend to directly influence the resulting MOF structures, providing some degree of design control. However, MOF synthesis is not a precise art; while a rational choice of precursors is likely to favor the formation of a given MOF with desired structure and function, it does not guarantee a specific outcome. For example, a linear organic linker, such as the ubiquitous 1,4-benzenedicarboxylate, favors straight linkages between adjacent metal ions (Scheme 7.1, left), while the inherent trigonal geometry of the 1,3,5-substituted trimesate ligand is more likely to induce trigonal (or hexagonal) symmetry elements into the resulting material (Scheme 7.1, right).[20,21]

The abundance of recent research performed in this expanding field—driven primarily by the attractiveness of MOFs for applications in molecular storage and separations—has also led to the realisation of a number of other potential applications, such as drug delivery,[22,23] catalysis,[24–26] and the focus of this chapter: chemical sensing. The following is a survey of recent research progress using MOFs as chemical sensors *via* a number of innovative synthetic and analytical approaches. It is difficult to overstate the need for improved chemical sensors that can achieve lower limits of detection and tolerate a range of chemical environments. The demand for reliable, affordable and portable sensors supports the development of many critical technologies that range from medicine to environmental pollution control,[27] and from food safety monitoring to the detection of illicit substances and chemical weapons.[28]

Chemical sensors provide analytical information about the chemical composition of a gaseous or liquid environment in a two-step process.[29–31] First, the target analyte species physically interacts with selective receptor sites in the sensor material, thus inducing a physical change in the state of the sensor. The analyte–sensor interactions should be reversible and can range from weak intermolecular interactions to the formation of strong chemical bonds. Second, the induced change in the sensor is converted into a measurable output signal. A number of detection methods are commonly employed to generate a diagnostic signal output (*e.g.*, electrochemical, mechanochemical, non-dispersive infrared (NDIR), surface plasmon resonance (SPR)). To date, the most popular mechanism investigated for sensing by MOFs has relied on monitoring the modulation of the solid-state luminescence as a function of gas or liquid analyte interactions. A smaller number of examples have also focused on mechanically-induced signaling. These systems are each carefully designed to take advantage of certain physical properties that are unique to MOFs, which allow for a broad scope of chemical detection.

We shall begin by describing the mechanisms of luminescence applicable to MOFs, which can be harnessed for sensing applications. In an attempt to better understand how MOFs function as effective sensors, we will then present an in-depth survey of recent literature reports describing applications in the detection of a range of molecular and ionic analytes, as well as changes in local physical properties such as temperature and pH. This is followed by a discussion of emerging examples of mechanical-based MOF sensors.

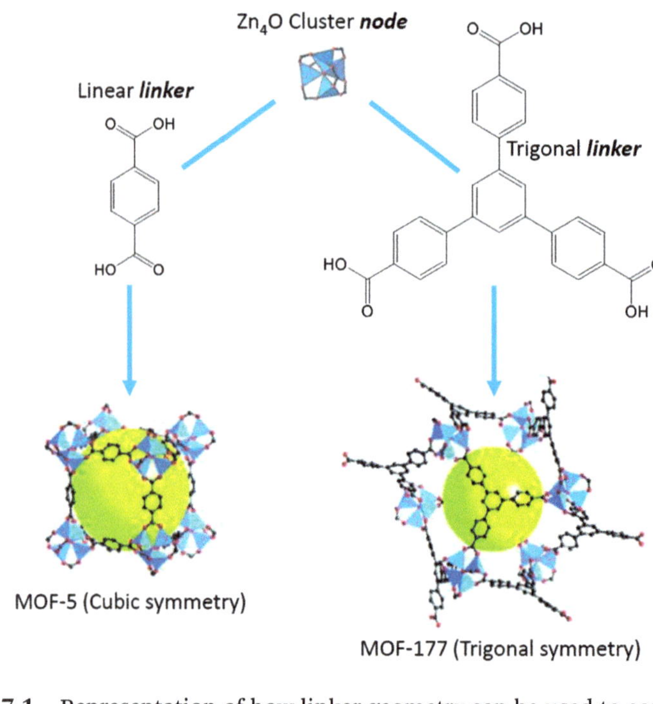

Scheme 7.1 Representation of how linker geometry can be used to control resulting MOF topology: (left) a linear linked used to generate a MOF with cubic symmetry; (right) a trigonal linker used to generate a MOF with corresponding trigonal symmetry. Adapted with permission from ref. 20. Copyright 2005 American Chemical Society. Adapted with permission from ref. 21. Copyright 2004 Nature Publishing Group.

Consideration will also be made to some of the current challenges and potential solutions related to the implementation of MOF sensors as practical devices, and how these compare to established analytical techniques.

7.1.1 General Mechanisms of Luminescence Applicable to MOFs

Luminescence is the general term used to describe the process of light emission from a cold body, *i.e.*, when photons are produced by processes other than heating.[32,33] Photoluminescence is the process by which light is emitted from a substance promoted by the prior absorption of incoming photons, and is most relevant to the use of luminescent MOFs as sensor materials. Photoluminescence processes are usually further distinguished as occurring *via* fluorescent or phosphorescent pathways (Figure 7.1).[34]

Fluorescence is the process by which an incoming photon ($h\nu_{ex}$) excites an electron in a molecule to an excited state (eqn (7.1)), followed by rapid (<10 ns) relaxation of that electron back to its ground state, accompanied by emission of a secondary fluorescent photon ($h\nu_F$) (Figure 7.1, left, and eqn (7.2)). Notably, there is no change in the electron spin multiplicity during this process, which remains a singlet (S). The energy of the emitted photon is lower than that of the incoming photon due to non-radiative (vibrational) energy losses; the fluorescence process is commonly summarized as:

$$\text{Excitation:} \quad S_0 + h\nu_{ex} \rightarrow S_1 \tag{7.1}$$

$$\text{Emission:} \quad S_1 \rightarrow S_0 + h\nu_F + \text{heat} \tag{7.2}$$

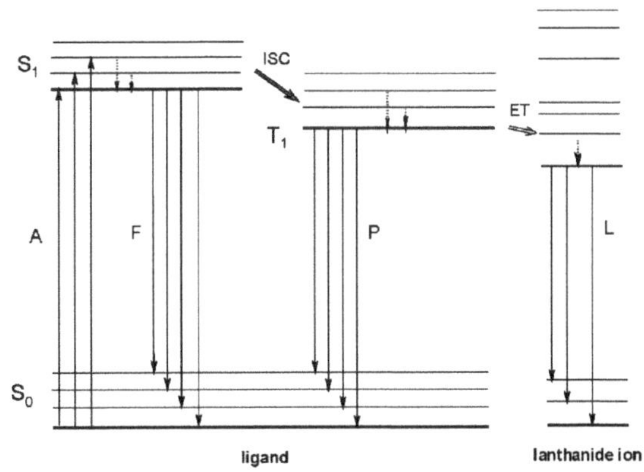

Figure 7.1 Schematic representation of photophysical processes in lanthanide(III) complexes (antenna effect). Abbreviations: A = absorption; F = fluorescence; P = phosphorescence; L = lanthanide-centered luminescence; S = singlet; T = triplet; ISC = intersystem crossing; ET = energy transfer. Full vertical lines indicate radiative transitions; dotted vertical lines indicate non-radiative transitions. Reprinted with permission from ref. 34. Copyright 2009 American Chemical Society.

In contrast, phosphorescence is a much slower emissive process that can occur over a period of microseconds, seconds, minutes or even hours (Figure 7.1, centre).[35] The initial excitation is the same as that shown in eqn (7.1); then, instead of direct relaxation back to the ground state, the excited electron undergoes an intersystem crossing event (ISC; Figure 7.1, centre, and eqn (7.3)) in which the electron can access a state of higher spin multiplicity (usually a triplet state, T_1, by spin reversal). The resulting T_1 electronic configuration must now relax back to an S_0 ground state; the relaxation is slower than fluorescence because a $T \rightarrow S$ transition is quantum mechanically forbidden. The relaxation is accompanied by emission of a phosphorescent photon ($h\nu_P$), which is typically of significantly lower energy than the incoming photon (eqn (7.4)):

$$\text{ISC:} \quad S_1 \rightarrow T_1 \tag{7.3}$$

$$\text{Emission:} \quad T_1 \rightarrow S_0 + h\nu_P + \text{heat} \tag{7.4}$$

A wide variety of compounds and materials exhibit photoluminescence because they have atomic (or multi-body) centres with electronic structures that are readily excited by incoming photons ($h\nu_{ex}$), having energies comparable to the S_0–S_1 gap that occurs in the UV/visible portion of the electromagnetic spectrum. The luminescence efficiency of a given compound or material, otherwise known as its emissive quantum yield (φ), is a simple measure of the efficiency of a photoluminescent material based on the ratio of the number of photons emitted *versus* the number of photons absorbed. There are many factors that are known to affect the value of φ, as well as other related properties, such as the excited state lifetime, τ. These are beyond the scope of this discussion, but several more elegant and detailed descriptions of chemical photoluminescence are available elsewhere.[28,36–39] Coordination compounds and organometallic complexes are often employed in photoluminescent devices because their electronic energy levels can be finely tuned by synthetic tailoring of the ligands, metals and coordination modes. This allows not only for optimisation of φ, but also tunability of the excitation and emission wavelengths, and the lifetimes of the excited states.

With the above considerations in mind, we are in a position to discuss a number of discernible reasons why MOF materials should make good chemical sensors. The crystalline, periodic nature of MOFs results in structurally well-defined chemical moieties that are immobilized at regular intervals within the polymer networks. It is relatively straightforward to install a wide range of both organic and inorganic photoluminescent species into MOFs. As a result, the luminescence properties (φ, τ, $h\nu_{ex}$, $h\nu_F$, *etc.*) should be more tunable, more predictable, and more easily optimisable in comparison to isolated molecular structures in solution. Furthermore, the inherent microporosity and correspondingly high surface areas of MOFs make possible a number of additional opportunities that are not readily available to traditional amorphous inorganic complexes.[28] MOFs are generally obtained as microcrystalline powders with individual crystallites that range between microns to millimetres in size. Therefore, the number of luminescent sensor receptor sites *inside* a given crystallite (and thus only accessible to analytes *via* the micropore structure) will always greatly dominate the number of equivalent sites on the crystallite exteriors. It follows that the MOF pore structures can be exploited to control which analytes are able to access the crystallite interiors—for example, by using a given pore structure as a size- or chemically-selective filter—so that only certain analytes are permitted access to the pores, resulting in a quantifiable sensor response. In this way, MOF sensors could be programmed with molecular specificity for diagnostic applications of complex mixtures of analytes, without the need for prior separation or purification steps. Microcrystalline MOFs are also mechanically robust, thus allowing for flexible processability into various device formats.

Photoluminescence quenching results in a reduction in the overall luminescence intensity of a given material. Quenching is generally viewed as an unwanted effect in molecular systems. Quenching can be broadly defined as any non-radiative relaxation pathway that results in a reduction of the overall fluorescent and phosphorescent quantum yields. There are many causes of quenching. One common example is vibrational coupling of an excited state with neighboring molecules (*e.g.*, solvents), which results in non-radiative decay of the excited state.[40] Another common example is concentration-dependent quenching in the solid-state, caused by *inter*-molecular energy transfer between closely-packed molecules. However, quenching can be exploited in MOFs for molecular sensing applications because the observed reduction of the luminescence intensity (or emission lifetime) may be proportional to the presence of a known quencher—thus providing a diagnostic signal.

All of these potential virtues—and others—are explored in the following pages through recent literature examples. First, it is appropriate to consider in more depth the types of chemical moieties that are useful in the construction of luminescent materials, and how these can be deliberately combined to create luminescent MOFs with multifunctional sensing capabilities.

7.1.1.1 Organic (Linker)-Based Emission

The majority of known MOFs are constructed using a variety of π-conjugated organic molecules with multiple substituents that are exploited to coordinate to metal ions. Benzene-based linkers and poly(phenyl) derivatives are the most common examples, although there are also many examples of linkers based on π-conjugated heterocycles (*e.g.*, pyridyls and azolates).[41–43] The primary reason for this reliance on unsaturated organic molecules as MOF linkers is due to their inherent rigidity and lack of conformational freedom; these attributes aid in the formation of correspondingly rigid polymer structures with largely predictable topologies. It is somewhat serendipitous that organic molecules with π-bonds are also the most ideal electronic structures for photoluminescence applications in the visible portion of the electromagnetic spectrum. Electrons in molecular orbitals that are dominated by π-conjugated systems are located further from the positively charged nuclei and are also more delocalized than electrons in σ-bonds; they are consequently more easily excited by photons of lower energy.[44] The lowest energy transitions in unsaturated C=C bonded systems involve both $n \rightarrow \pi^*$ and $\pi \rightarrow \pi^*$ transitions that are achieved with photons of $\lambda > 200$ nm (<600 kJ mol^{-1}), which corresponds to the ultraviolet-visible region of the electromagnetic spectrum. The presence of extended π-conjugated bonds (*e.g.*, aromatics) induces a red shift and allows for excitation with photons of even lower energy. The major photoluminescent pathways accessible to organic systems have already been described in Figure 7.1. Although fluorescence is the most common relaxation pathway observed for simple conjugated organic molecules, the associated quantum yields are characteristically low compared to charge-transfer processes.[45] Therefore, the organic components in MOFs are not usually able to produce luminescence intensities suitable for sensing applications alone. Instead, the light-harvesting capabilities of the conjugated π-systems need to be carefully united with other components of a MOF to generate a viable sensor output.

7.1.1.2 Metal (Node)-Based Emission and the Antenna Effect

In isolation, metal ions are not commonly luminescent. Only lanthanide ions and a few uranyl ($[UO_2]^{2+}$)-based complexes are luminescent.[46] The lanthanides, which generally exist as trivalent cations, have their own unique emission spectra. Only Sm(III), Eu(III), Tb(III) and Dy(III) display practically useful luminescence in the visible region of the electromagnetic spectrum; most other Ln(III) ions are luminescent to varying degrees in the near-infrared (NIR) spectral region; Gd(III) is emissive in the UV region.[28] Lanthanide-based luminescence has attracted much recent attention in advanced optics, partly because the spectral properties of Ln(III) ions are well defined and are not easily perturbed by external forces. This is because the 4f (valence) electrons that are involved in the optically-active transitions are contracted inside the radius of (and therefore shielded by) the filled $5s^2$ and $5p^6$ orbitals. As a result, Ln(III) ions do not engage in any substantial covalent bonding with ligands;[47] although they can be incorporated into a variety of charged ligands to form complexes, the stabilizing interactions are predominantly ionic and therefore the energy levels of the 4f-electrons remain largely unperturbed. This leads to inner shell 4f–4f transitions that are characteristically very sharp and easily identifiable (Figure 7.2).[48]

Unfortunately, lanthanide luminescence is generally very weak for isolated Ln(III) ions because f–f transitions are classically forbidden. This issue can be overcome by coupling the Ln(III) ions to secondary organic (ligand) species that can facilitate energy transfer—a process commonly referred to as

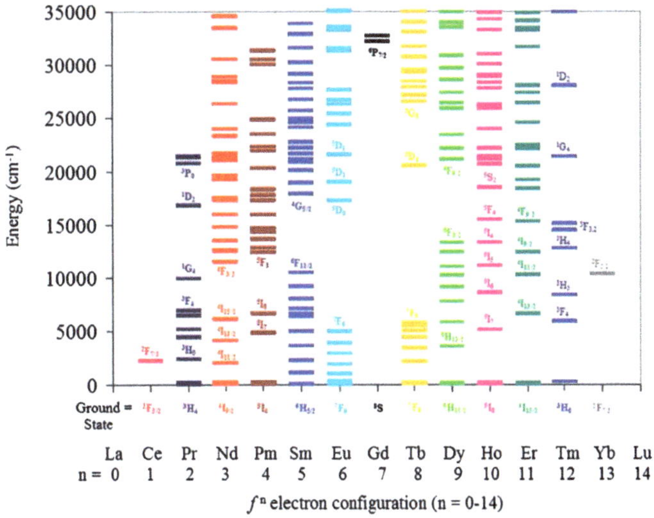

Figure 7.2 A summary of electronic excited-state energy levels for Ln(III) ions. Reprinted with permission from ref. 35. Copyright 2012 American Chemical Society.

the antenna effect. The mechanism of this effect in its most common form is summarized as follows (see Figure 7.1): first, light is absorbed by a conjugated organic group (ligand) in close proximity to a Ln(III) ion to generate a T_1 excited state *via* the process of ISC (*vide supra*); then, the energy of the excited state is transferred directly to the Ln(III) ion, resulting in promotion of an electron to an excited state of the Ln(III) ion, instead of resulting in ligand-based phosphorescence (ET; Figure 7.1); finally, the 4f-centred excited state may then undergo radiative decay to the ground state.[35] The efficiency of this specific type of ligand-to-metal energy transfer is dependent on the relative energetic positions of the ligand and Ln(III) ion excited states, as well as strong vibronic coupling between the two components. Therefore, it is important to ensure a good energy match between the ligand and Ln(III) ion excited states. Indeed, the development of improved Ln(III)-based organic materials is a topical field in its own right.[49] Specifically regarding MOF-based chemical sensors, the antenna effect presents an ideal opportunity to create more highly luminescent materials because Ln(III) ions are also able to act as the inorganic nodes. Classically, MOF chemistry has relied most heavily on the use of 3d-based metal ions as nodes, since they have somewhat predictable coordination geometries and numbers. The preparation of MOFs with desired structures using Ln(III) ions presents some additional challenges, due to the variable coordination numbers (6–12) and less defined geometries adopted by Ln(III) ions in complexes, which can result in the formation of denser, less porous materials.[50] However, recent research highlighted below has shown that a careful choice of precursors can still result in stable, highly porous materials with associated luminescence quantum yields that compare to the best known molecular complexes.

7.1.1.3 Up-Conversion Emission

Up-conversion (UC) is a non-linear anti-Stokes process that converts two or more low-energy NIR photons into a single emitted photon of higher energy. This additive process is useful for generating UV/visible light for practical applications. The numerous excited state energy levels available to Ln(III) ions (Figure 7.2) mean that mixed-Ln systems are ideal candidates to achieve UC. There are several energy transfer processes that can result in UC emission, which are summarized in Figure 7.3.[51]

The simplest mechanism by which UC can be achieved is by the simultaneous absorption of two NIR photons by a single Ln(III) ion (ESA; Figure 7.3). However, this direct process suffers from low quantum yields. Alternative pathways (Figure 7.3) involve cooperative behaviour of (at least) two independent Ln(III) ions and tend to achieve more efficient UC emission intensities. These examples involve some form of metal-to-metal energy transfer from the excited state of one Ln(III) to another.[36] The NIR absorption efficiency of Ln(III)-based UC materials can be maximized by co-doping of NIR emitting lanthanides (*e.g.*, Pr(III), Nd(III), Ho(III), Er(III), Tm(III), Yb(III)) in idealized relative ratios to achieve optimal sensitisation.[52] Since the coordination chemistry of the Ln(III) ions is largely insensitive to the ligand environment (owing to a lack of any substantial covalency), and since the ionic radii of the Ln(III) ions decreases very gradually across the 4f series, it is relatively trivial to prepare mixed-Ln(III) materials. This applies to Ln(III)-based MOFs, although UC MOFs have not yet attracted a great deal of attention.[53]

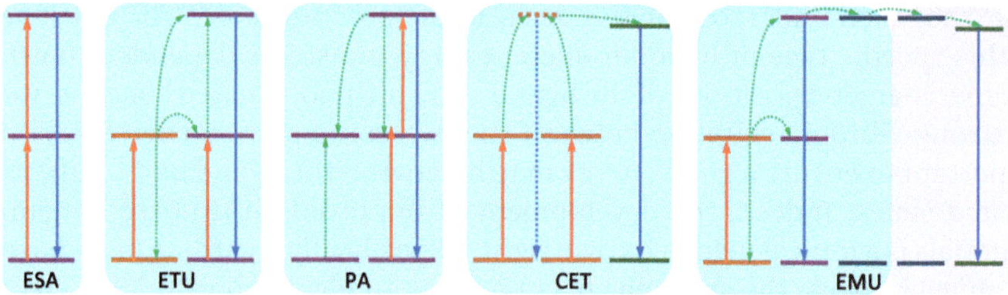

Figure 7.3 Schematic representations of UC processes, where upward (red) and downward (blue) full arrows stand for direct excitation processes and radiative emission processes, respectively. Dashed arrows (green) represent energy transfer processes. Note that non-radiative relaxations are not given here for clarity. Different kinds of luminescent centers are shown with different colored energy levels, and different regions are highlighted with different background colors. Abbreviations: ESA = excited state absorption; ETU = energy transfer up-conversion; PA = photon avalanche; CET = cooperative energy transfer; EMU = energy migration-mediated up-conversion. Adapted from ref. 51 with permission.

7.1.1.4 Charge-Transfer Luminescence

The luminescence processes discussed so far involve some kind of net energy transfer. Charge-transfer (CT) processes are generally classified as involving formal electron transfer from a donor species to an acceptor species. In MOFs, CT may conceivably occur by an *inter*-ligand mechanism (*inter*-ligand charge transfer, or ILCT), or by electron transfer between predominantly metal-based molecular orbitals and predominantly ligand-based molecular orbitals (LMCT and MLCT, depending on the direction of electron movement). The latter cases are most commonly observed in metal–organic complexes and MOFs, and usually result in a formal (and reversible) change of oxidation state of the metal ion. MLCT is prevalent for metals in low oxidation states and results in oxidation, while the corresponding reductive LMCT process occurs more often in materials that contain high oxidation state metals and ligands with available nonbonding electrons.[54]

Most commonly, LMCT is observed in MOFs that employ d^{10} metal centres (*e.g.*, Zn(II) or Cd(II)) and conjugated organic linkers, although LMCT have also been observed in mixed d-metal MOFs.[55–58] The transfer of a 4f electron into a 5d sub-shell to initiate luminescence has also been observed.[59] MLCT is somewhat less commonly observed in MOFs because the transfer of electrons from a metal centre into energetically-low-lying ligand π-orbitals is usually less favorable. However, since all CT absorption processes involve electronic transitions that are spin-allowed, their molar absorptivity coefficients can be very large (ε_{max} = 1000–50 000 $dm^3\ mol^{-1}\ cm^{-1}$) in comparison to the energy transfer-based luminescent mechanisms, which are formally Laporte-forbidden.[54] LMCT or MLCT luminescence may also compete with ILCT-based luminescence, resulting in a more complex series of emission bands.[60]

7.1.1.5 Vapochromism

Vapochromism is a distinct and simple detection mechanism that is accessible to certain MOFs. In general, a vapochromic substance exhibits a change in its characteristic absorption wavelength upon exposure to certain organic vapours or gases. This change is perceived by the naked eye as a change in the physical colour of the substance. Such materials have obvious applications in the sensing of volatile organic compounds (VOCs). The phenomenon of luminescent vapochromism (or vapoluminescence) has been widely studied for coordination complexes, and is therefore directly applicable to MOFs.[61] Luminescent vapochromism in MOFs is commonly associated with CT, which induces a large change in the net dipole moment of a chromophore upon excitation, due to the inherently charge-separated nature of the excited state. When the excited state of a given chromophore has a larger dipole moment than the ground state, it is favorably stabilized by polar solvents, resulting in a red shift; a blue shift is observed for the converse scenario.[2]

These visible changes in colour can be easily analytically measured and monitored, making this mechanism highly convenient for potential applications in devices. For example, a visible colour change was observed in the work of Takashima *et al.*[62] following the adsorption of different aromatic molecules inside a Zn(II)-based MOF (Figure 7.4).

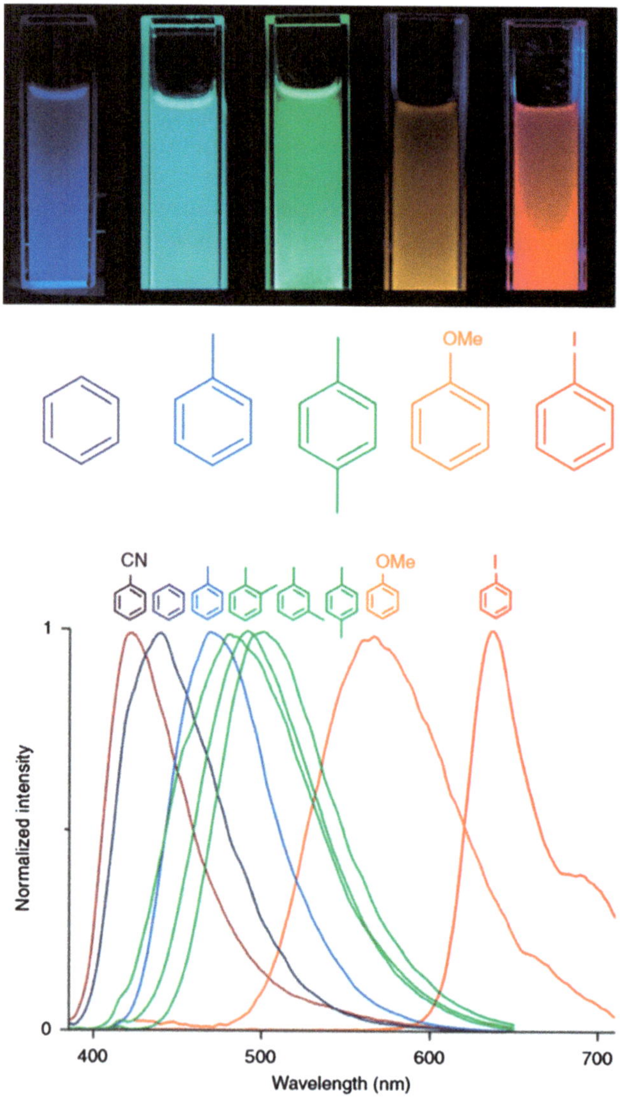

Figure 7.4 Multicolour luminescence of **1**⊃VOCs. (Top) The resulting luminescence of crystal powders of **1**, suspended in each VOC liquid after excitation at 365 nm using a commercial ultraviolet lamp; (bottom) height-normalized luminescent spectra of **1**⊃VOCs after excitation at 370 nm. Adapted with permission from ref. 62. Copyright 2011 Nature Publishing Group.

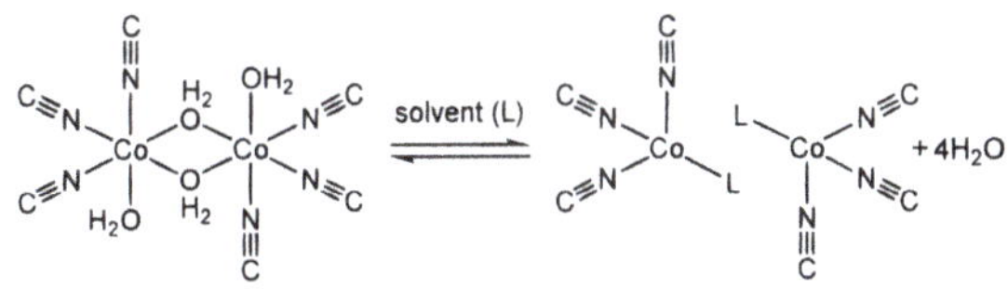

Scheme 7.2 Depiction of a vapochromic shift in a Co(II) complex, where the non-cyanide ligands (L) can be either water or solvent molecules. Reprinted with permission from ref. 63. Copyright 2000 American Chemical Society.

Figure 7.5 Visible colour changes of the coordination spheres in Scheme 7.2. Powder samples of compounds **1** (top) and **3** (bottom) deposited on filter paper and doused with selected solvents: water (as-prepared), tetrahydrofuran, and diethyl ether, from left to right, respectively. The apparent color changes are from orange to red-violet and blue-violet for compound **1**, and from orange to green and blue for compound **3**. Reprinted with permission from ref. 63. Copyright 2000 American Chemical Society.

An alternative route toward achieving luminescent vapochromic feedback in MOFs has also been demonstrated: since MOFs are essentially infinite networks of coordination complexes, manipulation of the metallic coordination sphere (or the node) *via* the introduction of a guest species can also result in vapochromic behaviour (Scheme 7.2 and Figure 7.5). This has been demonstrated in both older and newer literature in the field of MOFs and will likely continue being of interest for applications in sensing.[63,64]

7.1.1.6 Guest-Induced Luminescence

As a convenient alternative to the preparation of intrinsically emissive MOFs, it is possible to introduce lumophores into pre-formed, non-emissive MOFs *via* the incorporation of secondary emissive species. There are several strategies by which this can be achieved. For example, post-synthetic modification is a powerful strategy that has attracted significant recent interest,[65] and can enable a broader range of MOFs to be converted into potential sensors. A diverse range of post-synthetic modifications have been demonstrated for MOFs, such as organic modification in which covalent bonds are formed or broken,[66] or coordination chemistry in which secondary (non-nodal) metal species are installed into vacant linker sites.[67] Another common example of luminescence functionalisation of originally non-emissive MOFs involves the adsorption of luminescent 'guest' species within the pores. In addition, the incorporation of emissive secondary species into an already emissive MOF provides a means to alter the resulting emission profile. MOFs with larger pores are most well-suited to modifications with known chromaphoric sensitizers (*e.g.*, organic dyes or Ln(III)-based complexes). Also, the structural rigidity and periodic order of MOFs helps to enhance the overall luminescent properties, as well as to prevent guest molecules from undergoing guest–guest quenching interactions.[68,69]

The most common and simplest example of guest-induced luminescence is the direct incorporation of an emissive species into a MOF. This post-synthetic method has obvious advantages over other synthetic routes because it is applicable to a wide range of known MOFs. Thus, a designed-for-application material is, in theory, easier to obtain. The types of emissive guest species that have been studied in MOFs include conjugated organic lumophores, simple Ln(III) ions and complexes thereof. Regardless of the nature of the emissive guest species, it is usually found that the individual emissive properties of the guest (and framework, if applicable) remain relatively unchanged upon incorporation. Lumophore complexation and the resulting electronic interactions in host–guest systems have been thoroughly investigated outside of MOFs.[70] Although the introduction of secondary guest molecules is most commonly targeted at overall emission enhancement, the introduction of certain guest molecules can also be used to quench the luminescence of an emissive MOF.[71] The incorporation of lanthanides into molecular coordination complexes is a fairly straightforward technique to induce (or enhance) luminescence, but it is still a relatively new technique in the field of MOF chemistry.[72]

7.2 MOF Luminescence Sensing

We will now provide an in-depth discussion of the research advances in MOF luminescence sensing throughout the past decade. Each section includes the motivation behind the specific sensing application, a brief summary of the published materials to date, as well as specific examples and mechanisms of MOF luminescence sensing. We begin with the detection of smaller, charged species (including anions, cations, and pH sensing). This is followed by a survey of MOFs capable of molecular sensing (including gases, vapours and more complex biomolecules), and concludes with examples of temperature sensing by luminescent MOFs. This examination of the literature is not meant to be a fully comprehensive review, but aims to highlight the ability of MOFs to behave as advanced sensing materials.

7.2.1 Anion Sensing

There is a great deal of interest in the detection of anions in solution because anions play a variety of important roles in chemical, biological and environmental settings. The primary mechanism of anion detection involves ion exchange between an ionic MOF sensor and an anionic analyte species. The displacement of parent anions in the MOF by analyte anions produces a measurable change in the photoluminescence emission properties of the MOF. This form of guest-induced luminescence modulation therefore primarily involves ionic/electrostatic MOF···analyte interactions within the pores, although examples of anion detection based on hydrogen bonding interactions toward charge-neutral MOFs also exist. There are several examples of the use of MOFs to detect common small anions, including halides,[27,73,74] N_3^-,[74,75] CO_3^{2-},[74] SCN^-,[74,75] ClO_4^-,[75] PO_4^{3-},[76] CrO_4^{2-},[77,78] $Cr_2O_7^{2-}$,[79] $N(CN)_2^-$,[75] and polyoxometalates,[80] among others. One of the earliest examples of anion luminescence sensing in MOFs involves the detection of F^- in methanol by a Tb(III)/1,3,5-benzenetricarboxylate (BTC)-based material, MOF-76.[73] In this example, the neutral MOF was loaded with a range of Na(I) salts to incorporate halides, CO_3^{2-} or SO_4^{2-} anions inside the pores. The resulting materials were able to detect increasing F^- analyte concentration as a function of the degree of luminescence quenching, caused by vibrational coupling to methanol O–H oscillators. The analyte anions induced $CH_3O{-}H\cdots F^-$ hydrogen bonds, which impeded quenching within the system. A second example of F^- sensing involves the luminescence quenching of a mixed Eu(III)/Tb(III) MOF, when subjected to an increasing concentration of NaF in aqueous solution.[27] However, solutions of NaCl, NaBr, or NaI produced no measurable change in the observed luminescence intensity of this material, making it an excellent candidate as a selective sensor of F^- over all other halide ions.

Ghosh and co-workers have reported an example of tunable luminescence within a cationic MOF (**1**) composed of Zn(II)-based nodes and a neutral tetradentate Schiff-base ligand, which incorporates NO_3^- counter ions in the pores as a necessity of framework charge-balance.[75] The luminescence of MOF **1** arises due to $n \rightarrow \pi^*$ and $\pi \rightarrow \pi^*$ intra-ligand transitions, and the framework showed significant flexibility in the solid state. Anion exchange of **1** with different NaX precursors produced four subsequent materials, denoted as $\mathbf{1}\supset ClO_4^-$, $\mathbf{1}\supset N(CN)_2^-$, $\mathbf{1}\supset N_3^-$ and $\mathbf{1}\supset SCN^-$. Compared to the un-exchanged **1** (containing only NO_3^- anions), the luminescence intensities of $\mathbf{1}\supset N(CN)_2^-$ and $\mathbf{1}\supset ClO_4^-$ were enhanced by 70.72 and 84.14%, respectively, while $\mathbf{1}\supset N_3^-$ and $\mathbf{1}\supset SCN^-$ were quenched by 29.53 and 80.36%, respectively (Figure 7.6).

The authors attributed these unusual anion-specific effects to the coordination behaviour of each. Specifically, N_3^- and SCN^- were found to ligate to Zn(II) sites in the MOF, causing a dramatic structural deformation that

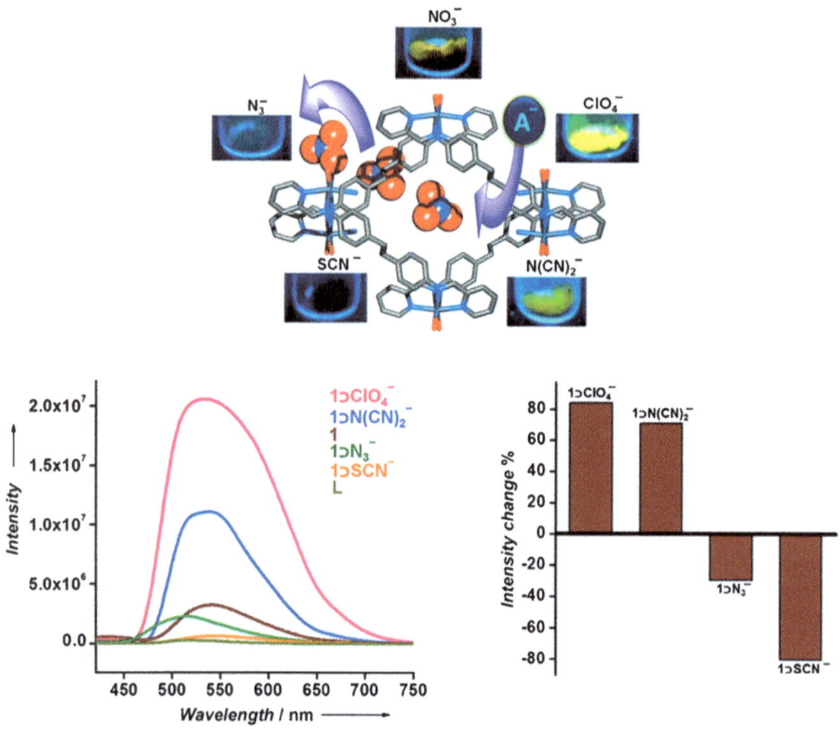

Figure 7.6 (Top) Effects of anion exchange on the solid state luminescence properties of **1**; (bottom left) luminescence intensity of different anion-exchanged samples; (bottom right) bar diagram showing luminescence enhancement and quenching of anion-exchanged samples with respect to **1**. Adapted with permission from ref. 75. Copyright 2013 John Wiley and Sons.

occurred in a single-crystal-to-single-crystal (SCSC) fashion. This structural change reduced the extent of intra-ligand interactions, ultimately reducing the luminescence intensity of the MOF. In contrast, the more weakly binding anions ClO_4^- and $N(CN)_2^-$ did not directly ligate to the Zn(II) centres but instead simply occupied different sites in the lattice compared to the displaced NO_3^- anions, resulting in stronger intra-ligand interactions and a higher overall luminescence intensity.

7.2.2 Cation Sensing

Compared to anion sensing, cation sensing encompasses a potentially much wider range of analytes, ranging from a simple proton to complex metal ions. For this reason, cation sensing will be discussed in two sections: pH sensing and metal ion sensing. pH sensing is of utmost importance in the physiological sciences, as well as in countless chemical and industrial processes.[81] Similarly, the determination of the presence, concentration, and identity of metal ions is also vital to numerous biological applications, industrial processes, and in environmental waste remediation.[82,83]

7.2.2.1 pH Sensing

The field of luminescence-based pH sensing in MOFs is still in its infancy as there are only three examples of the aforementioned materials to date.[84–86] In 2009, Rocha and Corma reported a self-calibrating pH sensor that showed a linear response in the pH range 5 to 7.5, which is highly relevant for biological systems.[84] ITQMOF-3-Eu was formed using 1,10-phenanthroline-2,9-dicarboxylic acid and Eu(III) centres, and the resulting framework contained two crystallographically-unique Eu(III) sites. When excited at 350 nm and 300 K, the emission spectrum showed two discrete transitions at 579.0 and 580.7 nm. These lines were assigned to the non-degenerate $^5D_0 \rightarrow {}^7F_0$ transitions for the two metal sites Eu2 and Eu1, respectively. However, when subjected to an increase in pH from pH 5 to pH 7.5, the line at 579.0 nm increased in intensity due to small changes induced in the coordination geometry of Eu2, while the peak at 580.7 nm was unchanged (Figure 7.7). By considering the ratio of luminescence intensities for the two lines (Eu2–Eu1), a linear pH response was observed, showing that the internally-calibrated ITQMOF-3-Eu is a good candidate for pH sensing in biological applications.

The second example of pH sensing by a luminescent MOF was reported by Zhou and co-workers.[85] Their material, named PCN-225, is composed of tetrakis(4-carboxyphenyl) porphyrin linkers and Zr(IV) nodes. PCN-225

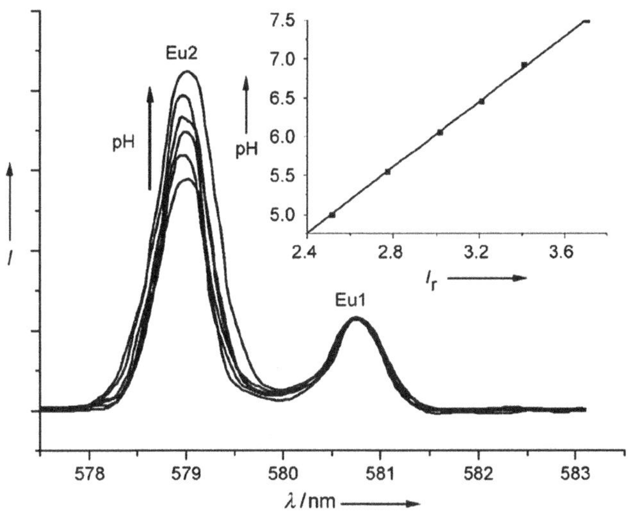

Figure 7.7 Intensity variation of the Eu2 $^5D_0 \rightarrow {}^7F_0$ transition from high (pH 7.5) to low (pH 5) pH values; the inset shows the linear variation of I_r (Eu2/Eu1) with the pH value. Reprinted with permission from ref. 84. Copyright 2009 John Wiley and Sons.

was found to possess excellent chemical stability in aqueous solution and achieved an impressive sensing range between pH = 1–11. Excitation of the MOF at 415 nm produced an emission band at 725 nm, which was assigned to intra-ligand transitions within the lattice. The mode of pH sensing in PCN-225 is by reversible protonation/deprotonation of the N-sites in the pyrrole rings of the porphyrins, which in turn affects the extent of ligand π-conjugation. An increase in pH produced enhanced emission and was accompanied by a slight blue shift at pH values above 7, caused by complete deprotonation of the imino groups.

Thirdly, Bradshaw *et al.*[86] have recently reported two amine-modified versions of the Zr(IV)-based MOF UiO-66, which also exhibited pH-dependent luminescence. Under basic conditions, UiO-66-NH_2 displayed a ligand-centred emission band at 428 nm (by excitation at 350 nm) due to $n \rightarrow \pi^*$ transitions of the free amine groups, the intensity of which increased with increasing pH. However, the lattice of UiO-66-NH_2 underwent irreversible structural changes above pH 9. To increase the stability of the material, post-synthetic modification of UiO-66-NH_2 was carried out by reaction with $NaNO_2$ followed by 1-methylindole to produce the diazo-functionalized version, UiO-66-N=N-ind. This material exhibited improved stability up to pH 12. Upon excitation at 350 nm, it also displayed a pH-dependent emission band *ca.* 428 nm. The luminescence intensity of UiO-66-N=N-ind was lower than that of the precursor MOF UiO-66-NH_2, but its sensitivity was substantially better, resulting in a limit of detection on the μM scale.

7.2.2.2 Metal Ion Sensing

There have been numerous reports of luminescence-based MOF sensors for the detection of s-block, d-block and f-block metal ions—the majority of which are comprised of Ln(III)-based metal nodes and utilize the antenna effect to achieve Ln(III) sensitisation. Sensing of Cu(II) ions in solution has been studied by several groups and has focused on the use of luminescent Tb(III)-MOFs,[78,87,88] Eu(III)-MOFs,[89–92] and a Eu(III)-doped Zn(II)-MOF,[72] among others.[93,94] There have also been examples of the detection of Mg(II),[95] Mn(II),[18] Fe(II),[96] Fe(III),[97–100] Co(II),[72] Zn(II),[101] Hg(II),[102,103] Tb(III),[104,105] and Dy(II).[105] One of the earliest accounts of cation luminescence sensing was reported by Zhao *et al.*[101] in 2004. They prepared a mixed d–f Mn(II)–Ln(III) (Ln = Eu (**1**), Tb (**2**)) MOF using pyridine-2,6-dicarboxylate linkers that gave a material with large hexagonal channels. Excitation of both **1** (λ_{ex} = 287 nm) and **2** (λ_{ex} = 296 nm) produced characteristic Eu(III) and Tb(III) emission spectra, respectively, which showed a marked response to the addition of guest (analyte) metal cations inside the channels. The luminescence intensity became enhanced by the presence of Zn(II) but was quenched by a range of other 3d-M(II) ions, while the presence of the s-block cations Ca(II) and Mg(II) did not cause any measurable change. The MOFs **1** and **2** were also able to detect the presence of low concentrations of Zn(II) ions in mixtures of Ca(II) and Mg(II).

The sensing of toxic heavy metal ions is important in environmental remediation. Chen and coworkers have reported the highly selective 'turn-on' sensing of aqueous Hg(II) ions by coordination polymer nanoparticles (CPNPs) composed of adenine, dipicolinic acid (DPA), and Tb(III), designated accordingly as Ad/Tb/DPA CPNPs.[103] Excitation of these nanomaterials at 270 nm gave typical Tb(III) emission bands at 490, 545, 584, and 620 nm caused by intra-molecular ligand-to-metal energy transfer. The sensor displayed a clear selectivity towards Hg(II) ions: no discernible changes in luminescence behaviour were observed in the presence of a range of common cations, including 3d-M(II) and M(III) ions, Ag^{I}, Pb^{II}, or Mg^{II}; however, addition of 100 nM Hg(II) ions caused a five-fold increase in the luminescence intensity (Figure 7.8).

The origin of this impressive selectivity was due to actual coordination of Hg(II) ions into the MOF structure by chelation between neighboring adenine-*N*-atoms and DPA carboxylic acid groups (Figure 7.9). This physical intercalation was thought to suppress the intra-molecular photo-induced electron transfer from adenine groups to DPA groups while concomitantly enhancing electron transfer from DPA groups to Tb(III) sites, thus producing an overall increase in the emission intensity of the material.

The detection limit of Ad/Tb/DPA CPNPs for Hg(II) ions in water (0.2 nM) and their near-linear luminescence response in the 0.2–100 nM range suggest that they would be effective for Hg(II) sensing in drinking water, for which the maximum permitted level is 10 nM.[106]

Tb(III)-based MOF sensors are popular, due in part to the fact that the luminescence quantum yields of Tb(III) complexes are commonly higher than for other Ln(III) ions and emission occurs in the visible region of the electromagnetic spectrum, thus enabling simple detection at higher levels of resolution. While the previous two examples rely on cation sensing by MOFs composed

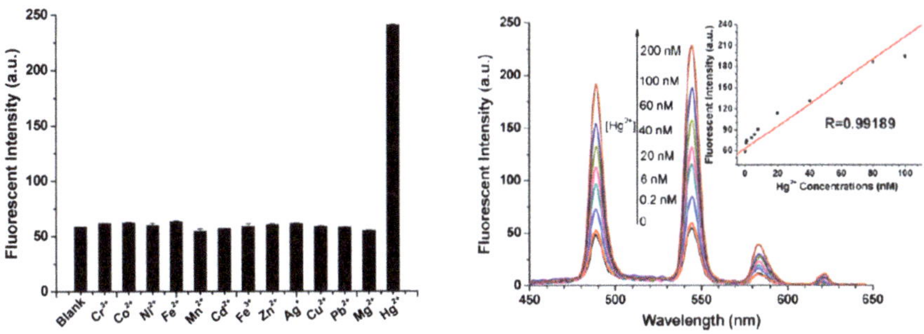

Figure 7.8 (Left) Effect of various metal ions (1 μM) on the fluorescence intensity of Ad/Tb/DPA CPNPs at 545 nm; (right) fluorescence emission spectra of Ad/Tb/DPA CPNPs in the presence of different concentrations of Hg(II) solution (0, 0.2, 6, 20, 40, 60, 100, 200 nM). Inset: linear relationship between the fluorescence intensity of Ad/Tb/DPA CPNPs at 545 nm and Hg(II) concentration. Adapted with permission from ref. 103. Copyright 2012 American Chemical Society.

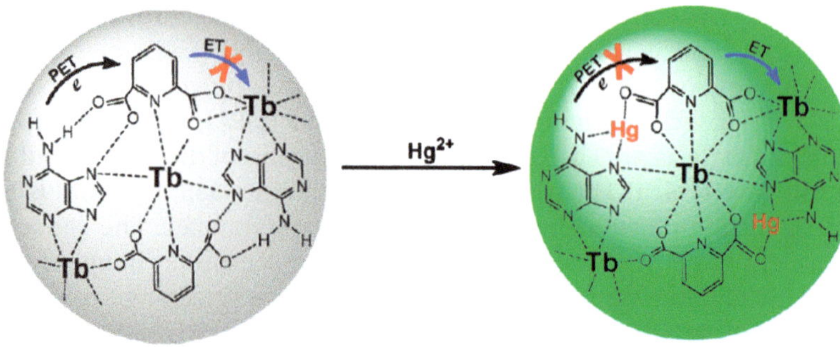

Figure 7.9 Coordination polymer nanoparticle Ad/Tb/DPA for sensing of Hg(II) by photoinduced electron transfer. Reprinted with permission from ref. 103. Copyright 2012 American Chemical Society.

of Tb(III) ions, the next example involves the selective sensing of Tb(III) ions over other Ln(III) cations by a Zn(II)-MOF. Liu *et al.*[104] have reported the selective sensing of Tb(III) ions by MOF–COOH, which was formed by reaction of 1,1′:2,1″-terphenyl-4,4′,4″,5′-tetracarboxylic acid (H_4ttca) and 4,4′-bipyridine with Zn(II). The H_4ttca ligand exhibited a strong emission band at 394 nm upon excitation at 339 nm, attributed to $\pi^* \rightarrow \pi$ and/or $\pi^* \rightarrow n$ transitions. However, the resulting MOF–COOH exhibited only a very weak, broad emission band between 400–500 nm (λ_{ex} = 339 nm), due to inefficient energy transfer from the ligand to the Zn(II) nodes and quenching by O–H vibrations of the free carboxylic acid groups in the framework. To probe the sensing abilities of MOF–COOH, the protons of the free carboxylic acid groups were exchanged with Na(I) ions, followed by subsequent exchange with Sm(III), Eu(III), Dy(III) and Tb(III) to produce Na^+@MOF–COO^-, Sm^{3+}@MOF–COO^-, Eu^{3+}@MOF–COO^-, Dy^{3+}@MOF–COO^-, and Tb^{3+}@MOF–COO^-, respectively. The luminescence intensity of Na@MOF–COO^- was significantly enhanced compared to that of MOF–COOH, and while Sm@MOF–COO^-, Eu@MOF–COO^-, and Dy@MOF–COO^- also displayed similar emission profiles, excitation of Tb@MOF–COO^- produced the distinctive $^5D_4 \rightarrow {}^7F_6$ (490 nm), $^5D_4 \rightarrow {}^7F_5$ (544 nm), $^5D_4 \rightarrow {}^7F_4$ (585 nm), and $^5D_4 \rightarrow {}^7F_3$ (621 nm) transitions expected for the Tb(III) ion (Figure 7.10). In this instance, the MOF-Tb(III) analyte energy transfer was more efficient than the intra-ligand emission of the MOF–COOH framework, allowing for the effective sensitisation of Tb(III), but not for the other Ln(III) ions studied.

7.2.3 Small Molecule Sensing

There are an increasing number of accounts of luminescence sensing in MOFs for the detection of small molecules such as O_2, H_2O, small organics, solvents, and nitroaromatics. However, there have been fewer reports of luminescence sensing of some other common small molecules that are otherwise of relevance to the MOF community. Luminescence sensing of CO_2

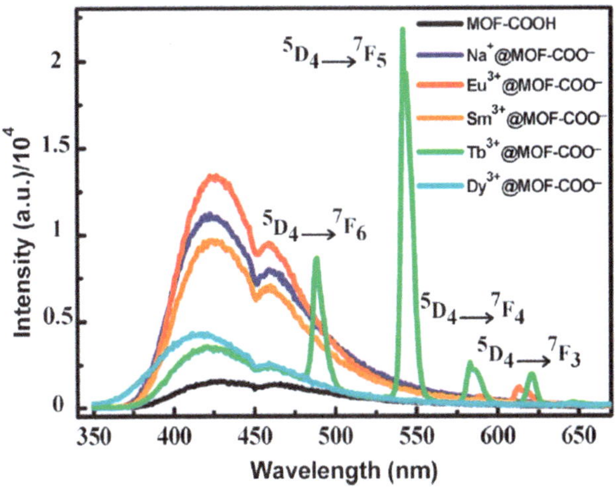

Figure 7.10 Photoluminescence of MOF–COOH, Na^+@MOF–COO$^-$, and Ln^{3+}@MOF–COO$^-$ (Ln = Sm(III), Dy(III), Eu(III), or Tb(III)). Reproduced from ref. 104 with permission.

is perhaps the most obvious example, since CO_2 storage by MOFs continues to attract a great deal of attention. A Zn(II)-MOF has been reported for the detection of low partial pressures of CO_2,[107] and Ag(0) nanoparticles coated with a Cu(II)-MOF have been shown to exhibit a 14-fold signal enhancement *versus* unmodified Ag(0) nanoparticles upon exposure to CO_2 gas.[108] There are a handful of examples of NH_3 sensing by Mg(II)-, Zn(II)-, and Tb(III)-based MOFs.[78,109,110] Luminescence sensing of molecular iodine by Cd(II)- and Zn(II)-MOFs has also been demonstrated.[111–113] Two highly selective and sensitive NO sensors have been described, based on Cu(II)- and Eu(III)-based MOFs, which can detect NO in aqueous solutions and living cells.[114]

7.2.3.1 O_2 Sensing

The accurate detection of O_2 is important in a wide array of fields, including biology, biochemistry, environmental analysis, food science, materials science, medicine and industry.[115] Molecular oxygen is known to be a very effective photoluminescence quencher *via* resonance energy transfer.[116] The quantitative detection of O_2 by the reduction in luminescence intensity of a given sensor is therefore a convenient method of detection applicable to MOFs. There have been several accounts of heterobimetallic MOFs for this purpose, including Ir(III)/Zn(II),[117–119] Ir(III)/Zr(IV),[120] Ru(II)/Zn(II),[121] and Ru(II)/Zr(IV),[120] as well as monometallic systems with Zn(II),[122] Eu(III),[123] and Tb(III).[123,124] For example, Lin *et al.*[120] have reported three different materials based on Ir(III)- or Ru(II)-doped UiO-67. Excitation at 420 nm produced emission maxima at 599, 593, 614 nm for these three MOFs, each ascribed to the formation of a ^{1}MLCT excited state, followed by intersystem crossing to a

^{3}MLCT state and eventual phosphorescent radiative relaxation to the ground state. These MOFs all exhibited exceptional sensing reversibility and phosphorescence quenching of up to 65% at 0.8 atm O_2.

In a related example, Sheu and coworkers have developed another mixed-metal system with four Ir(II)-containing MOFs by the reaction of $Ir(ppy)_2(H_2dcbpy)PF_6$ (H_2dcbpy = 4,4′-dicarboxy-2,2′-bipyridine) with $M(ClO_4)_2 \cdot xH_2O$ (M = Zn(II), **1**; Cd(II), **2**; Co(II), **3**; Ni(II), **4**).[119] Excitation at 406 nm produced emission maxima at 613, 595, 587, and 600 nm (for **1–4**, respectively), which were also due to phosphorescence. The absolute emission quantum yields of **1–4** were reported to be 0.274, 0.193, 0.001, and 0.002, respectively, and correlate to the d^n configurations of the M^{II} nodes. MOFs **1** and **2** were based on d^{10} metals, which provide enhanced structural rigidity to the framework lattices and limited the amount of non-radiative decay pathways within each material. Meanwhile, MOFs **3** and **4**, prepared with d^7 and d^8 metals, suffered from a greater proportion of non-radiative decay pathways. Materials **1** and **2** were subjected to varying concentrations of O_2 in N_2 ranging from 0 to 100% O_2; the observed emission intensities are shown in Figure 7.11. The quenching efficiencies of **1** and **2** at 1 atm O_2 were 46 and 74%, respectively, and repeated cycles of 100% O_2 and 100% N_2 showed a high degree of reversibility with *a* > 75% intensity recovery after each cycle.

An example of O_2 luminescence sensing by a noble-metal free MOF has been reported by Chen and coworkers that showed exceptional sensing selectivity and reversibility.[122] MAF-X11 is a highly symmetric Zn(II)-based MOF with a distorted cubic pore structure that was obtained by reaction of two types of linear linker groups, 2-aminoterephthalic acid and

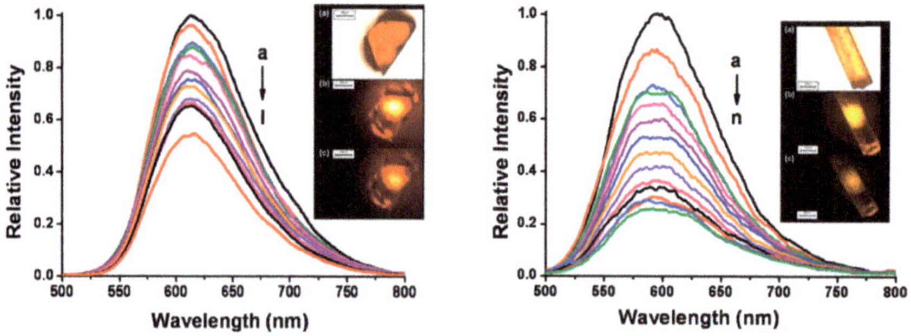

Figure 7.11 (Left) Response of **1** in single crystal under different mole fraction of $O_{2(g)}$ in nitrogen (a) 0, (b) 0.06, (c) 0.14, (d) 0.17, (e) 0.21, (f) 0.33, (g) 0.40, (h) 0.47, (i) 0.57, (j) 0.61, (k) 0.63, and (l) 1.0, λ_{ex} = 406 nm, inset: (a) color appearance of **1**, (b and c) luminescence changes of **1** with 100% N_2 and 100% O_2, respectively. Scale bar is 100 μm; (right) response of **2** in single crystal under different mole fraction of $O_{2(g)}$ in nitrogen (a) 0, (b) 0.05, (c) 0.10, (d) 0.15, (e) 0.20, (f) 0.25, (g) 0.30, (h) 0.40, (i) 0.50, (j) 0.60, (k) 0.70, (l) 0.8, (m) 0.9, and (n) 1.0, λ_{ex} = 406 nm. Inset: (a) color appearance of **2**, (b and c) luminescence changes of **2** with 100% N_2 and 100% O_2, respectively. Scale bar is 100 μm. Adapted from ref. 119 with permission from The Royal Society of Chemistry.

3,3′,5,5′-tetramethyl-4,4′-bipyrazole. The MOF was photoexcited (λ = 345 nm) such that the intensity of the corresponding broad fluorescence emission *ca.* 470 nm could be monitored under a range of O_2 pressures. At 1 bar O_2, the emission was quenched by 96.5%, which is higher than any previously reported O_2-sensing MOF and compares favorably to O_2 sensors composed of noble metal-containing complexes. When MAF-X11 was subjected to repeated cycles of exposure to 1 bar O_2 followed by evacuation, a rapid and highly reversible signal response was obtained. The material exhibited a similar reversibility using air, resulting in a quenching efficiency close to that observed for pure O_2 at 0.21 atm, indicating that MAF-X11 was highly selective for O_2 over other gas components in regular air.

7.2.3.2 H_2O Sensing

Humidity sensing is important for a variety of domestic applications and in meteorological science.[125] The majority of H_2O luminescence-based MOF sensors reported in the literature to date are composed of Eu(III),[126–130] Tb(III),[126,128] Sm(III),[127] or Dy(III).[131] H_2O is a strong quencher of photoluminescence because O–H oscillators facilitate efficient resonance energy transfer—particularly for excited Ln(III) species.[132] This presents both an opportunity and a challenge for H_2O sensing by Ln(III)-MOFs: while the presence of H_2O as an analyte should produce a measurable and quantitative suppression of the observed luminescence intensity, H_2O is also a good ligand for Ln(III) ions. Therefore, it is necessary to prepare Ln(III)-based MOF sensors that are not already highly coordinated (and therefore fully quenched) by H_2O, or that contain labile H_2O ligands that can be removed under facile conditions that do not result in decomposition of the MOF. For example, a Tb(III)-based MOF prepared using a triarylphosphine oxide ligand (referred to as PCM-15) has been shown to undergo a highly reversible exchange of H_2O ligands upon evacuation at 150 °C to provide an activated sensing material with well-defined, coordinatively-unsaturated Tb(III) centres.[110,133] PCM-15 retained its bulk crystallinity and porosity upon dehydration, exhibiting a two-fold increase in the luminescence intensity, which was quantitatively diminished upon re-exposure to humid air.

Another example of Tb(III)-based humidity sensing is provided by the work of Yu *et al.*,[128] who obtained two isostructural MOFs by reaction of pyridyl-4,5-imidazole dicarboxylic acid, oxalic acid, and $Ln(NO_3)_3$ (Ln = Tb(III), **1**; Eu(III), **2**). Excitation of the as-synthesized material **1** produced characteristic Tb(III) emission bands at 487, 542, 581, and 620 nm, while excitation of **2** gave similarly characteristic Eu(III) emission bands at 589, 612, 653, 699 nm. The MOFs were then activated by removal of uncoordinated H_2O, but notably not the Ln(III)-coordinated OH_2 groups, to yield the partially dehydrated analogues **1′** and **2′**. These materials were subsequently placed in an enclosed chamber with relative humidity held constant at 5, 30, 48, 75.8, and 85%. The luminescence responses were monitored at regular intervals between 1 and 24 h, which revealed *increasing* luminescence intensity with

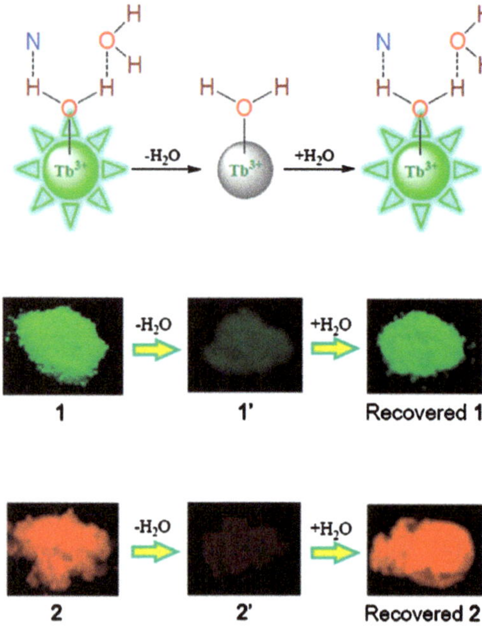

Figure 7.12 (Top) Proposed sensing mechanism; (center) the samples of **1**, **1**′, and recovered **1** illuminated with a 254 nm laboratory UV light; (bottom) the samples of **2**, **2**′, and recovered **2** illuminated with a 254 nm laboratory UV light. Adapted from ref. 128 with permission.

increasing relative humidity. This somewhat counterintuitive trend was explained by examination of the crystal structures of **1** and **2**. Prior to activation, uncoordinated H_2O molecules were H-bonded to the coordinated OH_2 ligands within the framework, thus perturbing vibrational coupling to the Ln(III) centres. However, removal of these guest H_2O molecules released the O–H oscillators of the coordinated OH_2 molecules, resulting in increased quenching (Figure 7.12).

An example of H_2O sensing by a non-Ln(III) MOF is provided by Chou *et al.*,[134] who developed a Zn(II)-based material based on both 1,2-bis(4-pyridyl)ethane (dpe) and 1,4-benzenedicarboxylate linkers. The hydrated framework exhibited a single emission band at 470 nm upon excitation at 325 nm due to a π*–π excimer-like transition between neighboring dpe ligand moieties. Dehydration of the sample produced a red-shifted, broad emission band at 500–550 nm, along with a remarkable 100-fold increase in the overall luminescence intensity.

7.2.3.3 Organic Molecule Sensing

Selective sensing of organics is a very broad domain given the almost endless range of potential analytes. Molecular sensors have become commonplace in many mainstream applications, but new and improved sensors are continually sought. The potential use of MOFs for specific, targeted molecular sensing applications has begun to be investigated in earnest. Pertinent recent examples in the literature range from the sensing of simple organic molecules, such as CS_2 and alkyl thiols by mixed In(III)–Ln(III) MOF thin films,[135,136] to the detection of more complex functionalized aromatics, such as melamine and acetoguanamine by a riboflavin-chelated Ag(I)-MOF.[137] In further examples, dipicolinic acid has been detected using mixed-Ln(III) MOFs,[138] and methyl viologen and methylene blue detection was demonstrated using a mixed Ru(II)–Zn(II) material.[139] Glucose molecules have been detected in human serum by a heterobimetallic Ir(III)/Zn(II) system,[118] and an Ag(I)-functionalized Zn(II)-MOF has been developed for selective olefin sensing.[140,141] In an example of chiral sensing, Lin *et al.*[142] have reported the enantioselective fluorescence sensing of four different amino alcohols by a chiral Cd(II)-MOF composed of carboxylated BINOL linkers. In the following section, we will highlight a number of examples that show how the tunable and versatile chemical properties of MOFs may be exploited to obtain selective sensors for a diverse range of molecular analytes.

7.2.3.3.1 Solvents. There are several reports of sensing of commonplace solvents by Zn(II)-based MOFs, including methanol,[107,143–145] ethanol,[14,107,146] isopropanol,[144] acetone,[14,93] benzene,[14,107,143] toluene,[14] pyridine,[143] and dimethylamine.[147] Shao *et al.*[14] developed a mixed Na(I)–Zn(II) MOF whose luminescence was enhanced by exposure to toluene or benzene, attributed to electron transfer from the higher energy π*-antibonding orbital of the analyte species to the conduction band of the framework. In contrast, the luminescence was quenched by coordinating solvents such as ethanol or acetone that could form dative donor interactions to metal ions within the framework, thus becoming more strongly vibrationally coupled *via* C–O and O–H oscillators. Zhu and coworkers have described the 'turn-on' sensing of methanol in ethanol by a Zn(II)-MOF composed of 4-(2-carboxyphenoxy) isophthalate linkers.[145] The intensity of the emission band at 404 nm (λ_{ex} = 304 nm) in this material displayed a linear relationship *versus* methanol concentration, and the high sensitivity of the material allowed for detection as low as 2×10^{-7}% methanol by volume.

There are an increasing number of accounts of solvent luminescence sensing by Ln(III)-MOFs, namely those based on Eu(III),[27,91,92,148–151] Tb(III),[18,27,78,110,151–154] Yb(III),[155] and some mixed Eu(III)/Tb(III),[27,151] Tb(III)/Gd(III),[156] and Yb(III)/Gd(III) systems.[156] Ng and coworkers recently reported a highly sensitive and selective methanol Tb(III)-MOF sensor composed of mixed 5-aminoisophthalate and 1,4-benzenedicarboxylate linkers.[153] The intensities of the typical Tb(III) emission bands at 490, 546, 586, and 624 nm (λ_{ex} = 341 nm) were monitored upon exposure to ten different solvents: methanol, ethanol, 1-propanol, 2-propanol, 1-butanol, acetone, acetonitrile, chloroform, ethyl acetate, and tetrahydrofuran. Only methanol caused a drastic decrease in luminescence intensity, while the nine other solvents produce little change. Exposure of the MOF to a 7 vol% solution of methanol quenched the emission by 96% (Figure 7.13, left). This high selectivity was attributed to the interaction of methanol with the ligands and/or open metal sites of the framework and the subsequent interruption of energy transfer from the ligand to the Tb(III) atoms. The selectivity was further probed by exposing the material to mixtures of methanol and other solvents, as well as all ten solvents combined as a mixture. The luminescence responses were virtually identical to that of the methanol-only solution, which indicates that the presence of other solvents has little to no effect on the high sensitivity and selectivity of the material (Figure 7.13, right).

Some pertinent examples of acetone sensing by luminescent MOFs serve to underline how the sensor signaling may be achieved by either luminescence quenching or enhancement—commonly referred to as 'turn-off' and 'turn-on' sensing, respectively. Chen *et al.*[155] reported the first near-infrared (NIR) luminescent MOF composed of biphenyl-3,4′,5-tricarboxylate linkers and Yb(III) nodes for the turn-off sensing of acetone. Upon excitation at 326 nm, the Yb(III)-MOF exhibited a single characteristic band for an expected $^2F_{5/2} \rightarrow {}^2F_{7/2}$ transition at 980 nm, which decreased in intensity linearly as a function of acetone concentration. Another Tb^{III}-MOF composed of biphenyl-4,4′-dicarboxylate and 3,3′-dicarboxylate-2,2′-dipyridine was shown to undergo almost complete quenching at a concentration of 21% acetone by volume.[78] Conversely, Ma *et al.*[18] reported the turn-on acetone sensor Cd(II)/Tb(III)-MOF, formed from the polymerisation of Cd(II) with pre-formed metalloligands composed of Tb(III)–oxydiacetic acid complexes. This material showed luminescence intensity enhancement upon exposure to acetone and an associated quantum efficiency increase from 57.6 ± 0.2% for the analyte-free MOF to 95.2 ± 0.2% in the presence of acetone.

An interesting series of Eu(III)/Tb(III) heterobimetallic MOFs called TCMs based on 4,4′,4″-(((2,4,6-trimethylbenzene-1,3,5-triyl)-tris(methylene))-tris-(oxy))tribenzoate (TCM) linkers was recently reported,[151] in which the

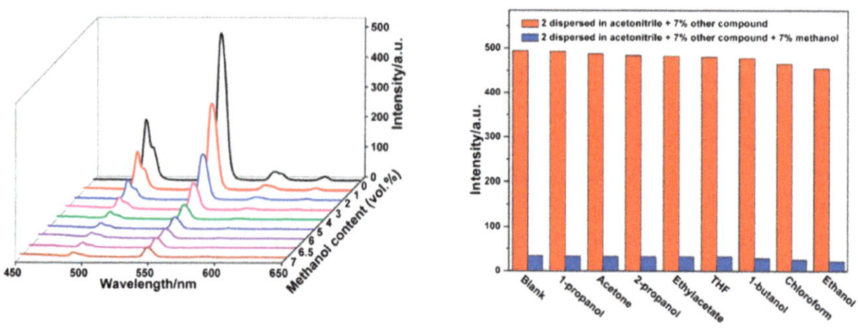

Figure 7.13 (Left) The photoluminescence spectra of **2**/acetonitrile emulsion in the presence of various amounts of methanol upon excitation at 341 nm; (right) the $^5D_4 \rightarrow {}^7F_5$ transition intensities of **2** dispersed into acetonitrile on the addition of different organics (red) and subsequent addition of methanol (blue) when excited at 341 nm. Adapted from ref. 153 with permission.

Eu–Tb ratio was systematically varied as a means to broadly tune the emission wavelength, allowing for multiple analyte sensing by a single MOF. This approach is most commonly referred to as MOF 'barcoding'.[157] The emission profiles of seven Eu_xTb_{1-x}-TCMs (x = 0, 0.1, 0.25, 0.35, 0.4, 0.65, 1) were compared upon excitation at 285 nm, and displayed continual tuning of the emission colours from red (Eu, $Eu_{0.65}Tb_{0.35}$, $Eu_{0.4}Tb_{0.6}$) to orange ($Eu_{0.35}Tb_{0.65}$), yellow ($Eu_{0.25}Tb_{0.75}$), and green ($Eu_{0.1}Tb_{0.9}$, Tb) (Figure 7.14).

The ratio of the intensities of the $^5D_0 \rightarrow {}^7F_2$ (Eu(III); 614 nm) and $^5D_4 \rightarrow {}^7F_5$ (Tb(III); 545 nm) transitions of $Eu_{0.35}Tb_{0.65}$-TCM were then monitored upon exposure to benzene, styrene, benzyl alcohol, nitrobenzene, chlorobenzene, aniline, methanol, and DMF (Figure 7.15).

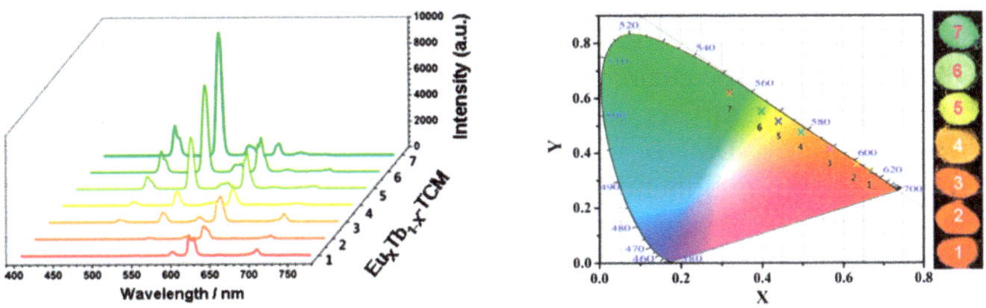

Figure 7.14 (Left) Emission spectra of Eu_xTb_{1-x}-TCM (x = 1 (1), 0.65 (2), 0.4 (3), 0.35 (4), 0.25 (5), 0.1 (6), 0 (7)) in the solid-state at room temperature (λ_{ex} = 285 nm); (right) CIE chromaticity coordinates and the corresponding optical photographs for Eu_xTb_{1-x}-TCM excited at 285 nm. Adapted with permission from ref. 151. Copyright 2014 American Chemical Society.

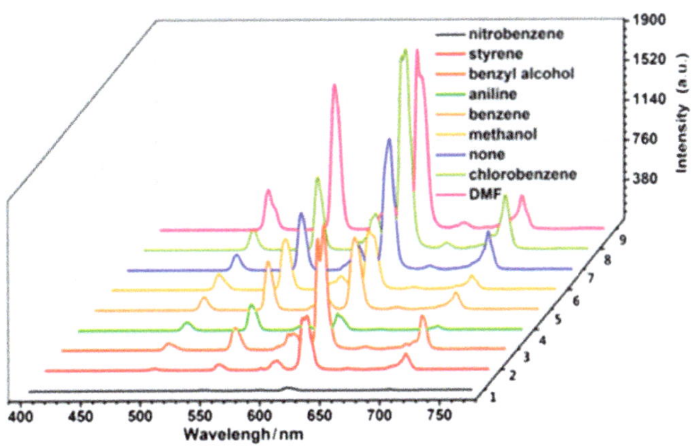

Figure 7.15 Emission spectra for $Eu_{0.35}Tb_{0.65}$-TCM after adsorption of nitrobenzene (1), styrene (2), benzyl alcohol (3), aniline (4), benzene (5), methanol (6), none (7), chlorobenzene (8), and DMF (9) molecules excited at 285 nm. Reprinted with permission from ref. 151. Copyright 2014 American Chemical Society.

Each solvent selectively quenched or enhanced the Eu(III) and/or Tb(III) luminescence to a different degree, resulting in further tuning of the overall emission wavelength, which was visually manifested as an instantaneous change in emission colour. From this data, a fingerprint of each solvent was created that led to the successful identification of each molecule. Further studies allowed for the differentiation between three isomers of xylene, four halobenzenes, and benzene, toluene, and ethylbenzene.

7.2.3.3.2 Nitroaromatics and Explosives. The rapid detection of nitroaromatics and related high-energy materials is increasingly important for safety and security on many levels. Existing detection methods employed in high-throughput screening processes, such as in airports, are not only expensive and cumbersome, but also lack the required levels of sensitivity.[158] Examples of nitroaromatic molecule sensing by luminescent MOFs are rapidly increasing in number and include the detection of a range of pertinent targets, including nitromethane (NM),[159] 2,3-dimethyl-2,3-dinitrobutane (DMNB),[160,161] nitrobenzene (NB),[161–166] 2-nitrotoluene (2-NT),[161,164,165] 4-nitrophenol (NP),[167] 4-nitroaniline (NA),[167] *m*-dinitrobenzene (*m*-DNB),[161,164–166] *p*-dinitrobenzene (*p*-DNB),[161,164–166] 2,4-dinitroaniline (DNA),[167] 2,4-dinitrotoluene (2,4-DNT),[159–161,164–166] 2,4-dinitrophenylhydrazine (DNPH),[167] 2,4,6-trinitrotoluene (TNT),[159,166,168] and 2,4,6-trinitrophenol, or 'picric acid' (PA).[167] Li and coworkers have designed two potential sensor materials based on the reaction of bis-(3,5-dicarboxyphenyl)terephthalamide (**L**) with Cd(II) (**1**) and Zn(II) (**2**) precursors.[165] The emission spectra for **L**, **1**, and **2** (λ_{ex} = 330 nm) exhibited strong bands at 458, 460 and 475 nm, respectively, confirming that the luminescence from MOFs **1** and **2** occurred solely due to ligand-centred electronic excitations. The luminescence intensities of **1** and **2** were quenched by ≥90% upon exposure to 2000 ppm *p*-DNB, NT, 2,4-DNT, *m*-DNB, and NB. Addition of methanol, ethanol, 1-propanol, 2-propanol, DMF, THF, DCM, acetone, benzene, toluene, or phenylmethanol had virtually no effect on these intensities, showing that these MOFs could be employed as highly selective sensors for the detection of trace amounts of nitroaromatic compounds even when diluted by solvents and other aromatic molecules.

There have also been numerous reports of nitroaromatics sensing by Eu(III)-MOFs for the detection of NM,[169] DMNB,[169] NB,[169–171] 2-NT,[169,172] 4-NT,[169,172] *o*-DNB,[169] *m*-DNB,[169] 2,4-DNT,[169–172] 2,6-DNT,[169,171,172] 3,4-DNT,[169] TNT,[170,171] and PA.[99,171] For example, Huang *et al.*[99] prepared a Eu(III)-MOF using 9,9-dimethylfluorene-2,7-dicarboxylate linkers that exhibited approximately 92% luminescence quenching in the presence of PA, but significantly less quenching was induced by NM, DMNB, *m*-DNB, 2-NT, 4-NT, 2,4-DNT, or 2,6-DNT. A nanosized Tb(III)-MOF composed of BTC linkers was prepared by a rapid ultrasound-assisted method, and has shown suitability for selective PA detection without interference from other common nitroaromatics.[173] The addition of 90 μM PA quenched the luminescence of the typical Tb(III) emission bands by 75% or more. However, there was little or no luminescence

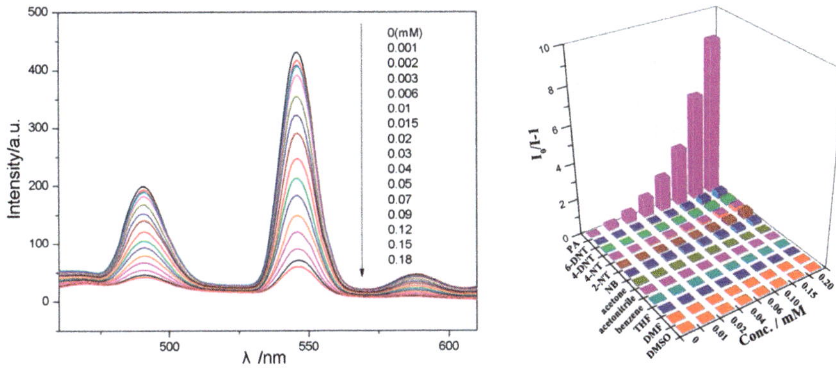

Figure 7.16 (Left) Variation of emission spectra of $[Tb(1,3,5\text{-}BTC)]_n$ nanocrystals with different concentrations of PA. The excitation wavelength is 324 nm; (right) luminescence quenching results of nitroaromatic explosives and commonly found interferents, including acetone, acetonitrile, benzene, THF, DMF, and DMSO, to the emission of the as-prepared $[Tb(1,3,5\text{-}BTC)]_n$ nanocrystals at various concentrations in ethanol. Adapted from ref. 173 with permission.

quenching in the presence of NB, 2-NT, 4-NT, 2,4-DNT, and 2,6-DNT, as well as acetone, acetonitrile, benzene, THF, DMF, or DMSO (Figure 7.16).

This high degree of selectivity was primarily attributed to the lower energy LUMO level in PA compared to the other nitroaromatics, which enabled the excited electrons in $[Tb(1,3,5\text{-}BTC)]_n$ to transfer more easily to the PA molecules. It was also postulated that stronger hydrogen bonding interactions between PA and the framework would facilitate more effective MOF–PA donor–acceptor charge and/or electron transfer. Other examples of selective sensors include Cd(II)-MOFs for the detection of PA,[174,175] Li(I)- and Mg(II)-MOFs for the detection of NB,[135,176] and an In(III)-MOF for the detection of NB, *m*-DNB, and 2,4-DNT.[177]

7.2.4 Sensing of Biomolecules

In recent years, there has been great interest in the luminescence sensing of biomolecules with masses in the kDa range, especially for the non-invasive detection of particular disease markers. The micropores of MOFs are generally too small to allow for adsorption of large biomolecule analytes; however, MOFs with lower framework connectivity (*e.g.*, 2-D layered MOFs) can be functionalized by a process of exfoliation of individual sheets. For example, the 2-D Cu(II)-based MOF *N*,*N'*-bis(2-hydroxyethyl)dithiooxamidatocopper (H_2dtoaCu)[178–180] was functionalized for the *in vitro* detection of duplex DNA strands by first adsorbing a triplex-forming oligonucleotide (TFO) labeled with a fluorescent fluorescein amidite dye (FAM) between the layers (Figure 7.17).

After functionalisation, emission from the FAM–TFO strands was already quenched due to photo-induced electron transfer from the MOF to a vacancy

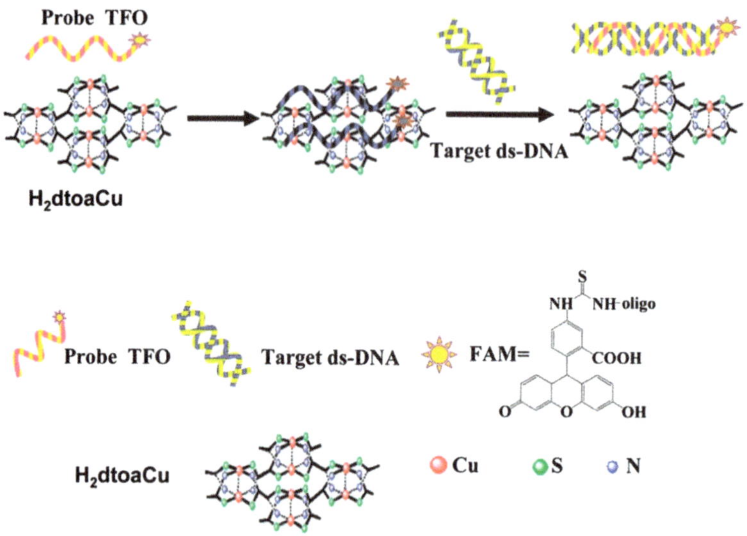

Figure 7.17 Mechanism for the ds-DNA fluorescent biosensor based on H_2dtoaCu. Reproduced from ref. 178 with permission.

in the HOMO of the fluorophore, formed by the promotion of an electron from the HOMO to the LUMO upon photoexcitation. However, in the presence of the target DNA analytes, fluorescence was turned back on due to the formation of Hoogsteen hydrogen bonds between bases in the DNA strands and the TFO receptors. A detection limit of 1.3 nmol L^{-1} was obtained. Chen and coworkers have utilized the same TFO-functionalised H_2dtoaCu MOF to successfully detect double-stranded DNA of the polypurine tract sequence of HIV-1 RNA, which plays a pivotal role in the life cycle of the HIV-1 virus.[178] H_2dtoaCu has also been utilized to detect oligonucleotide sequences of wild-type Hepatitis B,[180] the HIV-1 U5 long terminal repeat sequence,[179] and thrombin,[179] as well as antibodies of the H_5N_1 flu virus in human serum samples.[181]

Li *et al.*[165] have designed analogous Zn(II) and Cd(II)-based MOF sensors based on the polymerisation of bis(3,5-dicarboxyphenyl)terephthalamide. FAM-labeled probe DNA was first adsorbed by the MOFs, and upon introduction of the target DNA, the luminescence of the system was restored. These materials have been used to detect the presence of target DNA at concentrations as low as 0.05 nM, and both displayed a high degree of selectivity for the target DNA strands down to a single mismatched nucleotide.

7.2.5 Temperature Sensing

Temperature sensing by monitoring the changes in the luminescence intensity of MOFs has been reported in a few monometallic systems comprised of Al(III),[182] Cd(II),[183] and Zn(II),[163] as well as for some heterobimetallic Ln(III) materials.[184–188] Wu and coworkers[183] developed two structurally distinct 2-D Cd(II)-MOFs using the same organic linker, *N*,*N*′,*N*″-1,3,5-triazine-2,4,6-triyl tris(4-aminomethylbenzoic acid) (H_3TTAA), that has a C_3N_6 core. The ligand itself is highly luminescent, exhibiting strong blue emissions and temperature-dependent quantum yields. The absolute fluorescence quantum yields of the most highly luminescent MOF was 40.3% at 298 K, which increased to 74.0% upon cooling to 77 K accompanied by a slight blue shift of 8 nm. The substantial increase in fluorescence was achieved by a reduction of non-radiative decay of the intra-ligand excited state (π–π*) due to the increased rigidity of the ligand at lower temperature.

A mixed-metal Eu(III)/Tb(III)-MOF obtained using 2,5-dimethoxy-1,4-benzenedicarboxylic acid (H_2DMBDC) also exhibited temperature-dependent luminescence between 10 and 300 K.[184] The emission spectrum of $Eu_{0.0069}Tb_{0.9931}$–DMBDC (λ_{ex} = 355 nm) contained a hybrid of the characteristic $^5D_4 \rightarrow {}^7F_J$ (J = 6, 5, 4, 3; Tb(III)) and $^5D_0 \rightarrow {}^7F_J$ (J = 1, 2, 3, 4; Eu(III)) transitions. The intensities of the $^5D_4 \rightarrow {}^7F_5$ (*ca.* 545 nm) and $^5D_0 \rightarrow {}^7F_2$ (*ca.* 613 nm) transitions were monitored from 50–300 K in 50 K increments, revealing an increase in the intensity of the Tb(III) band with decreasing temperature but a concomitant decrease in the intensity for Eu(III). The ratio of the intensities (I_{Tb}–I_{Eu}) followed a linear correlation with temperature between 50 and 200 K. Hasegawa *et al.*[185] have reported another mixed Eu(III)/Tb(III)-MOF composed of 4,4′-bis(diphenylphosphoryl) biphenyl (dpbp) linkers that operates in a similar manner: $[Tb_{0.99}Eu_{0.01}(hfa)_3(dpbp)]_n$ (hfa = hexafluoro acetylacetonato) exhibited identical temperature-dependent fluorescence intensities of the $^5D_4 \rightarrow {}^7F_5$ (543 nm; Tb(III)) and $^5D_0 \rightarrow {}^7F_2$ (613 nm; Eu(III)) transitions between 200 and 450 K. The emission colour of the MOF changed from green to yellow, to orange, and finally to red at 250, 300, 350, and 400 K, respectively.

Monge and co-workers have presented an advanced example of emission colour tuning and temperature-dependent luminescence sensing with the development of multimetallic Eu(III)/Tb(III)/Gd(III)-based MOFs comprising 3,5-disulfobenzoate (DSB) linkers.[188] The isostructural monometallic MOFs (Gd–DSB, Eu–DSB, Tb–DSB) gave blue, red, or green emissions, respectively, when excited 10 K (λ = 270 nm). The blue-coloured emission from Gd–DSB was due solely to phosphorescence emission of the DSB ligand (since Gd(III) is a non-emissive species). The emission spectra of $Eu_{0.02}Gd_{0.98}$–DSB, $Tb_{0.02}Gd_{0.98}$–DSB, and $Eu_{0.05}Tb_{0.09}Gd_{0.86}$–DSB obtained at various temperatures between 10 and 300 K resulted in a broad range of emission colours. The calculated CIE chromaticity coordinates of each emission spectrum are shown in Figure 7.18 at the specified temperatures.

The luminescence intensities of $Eu_{0.02}Gd_{0.98}$–DSB and $Tb_{0.02}Gd_{0.98}$–DSB were greatly enhanced relative to those of Eu–DSB and Tb–DSB, respectively, due to the resonance energy transfer from the ligand triplet state in addition to energy transfer from the $^6P_{7/2}$ level of Gd(III) to Eu(III) and Tb(III). Additionally, because the intensity of the blue emission from Gd–DSB was highly sensitive to temperature from 10 to 300 K, doping the material with either or

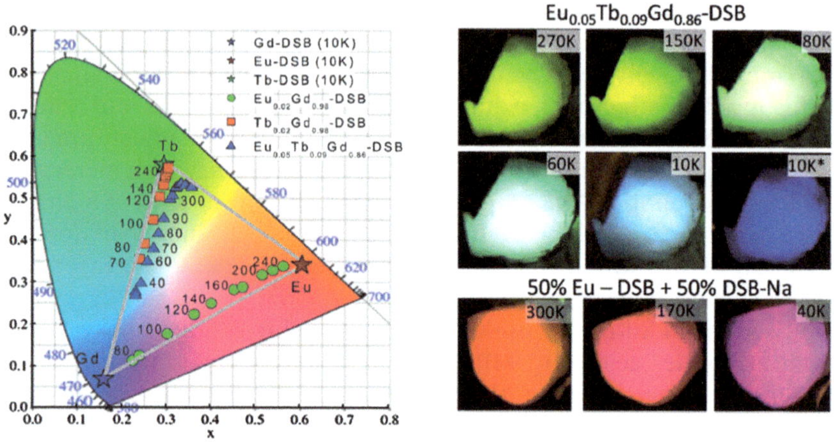

Figure 7.18 (Left) CIE chromaticity diagram showing the temperature dependence of the (x, y) color coordinates calculated to the corrected emission spectra of $Eu_{0.02}Gd_{0.98}$–DSB, $Tb_{0.02}Gd_{0.98}$–DSB, and $Eu_{0.05}Tb_{0.09}Gd_{0.86}$–DSB; (right) photographs of the pellets at different temperatures. The image labelled 10 K*, recorded with a time delay from the excitation, corresponds to the colour of the long lived triplet emission. Adapted from ref. 188 with permission.

both of the other Ln(III) ions produced a broad range of emissive colours that could be fine-tuned by small variations in temperature.

7.3 Electromechanical Sensors

The last part of this chapter addresses chemical sensing by MOFs in which the sensor feedback mechanism relies on mechanical signaling instead of some form of luminescence modulation. Although the majority of research to date has focused on the latter, some very interesting recent reports have shown how MOF-functionalized devices might be exploited to detect changes in local chemical environment *via* electromechanical effects. The main objective is to incorporate MOFs into known sensor technologies in order to exploit their inherently high porosity, extensive chemical tunability and adsorption selectivity as a means to increase overall sensitivity and analytical capacity.[5] For example, existing electromechanical sensors, such as quartz crystal microbalance (QCM)[189] and surface acoustic wave (SAW)[190] devices, rely on measurable changes in physical variables caused by surface-based chemical interactions. Present limitations, such as overall signal intensity or limited applicability to particular chemical analytes, may be overcome if MOFs can be incorporated into these devices to act as analyte-specific signal transducers.

Thus far, efforts in this field have primarily focused on the synthesis and application of MOF materials as bulk microcrystalline powders and their subsequent testing in model devices. In order to take full advantage of MOFs as electromechanical sensors for analytical applications, the development of MOF-based thin films or coatings is required, since this would permit MOFs to be directly incorporated into existing devices.[191] There are a number of inherent difficulties regarding the assembly of MOFs as thin films; therefore, before we discuss specific examples of these sensing materials, it is appropriate to explore some recent developments in MOF thin film growth.

7.3.1 Preparation of MOF Thin Films

When investigating the development of new types of thin films, it is imperative to consider some important physical attributes that will produce a well-functioning substrate, including controllability of film thickness, uniformity of microstructure, film smoothness, and crystallite orientation.[192] The synthesis of MOF thin films has proven difficult by traditional (solvothermal) means due to an inherent lack of control over morphology,[147] film thickness,[193] and surface smoothness.[194] Of the most common thin film preparation techniques employed industrially, (*e.g.*, chemical vapour deposition (CVD), atomic layer deposition (ALD), chemical solution deposition (CSD)), CSD is considered the fastest, simplest, and cheapest technique.[195] However, each of these techniques presents different practical difficulties when applied to MOF thin film growth, and as such, no preferred protocol appears to have been defined at this time.

The main sub-classes of MOF thin films are discussed in the following sections, as well as some innovative techniques that may enable the preparation of high-quality films for application in surface-based electromechanical sensing. An important distinction between types of MOF thin films concerns whether the film is prepared as a simple polycrystalline film, or by surface-attachment of MOF crystallites (SURMOFs; see later). A polycrystalline film or membrane is an assembly of irregularly-oriented crystallites obtained by deposition onto a substrate; this simple approach typically results in a thick, disordered film that has similar bulk characteristics to the unsupported MOF material. Meanwhile, SURMOFs are commonly thinner MOF multilayers with regularly oriented crystallites that produce more smooth surfaces by the process of controlled film growth.[196]

7.3.1.1 *Substrates for Thin Film Growth*

A substrate, in this context, is simply a material that acts as a support for the MOF thin film growth (Figure 7.19). Substrates add rigidity to the resulting film and can also introduce novel chemical or mechanical attributes that are useful in the desired applications. Examples of substrates that have been used for MOF thin film growth include ordered metal surfaces (*e.g.*, Au,[197] Cu[198]), carbon-based materials,[199] and common oxides (*e.g.*, alumina,[200] silica[201]), amongst others. These types of substrates can be obtained in several bulk forms, including meshes,[198] wafers,[146] porous solids,[202] or slides.[203] They are also compatible with existing sensor devices such as QCMs and SAWs

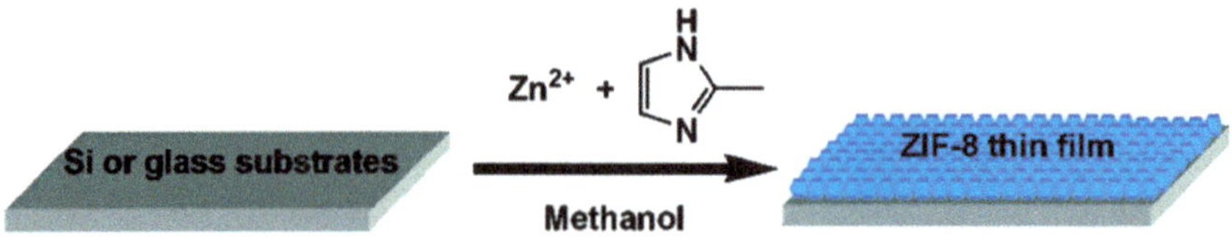

Figure 7.19 Synthesis of a ZIF-8 thin film on a Si or glass substrate. Adapted with permission from ref. 146. Copyright 2010 American Chemical Society.

(see later). In order for MOF crystals to successfully graft to the surface of a given substrate, a substrate with terminal chemical moieties that maximize the resulting MOF–substrate interactions should be chosen. For example, MOFs are commonly synthesized using carboxylated organic linker groups that facilitate strong metal–carboxylate bonds. Therefore, a carboxylic acid-terminated substrate should promote strong surface tethering of pre-formed MOF crystallites through the formation of formal coordination bonds.

An alternative means to functionalise the substrate surface involves the use of self-assembled monolayers (SAMs) as directing agents for the subsequent adsorption or growth of chemically reactive over-layers. SAMs are highly ordered monolayers that allow for the functionalisation of a given surface with particular chemical moieties.[204] For example, an ordered Au(111) surface can be functionalised with abundant carboxylic acid groups by reaction with appropriate long-chain thiocarboxylic acids.[205] The resulting exposed carboxylic acid termini facilitate subsequent functionalisation with the target sensor species. SAMs have been studied extensively in the functionalisation of a range of inorganic surfaces with organic species.[206] The adsorption of large molecules (*e.g.*, biomolecules) onto thin films can be rendered a highly controllable process using chemically-compatible SAMs.[207] This strategy is useful for the growth of MOF thin films because SAMs can be chosen to favour the organised adsorption of MOF crystallites *via* the formation of strong MOF–substrate interactions.[194]

7.3.1.2 Common Techniques for Polycrystalline Thin Film MOF Growth

7.3.1.2.1 Solvothermal Synthesis. Despite some of the property-related drawbacks associated with polycrystalline thin films, they remain attractive due to their simplicity. Conveniently, the synthetic procedure originally reported for the formation of a given MOF can usually be directly applied for polycrystalline thin film growth. Under regular solvothermal synthesis conditions, a solution containing the MOF precursors and other additives is heated (typically to temperatures in the 100–250 °C range) directly in the presence of the pre-functionalized substrate. This induces MOF crystallisation and deposition in a single-step process and typically produces film thicknesses on the order of micrometres, which is tolerable in systems that

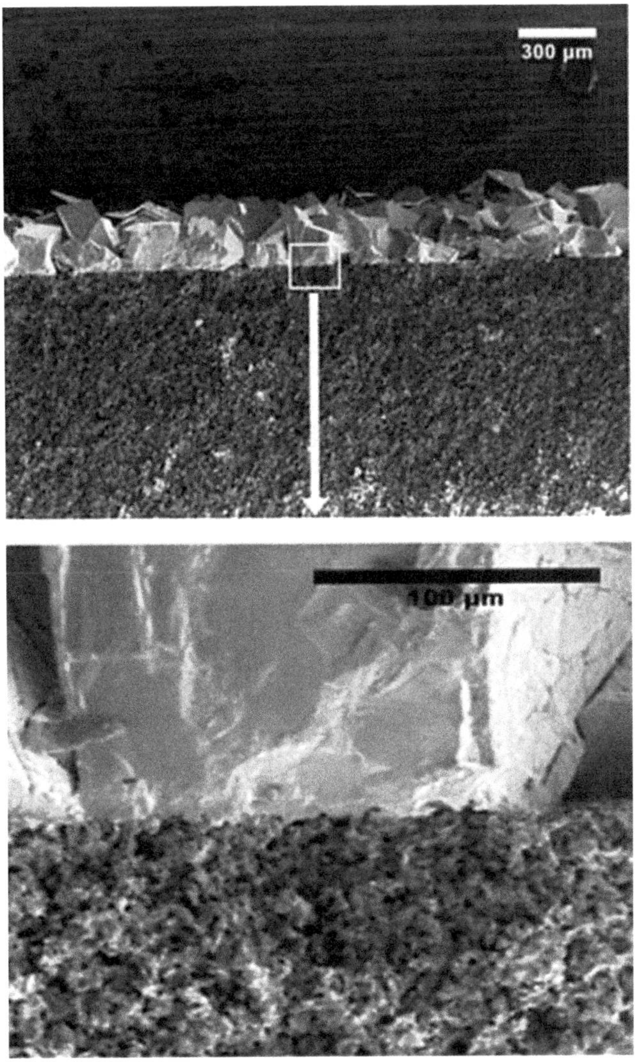

Figure 7.20 (Top) Profile of $Mn(HCO_2)_2$ crystals grown on a graphite support disc after two *in situ* crystallisations; (bottom) the magnified image shows intergrowth between the carbon support and $Mn(HCO_2)_2$. Reprinted with permission from ref. 209. Copyright 2007 John Wiley and Sons.

will take advantage of bulk material attributes (*i.e.*, porosity, sorption selectivity).[196,208] In cases where MOF crystallisation is rapid, dip-coating may also be employed. An example of non-oriented film growth is provided in Figure 7.20.[209] A thin film is easily accomplished by this method *via* repeated submersion of the substrate in fresh mother liquor solutions over multiple cycles until the desired film thickness is obtained.[210]

7.3.1.2.2 Room Temperature Synthesis. MOF thin films are most commonly synthesized by heating (*e.g.*, solvothermal,[211] microwave,[212] ultrasonic[213]). However, these types of methods may limit the choice of substrate because many are incompatible with harsher reaction conditions and suffer degradation;[214] examples include temperature-sensitive substrates and integrated devices (*i.e.*, gold surfaces or microcantilever devices). Substrates and MOF reagents may also undergo unwanted side reactions at elevated temperatures. In these instances, modified low-temperature synthetic routes must be established. Some pertinent advances in MOF thin film growth at room temperature, reported by Hermes and co-workers, involve the growth of MOF-5 on a carboxylic acid-terminated SAM (Figure 7.21).[215] The premise of the low-temperature method as applied to MOFs requires separation of crystallite nucleation and growth phases. This is accomplished by first preparing a MOF precursor solution, followed by a short heating period to initiate nucleation in the absence of the intended substrate. The solution is then cooled and filtered before the substrate is introduced.[196]

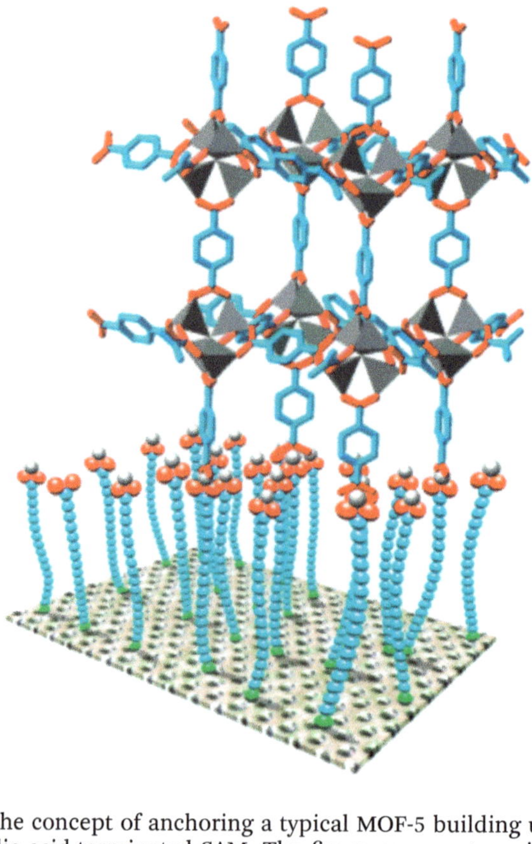

Figure 7.21 The concept of anchoring a typical MOF-5 building unit to a carboxylic acid-terminated SAM. The figure represents a simplified model, not excluding alternative possibilities of the linkage. Reprinted with permission from ref. 215. Copyright 2005 American Chemical Society.

7.3.1.3 SURMOFs

SURMOFs (surface-attached crystalline oriented MOFs), as the name suggests, are films of MOFs that have been grown on a substrate in a highly uniform and oriented fashion.[216] Although there have been several attempts to make highly oriented thin films, these have proven difficult to synthesise due to lack of control over the nucleation step. Two techniques have been employed in an attempt to circumvent this issue, namely the Langmuir–Blodgett method (LBM) and liquid-phase epitaxy (LPE); these have resulted in the most ordered crystalline MOF thin films reported to date.[217]

7.3.1.3.1 The Langmuir–Blodgett Method Applied to SURMOFs. The LBM is by far the oldest technique known for oriented thin film growth, with publications dating back to the early 20th century.[218] SURMOFs were first reported by a modified version of the LBM method, developed by Kitagawa and co-workers in 2010.[219] Their process involved a Langmuir–Blodgett layer-by-layer deposition technique that was compatible with standard MOF reagents and synthesis conditions. In the published example, the organic linker precursors were dissolved in an organic phase and then spread onto an aqueous sub-phase solution of the metal precursor ($CuCl_2$) in a Langmuir–Blodgett trough. This led to the formation of a single 2D layer of the MOF. The layer was then transferred on to a solid substrate (Si(100)) by dipping, followed by washing and drying steps to remove any unreacted precursors that could damage the film uniformity (Figure 7.22). Repeated cycles of this treatment allowed for a high degree of control over uniformity along the plane of the support. The overall smoothness and uniformity of MOF films obtained using this method was greatly improved in comparison to polycrystalline-based films. Since 2D layers of precursors tend to naturally order themselves based on van der Waals forces and/or π-stacking interactions, this technique is well-suited to the types of highly oriented precursors commonly employed in MOF synthesis.[219]

7.3.1.3.2 Liquid-Phase Epitaxy Applied to SURMOFs. LPE is a newer method for thin film growth that dates back to the early 1990s and was originally designed for polyelectrolytes.[220] LPE involves cycles of alternating submersion of a substrate into solutions of two or more precursor components. In the case of SURMOF assembly, this commonly entails alternating exposures of the substrate to solutions containing the organic ligand(s) and the metal salt(s). It is also possible to include washing steps between dips, if necessary. This layer by layer addition is visually demonstrated in Figure 7.23.

The LPE technique has been commonly used in conjunction with SAM-modified substrates in order to further improve the regularity of the film growth. Early attempts to apply LPE to the bottom-up growth of MOF-functionalized layers was hampered by necessary reaction conditions (*e.g.*, −60 °C and 6–8 h immersion times).[221] However, more recent reports have shown that a MOF thin film can be successfully synthesized by LPE at room temperature and

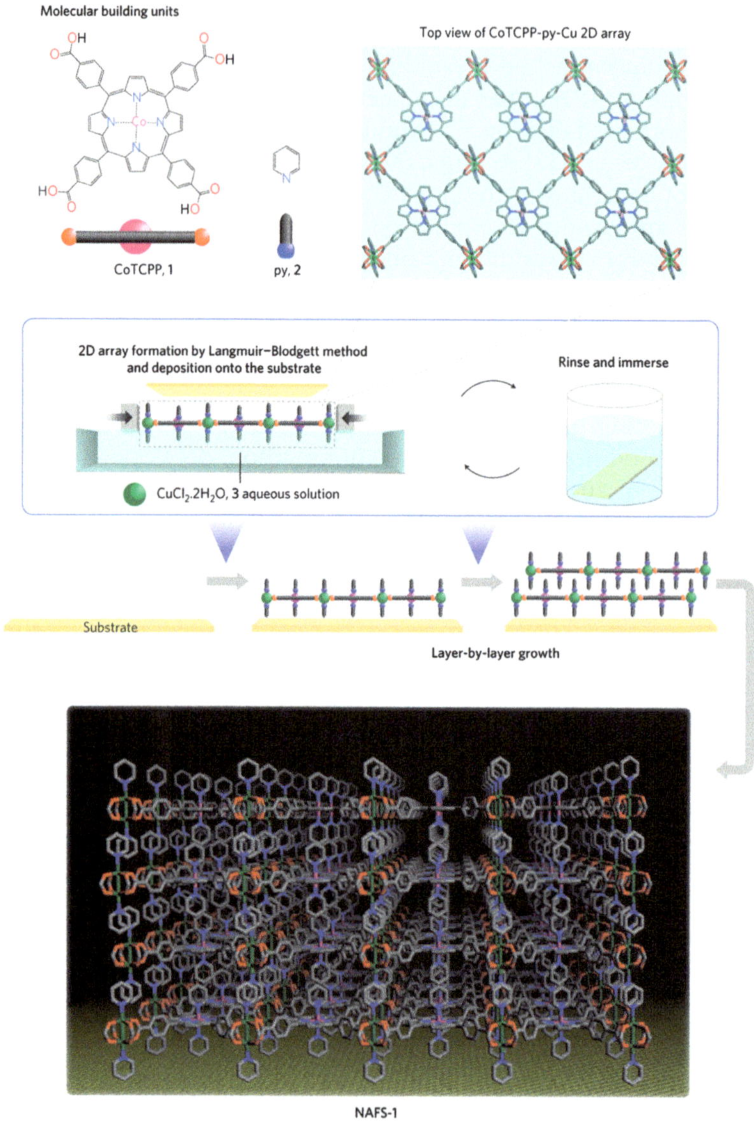

Figure 7.22 The solution mixture of CoTCPP (**1**) and pyridine (**2**) molecular building units is spread onto an aqueous solution of $CuCl_2{\cdot}2H_2O$, **3**, in a Langmuir trough. Pressing the surface with barrier walls leads to the formation of a copper-mediated CoTCPP 2D array (CoTCPP-py-Cu) (Langmuir–Blodgett method). Surface-pressure/area (π–a) isotherms were measured to evaluate the molecular area. The 2D arrays are deposited onto the substrate by the horizontal dipping method at room temperature. The substrate is then immersed in the pure solvent to remove excess starting materials or physisorbed components and dried. The repetitive process of successive sheet deposition and rinsing/solvent immersion leads to the sequential layer-by-layer growth of NAFS-1 with any desired thickness. C atoms are shown in grey, N in blue, O in red, Co(II) ions in pink and Cu(II) ions in green. Reprinted with permission from ref. 219. Copyright 2010 Nature Publishing Group.

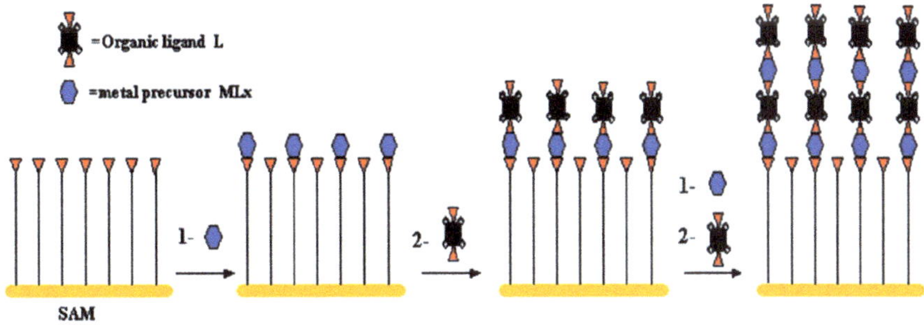

Figure 7.23 Schematic diagram for the step-by-step growth of the MOFs on a SAM by repeated immersion cycles, first in a solution of metal precursor and subsequently in a solution of organic ligand. Here, the scheme simplifies the assumed structural complexity of the carboxylic acid coordination modes. Reprinted with permission from ref. 222. Copyright 2007 American Chemical Society.

by employing submersion times of only 0.5–1.5 h.[222] These materials can be easily incorporated into analytical devices, such as QCMs.[196]

7.3.2 MOF Thin Films and QCMs

A quartz crystal microbalance (QCM) is a highly sensitive analytical tool that measures mass per unit area down to the nanogram scale. It operates by sensing changes in the resonant frequency of a quartz crystal acoustic resonator due to the presence of an analyte at its surface, which affects the apparent thickness of the resonator.[223] Interest in MOF-based QCMs has increased over the past several years.[224–226] MOFs are attractive materials to functionalise quartz for selective sensing applications due to the guest adsorption properties of the MOF. MOF thin films have been employed as QCM substrates both for selective sensing applications and for the direct analysis of the MOFs themselves.[227,228] An operational QCM consists of many components, the most important of which is a quartz single crystal that is polished into a very thin wafer (Figure 7.24).

Quartz is chosen due to its superior chemical, electronic and mechanical properties as a piezo-electric material. Electrodes are attached to either side of the quartz wafer to induce an oscillating electric field; then, perturbations to the oscillating field are induced by the analyte. The changes in field oscillation are detected and converted into a diagnostic measurement of mass per surface area.[229] Typically, if a particular chemical analysis requires the use of a substrate other than quartz, a secondary thin film is deposited or grown directly on top of the quartz wafer. To do so, a SAM is often employed to functionalise the quartz surface. It is therefore possible to deposit an ordered layer of MOF crystallites onto quartz using the deposition strategies described above. For example, a hydroxyl-functionalised quartz crystal

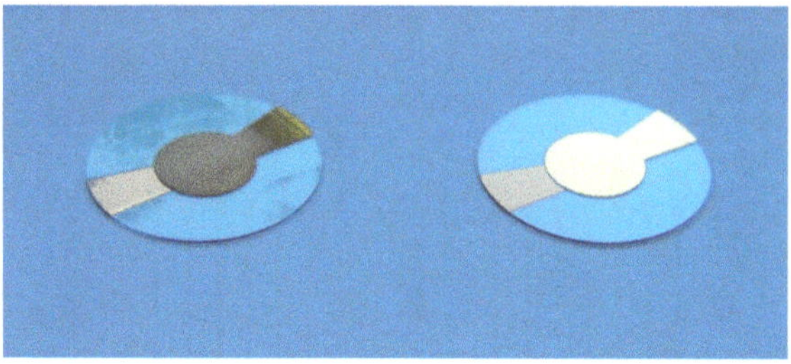

Figure 7.24 (Left) QCM-devices after growth of $Cu_3(BTC)_2(H_2O)_3{\cdot}xH_2O$: opaque layer on the SAM-modified gold electrode; (right) reference chip without SAM on the gold surface. Reprinted with permission from ref. 225. Copyright 2008 Elsevier.

allows for the direct grafting of hydroxyl- or carboxyl-decorated MOFs, which are able to undergo condensation reactions.[2]

In 2012, Khoshaman and co-workers published a study on the detection of VOCs using a QCM that employed a MOF thin film of HKUST-1 ($[Cu_3BTC_2(H_2O)_3]_n$).[230] The thin film was applied to a quartz wafer *via* two different processes: drop-casting and electrospray. These two contrasting methods were studied in order to allow for a comparison of the overall performance of a non-uniform MOF film (obtained by drop-casting) *versus* a much more uniform MOF thin film (obtained by electrospray deposition). Both substrates were employed in an electronically-monitored QCM arrangement in the presence of different VOC analytes (acetone, tetrahydrofuran and isopropanol at vapour pressure concentrations of 25, 23, and 5%, respectively). The electrospray-prepared MOF sensor achieved a response with approximately four times the sensitivity compared to the drop-casted MOF substrate, indicating superior performance of the more highly-ordered film. Although the VOCs detected in this study were at high relative concentrations, the researchers calculated that the best-performing substrate could achieve detection down to 2–50 ppm.

In a more recent example, Hirai *et al.*[231] employed a multi-step layering process in which a gold-coated QCM substrate was modified with a SAM consisting of 16-mercaptohexanedecanoic acid ($HS(CH_2)_{12}CO_2H$) groups. The resulting carboxylic acid-terminated surface was subsequently exposed to solvothermal reactions in the presence of precursors used to form the MOF Zn-CID-5. This LPE process resulted in Au surfaces decorated with bundles of Zn-CID-5 crystals ranging between 50 and 100 μm in size. Smaller crystals (5–10 μm) were obtained using a more rapid microwave-assisted heating method. More importantly, it was possible to control the orientations of the MOF crystals relative to the Au substrate using chemical modulators, such as 4-phenylpyridine. Zn-CID-5 crystallites grown in the [0,0,1], [1,−1,−1] and [0,1,0] directions allowed for the MOF micropore channels to be presented to incoming analytes in different orientations relative to the substrate surface (Figure 7.25).

This was found to affect the rate of pore filling upon exposure of the device to VOCs. Each functionalised QCM wafer was used to measure the

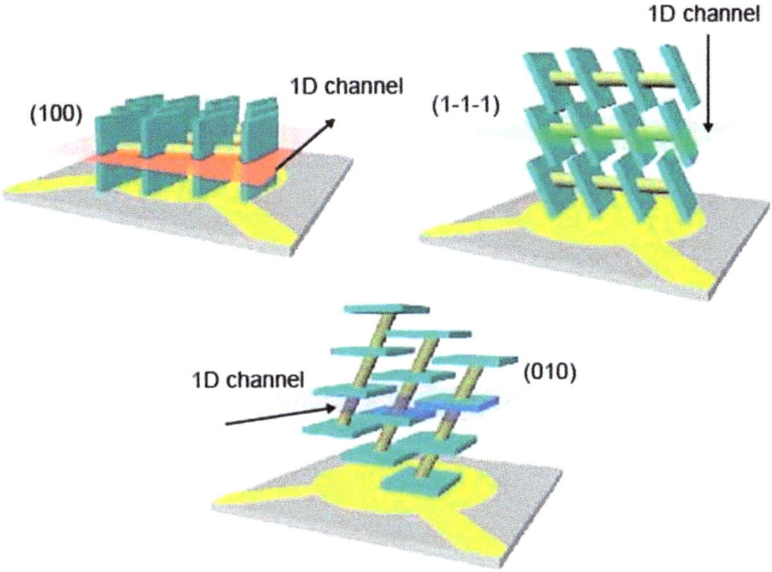

Figure 7.25 The channel orientation of (top left) [100], (top right) [1-1-1], and (bottom) [010] oriented samples. The 1D channel is running nearly parallel to the substrate for [100] and [010] oriented samples, but perpendicular to the substrate for the [1-1-1] oriented sample. Reproduced from ref. 231 with permission.

time-dependent methanol uptake for each orientation as a function of mass change. A comparison of the relationship between adsorption kinetics and MOF crystallite orientation is shown in Figure 7.26. This type of orientation-based control demonstrates great promise for such systems as highly tunable sensor devices.

7.3.3 Microcantilever-Based MOF Sensing Devices

Microcantilever (MCL) devices were originally employed in atomic force microscopy and have since been employed in many other areas, including biological sensing. An MCL is essentially a micron-sized diving board that is anchored at one end and oscillates with a regular frequency (Figure 7.27).[232] Similar to the action of a QCM, the oscillation frequency is modulated in response to an analyte that becomes adsorbed to its surface (Figure 7.28). Therefore, MOF-functionalised MCLs have the ability to offer higher overall adsorption capacities (which increases the device resolution) *versus* unfunctionalised MCLs, and may also offer the benefit of adsorption selectivity.

An MCL device is supported at one end by a piezo-resistive chip that can detect small changes in surface forces (*e.g.* stress/strain).[233] MCLs operate by adsorbing a target analyte, which induces a measurable change in the resonance frequency. A multi-layer MCL platform can be designed to

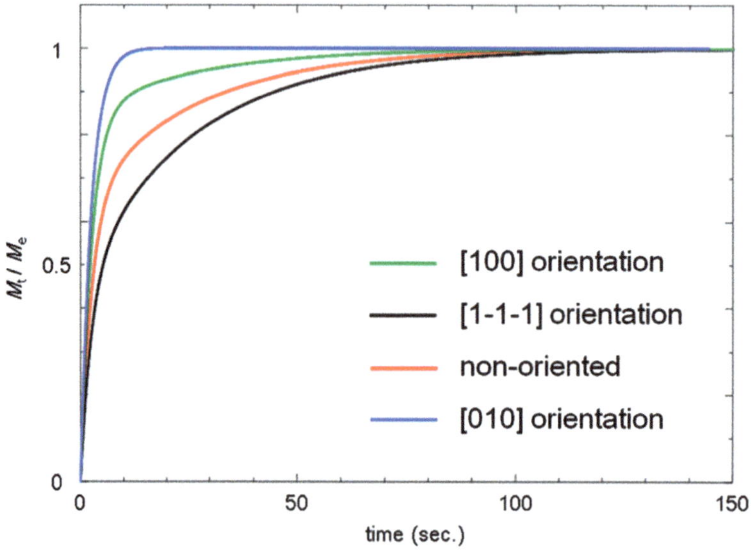

Figure 7.26 Time-dependent methanol uptakes of [100] orientation (green), [1-1-1] orientation (black), and non-oriented (red), [010] orientation (blue) at $P/P_0 = 0.4$. Reproduced from ref. 231 with permission.

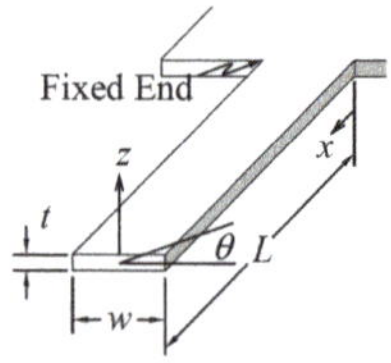

Figure 7.27 Microcantilever geometry and nomenclature. Abbreviations: t = thickness; w = width; L = length. Reprinted with permission from ref. 232. Copyright 2008 American Chemical Society.

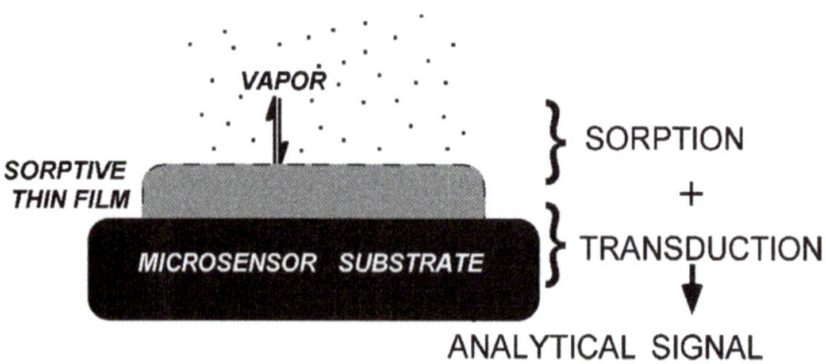

Figure 7.28 Sorption-based sensor. Reprinted with permission from ref. 233. Copyright 2008 American Chemical Society.

allow for quantifiable changes in the surface stress generated between the films.[234] The inherently small size of an MCL, coupled with its high sensitivity toward surface stress, can produce forces as small as ~10^{-4} N m^{-1}. In addition, when compared to QCM or SAW (see later) devices, MCLs operate with very low power consumption and are simpler in their design. However, the incorporation of MOFs into such devices remains relatively unexplored and, as discussed earlier, is hampered by obstacles associated with thin film growth.[235]

The first example of a MOF-incorporated MCL sensor appeared in the literature in 2008. Allendorf, Hesketh and co-workers demonstrated the use of the MOF HKUST-1 on an MCL for the detection of water vapour and VOCs.[236] Highly ordered thin films of HKUST-1 were assembled on a SAM-functionalised gold substrate. Twenty cycles of film growth *via* immersion in solvent resulted in a composite material with an approximately 100 nm thick MOF layer. Surprisingly, the hydrated form of the device produced time-dependent changes in the resistance response when exposed to vapours of MeOH, EtOH and H_2O; the responses of the latter (recorded in the range of 0.18–0.98 mol% H_2O in an inert gas dilutant) are shown in Figure 7.29.

The sensor was also able to detect CO_2 when dehydrated, achieved by heating at 50 °C under dry N_2 for 2 h. However, the dehydrated sensor showed little change in response to MeOH, EtOH, and H_2O. It was determined that the resistance responses from MeOH, EtOH, and H_2O, and the relatively weak response from CO_2, were induced by H-bonding *versus* coordination to the exposed metal sites of the dehydrated sample, respectively. By careful control over the hydration of the material, selectivity for various analytes could be achieved. Although all attributes of the systems investigated in this work indicate good feasibility for implementation, the incorporation of MOFs into

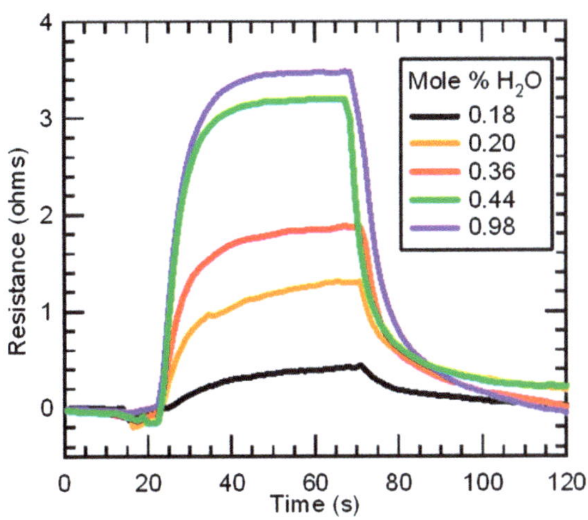

Figure 7.29 Temporal response of the cantilever piezoresistive sensor to water vapor diluted in N_2 (room temperature, 1 atm). Reprinted with permission from ref. 236. Copyright 2008 American Chemical Society.

MCL devices remains in the early stages of development. The same researchers have recently assessed the incorporation of a library of different MOFs into MCL sensors *via* a computational study of some relevant mechanical attributes, such as Young's Moduli and Poisson's ratios.[237]

7.3.4 MOF-Functionalised SAW Sensors

Surface acoustic wave (SAW) sensors have shown great potential in applications geared toward the measurement of ionic concentrations,[238] magnetic fields,[239] and vapour concentration.[240] These unique electromechanical devices detect the modulation of acoustic waves caused by the local external environment. Typically, some kind of piezo-electric material, such as $LiNbO_3$ or quartz, is employed as a substrate, and SAW sensors can utilise changes in amplitude, phase, or frequency to detect analytes.[241]

Examples of MOF-based SAW sensors are presently rare. In 2012, Allendorf *et al.* demonstrated the first employment of a MOF-SAW device for ultrasensitive humidity detection.[242] The MOF HKUST-1 was deposited onto a SiO_2 substrate as a thin film *via* a layer-by-layer (LBL) method similar to the LPE route employed in the MCL example discussed earlier. The LBL process took approximately 20 h in order to deposit a 100 nm film. A vector network analyser was used to measure experimental data and all findings were reported as phase shifts, normalized *versus* an uncoated SAW sensor. The SAW devices were then exposed to water vapour using a humidity generator. Different MOF film thicknesses were investigated and displayed increased phase shift responses with increasing humidity levels (Figure 7.30); the films tested in this work ranged from 70 to 350 nm in thickness (obtained *via*

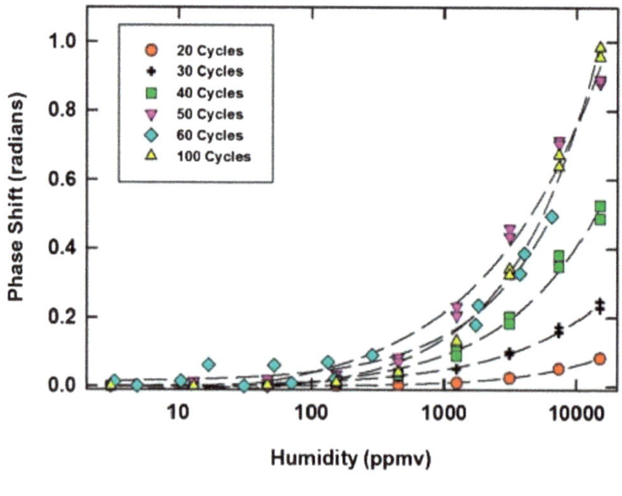

Figure 7.30 Response of Cu–BTC SAWs with a different number of coating cycles to various humidity levels. Reprinted with permission from ref. 242. Copyright 2012 American Chemical Society.

20–100 cycles). Full regeneration of the dehydrated MOF-bound substrate was easily accomplished through heating at 55–60 °C for 2 h. Experimental results showed the device could achieve sensitivities as low as 1.2 ng cm^{-2}, which is on the order of 200 times better sensitivity than that obtained using a standard 5 MHz QCM. Furthermore, the HKUST-1 films showed exceptional durability upon continuous exposure to water vapor, with relative humidity conditions high as 42.5%.

7.4 Summary and Outlook

MOFs have become well-established as advanced materials for applications in gas storage and molecular separations and have shown recent promise in other areas, including drug delivery and catalysis. Interest in MOFs as novel chemical sensors is a comparatively less mature and less explored field. Nonetheless, the research described in this chapter shows that MOFs hold much promise in this regard. Luminescence-based chemical detection by MOFs is attracting increasing interest. We discussed some key examples from the recent literature, as well as the design requirements for achieving an effective sensing material. MOFs can be designed to act as selective and sensitive sensors for pH, cations, anions, volatile organic compounds and nitroaromatics. It was also shown how the structural benefits and luminescence mechanisms of MOFs can be simultaneously harnessed to develop effective sensors for environmental, physiological and industrial applications. We then discussed the fundamentals of thin film technology and their application to MOF devices, along with a brief summary of current publications in the field of electromechanical sensors. These materials are advantageous in the areas of QCM, MCL, and SAW sensing and possess high degrees of precision and selectivity. However, research into MOFs as chemical sensors is still relatively undeveloped. The continuing need to identify novel sensors—coupled to the boundless chemical flexibility available to MOFs—will undoubtedly attract increasing attention to this arena in the near future.

Acknowledgements

The authors acknowledge the Welch foundation (Grant F-1738) and the Texas Higher Education Coordinating Board Norman Hackerman Advanced Research program for funding this work, and Dr Bradley J. Holliday (U. T. Austin) for helpful suggestions.

References

1. P. L. Feng, J. V. Branson, K. Hattar, G. Vizkelethy, M. D. Allendorf and F. Patrick Doty, *Nucl. Instrum. Methods Phys. Res., Sect. A*, 2011, **652**, 295.
2. L. E. Kreno, K. Leong, O. K. Farha, M. Allendorf, R. P. Van Duyne and J. T. Hupp, *Chem. Rev.*, 2011, **112**, 1105.

3. K. Koh, A. G. Wong-Foy and A. J. Matzger, *J. Am. Chem. Soc.*, 2009, **131**, 4184.
4. H.-L. Jiang, B. Liu, Y.-Q. Lan, K. Kuratani, T. Akita, H. Shioyama, F. Zong and Q. Xu, *J. Am. Chem. Soc.*, 2011, **133**, 11854.
5. H.-C. Zhou, J. R. Long and O. M. Yaghi, *Chem. Rev.*, 2012, **112**, 673.
6. D.-W. Lim, S. A. Chyun and M. P. Suh, *Angew. Chem., Int. Ed.*, 2014, **53**, 7819.
7. R. M. Kumar, J. V. Sundar and V. Subramanian, *Int. J. Hydrogen Energy*, 2012, **37**, 16070.
8. M. P. Suh, H. J. Park, T. K. Prasad and D.-W. Lim, *Chem. Rev.*, 2011, **112**, 782.
9. N. L. Rosi, J. Eckert, M. Eddaoudi, D. T. Vodak, J. Kim, M. O'Keeffe and O. M. Yaghi, *Science*, 2003, **300**, 1127.
10. A. Ö. Yazaydın, R. Q. Snurr, T.-H. Park, K. Koh, J. Liu, M. D. LeVan, A. I. Benin, P. Jakubczak, M. Lanuza, D. B. Galloway, J. J. Low and R. R. Willis, *J. Am. Chem. Soc.*, 2009, **131**, 18198.
11. R. Lyndon, K. Konstas, B. P. Ladewig, P. D. Southon, P. C. J. Kepert and M. R. Hill, *Angew. Chem., Int. Ed.*, 2013, **52**, 3695.
12. S. Ye, X. Jiang, L.-W. Ruan, B. Liu, Y.-M. Wang, J.-F. Zhu and L.-G. Qiu, *Microporous Mesoporous Mater.*, 2013, **179**, 191.
13. D. Dubbeldam, C. J. Galvin, K. S. Walton, D. E. Ellis and R. Q. Snurr, *J. Am. Chem. Soc.*, 2008, **130**, 10884.
14. C.-Y. Sun, X.-L. Wang, C. Qin, J.-L. Jin, Z.-M. Su, P. Huang and K.-Z. Shao, *Chem.–Eur. J.*, 2013, **19**, 3639.
15. L. Alaerts, C. E. A. Kirschhock, M. Maes, M. A. van der Veen, V. Finsy, A. Depla, J. A. Martens, G. V. Baron, P. A. Jacobs, J. F. M. Denayer and D. E. De Vos, *Angew. Chem., Int. Ed.*, 2007, **46**, 4293.
16. D. T. de Lill, A. de Bettencourt-Dias and C. L. Cahill, *Inorg. Chem.*, 2007, **46**, 3960.
17. R. Zhang, S. Ji, N. Wang, L. Wang, G. Zhang and J.-R. Li, *Angew. Chem., Int. Ed.*, 2014, **53**, 9775.
18. J. Ma, X. Huang, X. Song and W. Liu, *Chem.–Eur. J.*, 2013, **19**, 3590.
19. J.-W. Cheng, S.-T. Zheng, E. Ma and G.-Y. Yang, *Inorg. Chem.*, 2007, **46**, 10534.
20. J. L. C. Rowsell, J. Eckert and O. M. Yaghi, *J. Am. Chem. Soc.*, 2005, **127**, 14904.
21. H. K. Chae, D. Y. Siberio-Pérez, J. Kim, Y. Go, M. Eddaoudi, A. J. Matzger, M. O'Keeffe and O. M. Yaghi, *Nature*, 2004, **427**, 523.
22. P. Horcajada, T. Chalati, C. Serre, B. Gillet, C. Sebrie, T. Baati, J. F. Eubank, D. Heurtaux, P. Clayette, C. Kreuz, J.-S. Chang, Y. K. Hwang, V. Marsaud, P.-N. Bories, L. Cynober, S. Gil, G. Férey, P. Couvreur and R. Gref, *Nat. Mater.*, 2010, **9**, 172.
23. T. Kundu, S. Mitra, P. Patra, A. Goswami, D. Díaz Díaz and R. Banerjee, *Chem.–Eur. J.*, 2014, **20**, 10514.
24. D. Farrusseng, S. Aguado and C. Pinel, *Angew. Chem., Int. Ed.*, 2009, **48**, 7502.

25. J. Lee, O. K. Farha, J. Roberts, K. A. Scheidt, S. T. Nguyen and J. T. Hupp, *Chem. Soc. Rev.*, 2009, **38**, 1450.
26. W. Zhang, G. Lu, C. Cui, Y. Liu, S. Li, W. Yan, C. Xing, Y. R. Chi, Y. Yang and F. Huo, *Adv. Mater.*, 2014, **26**, 4056.
27. J.-M. Zhou, W. Shi, N. Xu and P. Cheng, *Inorg. Chem.*, 2013, **52**, 8082.
28. J. J. Perry IV, C. A. Bauer and M. D. Allendorf, *Metal-Organic Frameworks*, Wiley-VCH, Weinheim, 2011, ch. 12, pp. 267–308.
29. G. Korotcenkov, *Chemical Sensors*, Momentum Press, New York, 2011, vol. 6.
30. P.-I. Gouma, *Nanomaterials for Chemical Sensors and Biotechnology*, Pan Stanford Publishing, Singapore, 2009.
31. P. Gründler, *Chemical Sensors*, Springer, Berlin, 2010.
32. *Luminescence Applied in Sensor Science*, ed. L. Prodi, M. Montalti and N. Zaccheroni, Springer, 2013.
33. R. C. Ropp, *Luminescence and the Solid State*, Elsevier Science, Amsterdam, 2nd edn, 2004.
34. K. Binnemans, *Chem. Rev.*, 2009, **109**, 4283.
35. Y. Cui, Y. Yue, G. Qian and B. Chen, *Chem. Rev.*, 2012, **112**, 1126.
36. V. Balzani, *Supramolecular Photochemistry*, D. Reidel Publishing Company, Anacapri, 1987.
37. N. J. Turro, J. C. Scaiano and V. Ramamurthy, *Modern Molecular Photochemistry of Organic Molecules*, University Science Books, Sausalito, 2010.
38. *Supramolecular Photochemistry: Controlling Photochemical Processes*, ed. V. Ramamurthy and Y. Inoue, Wiley, Hoboken, 2011.
39. V. Ramamurthy and K. S. Schanze, *Solid State and Surface Photochemistry*, CRC Press, New York, 2000.
40. T. J. Penfold, S. Karlsson, G. Capano, F. A. Lima, J. Rittmann, M. Reinhard, M. H. Rittmann-Frank, O. Braem, E. Baranoff, R. Abela, I. Tavernelli, U. Rothlisberger, C. J. Milne and M. Chergui, *J. Phys. Chem. A*, 2013, **117**, 4591.
41. J. S. Seo, D. Whang, H. Lee, S. I. Jun, J. Oh, Y. J. Jeon and K. Kim, *Nature*, 2000, **404**, 982.
42. J. R. Stork, V. S. Thoi and S. M. Cohen, *Inorg. Chem.*, 2007, **46**, 11213.
43. A.-X. Zhu, Y. Liu and W.-X. Zhang, *Inorg. Chem. Commun.*, 2013, **30**, 88.
44. J. R. Albani, *Principles and Applications of Fluorescence Spectroscopy*, Wiley-Blackwell, Oxford, 2007.
45. E. V. Anslyn and D. A. Dougherty, *Modern Physical Organic Chemistry*, University Science, Sausalito, illustrated edn, 2005, ch. 16, pp. 935–1000.
46. A. P. Demchenko, *Introduction to Fluorescence Sensing*, Springer, Netherlands, 2008.
47. E. G. Moore, A. P. S. Samuel and K. N. Raymond, *Acc. Chem. Res.*, 2009, **42**, 542.
48. S. V. Eliseeva and J.-C. G. Bünzli, *Chem. Soc. Rev.*, 2009, **39**, 189.
49. Z. Liu, W. He and Z. Guo, *Chem. Soc. Rev.*, 2013, **42**, 1568.
50. J. Rocha, L. D. Carlos, F. A. Almeida Paz and D. Ananias, *Chem. Soc. Rev.*, 2011, **40**, 926.

51. H. Dong, L.-D. Sun and C.-H. Yan, *Chem. Soc. Rev.*, 2015, **44**, 1608.
52. D. Yang, P. Ma, Z. Hou, Z. Cheng, C. Li and J. Lin, *Chem. Soc. Rev.*, 2015, **44**, 1416.
53. C. Liu and B. Yan, *New J. Chem.*, 2015, **39**, 1125.
54. D. Shriver and P. Atkins, *Inorganic Chemistry*, W. H. Freeman and Company, New York, 4th edn, 2006, ch. 19, pp. 459–490.
55. H.-L. Jiang, B. Liu and Q. Xu, *Cryst. Growth Des.*, 2009, **10**, 806.
56. G.-H. Wang, Z.-G. Li, H.-Q. Jia, N.-H. Hu and J.-W. Xu, *CrystEngComm*, 2009, **11**, 292.
57. C. A. Bauer, T. V. Timofeeva, T. B. Settersten, B. D. Patterson, V. H. Liu, B. A. Simmons and M. D. Allendorf, *J. Am. Chem. Soc.*, 2007, **129**, 7136.
58. Z.-Z. Lu, R. Zhang, Y.-Z. Li, Z.-J. Guo and H.-G. Zheng, *J. Am. Chem. Soc.*, 2011, **133**, 4172.
59. P. Dorenbos, *J. Lumin.*, 2000, **91**, 91.
60. X. Wang, J.-Z. Chen and J.-H. Liu, *Cryst. Growth Des.*, 2007, **7**, 1227.
61. O. S. Wenger, *Chem. Rev.*, 2013, **113**, 3686.
62. Y. Takashima, V. M. Martínez, S. Furukawa, M. Kondo, S. Shimomura, H. Uehara, M. Nakahama, K. Sugimoto and S. Kitagawa, *Nat. Commun.*, 2011, **2**, 168.
63. L. G. Beauvais, M. P. Shores and J. R. Long, *J. Am. Chem. Soc.*, 2000, **122**, 2763.
64. G.-B. Li, L. Li, J.-M. Liu, T. Yang and C.-Y. Su, *Cryst. Growth Des.*, 2013, **13**, 1518.
65. S. M. Cohen, *Chem. Rev.*, 2011, **112**, 970.
66. A. J. Nuñez, M. S. Chang, I. A. Ibarra and S. M. Humphrey, *Inorg. Chem.*, 2013, **53**, 282.
67. A. M. Bohnsack, I. A. Ibarra, V. I. Bakhmutov, V. M. Lynch and S. M. Humphrey, *J. Am. Chem. Soc.*, 2013, **135**, 16038.
68. J. An, C. M. Shade, D. A. Chengelis-Czegan, S. Petoud and N. L. Rosi, *J. Am. Chem. Soc.*, 2011, **133**, 1220.
69. M. D. Allendorf, C. A. Bauer, R. K. Bhakta and R. J. T. Houk, *Chem. Soc. Rev.*, 2009, **38**, 1330.
70. L. Flamigni, A. M. Talarico, F. Barigelletti and M. R. Johnston, *Photochem. Photobiol. Sci.*, 2002, **1**, 190.
71. G. J. McManus, Perry, M. Perry, B. D. Wagner and M. J. Zaworotko, *J. Am. Chem. Soc.*, 2007, **129**, 9094.
72. F. Luo and S. R. Batten, *Dalton Trans.*, 2010, **39**, 4485.
73. B. Chen, L. Wang, F. Zapata, G. Qian and E. B. Lobkovsky, *J. Am. Chem. Soc.*, 2008, **130**, 6718.
74. Y.-Q. Chen, G.-R. Li, Z. Chang, Y.-K. Qu, Y.-H. Zhang and X.-H. Bu, *Chem. Sci.*, 2013, **4**, 3678.
75. B. Manna, A. K. Chaudhari, B. Joarder, A. Karmakar and S. K. Ghosh, *Angew. Chem., Int. Ed.*, 2013, **52**, 998.
76. H. Xu, Y. Xiao, X. Rao, Z. Dou, W. Li, Y. Cui, Z. Wang and G. Qian, *J. Alloys Compd.*, 2011, **509**, 2552.

77. P.-F. Shi, B. Zhao, G. Xiong, Y.-L. Hou and P. Cheng, *Chem. Commun.*, 2012, **48**, 8231.
78. J.-M. Zhou, W. Shi, H.-M. Li, H. Li and P. Cheng, *J. Phys. Chem. C*, 2014, **118**, 416.
79. X. Li, H. Xu, F. Kong and R. Wang, *Angew. Chem., Int. Ed.*, 2013, **52**, 13769.
80. Q.-B. Bo, H.-T. Zhang, H.-Y. Wang, J.-L. Miao and Z.-W. Zhang, *Chem.–Eur. J.*, 2014, **20**, 3712.
81. J. Han and K. Burgess, *Chem. Rev.*, 2010, **110**, 2709.
82. Y. Cui, B. Chen and G. Qian, *Coord. Chem. Rev.*, 2014, **273**, 76.
83. L. Cui, J. Wu and H. Ju, *Biosens. Bioelectron.*, 2015, **63**, 276.
84. B. V. Harbuzaru, A. Corma, F. Rey, J. L. Jorda, D. Ananias, L. D. Carlos and J. Rocha, *Angew. Chem., Int. Ed.*, 2009, **48**, 6476.
85. H.-L. Jiang, D. Feng, K. Wang, Z.-Y. Gu, Z. Wei, Y.-P. Chen and H.-C. Zhou, *J. Am. Chem. Soc.*, 2013, **135**, 13934.
86. J. Aguilera-Sigalat and D. Bradshaw, *Chem. Commun.*, 2014, **50**, 4711.
87. W. Yang, J. Feng and H. Zhang, *J. Mater. Chem.*, 2012, **22**, 6819.
88. L. Sun, Y. Li, Z. Liang, J. Yu and R. Xu, *Dalton Trans.*, 2012, **41**, 12790.
89. B. Chen, L. Wang, Y. Xiao, F. R. Fronczek, M. Xue, Y. Cui and G. Qian, *Angew. Chem., Int. Ed.*, 2009, **48**, 500.
90. Y. Xiao, Y. Cui, Q. Zheng, S. Xiang, G. Qian and B. Chen, *Chem. Commun.*, 2010, **46**, 5503.
91. Z. Hao, X. Song, M. Zhu, X. Meng, S. Zhao, S. Su, W. Yang, S. Song and H. Zhang, *J. Mater. Chem. A*, 2013, **1**, 11043.
92. Z. Hao, G. Yang, X. Song, M. Zhu, X. Meng, S. Zhao, S. Song and H. Zhang, *J. Mater. Chem. A*, 2014, **2**, 237.
93. S. Liu, Z. Xiang, Z. Hu, X. Zheng and D. Cao, *J. Mater. Chem.*, 2011, **21**, 6649.
94. K. Jayaramulu, R. P. Narayanan, S. J. George and T. K. Maji, *Inorg. Chem.*, 2012, **51**, 10089.
95. B. Zhao, X.-Y. Chen, Z. Chen, W. Shi, P. Cheng, S.-P. Yan and D.-Z. Liao, *Chem. Commun.*, 2009, 3113.
96. B. Liu, C. Sun and Y. Chen, *J. Mater. Chem. B*, 2014, **2**, 1661.
97. S. Dang, E. Ma, Z.-M. Sun and H. Zhang, *J. Mater. Chem.*, 2012, **22**, 16920.
98. Q. Tang, S. Liu, Y. Liu, J. Miao, S. Li, L. Zhang, Z. Shi and Z. Zheng, *Inorg. Chem.*, 2013, **52**, 2799.
99. X.-H. Zhou, L. Li, H.-H. Li, A. Li, T. Yang and W. Huang, *Dalton Trans.*, 2013, **42**, 12403.
100. Z. Xiang, C. Fang, S. Leng and D. Cao, *J. Mater. Chem. A*, 2014, **2**, 7662.
101. B. Zhao, X. Y. Chen, P. Cheng, D. Z. Liao, S. P. Yan and Z. H. Jiang, *J. Am. Chem. Soc.*, 2004, **126**, 15394.
102. Y.-M. Zhu, C.-H. Zeng, T.-S. Chu, H.-M. Wang, Y.-Y. Yang, Y.-X. Tong, C.-Y. Su and W.-T. Wong, *J. Mater. Chem. A*, 2013, **1**, 11312.
103. H. Tan, B. Liu and Y. Chen, *ACS Nano*, 2012, **6**, 10505.
104. J. Cao, Y. Gao, Y. Wang, C. Du and Z. Liu, *Chem. Commun.*, 2013, **49**, 6897.

105. J.-S. Qin, S.-R. Zhang, D.-Y. Du, P. Shen, S.-J. Bao, Y.-Q. Lan and Z.-M. Su, *Chem.–Eur. J.*, 2014, **20**, 5625.
106. http://water.epa.gov/drink/contaminants/basicinformation/mercury.cfm.
107. X.-L. Qi, R.-B. Lin, Q. Chen, J.-B. Lin, J.-P. Zhang and X.-M. Chen, *Chem. Sci.*, 2011, **2**, 2214.
108. L. E. Kreno, J. T. Hupp and R. P. Van Duyne, *Anal. Chem.*, 2010, **82**, 8042.
109. N. B. Shustova, A. F. Cozzolino, S. Reineke, M. Baldo and M. Dinca, *J. Am. Chem. Soc.*, 2013, **135**, 13326.
110. I. A. Ibarra, T. W. Hesterberg, J.-S. Chang, J. W. Yoon, B. J. Holliday and S. M. Humphrey, *Chem. Commun.*, 2013, **49**, 7156.
111. Q.-K. Liu, J.-P. Ma and Y.-B. Dong, *Chem. Commun.*, 2011, **47**, 7185.
112. Z.-J. Zhang, W. Shi, Z. Niu, H.-H. Li, B. Zhao, P. Cheng, D.-Z. Liao and S.-P. Yan, *Chem. Commun.*, 2011, **47**, 6425.
113. L. Chen, K. Tan, Y.-Q. Lan, S.-L. Li, K.-Z. Shao and Z.-M. Su, *Chem. Commun.*, 2012, **48**, 5919.
114. P. Wu, J. Wang, C. He, X. Zhang, Y. Wang, T. Liu and C. Duan, *Adv. Funct. Mater.*, 2012, **22**, 1698.
115. X. Wang and O. S. Wolfbeis, *Chem. Soc. Rev.*, 2014, **43**, 3666.
116. A. Ruggi, F. W. B. van Leeuwen and A. H. Velders, *Coord. Chem. Rev.*, 2011, **255**, 2542.
117. Z. Xie, L. Ma, K. E. deKrafft, A. Jin and W. Lin, *J. Am. Chem. Soc.*, 2010, **132**, 922.
118. K.-Y. Cheng, J.-C. Wang, C.-Y. Lin, W.-R. Lin, Y.-A. Chen, F.-J. Tsai, Y.-C. Chuang, G.-Y. Lin, C.-W. Ni, Y.-T. Zeng and M.-L. Ho, *Dalton Trans.*, 2014, **43**, 6536.
119. M.-L. Ho, Y.-A. Chen, T.-C. Chen, P.-J. Chang, Y.-P. Yu, K.-Y. Cheng, C.-H. Shih, G.-H. Lee and H.-S. Sheu, *Dalton Trans.*, 2012, **41**, 2592.
120. S. M. Barrett, C. Wang and W. Lin, *J. Mater. Chem.*, 2012, **22**, 10329.
121. X.-L. Qi, S.-Y. Liu, R.-B. Lin, P.-Q. Liao, J.-W. Ye, Z. Lai, Y. Guan, X.-N. Cheng, J.-P. Zhang and X.-M. Chen, *Chem. Commun.*, 2013, **49**, 6864.
122. R.-B. Lin, F. Li, S.-Y. Liu, X.-L. Qi, J.-P. Zhang and X.-M. Chen, *Angew. Chem., Int. Ed.*, 2013, **52**, 13429.
123. J. Demel, P. Kubat, F. Millange, J. Marrot, I. Cisarova and K. Lang, *Inorg. Chem.*, 2013, **52**, 2779.
124. Z. Dou, J. Yu, Y. Cui, Y. Yang, Z. Wang, D. Yang and G. Qian, *J. Am. Chem. Soc.*, 2014, **136**, 5527.
125. A. Tripathy, S. Pramanik, J. Cho, J. Santhosh and N. A. Abu Osman, *Sensors*, 2014, **14**, 16343.
126. W.-H. Zhu, Z.-M. Wang and S. Gao, *Inorg. Chem.*, 2007, **46**, 1337.
127. T.-F. Liu, W. Zhang, W.-H. Sun and R. Cao, *Inorg. Chem.*, 2011, **50**, 5242.
128. Y. Yu, J.-P. Ma and Y.-B. Dong, *CrystEngComm*, 2012, **14**, 7157.
129. H.-L. Sun, D.-D. Yin, Q. Chen and Z. Wang, *Inorg. Chem.*, 2013, **52**, 3582.
130. F.-N. Shi, M. L. Pinto, D. Ananias and J. Rocha, *Microporous Mesoporous Mater.*, 2014, **188**, 172.
131. S. Mohapatra, B. Rajeswaran, A. Chakraborty, A. Sundaresan and T. K. Maji, *Chem. Mater.*, 2013, **25**, 1673.

132. W. D. Horrocks and D. R. Sudnick, *Acc. Chem. Res.*, 1981, **14**, 384.
133. I. A. Ibarra, T. W. Hesterberg, B. J. Holliday, V. M. Lynch and S. M. Humphrey, *Dalton Trans.*, 2012, **41**, 8003.
134. C.-C. Wang, C.-C. Yang, W.-C. Chung, G.-H. Lee, M.-L. Ho, Y.-C. Yu, M.-W. Chung, H.-S. Sheu, C.-H. Shih, K.-Y. Cheng, P.-J. Chang and P.-T. Chou, *Chem.–Eur. J.*, 2011, **17**, 9232.
135. Z.-F. Wu, B. Tan, M.-L. Feng, A.-J. Lan and X.-Y. Huang, *J. Mater. Chem. A*, 2014, **2**, 6426.
136. Z. Dou, J. Yu, H. Xu, Y. Cui, Y. Yang and G. Qian, *Microporous Mesoporous Mater.*, 2013, **179**, 198.
137. T. Lee, M. H. Tsai and H. L. Lee, *Cryst. Growth Des.*, 2012, **12**, 3181.
138. W. J. Rieter, K. M. L. Taylor and W. Lin, *J. Am. Chem. Soc.*, 2007, **129**, 9852.
139. C. A. Kent, D. Liu, T. J. Meyer and W. Lin, *J. Am. Chem. Soc.*, 2012, **134**, 3991.
140. A. M. Marti, N. Nijem, Y. J. Chabal and K. J. Balkus, *Microporous Mesoporous Mater.*, 2013, **174**, 100.
141. A. M. Marti, S. D. Perera, L. D. McBeath and K. J. Balkus, *Langmuir*, 2013, **29**, 5927.
142. M. M. Wanderley, C. Wang, C.-D. Wu and W. Lin, *J. Am. Chem. Soc.*, 2012, **134**, 9050.
143. E. Y. Lee, S. Y. Jang and M. P. Suh, *J. Am. Chem. Soc.*, 2005, **127**, 6374.
144. Z. Zhang, S. Xiang, Q. Zheng, X. Rao, J. U. Mondal, H. D. Arman, G. Qian and B. Chen, *Cryst. Growth Des.*, 2010, **10**, 2372.
145. Z. Jin, H. He, H. Zhao, T. Borjigin, F. Sun, D. Zhang and G. Zhu, *Dalton Trans.*, 2013, **42**, 13335.
146. G. Lu and J. T. Hupp, *J. Am. Chem. Soc.*, 2010, **132**, 7832.
147. X. Zou, G. Zhu, I. J. Hewitt, F. Sun and S. Qiu, *Dalton Trans.*, 2009, 3009.
148. B. Chen, Y. Yang, F. Zapata, G. Lin, G. Qian and E. B. Lobkovsky, *Adv. Mater.*, 2007, **19**, 1693.
149. B. V. Harbuzaru, A. Corma, F. Rey, P. Atienzar, J. L. Jorda, H. Garcia, D. Ananias, L. D. Carlos and J. Rocha, *Angew. Chem., Int. Ed.*, 2008, **47**, 1080.
150. J.-J. Wu, Y.-X. Ye, Y.-Y. Qiu, Z.-P. Qiao, M.-L. Cao and B.-H. Ye, *Inorg. Chem.*, 2013, **52**, 6450.
151. C. Zhan, S. Ou, C. Zou, M. Zhao and C.-D. Wu, *Anal. Chem.*, 2014, **86**, 6648.
152. H. Li, W. Shi, K. Zhao, Z. Niu, H. Li and P. Cheng, *Chem.–Eur. J.*, 2013, **19**, 3358.
153. H.-M. Wang, H.-P. Liu, T.-S. Chu, Y.-Y. Yang, Y.-S. Hu, W.-T. Liu and S. W. Ng, *RSC Adv.*, 2014, **4**, 14035.
154. W. Ahmad, L. Zhang and Y. Zhou, *Photochem. Photobiol. Sci.*, 2014, **13**, 660.
155. Z. Guo, H. Xu, S. Su, J. Cai, S. Dang, S. Xiang, G. Qian, H. Zhang, M. O'Keeffe and B. Chen, *Chem. Commun.*, 2011, **47**, 5551.
156. S. Dang, X. Min, W. Yang, F.-Y. Yi, H. You and Z.-M. Sun, *Chem.–Eur. J.*, 2013, **19**, 17172.
157. K. A. White, D. A. Chengelis, K. A. Gogick, J. Stehman, N. L. Rosi and S. Petoud, *J. Am. Chem. Soc.*, 2009, **131**, 18069.

158. S. J. Toal and W. C. Trogler, *J. Mater. Chem.*, 2006, **16**, 2871.
159. C. Zhang, Y. Che, Z. Zhang, X. Yang and L. Zang, *Chem. Commun.*, 2011, **47**, 2336.
160. A. Lan, K. Li, H. Wu, D. H. Olson, T. J. Emge, W. Ki, M. Hong and J. Li, *Angew. Chem., Int. Ed.*, 2009, **48**, 2334.
161. S. Pramanik, C. Zheng, X. Zhang, T. J. Emge and J. Li, *J. Am. Chem. Soc.*, 2011, **133**, 4153.
162. Z. Zhang, S. Xiang, X. Rao, Q. Zheng, F. R. Fronczek, G. Qian and B. Chen, *Chem. Commun.*, 2010, **46**, 7205.
163. D. Ma, B. Li, X. Zhou, Q. Zhou, K. Liu, G. Zeng, G. Li, Z. Shi and S. Feng, *Chem. Commun.*, 2013, **49**, 8964.
164. D. Banerjee, Z. Hu, S. Pramanik, X. Zhang, H. Wang and J. Li, *CrystEngComm*, 2013, **15**, 9745.
165. G.-Y. Wang, C. Song, D.-M. Kong, W.-J. Ruan, Z. Chang and Y. Li, *J. Mater. Chem. A*, 2014, **2**, 2213.
166. L. Li, S. Zhang, L. Xu, L. Han, Z.-N. Chen and J. Luo, *Inorg. Chem.*, 2013, **52**, 12323.
167. I.-H. Park, R. Medishetty, J.-Y. Kim, S. S. Lee and J. J. Vittal, *Angew. Chem., Int. Ed.*, 2014, **53**, 5591.
168. B. Gole, A. K. Bar and P. S. Mukherjee, *Chem. Commun.*, 2011, **47**, 12137.
169. X. Zhou, H. Li, H. Xiao, L. Li, Q. Zhao, T. Yang, J. Zuo and W. Huang, *Dalton Trans.*, 2013, **42**, 5718.
170. H. Xu, F. Liu, Y. Cui, B. Chen and G. Qian, *Chem. Commun.*, 2011, **47**, 3153.
171. Y.-N. Gong, L. Jiang and T.-B. Lu, *Chem. Commun.*, 2013, **49**, 11113.
172. S.-B. Ding, W. Wang, L.-G. Qiu, Y.-P. Yuan, F.-M. Peng, X. Jiang, A.-J. Xie, Y.-H. Shen and J.-F. Zhu, *Mater. Lett.*, 2011, **65**, 1385.
173. J.-D. Xiao, L.-G. Qiu, F. Ke, Y.-P. Yuan, G.-S. Xu, Y.-M. Wang and X. Jiang, *J. Mater. Chem. A*, 2013, **1**, 8745.
174. L. Sun, H. Xing, J. Xu, Z. Liang, J. Yu and R. Xu, *Dalton Trans.*, 2013, **42**, 5508.
175. S. S. Nagarkar, B. Joarder, A. K. Chaudhari, S. Mukherjee and S. K. Ghosh, *Angew. Chem., Int. Ed.*, 2013, **52**, 2881.
176. T. K. Kim, J. H. Lee, D. Moon and H. R. Moon, *Inorg. Chem.*, 2013, **52**, 589.
177. Y.-S. Xue, Y. He, L. Zhou, F.-J. Chen, Y. Xu, H.-B. Du, X.-Z. You and B. Chen, *J. Mater. Chem. A*, 2013, **1**, 4525.
178. L. Chen, H. Zheng, X. Zhu, Z. Lin, L. Guo, B. Qiu, G. Chen and Z.-N. Chen, *Analyst*, 2013, **138**, 3490.
179. X. Zhu, H. Zheng, X. Wei, Z. Lin, L. Guo, B. Qiu and G. Chen, *Chem. Commun.*, 2013, **49**, 1276.
180. T. Ye, Y. Liu, M. Luo, X. Xiang, X. Ji, G. Zhou and Z. He, *Analyst*, 2014, **139**, 1721.
181. X. Wei, L. Zheng, F. Luo, Z. Lin, L. Guo, B. Qiu and G. Chen, *J. Mater. Chem. B*, 2013, **1**, 1812.
182. P. L. Feng, K. Leong and M. D. Allendorf, *Dalton Trans.*, 2012, **41**, 8869.

183. Q. Zhu, C. Shen, C. Tan, T. Sheng, S. Hu and X. Wu, *Chem. Commun.*, 2012, **48**, 531.
184. Y. Cui, H. Xu, Y. Yue, Z. Guo, J. Yu, Z. Chen, J. Gao, Y. Yang, G. Qian and B. Chen, *J. Am. Chem. Soc.*, 2012, **134**, 3979.
185. K. Miyata, Y. Konno, T. Nakanishi, A. Kobayashi, M. Kato, K. Fushimi and Y. Hasegawa, *Angew. Chem., Int. Ed.*, 2013, **52**, 6413.
186. A. Cadiau, C. D. S. Brites, P. M. F. J. Costa, R. A. S. Ferreira, J. Rocha and L. D. Carlos, *ACS Nano*, 2013, **7**, 7213.
187. X. Rao, T. Song, J. Gao, Y. Cui, Y. Yang, C. Wu, B. Chen and G. Qian, *J. Am. Chem. Soc.*, 2013, **135**, 15559.
188. R. F. D'Vries, S. Alvarez-Garcia, N. Snejko, L. E. Bausa, E. Gutierrez-Puebla, A. de Andres and M. Angeles Monge, *J. Mater. Chem. C*, 2013, **1**, 6316.
189. H. Matsumoto, M. Matsuoka, T. Iwasaki, S. Kinoshita, S. Iwamori and K. Noda, *J. Photopolym. Sci. Technol.*, 2009, **22**, 279.
190. D. Greve, T.-L. Chin, P. Zheng, P. Ohodnicki, J. Baltrus and I. Oppenheim, *Sensors*, 2013, **13**, 6910.
191. R. Ameloot, L. Stappers, J. Fransaer, L. Alaerts, B. F. Sels and D. E. De Vos, *Chem. Mater.*, 2009, **21**, 2580.
192. M. A. Allendorf, A. Bétard, R. A. Fischer, *Metal-Organic Frameworks*, Wiley-VCH, Weinheim, 2011, ch. 13, pp. 309–338.
193. C. Munuera, O. Shekhah, H. Wang, C. Wöll and C. Ocal, *Phys. Chem. Chem. Phys.*, 2008, **10**, 7257.
194. O. Shekhah, J. Liu, R. A. Fischer and C. Wöll, *Chem. Soc. Rev.*, 2011, **40**, 1081.
195. A. Demessence, C. Boissière, D. Grosso, P. Horcajada, C. Serre, G. Férey, G. J. A. A. Soler-Illia and C. Sanchez, *J. Mater. Chem.*, 2010, **20**, 7676.
196. A. Bétard and R. A. Fischer, *Chem. Rev.*, 2011, **112**, 1055.
197. C. Scherb, A. Schödel and T. Bein, *Angew. Chem., Int. Ed.*, 2008, **47**, 5777.
198. H. Guo, G. Zhu, I. J. Hewitt and S. Qiu, *J. Am. Chem. Soc.*, 2009, **131**, 1646.
199. Y. Yoo, Z. Lai and H.-K. Jeong, *Microporous Mesoporous Mater.*, 2009, **123**, 100.
200. J. Nan, X. Dong, W. Wang, W. Jin and N. Xu, *Langmuir*, 2011, **27**, 4309.
201. P. Horcajada, C. Serre, D. Grosso, C. Boissière, S. Perruchas, C. Sanchez and G. Férey, *Adv. Mater.*, 2009, **21**, 1931.
202. M. G. Schwab, I. Senkovska, M. Rose, M. Koch, J. Pahnke, G. Jonschker and S. Kaskel, *Adv. Eng. Mater.*, 2008, **10**, 1151.
203. R. Ameloot, E. Gobechiya, H. Uji-i, J. A. Martens, J. Hofkens, L. Alaerts, B. F. Sels and D. E. De Vos, *Adv. Mater.*, 2010, **22**, 2685.
204. A. Ulman, *Chem. Rev.*, 1996, **96**, 1533.
205. J. Madoz, B. A. Kuznetzov, F. J. Medrano, J. L. Garcia and V. M. Fernandez, *J. Am. Chem. Soc.*, 1997, **119**, 1043.
206. D. L. Allara and R. G. Nuzzo, *Langmuir*, 1985, **1**, 52.
207. E. Biemmi, C. Scherb and T. Bein, *J. Am. Chem. Soc.*, 2007, **129**, 8054.
208. D. Zacher, O. Shekhah, C. Wöll and R. A. Fischer, *Chem. Soc. Rev.*, 2009, **38**, 1418.

209. M. Arnold, P. Kortunov, D. J. Jones, Y. Nedellec, J. Kärger and J. Caro, *Eur. J. Inorg. Chem.*, 2007, 60.
210. M. Kubo, W. Chaikittisilp and T. Okubo, *Chem. Mater.*, 2008, **20**, 2887.
211. C. Livage, C. Egger and G. Férey, *Chem. Mater.*, 1999, **11**, 1546.
212. S. H. Jhung, J.-H. Lee, J. W. Yoon, C. Serre, G. Férey and J.-S. Chang, *Adv. Mater.*, 2007, **19**, 121.
213. L.-G. Qiu, Z.-Q. Li, Y. Wu, W. Wang, T. Xu and X. Jiang, *Chem. Commun.*, 2008, 3642.
214. J.-L. Zhuang, D. Ceglarek, S. Pethuraj and A. Terfort, *Adv. Funct. Mater.*, 2011, **21**, 1442.
215. S. Hermes, F. Schröder, R. Chelmowski, C. Wöll and R. A. Fischer, *J. Am. Chem. Soc.*, 2005, **127**, 13744.
216. B. Liu, M. Ma, D. Zacher, A. Bétard, K. Yusenko, N. Metzler-Nolte, C. Wöll and R. A. Fischer, *J. Am. Chem. Soc.*, 2011, **133**, 1734.
217. K. Otsubo, T. Haraguchi, O. Sakata, A. Fujiwara and H. Kitagawa, *J. Am. Chem. Soc.*, 2012, **134**, 9605.
218. K. B. Blodgett, *J. Am. Chem. Soc.*, 1934, **56**, 495.
219. R. Makiura, S. Motoyama, Y. Umemura, H. Yamanaka, O. Sakata and H. Kitagawa, *Nat. Mater.*, 2010, **9**, 565.
220. G. Decher, *Science*, 1997, **277**, 1232.
221. C. M. Bell, M. F. Arendt, L. Gomez, R. H. Schmehl and T. E. Mallouk, *J. Am. Chem. Soc.*, 1994, **116**, 8374.
222. O. Shekhah, H. Wang, S. Kowarik, F. Schreiber, M. Paulus, M. Tolan, C. Sternemann, F. Evers, D. Zacher, R. A. Fischer and C. Wöll, *J. Am. Chem. Soc.*, 2007, **129**, 15118.
223. B. Liu, O. Shekhah, H. K. Arslan, J. Liu, C. Wöll and R. A. Fischer, *Angew. Chem., Int. Ed.*, 2012, **51**, 807.
224. O. Zybaylo, O. Shekhah, H. Wang, M. Tafipolsky, R. Schmid, D. Johannsmann and C. Wöll, *Phys. Chem. Chem. Phys.*, 2010, **12**, 8093.
225. E. Biemmi, A. Darga, N. Stock and T. Bein, *Microporous Mesoporous Mater.*, 2008, **114**, 380.
226. M. Tu and R. A. Fischer, *J. Mater. Chem. A*, 2014, **2**, 2018.
227. X. Si, C. Jiao, F. Li, J. Zhang, S. Wang, S. Liu, Z. Li, L. Sun, F. Xu, Z. Gabelica and C. Schick, *Energy Environ. Sci.*, 2011, **4**, 4522.
228. V. Stavila, J. Volponi, A. M. Katzenmeyer, M. C. Dixon and M. D. Allendorf, *Chem. Sci.*, 2012, **3**, 1531.
229. M. R. Deakin and D. A. Buttry, *Anal. Chem.*, 1989, **61**, 1147A.
230. A. H. Khoshaman and B. Bahreyni, *Sens. Actuators, B*, 2012, **162**, 114.
231. K. Hirai, K. Sumida, M. Meilikhov, N. Louvain, M. Nakahama, H. Uehara, S. Kitagawa and S. Furukawa, *J. Mater. Chem. C*, 2014, **2**, 3336.
232. K. M. Goeders, J. S. Colton and L. A. Bottomley, *Chem. Rev.*, 2008, **108**, 522.
233. J. W. Grate, *Chem. Rev.*, 2008, **108**, 726.
234. A. M. Moulin, S. J. O'Shea and M. E. Welland, *Ultramicroscopy*, 2000, **82**, 23.
235. M. D. Allendorf, A. Schwartzberg, V. Stavila and A. A. Talin, *Chem.–Eur. J.*, 2011, **17**, 11372.

236. M. D. Allendorf, R. J. T. Houk, L. Andruszkiewicz, A. A. Talin, J. Pikarsky, A. Choudhury, K. A. Gall and P. J. Hesketh, *J. Am. Chem. Soc.*, 2008, **130**, 14404.
237. A. Venkatasubramanian, J.-H. Lee, V. Stavila, A. Robinson, M. D. Allendorf and P. J. Hesketh, *Sens. Actuators, B*, 2012, **168**, 256.
238. M. J. Vellekoop, G. W. Lubking, P. M. Sarro and A. Venema, *Sens. Actuators, A*, 1994, **44**, 249.
239. S. M. Hanna, *IEEE Trans. Ultrason. Ferroelectr.*, 1987, **34**, 191.
240. H. Wohltjen and R. Dessy, *Anal. Chem.*, 1979, **51**, 1458.
241. C. Caliendo, P. Verardi, E. Verona, A. D'Amico, C. D. Natale, G. Saggio, M. Serafini, R. Paolesse and S. E. Huq, *Smart Mater. Struct.*, 1997, **6**, 689.
242. A. L. Robinson, V. Stavila, T. R. Zeitler, M. I. White, S. M. Thornberg, J. A. Greathouse and M. D. Allendorf, *Anal. Chem.*, 2012, **84**, 7043.

CHAPTER 8

Anchoring Metallosupramolecular Materials on Solid Substrates: Specific Surface–Molecule Interactions and Self-Assembly

GIUSEPPINA PACE*[a] AND ARTUR R. STEFANKIEWICZ*[b,c]

[a]Center for Nano Science and Technology@PoliMi, Istituto Italiano di Tecnologia, Via Pascoli 70/3, 20133 Milano, Italy; [b]Faculty of Chemistry, Adam Mickiewicz University, Umultowska 89b, 61-614 Poznań, Poland; [c]Wielkopolska Center for Advanced Technologies, Umultowska 89c, 61-614 Poznań, Poland
*E-mail: Giuseppina.Pace@iit.it, ars@amu.edu.pl

8.1 Introduction

Metallosupramolecular compounds (coordination complexes) have a wide variety of molecular structures, and display interesting optical, electrochemical and magnetic properties. It is possible to interact with each of these properties using the appropriate external optical, electrical, chemical or magnetic stimuli. Introducing supramolecular chemistry concepts[1] into the design of metallosupramolecular compounds enriches their chemistry with dynamic processes such as self-organization and self-repair. Of all the possible factors

RSC Smart Materials No. 15
Functional Metallosupramolecular Materials
Edited by John G. Hardy and Felix H. Schacher

Published by the Royal Society of Chemistry, www.rsc.org

that can affect the properties of such metallosupramolecular complexes, surface interactions are some of the most challenging to investigate. However, when considering the integration of a metallosupramolecular molecule into sensors, optoelectronics and molecular electronics devices, it is important to identify surface-/molecule-specific interactions that can enable the control of the desired device performance.

Molecules absorbed on solid substrates can undergo structural rearrangement; new electronic configurations can arise from the new intermolecular packing at the surface and from the convolution between molecular and surface electronic states.[2] In this last case, new hybrid states can be formed and both molecular as well as surface electronic states can be strongly affected, an example being metal complexes that can undergo redox reactions on a surface.[3]

The design of metallosupramolecular compounds that can self-assemble on solid substrates enables the adoption of bottom-up manufacturing approaches, making it very appealing for technological applications. A popular strategy for achieving controlled adsorption on a solid surface is to embed specific functional groups, which can foster weak intermolecular interactions such as van der Waals forces, hydrogen bonding, π–π and extended metal coordination interactions. In this chapter, we will also see that metal–ligand coordination interactions can be exploited to grow hierarchical assemblies leading to the formation of two- and three-dimensional crystals on solid substrates (Section 8.4).

The purpose of this chapter is to describe the molecular design principles so far exploited to self-assemble metallosupramolecular compounds on surfaces. Particular attention will be given to the main driving forces for surface binding and self-assembly and most importantly, to how molecule–surface interactions affect the properties of the adsorbate.

8.2 Electronic Structure of Metallosupramolecules on Surfaces

The huge flexibility offered by supramolecular chemistry allows the design of metal complexes with different architectures such as molecular grids,[4,5] molecular cages,[6] and coordination polymers[7,8] which are very stable in solution. In Section 8.4 we will describe the specific case of surface metal–organic frameworks (SURMOFs), which are a class of metal coordination polymers that can self-assemble on solid substrates.

Very complex architectures can be found for metallosupramolecular polymers and metallodendrimers, which will not be described in the current chapter. The reader is referred to the specific literature for more details on this topic.[9–12]

The complex geometry found for such molecules very often hampers their packing at surfaces. A common strategy used to make those compounds likely to self-assemble on surfaces is to introduce specific functional groups onto the ligand that favor intermolecular interactions between the adsorbed molecules on a substrate. Intermolecular packing of an organic compound can be favored by, for example, van der Waals interactions between long alkyl-chain substituents, hydrogen bonding and π–π interactions. Specific molecule–surface interactions can also foster the assembly process, such as π-interactions between planar aromatic groups and hydrophobic surfaces, or covalent bonding through specific anchoring groups such as thiols on Au or carboxylic acids on oxide surfaces (*e.g.*, Al/Al_yO_x, TiO_2, ZnO).[13,14]

A useful tool to investigate the electronic structure of metallosupramolecules adsorbed on conductive substrates is scanning tunneling microscopy (STM). STM enables the investigation of the electron density of organic compounds adsorbed on atomically flat and conductive substrates, also at the single molecule level.

As already mentioned, metallosupramolecular chemistry allows exotic molecular structures such as molecular capsules/cages[15] and molecular grids[9–11] to be designed. A recent example of molecular cages assembled on a solid surface is provided in the work of Mahimaidoss *et al.*, who synthesized hybrid vanadate capsules of different dimensions.[16] Stable aqueous solutions of the V-capsules as sodium salts could be deposited on Au(111) surfaces *via* spin-coating. The similarity of the array obtained to that of planes within the crystal lattice suggested that the Na^+ counter ions were located at the extreme lobes of the capsules, inducing hydrophilic intermolecular interactions between capsules to occur within a plane parallel to the Au(111) surface. Such an arrangement of the counter ions should reduce the unfavorable repulsive interactions with the hydrophobic Au surface, thereby increasing the probability of adsorption.

The choice of the metal ion can also favour particular assemblies. A face-on arrangement of the metallosupramolecular complex can be favoured by a metal ion, such as Pt(II), which adopts square-planar coordination. Li *et al.*[17] showed that by using a Pt(II) complex of planar aromatic ligands they could achieve a face-on configuration of the Pt-complex with respect to the Au(111) surface. The flat aromatic moieties further stabilized the planar configuration of the assembly due to favourable π interactions with the Au surface.

It is well known that alkanes form densely packed 2D assemblies on HOPG (highly ordered pyrolytic graphite) due to the hydrophobic interactions which favour commensurate chain alignments to give lamellar phases.[18] The stability induced by van der Waals forces between the interdigitated alkyl chains can be used to force a metal–organic structure to self-assemble onto the solid substrate.[19–21] This strategy is very often successfully employed to self-assemble organic compounds with a low affinity for hydrophobic substrates.[22,23]

Different examples of [2 × 2] metal grid architectures based on transition metal ions with octahedral coordination and assembled on solid substrate can be found in the literature.[24] This class of compounds has raised strong interest because of their optical, redox and magnetic properties in solution.[25–28] By adding long alkyl chains to the bis-hydrazone ligand, Pace *et al.* achieved the self-assembly of a [2 × 2] metallosupramolecular grid containing Co(II) onto a HOPG surface. Such a strategy overcame the repulsive interactions between the hydrophobic surface and the Co(II) cations and counter ions.[27] The 2D lattice parameters extracted from STM images (Figure 8.1) showed an edge-on configuration of the grid with respect to the substrate plane. This implies that an insulating layer of the alkyl chains was assembled between the HOPG surface and the metallosupramolecular core, thus enabling a reduced repulsion between the hydrophobic substrate and the polar species.

Semenov *et al.*[26] showed that by varying the substituent groups on a 4,6-bis(2′,2″-bipyridyl-6-yl)pyrimidine ligand, such as by simply replacing a hydrogen atom with a methylene group, the grid assembly on the HOPG surface could evolve from a face-on to an edge-on configuration (Figure 8.2).

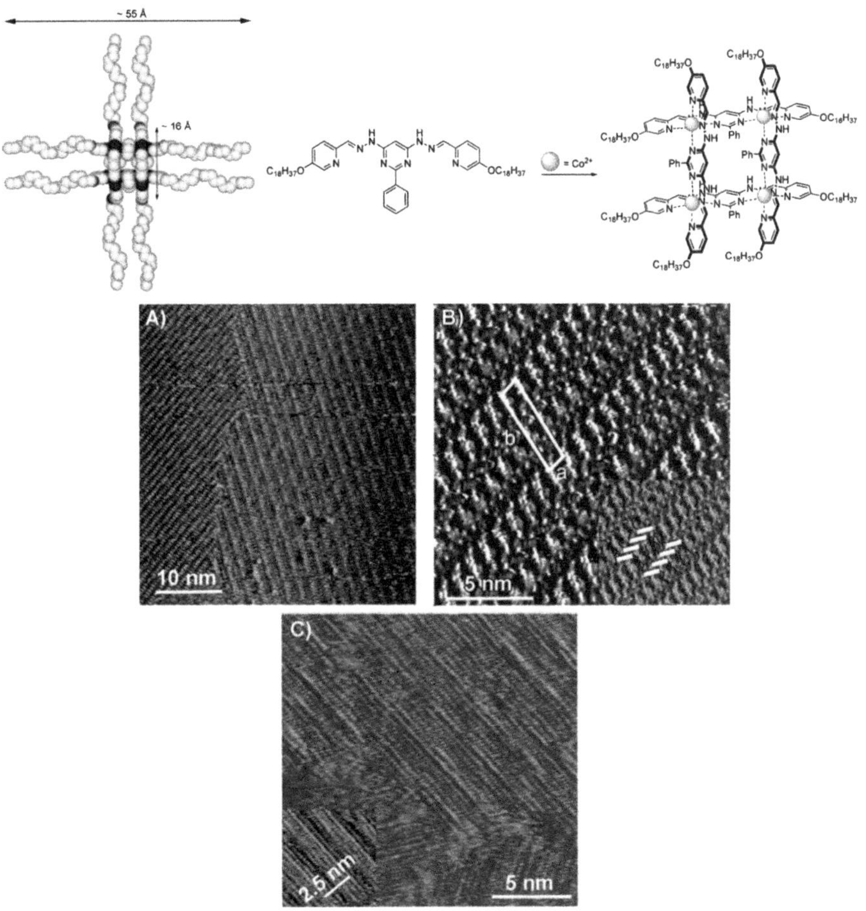

Figure 8.1 Bis-hydrazone ligand (L) and the molecular grid structure $[Co_4(L)_4]$. (A and B) Solid liquid interface STM images of $[Co_4(L)_4]$ assembled on HOPG substrate. Unit cell parameters: $a = 1.2$ nm, $b = 5.1$ nm; $\alpha = 70°$. (B) Inset: molecular sketch of the grid packing on HOPG. (C) STM image of $[Co_4(L_1)_4]$ assembled on a HOPG after annealing.[27]

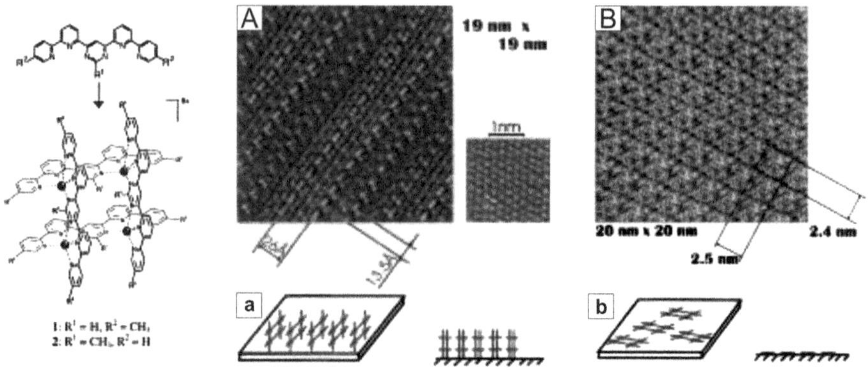

Figure 8.2 Left: molecular structure of the bis-terdentate ligands [2 × 2] grid type Co-complex $[Co_4L_4]^{8+}$. (a) STM height image of grid 1 with a sketch of the edge-on configuration on HOPG. (b) STM height image of grid 2 and a sketch of the face-on configuration.[26]

As noted in the Introduction, intermolecular interactions as well as surface–molecule interactions can affect molecular properties. Detailed studies of metallosupramolecular complex–surface interactions as obtained from single molecule studies have resulted in important findings.[29,30] We discuss later a few examples of how the molecular electronic structure is modified by interaction with conductive surfaces.

In STM imaging, the tunneling current is a function of the probability of coupling the electronic states of the scanning tip–molecule–substrate, so that the molecular orbitals involved in the conduction channel can be imaged. Though STM provides atomic resolution, it is very difficult to identify the specific interactions between the overlying molecule and the substrate atomic layer. Since a 2D-gas-like phase exists at room temperatures, ultra-high vacuum (UHV) and very low temperature (a few Kelvin) conditions are required, in order to freeze the diffusion of molecules at the surface. Despite the extreme experimental conditions, important insights into surface–molecule interactions can still be gained. The most stable configuration of the absorbed molecules is a thermodynamic minimum, which can require a lattice reconstruction of the underlying substrate, *e.g.* in the case of the Au(111) surface herringbone structure, a notable reconstruction occurs upon thiol binding. Since uncoordinated surface atoms are unstable with respect to the bulk atoms (because of the lack of a complete coordination sphere), surface sites at a higher surface energy are the most favorable for coordination of the adsorbate. Heinrich *et al.*[31] determined the relative orientation of a metal complex on a Cu(111) substrate through the deposition of co-adsorbates, such as CO molecules, which define the substrate's lattice orientation. Following this procedure, the presence of a commensurate pattern of Co-phthalocyanine (CoPc) on Cu(111) could be defined indirectly. The binding site for CoPc was therefore deduced, and the Co atom was found to be centred between two Cu surface atoms. Differential conductance (dI/dV)

mapping showed that the main conductance channel was localized over the Co atom at negative bias. Conductive channels could be delocalized also over the N atom bound to the Co centre and over the pyrrole rings at an even lower negative bias (−2.4 V). This finding revealed the specific interaction between the p orbitals of the nitrogen atom or the benzene ring with the d_{z^2} orbitals of the copper atoms (Figure 8.3).

Transition metals can confer magnetic properties to the metal complex system but these can be environmentally sensitive, so it is important to establish that properties defined through solution studies are maintained upon surface adsorption. This is the case, for example, for single molecule magnets (SMMs),[32,33] where the specific magnetic properties have to be preserved upon interaction with the substrates. A well-studied case is terbium bis-phthalocyanine, where the lanthanide metal ion is sandwiched between two Pc units. The Pc units can screen the metal core from direct interaction with the metal substrate. Such molecular screening reduces the probability of undesired redox reactions or weak metal ion–substrate coupling interactions, responsible for the change of the spin-electronic configuration of the pristine material. Spin polarized STM enabled the imaging of the spin-splitting[34,35] and X-ray magnetic circular dichroism measurements enabled resolution of the spin character of the molecular orbitals.[36–38]

The examples presented in this section illustrate how important specific substrate–molecule interactions are, which have to be taken into account carefully when designing new molecular structures. Gaining full control over those interactions represents a big challenge prior to the implementation of metallosupramolecular complexes in actual devices. A few examples of such systems in functional electronic devices are presented in Section 8.5. In Sections 8.3 and 8.4, we will focus on the strategies developed so far for the controlled growth of 2D and 3D assemblies of metal complexes on surfaces.

8.3 Surface-Confined Supramolecular Coordination Chemistry

In all previous examples of metal–organic compounds assembled on surfaces, we encountered molecules whose metal complex structures were stable in solution and their deposition on surfaces was conducted after their synthesis in solution. Assembly on surfaces could then be driven by substrate–molecule and molecule–molecule interactions.

However, metallosupramolecular chemistry offers an additional flexibility due to the favoured coordination reaction between a well-chosen ligand and metal ion, which can be forced to occur directly on a surface (*in situ*). An example of surface-confined coordination chemistry[39] is given by Écija *et al.*, who observed tessellation of Ag(111) substrates *via in situ* surface-driven coordination chemistry.[40] Sequential deposition of an organic ligand and coordinating metal source was achieved by evaporating the reactants under UHV conditions. This assembly process is strongly driven by the surface

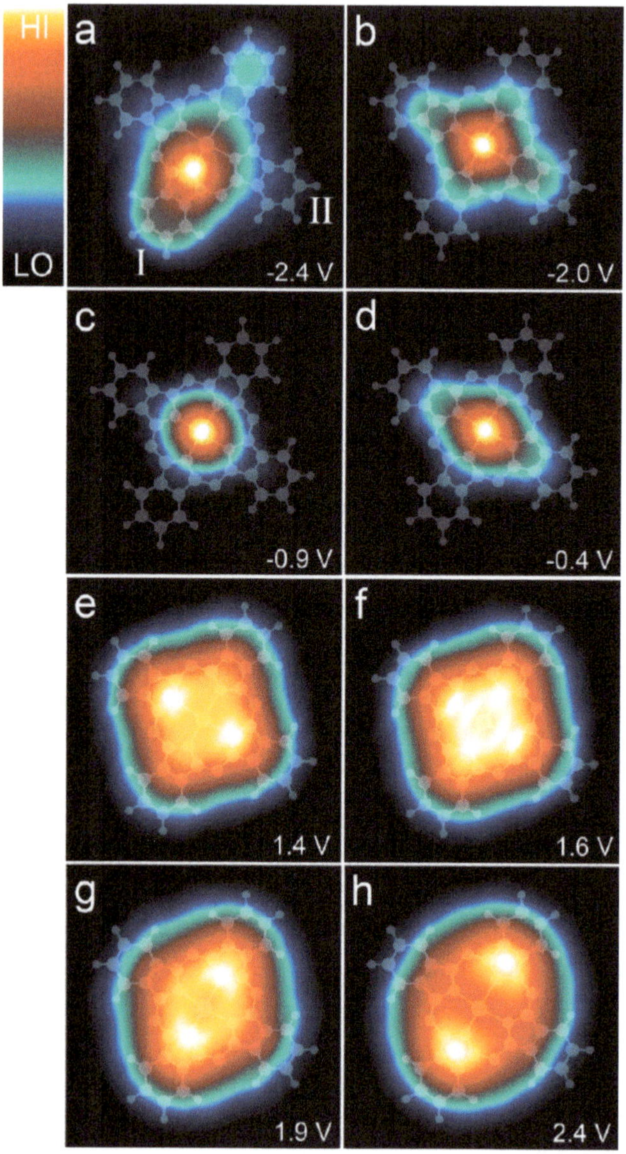

Figure 8.3 Differential conductance maps (dI/dV) of CoPc (20 × 20 Å) physisorbed on Cu(111) acquired at constant height and at a bias of (a) −2.4 V, (b) −2.0 V, (c) −0.9 V, (d) −0.4 V, (e) 1.4 V, (f) 1.6 V, (g) 1.9 V and (h) 2.4 V. The electron density is dependent on the applied bias: higher conductivity regions appear in red and lower conductivity regions in blue. The depicted structure of CoPc is simply a guide to assist readers to localize metal and ligand atoms on the electron density map.[31]

interactions with the pre-evaporated ligand, and the final 2D metal complex array can feature assembled architectures very different from the one formed from solution deposition or existing in the solid state (Scheme 8.1).[39] The final assembly is strongly affected by the values of the atomic lattice constants, the crystalline orientation, atomic steps and surface atomic reconstruction.[41]

Supramolecular assembly on surfaces features the same advantages of self-recognition, self-selection, self-repair and dynamic self-(re)organization as observed for supramolecular assembly processes in solution. This leads to thermodynamic control of the reaction and the most stable configuration on the surface is the one most likely to be formed. In surface coordination chemistry, the molecular building blocks often display carboxylates and nitrogen groups (as in pyridyl, hydrazine and terpyridine linkers) that are strongly bound to transition metal ions such as Fe(II), Cu(II) and Co(II). A wide variety of self-assembled supramolecular complexes have been found to adsorb on solid substrates *via* surface-confined coordination chemistry.[39] Owing to the high flexibility in molecular design and synthesis offered by coordination chemistry, different 2D-arrays can be achieved by simply varying the molecular length, the metal coordination centres and the relative orientation of the coordinating groups. We highlight a few such beautiful surface architectures in Figures 8.4 and 8.5. Metal–organic honeycomb networks formed from ditopic dicarbonitrile molecular building blocks and Co metal ions on Ag(111) are shown in Figure 8.4.[42,43] By varying the number of the phenyl rings composing the ditopic ligand, it is possible to tune the cavity size of the formed "nanomesh" array. In this 2D-assembly, three ligands coordinate to the Co centers, something not common in the coordination configuration of cobalt–carbonitrile complexes found in the bulk. This illustrates the role played by the Ag(111) surface in driving the self-assembly. Following a combinatorial approach, Langner *et al.*[44] were able to build a small library of size-tunable, rectangular nanomesh patterns by assembling bi-component solution ligands coordinating Fe(II) ions (Figure 8.5).

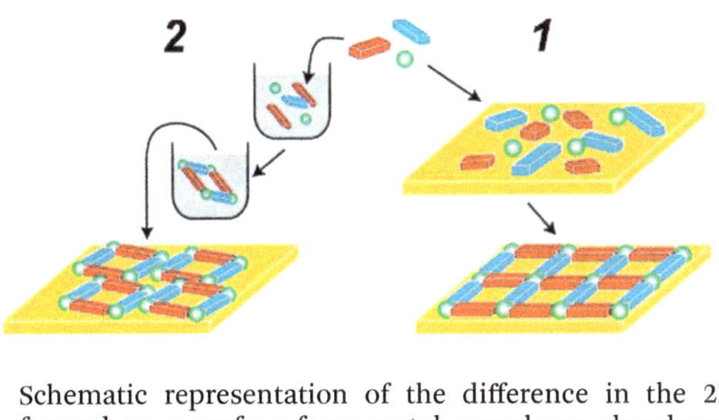

Scheme 8.1 Schematic representation of the difference in the 2D-assemblies formed on a surface from metal complexes already synthesized in solution (**2**) or from direct assembly of metal and ligand (*in situ* assembly) (**1**). The presence of the specific substrate ligand and metal interactions can drive the assembly on the surface. The final assemblies are characterized by a different stoichiometry.[39]

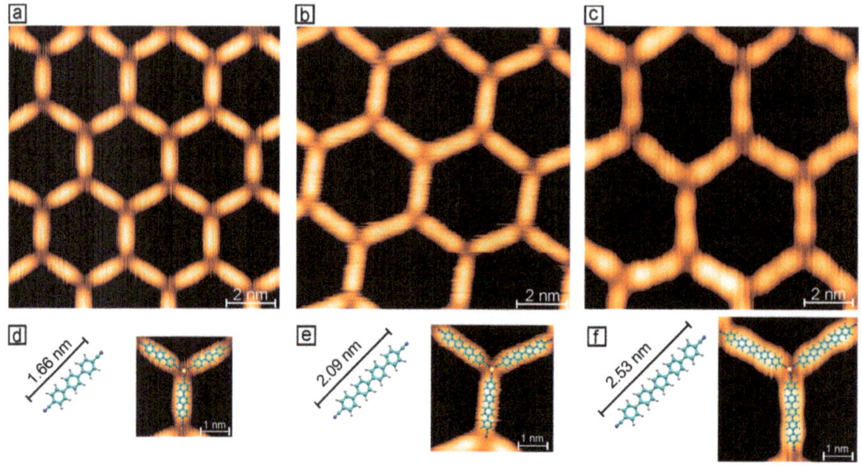

Figure 8.4 STM images showing the result of Co-directed assembly of ditopic dicarbonitrile (CN) polyphenyl (Ph); (a) NC–Ph_3–CN, (b) NC–Ph_4–CN, and (c) NC–Ph_5–CN, respectively on a Ag(111) surface. (d–f) Molecular structure superimposed on the experimental data. The threefold Co-carbonitrile coordination motif as found in (a)–(c) is represented in yellow (cobalt centre); turquoise (carbon); white (hydrogen); blue (nitrogen).[42]

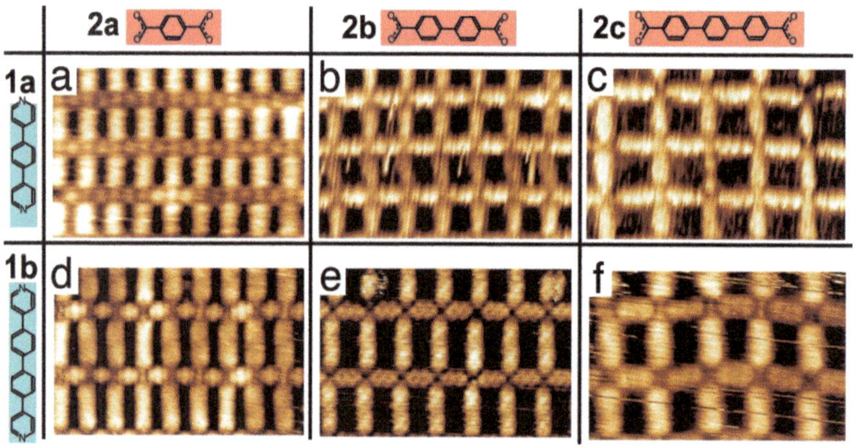

Figure 8.5 Combinatorial approach to the surface directed synthesis of 2D arrays. STM images of the pattern found for the combination of bipyridine functionalized ligands (**1a** and **1b**) with and bis-carboxylic acid functionalized ligands (**2a**, **2b**, and **2c**). The ligands bind to two Fe metal coordination centres (9.4 nm × 6.0 nm). Varying the phenyl chain length provides different lattice parameters: (1.1 × 1.8) nm (a), (1.5 × 1.8) nm (b), (1.8 × 1.8) nm (c), (1.1 × 2.3) nm (d), (1.5 × 2.3) nm (e), and (1.8 × 2.3) nm (f).[44]

Comprehensive studies carried out on the dynamic assembly of phenyl dicarboxylic acids on different metal substrates and various metal centres have shown that the interplay between the surface template effect and metal–ligand interactions can drive the assembly of completely different structures. If the metal-to-ligand binding overcomes the templating effect of the metal substrate, as for the case of Fe and Co-terephthalate assemblies on Au(111), a non-commensurate assembly will be formed and small 2D crystals appear.[45] Conversely, pyridine–Cu coordination is weaker and the substrate template effect dominates, favoring the formation of a commensurate lattice extending over a large area (few microns).[46,47] It is then clear that by matching the symmetry of the underlying substrate with the geometry of the most favoured metal-to-ligand configuration, it is possible to form 2D assemblies that can extend over a very long range, as in the case of *para*-hexaphenyl-dicarbonitrile and Co complex nanomeshes formed on an Ag(111) surface.[48]

So far, the examples presented in this section have mostly dealt with systems grown by evaporating the metal and ligand within UHV chambers. In addition, within the framework of molecular tectonics,[49] it is possible to find examples of surface-confined coordination chemistry starting from solutions of ligand (tectons) and metal ions. An example is the growth of metal–organic complexes at a liquid–solid interface by post-deposition of the metal ion on a pre-adsorbed terpyridine ligand.[49]

The ease of formation of such 2D nanomeshes led researchers to accommodate molecules in the hosting cavity.[50] An example is presented in Figure 8.6 where fullerene, cystine or phenylalanine molecules were found to bind at the pore sites of a Fe–trimesic acid complex nanomesh.[51]

Surface coordination chemistry has certainly been mostly driven by academic curiosity but it has enabled the discovery of the principles that underpin ligand–metal assembly on surfaces. The fine balance of weak interactions that drive the metallosupramolecular assembly enables the formation of 2D networks whose order can extend over the mesoscopic range. This is possible

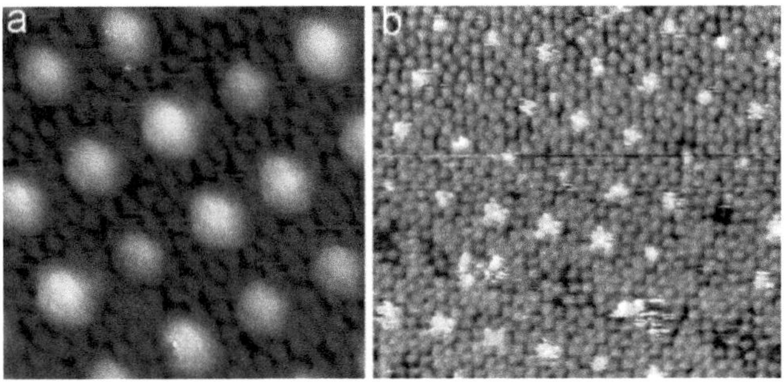

Figure 8.6 (a) STM image (14 nm × 14 nm) showing the binding of single C_{60} molecules into the nano-cavities network formed by trimesic acid molecules coordinating Fe atoms and self-assembled on Cu(100). (b) STM image (25 nm × 22 nm) showing the binding of phenylalanine molecules into the nanopores.[51]

thanks to the self-recognition and self-repairing processes intrinsic in supramolecular chemistry. Mastering such interactions has led to the production of nanomeshes where size tunable cavities can be prepared, which can host a variety of reactants. This is an example of how such an approach to metal–organic chemistry can foster new discoveries and opportunities. Such cavities can be employed to isolate a wide variety of guest species, therefore enabling studies at the single molecule level. The prospective applications of such an approach have yet to be explored further.

8.4 Hierarchical Assembly of Metal–Organic Frameworks (MOFs)

Surface-confined coordination chemistry has shown the possibility of *in situ* coordination chemistry, where complexes are formed directly on a solid substrate. The surface interface is therefore the reaction medium for the ligand–metal ion coordination chemistry. The final 2D array of the metal–organic complex implies a 2D crystal growth. However, surface-confined synthesis has been successfully employed to grow inorganic–organic hybrid materials in the third dimension also by using wet processes.[52]

This hierarchical metal–organic assembly at surfaces belongs to a larger family of metal–organic compounds termed metal–organic frameworks (MOFs; also termed coordination polymers, see Chapter 7) that are hybrid inorganic–organic solid state compounds.[53–55] Among these frameworks, the most interesting to the purpose of this chapter are surface metal–organic frameworks (SURMOFs), which are substrate-supported highly oriented MOFs.[56,57] Through a layer-by-layer deposition on an appropriately chosen self-assembled monolayer (SAM)-functionalized substrate, it is in fact possible to grow 2D and 3D crystalline metal–organic structures directly on surfaces. SAMs can form on various substrates, examples being thiolated molecules on gold, silane and carboxylic groups on silicon oxides and other oxide substrates.[13] In order to foster the crystalline growth of a SURMOF, these SAMs have to expose to the solution interface appropriate functional groups able to specifically coordinate the metal ions. Typically, these are carboxylates, phosphonates or pyridine groups to which a variety of metal ions can bind. After SAM–metal binding is completed, the next step is the exposure to a solution containing the coordinating ligand. This process is reiterated to grow crystalline SURMOFs; this cycling procedure is shown in Scheme 8.2.

This process has been termed liquid-phase epitaxy (LPE) since such quasi-epitaxial step-growth leads to the formation of crystalline layers on solid substrates due to the template effect of the nucleation sites offered by the SAMs' functional groups.[59] Such a template effect has been exploited for the growth of enantioselective SURMOFs[60] and to drive specific crystal orientation growth.[61] Heteroepitaxial growth has also been shown by the formation of layered-crystal component SURMOFs where two lattice-mismatched

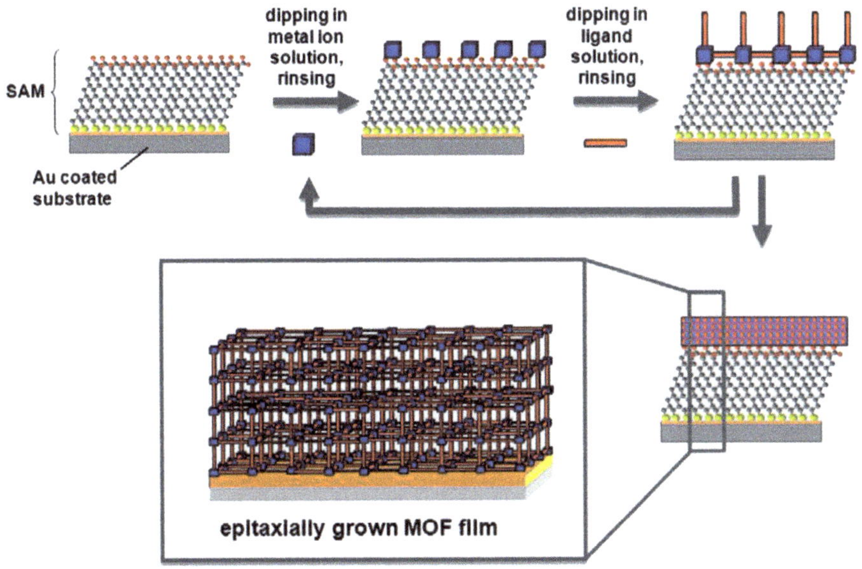

Scheme 8.2 Scheme of the hierarchical self-assembly of metal ions and ligands from solutions for the fabrication of SURMOF films. The starting substrate is a preformed SAM-terminated gold surface followed by the epitaxial layer-by-layer growth of the SURMOF. The crystal size is defined by the number of coating cycles.[58]

Cu-based MOFs could be sequentially grown on the same surface.[62] Controlled micro- and mesopores can be designed by the proper choice of ligands and metal ions, opening prospective applications in gas storage and separation,[63] catalysis,[64] delivery of therapeutic agents[65] and sensor devices.[66–68] Furthermore, the selective growth on functionalized SAM surfaces allows patterning[69] of the surface through microcontact printing,[70] lithography, nanografting,[71] pen-lithography,[72] inkjet printing[73] and spray techniques.[74]

In Figure 8.7, we show an example of a HKUST-1 single crystal MOF selectively grown on patterned substrates. Pen-type lithography was used to fabricate the SAM pattern as well as to locally release the HKUST-1 solution. Such a strategy allowed the growth of a single crystal per SAM patterned site. In Figure 8.8, another example of substrate patterning is shown. The fabrication process employed atomic force microscopy (AFM) as a nanolithographic tool for the selective deposition of specifically-functionalized SAMs, while spray coating was used for the layer-by-layer growth of the HKUST-1 crystals.[71]

An exciting prospect for MOFs was recently demonstrated by Talin *et al.*[75] who infiltrated the host cavities available in the binuclear Cu(II) paddlewheel MOF structure $Cu_3(BTC)_2$ (also known as HKUST-1; BTC, benzene-1,3,5-tricarboxylic acid), with the semiconducting TCNQ (7,7,8,8-tetracyanoquinododimethane), which could impart semiconducting properties to the MOF (Figure 8.9).

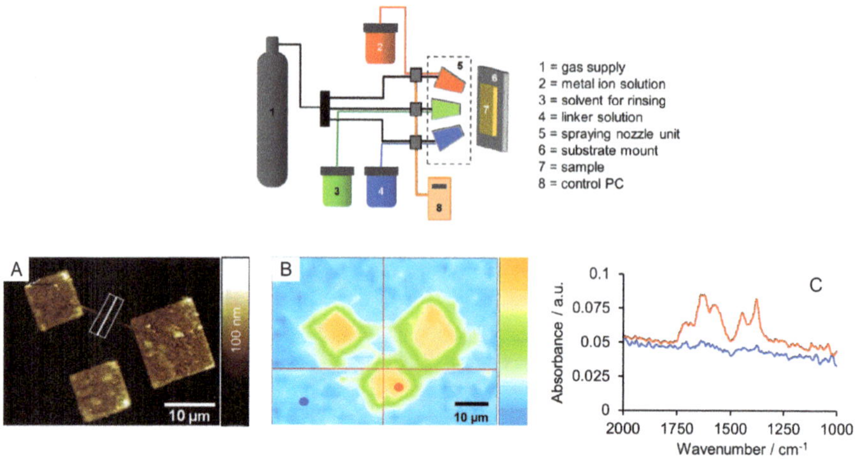

Figure 8.7 (a) Schematic showing the fabrication of printed SURMOFs patterns. (b) Optical micrograph of the droplet arrays; the drop spacing is 10 μm. (c) Scanning electron microscopy image of the resulting MOF of HKUST-1.[72]

Such patterned and functionalized surfaces can be integrated into a variety of electronic devices, making them appealing for further technological applications in optoelectronics and sensing.

8.5 Metallosupramolecular Complexes in Devices

Essentially, full comprehension of the self-assembly processes of metal–organic complexes on surfaces has been achieved. Proper metal and ligand choice can be used to drive the assembly towards the desired morphology. A functional assembly obtained through layer-by-layer deposition of a terpyridine ligand and Ru(II) source has been described by Gupta *et al.* (Scheme 8.3).[76] The authors developed a surface-confined growth of hybrid semiconductor/redox-active molecule materials from solution precursors.

Another striking example comes from the joint work of Anderson, Lambert and Nichols in which short chains of metalloporphyrin molecules have been employed to mediate electron transport over quite significant distances of 5–10 nm (Figure 8.10).[77]

The authors used a variable-temperature STM technique to investigate the temperature-dependence conductivity of oligo-metalloporphyrin single-molecule junctions (with one, two and three porphyrin units), and also examined the mechanism of charge transport with both an analytical model and *ab initio* transport computations based on density functional theory (DFT). The results obtained showed that the conductance process strongly depends on both the temperature (a temperature range of 25–100 °C was investigated) and to a lesser extent on the number of porphyrin units between the gold leads. The values obtained from both of these factors led

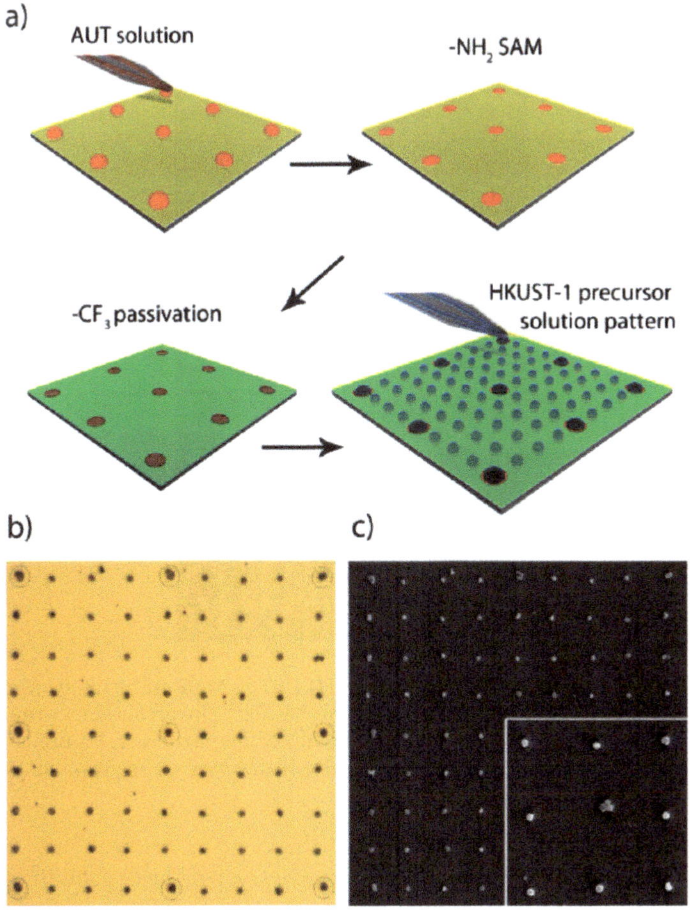

Figure 8.8 Scheme of the spray coating technique employed. (a) AFM image of HKUST-1 crystals grown by spray coating on a nanolithographically pre-patterned SAM substrate of 3-mercaptopropionic acid SAM in a 1-decanethiol SAM matrix. (b) Fourier transform-IR (FT-IR) imaging showing the selective growth of the HKUST-1 crystal. (c) FT-IR spectra acquired inside (red) and outside (blue) the patterned area (the bands between 1420 cm^{-1} and 1300 cm^{-1} are assigned to the C–C vibration of the aromatic ring of the crystal).[71]

to the conclusion that the electron conductivity is consistent with phase-coherent tunneling through the whole molecular junction rather than a thermally-assisted incoherent (hopping) mechanism. This type of conjugated metallosupramolecular architecture is very promising as a candidate for application in the field of nanotechnology as a new type of photovoltaic device.

Construction of functional metallosupramolecular thin films and its use in the fabrication of lateral nanoscale devices have been described by Lin and

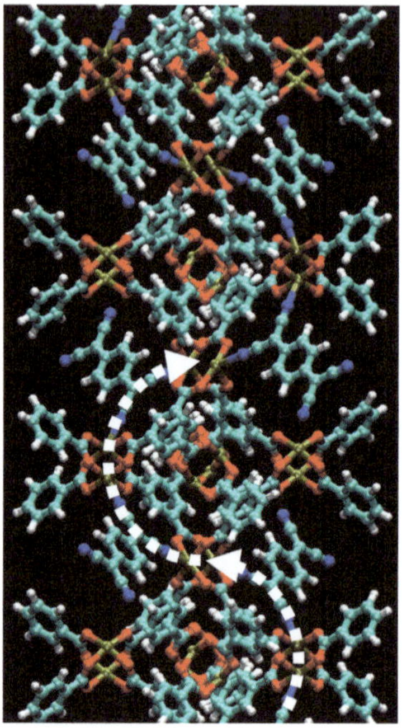

Figure 8.9 Molecular structure of the HKUST-1 MOF infiltrated with TCNQ (7,7,8,8-tetracyanoquinododimethane).[75] The arrow shows the possible electron conductive path in the molecular structure. Figure courtesy of M. Foster/Sandia National Labs (http://www.nist.gov/cnst/20131205_mof.cfm).

Kagan.[78] The authors utilized the progressive layer-by-layer growth of metal–metal bonded compounds to generate a new system of inorganic coordination complexes anchored on a gold substrate (Scheme 8.4).

The formation of the metallosupramolecular thin film was monitored by electronic spectroscopy whereas its thickness (average 1.5 nm) was evaluated by atomic force microscopy (AFM). More importantly, cyclic voltammetry showed that this system is redox active (two one-electron reversible oxidation processes of the Rh metal centers). Utilizing the layer-by-layer directed assembly method of this system, the authors produced a lateral nanodevice, which preserved interesting redox properties. This kind of approach opens the door for the production of new types of supramolecular devices possessing fascinating physicochemical properties of different kinds depending on the type of metallic center employed.

Gao and Sun employed a supramolecular self-assembly method to obtain a functional water splitting device based on a photoactive anode on the surface of nanostructured TiO_2.[79] Their system consisted of two metallosupramolecular Ru complexes **1** (photosensitizer) and **2** (catalyst) and a Zr^{4+} ion as a linkage to connect **1** and **2** in the photoanode, as shown in Scheme 8.5.

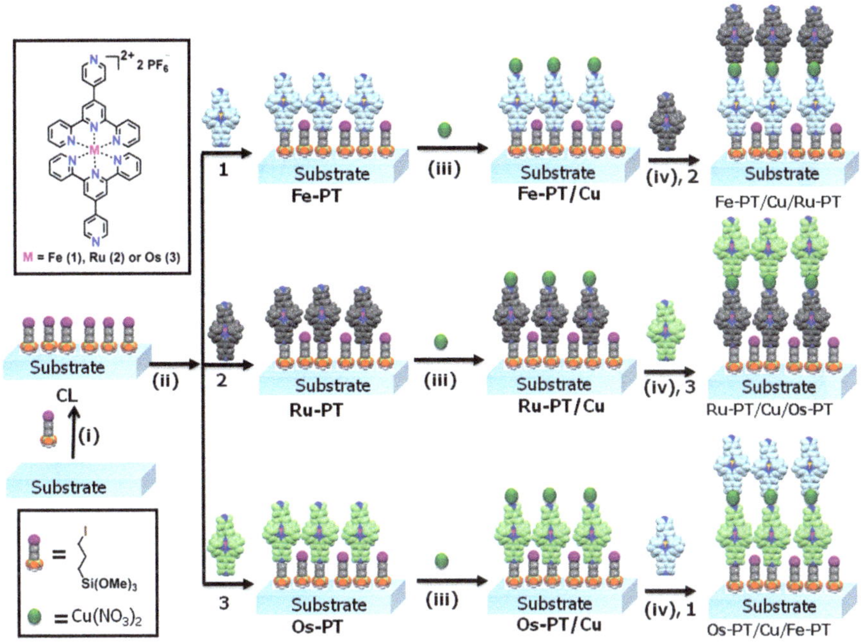

Scheme 8.3 Scheme of hierarchical self-assembly of metal ion and ligand for the fabrication of a Ru-based SURMOF: (i) immobilization of 3-iodo-*n*-propyl-trimethoxysilane on a SiO*x* substrate to form a coupling layer (CL) serving as a primary template, (ii) quaternization of the pendant pyridyl group of Fe–PT, Ru–PT or Os–PT to CL to form a secondary template layer, (iii) coordination of copper nitrate with the pyridine-terminated template layer to form a Cu-terminated tertiary template layer, and (iv) coordination of Ru–PT, Os–PT and Fe–PT with the Cu-terminated template layer to form the hetero-metallic molecular dyads.[76]

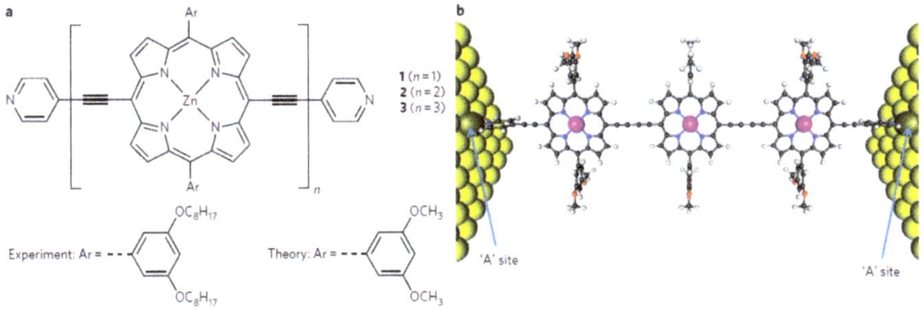

Figure 8.10 (a) Chemical structure of the conjugated Zn(II) metalloporphyrin molecules used in electron transport. (b) Schematic representation of the metallosupramolecular device between gold leads.[77]

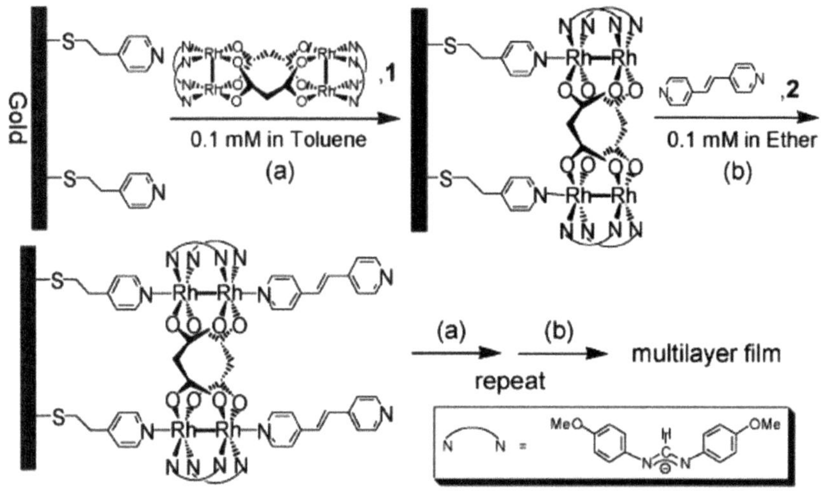

Scheme 8.4 Schematic representation of the layer-by-layer growth of metal–metal bonded structures.[78]

Upon combining all three components, a new metallosupramolecular assembly formed *via* coordination of the Zr^{4+} ion by the phosphate and carboxylate groups of **1** and **2**, respectively. Electrochemical measurements (CV) on this system showed two separate oxidation processes of the Ru metal centers (Ru^{II}/Ru^{III}) in **1** and **2**. Due to the higher oxidation potential of the Ru center in **1** (1.40 V) than the onset potential for catalytic water oxidation (1.20 V), this assembly was used for visible light-driven water splitting. Although more detailed photoelectrochemical analysis showed relatively low values of incident photon to current conversion efficiency (IPCE) and photocurrent density, this metallosupramolecular photoanode still offers an interesting new platform for constructing "monolayer" molecular devices using a supramolecular self-assembly method.

In work from Itaya and co-workers, adlayers of Co(III) tetraphenylporphyrin were prepared on Au electrodes and examined for axial coordination of an O_2 molecule (Figure 8.11). Interestingly, it was found that these metallosupramolecular architectures adsorbed on the Au surface had a higher O_2 binding affinity than native myoglobin. The assembly of Co(III) tetraphenylporphyrin on both Au(111) and Au(100)–(hex) surfaces was performed and led to the formation of well-ordered nanobelt arrays.[80,81] The deposition process was repeated with the O_2-adducted metalloporphyrin but the ordered arrays were observed only on the Au(100)–(hex) surface, thus suggesting a strong dependence of this process on the crystallographic orientation of Au.

The authors explained this observation by suggesting that underlying Au atoms can act as axial ligands, facilitating the formation of the O_2-adducted metalloporphyrin. Analysis of the STM images showed a mixture of bright and dark spots corresponding to metalloporphyrin macrocycles with and without coordinated O_2 molecules, respectively. Time-dependent STM images showed

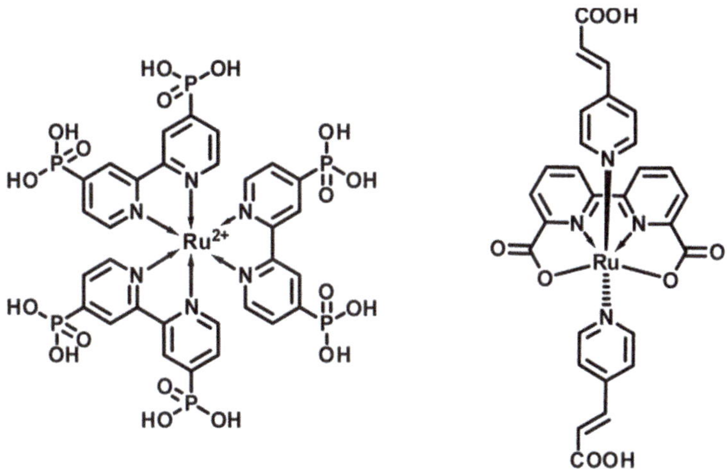

1 **2**

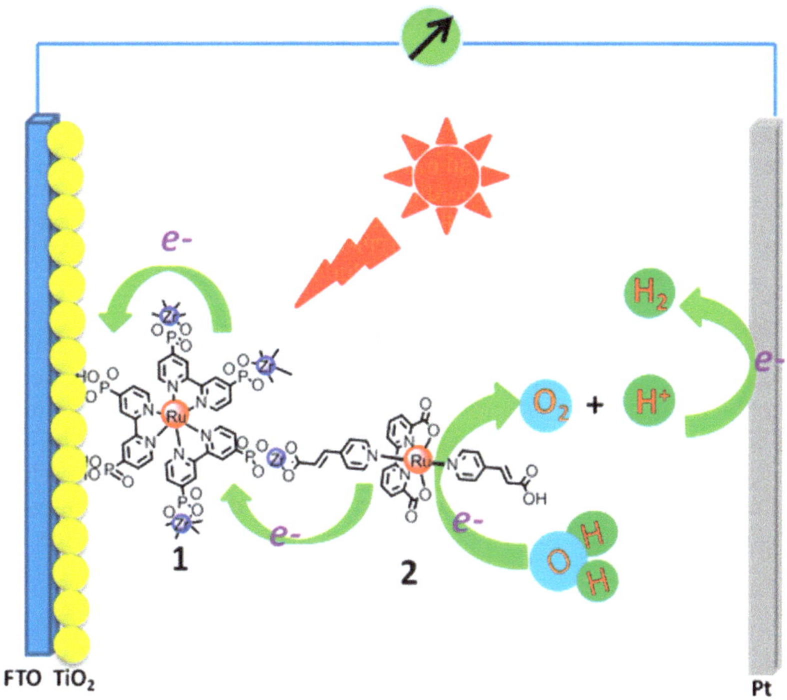

Scheme 8.5 Schematic representation of the photoelectrochemical cell (PEC) with a photoanode assembled from a Zr^{4+} ion, metallosupramolecular complexes **1** and **2** on the nanostructured TiO_2 and a passive Pt cathode, for visible light-driven water splitting in an aqueous solution.[79]

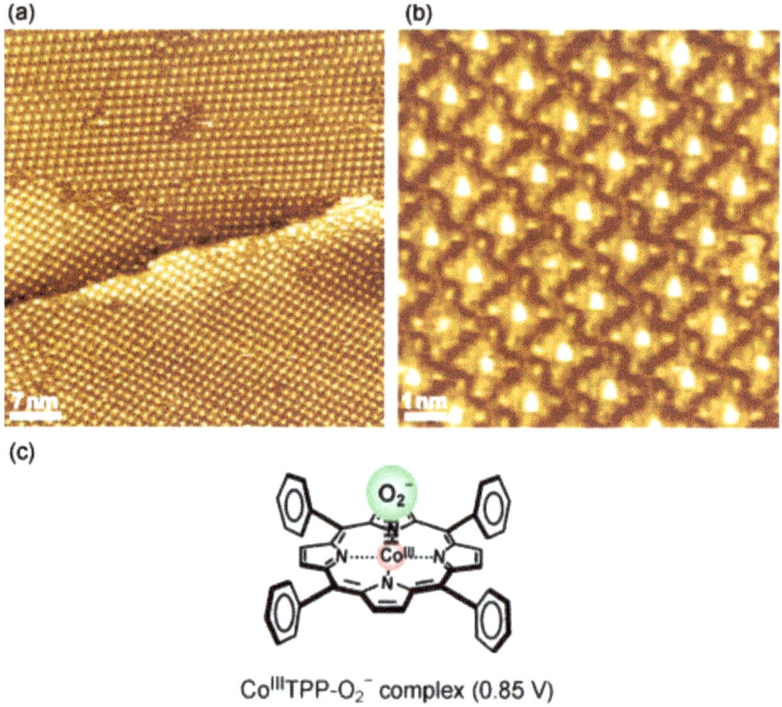

Figure 8.11 (a) Large-scale (50×50 nm^2) and (b) high-resolution (9×9 nm^2) STM images of the O_2-coordinated Co(III) tetraphenylporphyrin on an Au(111) surface. (c) Schematic representation of the proposed model of the O_2-coordinated Co(III)TPP complex in the highly ordered CoTPP array on Au(111) in an acidic solution.[80]

that the coordinated O_2 molecules were highly mobile, having the ability to move from one "picket-fence" to another. Furthermore, the investigated system enabled the gradual electrocatalytic reduction of the oxygen molecules in the cavities of the Co(III) porphyrin seen by the disappearance of the bright spots in the STM images, thus indicating that all the O_2 had been released into solution. This final experiment shows that the described metallosupramolecular system could function as a nanoscale oxygen storage device.

8.6 Conclusion

Metallosupramolecular complexes with interesting electronic, spin and optical states pave the way for the production of new technology. The design of the ligand and choice of metal ion center allow researchers to engineer in a wide variety of functionalities, perhaps most importantly of all their incorporation into solid state structures. In this chapter, we have discussed the main strategies employed so far to confine metallosupramolecular complexes on surfaces, focusing on their specific interaction with the substrate, and highlighting some examples of operational devices.

Acknowledgements

We would like to thank Prof. Jack Harrowfield for his suggestions and corrections. ARS thanks NCBR (grant LIDER/024/391/L-5/13/NCBR/2014) for financial support.

References

1. J. M. Lehn, *Proc. Natl. Acad. Sci. U. S. A.*, 2002, **99**, 4763.
2. G. Heimel, S. Duhm, I. Salzmann, A. Gerlach, A. Strozecka, J. Niederhausen, C. Burker, T. Hosokai, I. Fernandez-Torrente, G. Schulze, S. Winkler, A. Wilke, R. Schlesinger, J. Frisch, B. Broker, A. Vollmer, B. Detlefs, J. Pflaum, S. Kera, K. J. Franke, N. Ueno, J. I. Pascual, F. Schreiber and N. Koch, *Nat. Chem.*, 2013, **5**, 187.
3. M. K. Samantaray, J. Alauzun, D. Gajan, S. Kavitake, A. Mehdi, L. Veyre, M. Lelli, A. Lesage, L. Emsley, C. Coperet and C. Thieuleux, *J. Am. Chem. Soc.*, 2013, **135**, 3193.
4. J. G. Hardy, *Chem. Soc. Rev.*, 2013, **42**, 7881.
5. H. Hofmeier and U. S. Schubert, *Chem. Soc. Rev.*, 2004, **33**, 373.
6. C. Wieser, C. B. Dieleman and D. Matt, *Coord. Chem. Rev.*, 1997, **165**, 93.
7. S. Leininger, B. Olenyuk and P. J. Stang, *Chem. Rev.*, 2000, **100**, 853.
8. G. R. Whittell, M. D. Hager, U. S. Schubert and I. Manners, *Nat. Mater.*, 2011, **10**, 176.
9. U. S. Schubert and C. Eschbaumer, *Angew. Chem., Int. Ed.*, 2002, **41**, 2893.
10. P. R. Andres and U. S. Schubert, *Adv. Mater.*, 2004, **16**, 1043.
11. D. G. Kurth and M. Higuchi, *Soft Matter*, 2006, **2**, 915.
12. A. Wild, F. Schlutter, G. M. Pavlov, C. Friebe, G. Festag, A. Winter, M. D. Hager, V. Cimrova and U. S. Schubert, *Macromol. Rapid Commun.*, 2010, **31**, 868.
13. A. Ulman, *Chem. Rev.*, 1996, **96**, 1533.
14. J. C. Love, L. A. Estroff, J. K. Kriebel, R. G. Nuzzo and G. M. Whitesides, *Chem. Rev.*, 2005, **105**, 1103.
15. M. Schmittel and V. Kalsani, *Funct. Mol. Nanostruct.*, 2005, **245**, 1.
16. M. B. Mahimaidoss, S. A. Krasnikov, L. Reck, C. I. Onet, J. M. Breen, N. Y. Zhu, B. Marzec, I. V. Shvets and W. Schmitt, *Chem. Commun.*, 2014, **50**, 2265.
17. S. S. Li, B. H. Northrop, Q. H. Yuan, L. J. Wan and P. J. Stang, *Acc. Chem. Res.*, 2009, **42**, 249.
18. M. Hibino, A. Sumi, H. Tsuchiya and I. Hatta, *J. Phys. Chem. B*, 1998, **102**, 4544.
19. X. H. Qiu, C. Wang, Q. D. Zeng, B. Xu, S. X. Yin, H. N. Wang, S. D. Xu and C. L. Bai, *J. Am. Chem. Soc.*, 2000, **122**, 5550.
20. Z. G. Wang, Q. D. Zeng, Y. B. Luan, X. J. Wu, L. J. Wan, C. Wang, G. U. Lee, S. X. Yin, J. L. Yang and C. L. Bai, *J. Phys. Chem. B*, 2003, **107**, 13384.
21. M. Koepf, J. A. Wytko, J. P. Bucher and J. Weiss, *J. Am. Chem. Soc.*, 2008, **130**, 9994.

22. S. De Feyter and F. C. De Schryver, *Chem. Soc. Rev.*, 2003, **32**, 139.
23. S. De Feyter and F. C. De Schryver, *J. Phys. Chem. B*, 2005, **109**, 4290.
24. M. Ruben, J. Rojo, F. J. Romero-Salguero, L. H. Uppadine and J. M. Lehn, *Angew. Chem., Int. Ed.*, 2004, **43**, 3644.
25. A. R. Stefankiewicz, G. Rogez, J. Harrowfield, A. N. Sobolev, A. Madalan, J. Huuskonen, K. Rissanen and J. M. Lehn, *Dalton Trans.*, 2012, **41**, 13848.
26. A. Semenov, J. P. Spatz, M. Moller, J. M. Lehn, B. Sell, D. Schubert, C. H. Weidl and U. S. Schubert, *Angew. Chem., Int. Ed.*, 1999, **38**, 2547.
27. G. Pace, A. Stefankiewicz, J. Harrowfield, J. M. Lehn and P. Sarnori, *ChemPhysChem*, 2009, **10**, 699.
28. U. Ziener, J. M. Lehn, A. Mourran and M. Moller, *Chem.–Eur. J.*, 2002, **8**, 951.
29. S. V. Aradhya and L. Venkataraman, *Nat. Nanotechnol.*, 2013, **8**, 399.
30. B. Hulsken, R. Van Hameren, J. W. Gerritsen, T. Khoury, P. Thordarson, M. J. Crossley, A. E. Rowan, R. J. M. Nolte, J. A. A. W. Elemans and S. Speller, *Nat. Nanotechnol.*, 2007, **2**, 285.
31. B. W. Heinrich, C. Iacovita, T. Brumme, D. J. Choi, L. Limot, M. V. Rastei, W. A. Hofer, J. Kortus and J. P. Bucher, *J. Phys. Chem. Lett.*, 2010, **1**, 1517.
32. D. N. Woodruff, R. E. P. Winpenny and R. A. Layfield, *Chem. Rev.*, 2013, **113**, 5110.
33. L. Bogani and W. Wernsdorfer, *Nat. Mater.*, 2008, **7**, 179.
34. T. Komeda, H. Isshiki, J. Liu, Y. F. Zhang, N. Lorente, K. Katoh, B. K. Breedlove and M. Yamashita, *Nat. Commun.*, 2011, **2**, 217.
35. M. Gonidec, R. Biagi, V. Corradini, F. Moro, V. De Renzi, U. del Pennino, D. Summa, L. Muccioli, C. Zannoni, D. B. Amabilino and J. Veciana, *J. Am. Chem. Soc.*, 2011, **133**, 6603.
36. J. Schwobel, Y. S. Fu, J. Brede, A. Dilullo, G. Hoffmann, S. Klyatskaya, M. Ruben and R. Wiesendanger, *Nat. Commun.*, 2012, **3**, 953.
37. M. Mannini, F. Pineider, P. Sainctavit, C. Danieli, E. Otero, C. Sciancalepore, A. M. Talarico, M. A. Arrio, A. Cornia, D. Gatteschi and R. Sessoli, *Nat. Mater.*, 2009, **8**, 194.
38. S. Stepanow, J. Honolka, P. Gambardella, L. Vitali, N. Abdurakhmanova, T. C. Tseng, S. Rauschenbach, S. L. Tait, V. Sessi, S. Klyatskaya, M. Ruben and K. Kern, *J. Am. Chem. Soc.*, 2010, **132**, 11900.
39. N. Lin, S. Stepanow, M. Ruben and J. V. Barth, *Top. Curr. Chem.*, 2009, **287**, 1.
40. D. Écija, J. I. Urgel, A. C. Papageorgiou, S. Joshi, W. Auwarter, A. P. Seitsonen, S. Klyatskaya, M. Ruben, S. Fischer, S. Vijayaraghavan, J. Reichert and J. V. Barth, *Proc. Natl. Acad. Sci. U. S. A.*, 2013, **110**, 6678.
41. T. Kudernac, S. B. Lei, J. A. A. W. Elemans and S. De Feyter, *Chem. Soc. Rev.*, 2009, **38**, 402.
42. U. Schlickum, R. Decker, F. Klappenberger, G. Zoppellaro, S. Klyatskaya, M. Ruben, I. Silanes, A. Arnau, K. Kern, H. Brune and J. V. Barth, *Nano Lett.*, 2007, **7**, 3813.
43. J. Reichert, M. Marschall, K. Seufert, D. Ecija, W. Auwarter, E. Arras, S. Klyatskaya, M. Ruben and J. V. Barth, *J. Phys. Chem. C*, 2013, **117**, 12858.

44. A. Langner, S. L. Tait, N. Lin, C. Rajadurai, M. Ruben and K. Kern, *Proc. Natl. Acad. Sci. U. S. A.*, 2007, **104**, 17927.
45. S. Clair, S. Pons, S. Fabris, S. Baroni, H. Brune, K. Kern and J. V. Barth, *J. Phys. Chem. B*, 2006, **110**, 5627.
46. S. L. Tait, A. Langner, N. Lin, S. Stepanow, C. Rajadurai, M. Ruben and K. Kern, *J. Phys. Chem. C*, 2007, **111**, 10982.
47. S. Stepanow, N. Lin, D. Payer, U. Schlickum, F. Klappenberger, G. Zoppellaro, M. Ruben, H. Brune, J. V. Barth and K. Kern, *Angew. Chem., Int. Ed.*, 2007, **46**, 710.
48. D. Kuhne, F. Klappenberger, R. Decker, U. Schlickum, H. Brune, S. Klyatskaya, M. Ruben and J. V. Barth, *J. Am. Chem. Soc.*, 2009, **131**, 3881.
49. M. W. Hosseini, *Coord. Chem. Rev.*, 2003, **240**, 157.
50. H. Spillmann, A. Kiebele, M. Stohr, T. A. Jung, D. Bonifazi, F. Y. Cheng and F. Diederich, *Adv. Mater.*, 2006, **18**, 275.
51. S. Stepanow, N. Lin, J. V. Barth and K. Kern, *Chem. Commun.*, 2006, 2153.
52. N. Tuccitto, V. Torrisi, M. Cavazzini, T. Morotti, F. Puntoriero, S. Quici, S. Campagna and A. Licciardello, *ChemPhysChem*, 2007, **8**, 227.
53. E. Redel, Z. B. Wang, S. Walheim, J. X. Liu, H. Gliemann and C. Woll, *Appl. Phys. Lett.*, 2013, **103**, 091903.
54. A. Carne-Sanchez, I. Imaz, M. Cano-Sarabia and D. Maspoch, *Nat. Chem.*, 2013, **5**, 203.
55. D. J. Tranchemontagne, J. L. Mendoza-Cortes, M. O'Keeffe and O. M. Yaghi, *Chem. Soc. Rev.*, 2009, **38**, 1257.
56. D. Zacher, R. Schmid, C. Woll and R. A. Fischer, *Angew. Chem., Int. Ed.*, 2011, **50**, 176.
57. D. Zacher, O. Shekhah, C. Woll and R. A. Fischer, *Chem. Soc. Rev.*, 2009, **38**, 1418.
58. H. Gliemann and C. Woll, *Mater. Today*, 2012, **15**, 110.
59. J. X. Liu, O. Shekhah, X. Stammer, H. K. Arslan, B. Liu, B. Schupbach, A. Terfort and C. Woll, *Materials*, 2012, **5**, 1581.
60. B. Liu, O. Shekhah, H. K. Arslan, J. X. Liu, C. Woll and R. A. Fischer, *Angew. Chem., Int. Ed.*, 2012, **51**, 807.
61. B. Liu, M. Tu and R. A. Fischer, *Angew. Chem., Int. Ed.*, 2013, **52**, 3402.
62. M. Tu and R. A. Fischer, *J. Mater. Chem. A*, 2014, **2**, 2018.
63. Z. Y. Gu, C. X. Yang, N. Chang and X. P. Yan, *Acc. Chem. Res.*, 2012, **45**, 734.
64. S. T. Meek, J. A. Greathouse and M. D. Allendorf, *Adv. Mater.*, 2011, **23**, 249.
65. J. Della Rocca, D. M. Liu and W. B. Lin, *Acc. Chem. Res.*, 2011, **44**, 957.
66. G. Lu and J. T. Hupp, *J. Am. Chem. Soc.*, 2010, **132**, 7832.
67. U. Mueller, M. Schubert, F. Teich, H. Puetter, K. Schierle-Arndt and J. Pastre, *J. Mater. Chem.*, 2006, **16**, 626.
68. H. Furukawa, K. E. Cordova, M. O'Keeffe and O. M. Yaghi, *Science*, 2013, **341**, 974.
69. A. Carne, C. Carbonell, I. Imaz and D. Maspoch, *Chem. Soc. Rev.*, 2011, **40**, 291.
70. J. L. Zhuang, D. Ceglarek, S. Pethuraj and A. Terfort, *Adv. Funct. Mater.*, 2011, **21**, 1442.

71. T. Ladnorg, A. Welle, S. Heisser, C. Woll and H. Gliemann, *Beilstein J. Nanotechnol.*, 2013, **4**, 638.
72. C. Carbonell, I. Imaz and D. Maspoch, *J. Am. Chem. Soc.*, 2011, **133**, 2144.
73. J. L. Zhuang, D. Ar, X. J. Yu, J. X. Liu and A. Terfort, *Adv. Mater.*, 2013, **25**, 4631.
74. H. K. Arslan, O. Shekhah, J. Wohlgemuth, M. Franzreb, R. A. Fischer and C. Woll, *Adv. Funct. Mater.*, 2011, **21**, 4228.
75. A. A. Talin, A. Centrone, A. C. Ford, M. E. Foster, V. Stavila, P. Haney, R. A. Kinney, V. Szalai, F. El Gabaly, H. P. Yoon, F. Leonard and M. D. Allendorf, *Science*, 2014, **343**, 66.
76. T. Gupta, P. C. Mondal, A. Kumar, Y. L. Jeyachandran and M. Zharnikov, *Adv. Funct. Mater.*, 2013, **23**, 4227.
77. G. Sedghi, V. M. Garcia-Suarez, L. J. Esdaile, H. L. Anderson, C. J. Lambert, S. Martin, D. Bethell, S. J. Higgins, M. Elliott, N. Bennett, J. E. Macdonald and R. J. Nichols, *Nat. Nanotechnol.*, 2011, **6**, 517.
78. C. Lin and C. R. Kagan, *J. Am. Chem. Soc.*, 2003, **125**, 336.
79. X. Ding, Y. Gao, L. Zhang, Z. Yu, J. Liu and L. Sun, *ACS Catal.*, 2014, **4**, 2347.
80. S. Yoshimoto, K. Sato, S. Sugawara, Y. Chen, O. Ito, T. Sawaguchi, O. Niwa and K. Itaya, *Langmuir*, 2007, **23**, 809.
81. S. Mohnani and D. Bonifazi, *Coord. Chem. Rev.*, 2010, **254**, 2342.

CHAPTER 9

Metallosupramolecular Materials for Electronic Applications: Molecular Boolean Computation

BRIAN DALY[a], JUE LING[a] AND A. PRASANNA DE SILVA*[a]

[a]School of Chemistry and Chemical Engineering, Queen's University, Belfast BT9 5AG, Northern Ireland, UK
*E-mail: a.desilva@qub.ac.uk

9.1 Introduction

Ever since molecules were shown to possess logical properties,[1] the field of molecular Boolean computation has grown, also visible through the appearance of many reviews,[2–25] several book chapters[26–29] and books.[30–33] The recognition that any observable chemical interaction or reaction can be viewed as a molecular device that receives an input in order to produce an output was the bridge between (bio)chemical sciences and computer technology. It has become increasingly clear that small molecules can carry out information processing tasks of various kinds, especially in small biorelevant spaces that semiconductor devices find difficult if not impossible to access and operate from. The fact that biology relies on information handling by a host of atomic and molecular players means that there are many lessons

RSC Smart Materials No. 15
Functional Metallosupramolecular Materials
Edited by John G. Hardy and Felix H. Schacher

Published by the Royal Society of Chemistry, www.rsc.org

to be learned from the oldest information technology revolution of all—life itself. If they are biocompatible, even synthetic molecules can perform useful Boolean computation in living organisms. Of course, much of the background molecular Boolean computation occurs in the laboratory environment. We take this opportunity to discuss some recent developments in the field that have attracted our attention. The examples will be handled in order of increasing number of inputs and outputs to/from the molecular logic device.

9.2 One/Two-Input One-Output Logic

One of the simplest expressions of molecular logic is in sensing metal ions. This corresponds to metal ion-driven YES logic with fluorescence output. Metal-induced biasing of the competition between photoinduced electron transfer (PET) and fluorescent pathways is the underlying principle.[34–36] Though notable exceptions exist,[37] the most common metal-ligating atom employed is nitrogen. However, when they are metal-free, nitrogen lone electron pairs can undergo protonation, which limits the pH range at which the sensor can function. Vonlanthen and Finney examine sulfur (within thiourea derivatives **1** and **2**) as an alternative.[38] **1** shows a large Zn^{2+}-induced fluorescence enhancement factor of 20 (with some cross-reactivity with Cd^{2+}), though the affinity is weak ($\log\beta = 1.3$ in methanol). The affinity can be increased by adding additional sites for metal chelation. This means that the chelating conformer must be at least kinetically accessible, if not thermodynamically more stable. The use of a pyridyl group might appear to defeat the purpose of avoiding nitrogen lone electron pairs but its placement far from the fluorophore avoids any PET-type interaction so that pH-sensitivity is absent. The Hg^{2+}-affinity of **2** ($\log\beta = 5.8$ in methanol, *cf.* $\log\beta_{Zn2+} = 3.2$) arises much more from the pyridyl group than from the sulfur centre.

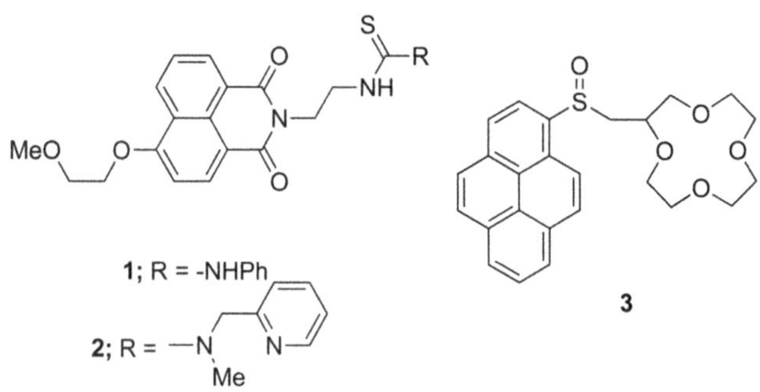

1; R = -NHPh

2; R =

3

Kathayat and Finney[39] enhance sulfur's role in sensing by bringing in a novel fluorescence signalling mechanism.[40] They use aryl sulfoxides, which have weak fluorescence because the excited states allow pyramidal inversion of the sulfoxide unit. The activation energy for these inversions is decreased by internal charge transfer (ICT)[34,41] from the sulfoxide sulfur to the aryl group in the planar excited state. This means that if the electron density of the sulfoxide is decreased *via* coordination of a metal ion, the ICT pathway will be avoided and fluorescence is enhanced. Although the metal affinity of sulfoxide oxygen is low, this can be improved by appending proximal ionophores. **3** is notable in being devoid of nitrogen and still displays a fluorescence quantum yield enhancement from 0.015 to 0.4 in the presence of Ca^{2+} in acetonitrile ($\log\beta_{Ca2+} = 5.7$).

Uchiyama and his co-workers have previously shown how temperature distribution within animal cells can be mapped[42] by exploiting thermally-driven random coil–globule transitions of poly(*N*-alkylacrylamide). Some of these show temperature-driven YES logic in their fluorescence output.[43–45] They recently tackled plant cells,[46] which thwart many attempts to introduce molecular sensors to the cytosol. Uchiyama's team succeeded by employing copolymers containing quaternary ammonium groups, **4**, which diffuse through the pores of the cell wall and then permeate the plasma membrane. The electrostatic interaction between the cationic polymer and the anionic phospholipids in the membrane appears to be responsible for the facile entry and maintenance in the cytosol. The benzofurazan fluorophore with donor(amino)–acceptor(sulfonamide) substituents leads to a pronounced water-sensitivity of the emission intensity and the fluorescence lifetime. The thermal transition of the polymer then confers temperature-sensitivity on the fluorescence parameters of **4**. Besides low cytotoxicity, the fluorescence lifetime of **4** is prolonged as the temperature increases: 6.2 ns at 15 °C and 8.6 ns at 35 °C. Since the temperature resolution is 0.09–0.78 °C in this temperature range, **4** is capable of thermal mapping of both plant and animal cells.

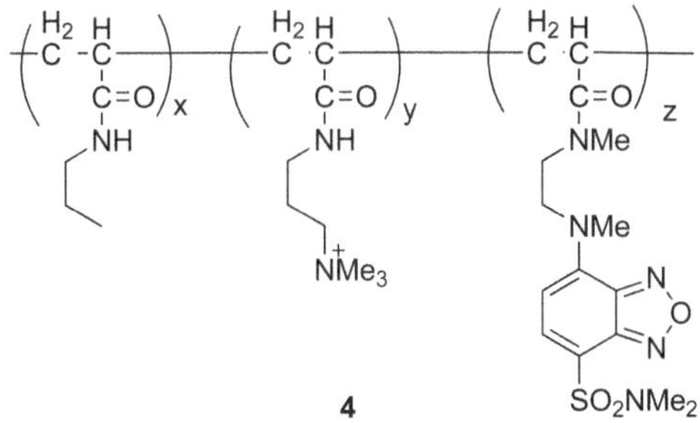

4

In spite of a controversial beginning,[47] molecular electronic switches are now well-established.[48] Interestingly, the work due of Rigaut, Chen and co-workers[49] contains switches with memory aspects. These can be understood as YES logic gates if the memory aspects are put aside or else these would be more precisely described as R–S latches.[50] Implantation of these devices into gaps fabricated into gold nanowires controls the flow of electrons across the gap. The molecule **5**/**6** contains the well-known dithienylethene (DTE) photochrome unit sandwiched between two electroactive ruthenium complexes; since the isomerization between the open and closed form of DTE is controlled by photochemistry, the DTE unit has in essence produced a molecular light-dependent resistor. The terminal sulfurs in **5**/**6** are protected with acetyl groups, which are cleaved when the sulfur terminals attach to the gold electrodes. Since photochromics can maintain their state after a light dose has been applied, they clearly fall into the category of R–S latches.[33] In the uncyclized form (**5**), the electron flow is disrupted but application of UV (366 nm) light cyclizes the DTE unit and in this conjugated form (**6**), the electron flow can occur across the linearly conjugated π system; the flow can then be disrupted by using red (700 nm) light. The authors have overcome the problem of repeated switching causing photodecomposition by showing that the cyclization can also be triggered electrochemically.

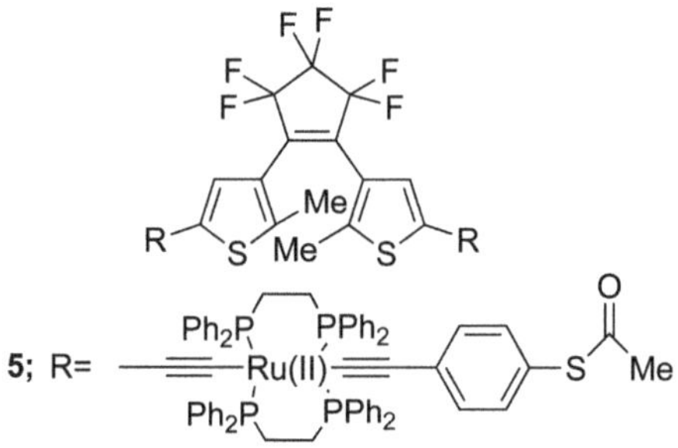

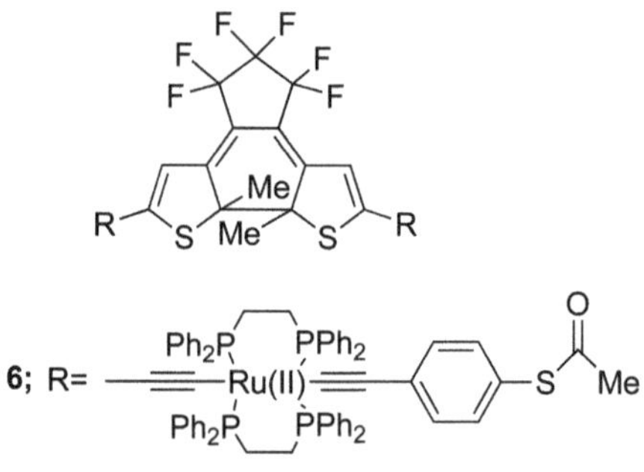

It is common practice in electronics textbooks to introduce AND logic gates as a series combination of simple switches.[51] Further elaboration of **5**/**6** to contain three ruthenium units and two DTE moieties has allowed demonstration of AND logic behaviour as each of the DTE units in the system respond to different frequencies of light and electrochemical potentials and can therefore be closed/opened selectively.

Unidirectional electron flow, which is arranged in semiconductor electronics with p–n junctions,[51] can also be organized in electrochemical situations with electrodes coated sequentially with two redox polymer layers to set up a redox potential gradient.[52,53] The inner polymer layer serves as a redox mediator but in one direction only. Interestingly, one pioneering paper on bilayer electrodes[52] also refers to 'truth tables' though the Boolean idea is not developed further. Recently, Mayer and colleagues reported the development of XOR logic using bilayer electrode concepts.[54] They used gold electrodes coated with hexadecanethiol as an insulator and incorporated 11-undecanethiol-1-ferrocene (E_{redox} = 0.34 V *vs.* standard calomel electrode) as a redox mediator. The insulator prevents direct electron transfer between the electrode and solution species which are chosen as $[Fe(II)(CN)_6]^{4-}$ (E_{redox} = 0.23 V) and $[Ir(IV)Cl_6]^{2-}$ (E_{redox} = 0.71 V). The latter serve as inputs and the modulus of the current is taken as the output.

$[Ir(IV)Cl_6]^{2-}$ can receive an electron from the mediator due to the redox potential gradient and the application of a negative potential (during cyclic voltammetry) to the electrode will replenish the reduced form of the mediator so that a large cathodic current is seen. The positive (reverse) potential results in no current as electron transfer from the reduced $[Ir(IV)Cl_6]^{2-}$ adjacent to the electrode to the oxidized form of the mediator is unfavoured. On the other hand, only a large anodic current is observed on application of a positive potential to a solution containing $[Fe(II)(CN)_6]^{4-}$. When both ions are present together in solution, electron transfer happens predominantly between the ions to produce reduced $[Ir(IV)Cl_6]^{2-}$ and oxidized $[Fe(II)(CN)_6]^{4-}$, and interfacial electron transfer from/to the electrode *via* the mediator from/to these two species is unfavoured. Thus the current at whatever potential is greatly reduced. Of course, no current flows when both $[Fe(II)(CN)_6]^{4-}$ and $[Ir(IV)Cl_6]^{2-}$ are absent. Mutual destruction of species was employed in the first XOR gate,[55] though later designs[56,57] overcame this situation. However, taking the modulus impresses a manual algebraic operation, which takes away from the exclusively molecular nature of the logic operation that is desired.

Room temperature phosphorescence (RTP), with advantages of long lifetimes and high signal-to-noise ratios, has been used in logic devices only rarely.[58–60] Tian, Ma and co-workers recently returned to it, but with an exciting new twist.[61] A β-cyclodextrin (CD) derivatized with an *E*-azobenzene (**7**) allows self-complexation of the long and thin π-system so that 1-bromonaphthalene (**9**) cannot be included in the CD cavity. However, application of a 360 nm light dose (input$_1$) isomerizes **7** to **8** so that a *Z*-azobenzene unit, which is too wide to be contained within the CD, is produced. Now, **9** can occupy the vacant cavity of **8**. Still, RTP at 528 nm cannot emerge until O_2 (input$_2$) is removed, since paramagnetic oxygen quenches the paramagnetic triplet excited state precursor of **9**'s phosphorescence. For logic analysis, we consider **9** as an integral part of the logic device rather than as an input since the observation of RTP output would be impossible in the absence of **9**. Then we have a light dose, O_2-driven INHIBIT(O_2) gate where O_2 is the disabling input.

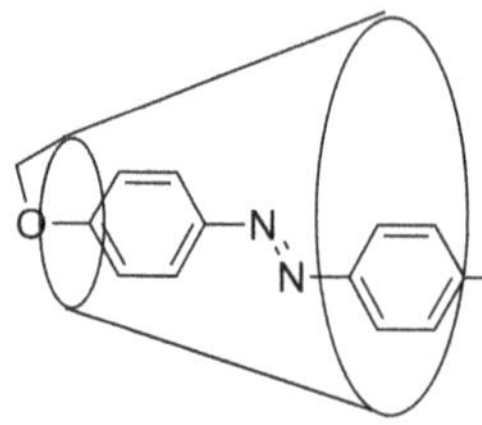

7

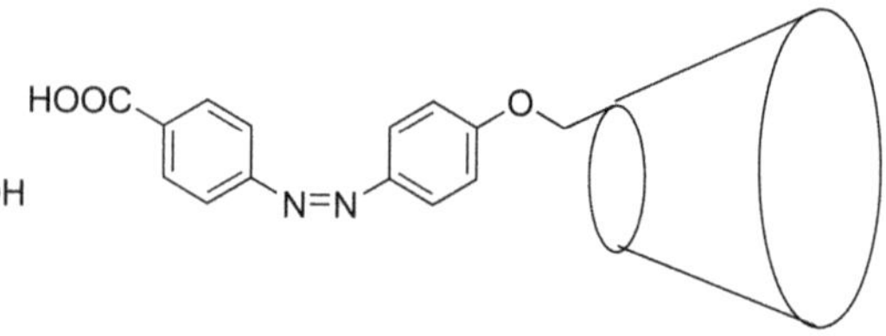

8

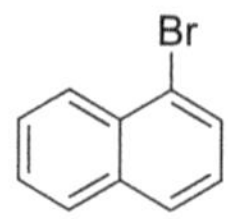

9

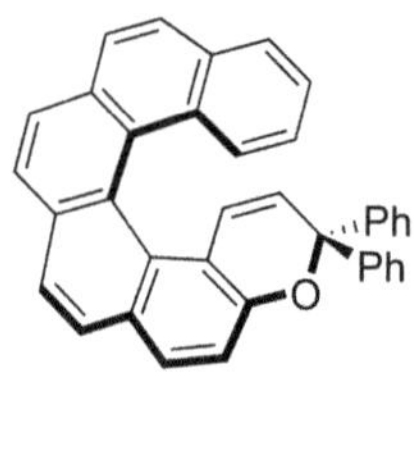

10

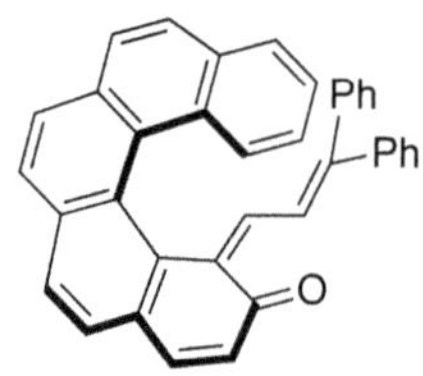

11

The chromene-*o*-quinone methide interconversion is not the most famous among photochromic reactions,[62] because of poor colour retention. However, Moorthy and his colleagues developed an interesting version, **10–11**, based on a helicene so that substantial intra-helix steric crowding prevents easy thermal reversion.[63] This is valuable as a molecular memory. Indeed, **11** has a half-life of 46 years at room temperature, though structural mutants with far shorter half-lives are also available. If we lay aside the memory aspects, a combinational logic operation can be uncovered by considering a 350 nm light dose and heat as inputs and the absorption of the *o*-quinone methide at 524 nm as the output. The heat input supplied is 9 hours at 100 °C or a shorter time at a higher temperature. Then we have **10** displaying a light dose and heat-driven INHIBIT(heat) logic where heat is the disabling input.

Following their recent success with calix[4]pyrrole-based INHIBIT gates,[64] Sessler and co-workers turned to NAND logic.[65] Calix[4]pyrrole with tetrathiafulvalene (TTF) extensions, **12**, and its partner, **13**, which has two pyridinium units on opposite ends, constitute the logic device. **12** usually exists in a 1,3-dialternate conformation and engages in π–π donor(TTF)–acceptor(pyridinium) interactions to form a 1 : 1 polymer in $CHCl_3$ solution. This leads to a charge-transfer absorption band at 650 nm. The absorbance at 690 nm is taken as the output because of the better clarity of the truth table. Anions such as I^- (input$_1$) bind to the N–H groups of **12** and force it into a cone conformation. Now, two cones of **12** encapsulate a single **13** with π–π donor–acceptor forces so that the CT absorption is preserved, albeit at 655 nm. On the other hand, cations such as Et_4N^+ (input$_2$) fit into the cone conformation of **12** (when enforced by I^-) by cation–π interactions and displace **13** so that the CT absorption disappears. Similar displacement does not occur from the 1,3-dialternate conformation of **12** in the absence of I^-, *i.e.* the CT absorption remains 'high'. Overall, this behaviour corresponds to NAND logic since the absorbance output becomes 'low' only when both inputs are 'high'.

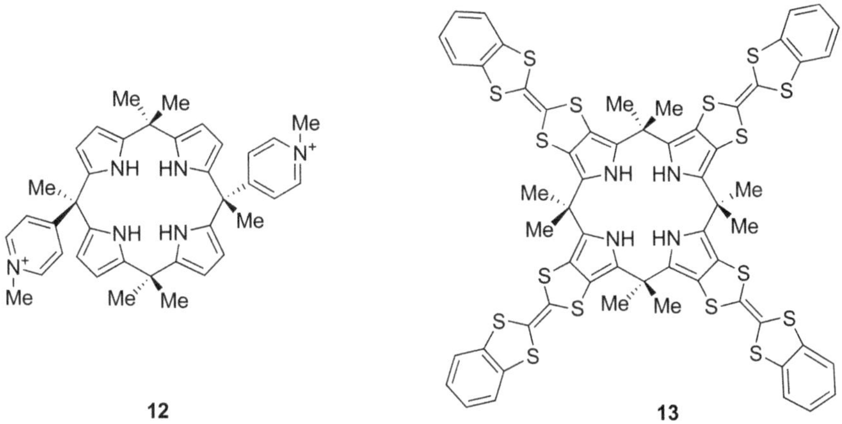

12 **13**

A 'high' binding constant, above a chosen threshold, as measured by isothermal titration calorimetry (ITC) can be considered as the output when lectin and Ca^{2+} inputs simultaneously encounter a glycodendrimer based on two β-cyclodextrins, which are heptasubstituted on the lower rim, binding to the two ends of **14**.[66] AND logic is found since Ca^{2+} enhances the carbohydrate–protein interaction. However, the 'high' and 'low' states of the outputs are sometimes relatively close. Such situations are ignored by binary logic and it is possible that important information is lost during a Boolean analysis. In the present instance, there is a significant carbohydrate–protein interaction (amounting to 35% of the 'high' binding constant) even in the absence of Ca^{2+}. Fuzzy logic[67,68] can be used to examine such cases although it is subjective in-so-much as the intermediate states 'weak', 'medium' and 'strong' are user-defined (just as binary thresholds are). Multi-valued logic can be applied similarly.[69–72] It is worth making a general point: a major reason for the growth of Boolean molecular science[30–33,73,74] is that avoiding information overload to permit incisive analyses and understanding is a sacrifice worth making in many instances.

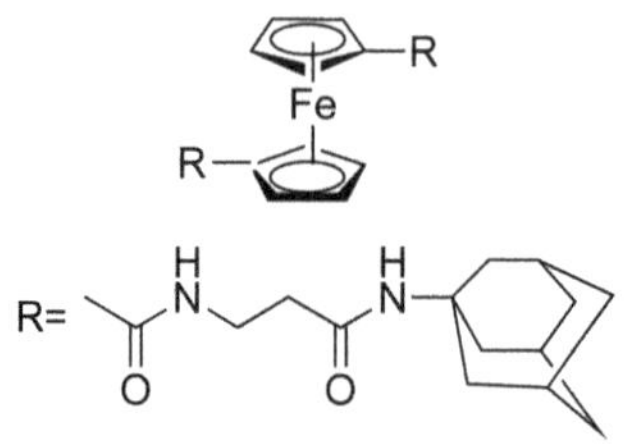

14

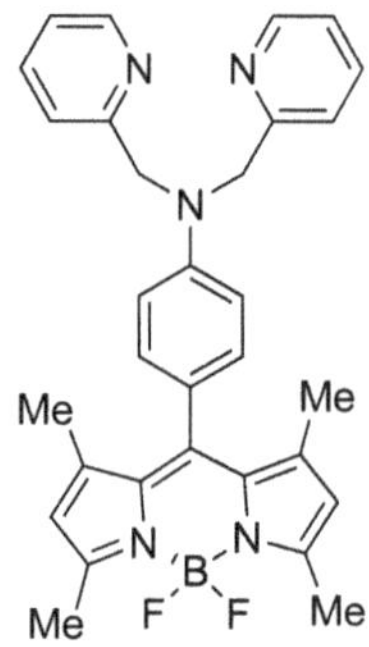

17

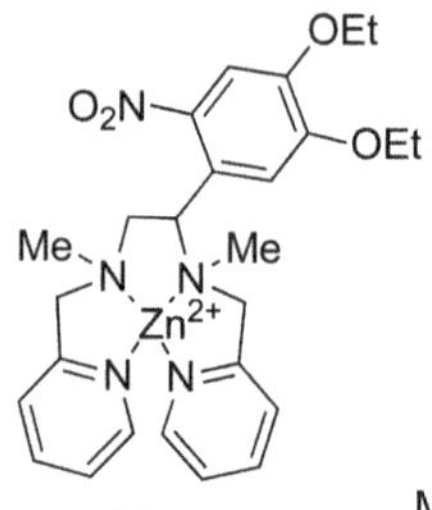

15

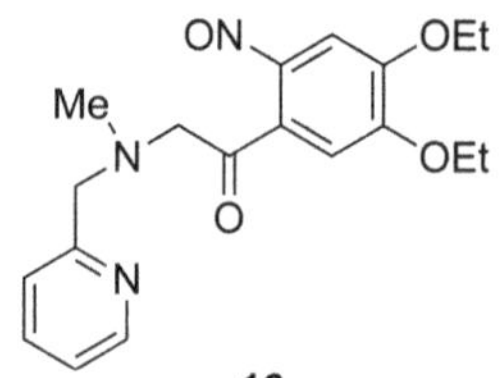

16

Short DNA fragments serve as the two primers in the polymerase chain reaction,[75–77] which is usually run with Taq polymerase. Each of the primers (forward and reverse) needs to hybridize smoothly with the 3'-end of its strand of the target which is to be amplified. Exponential DNA amplification needs both primers to succeed in this way. However, the primers can be chosen so that a thymine (T)–thymine mismatch creates a point mutation during hybridization and prevents the Taq polymerase from doing its job. Nevertheless, Hg^{2+} can selectively bridge opposing thymines by complexation so that the Taq-induced polymerization is reinstated. Ag^{+} similarly repairs cytosine (C)–cytosine mismatches. If the two primers are chosen to carry a mismatch each (T–T and C–C), exponential DNA amplification occurs only if Hg^{2+} and Ag^{+} are provided as the two inputs.[78,79] If, say, only Hg^{2+} was made available, only one of the two strands of the target would be hybridized with its primer without a point mutation. Thus, only linear amplification would occur. If a suitable threshold is chosen for the final observation technique (gel electrophoresis or fluorescence detection of double-strand DNA after 30 cycles), this would correspond to Hg^{2+}, Ag^{+}-driven AND logic behaviour.

The corresponding OR logic action[78,79] is arranged by creating two primers (instead of the one used previously) for each of the forward and reverse directions. Once hybridized, each of these two primers presents a point mutation arising from a mismatch (T–T and C–C respectively). Presentation of either Hg^{2+} or Ag^{+} inputs would bridge one of these mismatched pairs so that Taq-induced polymerization succeeds, though some of the product strands would

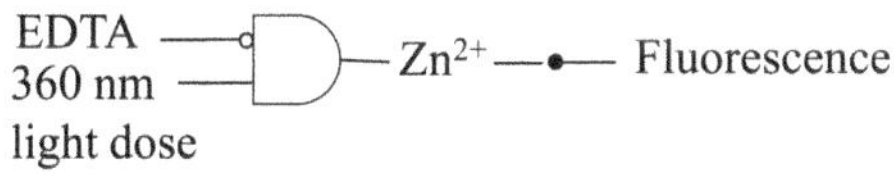

Figure 9.1 Logic gate array arising from **15** and **17**.

be slightly shorter than the target strand. Presentation of both Hg^{2+} and Ag^{+} would succeed similarly except that the full length target strand is produced. Such slight variations in strand length would not matter for our purpose.

Concatenation of logic gates is the basis of serial operations in electronic computation but the inhomogeneity of the input/output in molecular logic systems poses problems. Akkaya offers a way of cascading molecular gates using metal ions as the linker species[80] by tapping into the field of photo-uncaging of bioactive species.[81] Previous work by Raymo and Giordani had also used H^{+} as the linker species.[82,83] The first gate, complex **15**, is a form of caged Zn^{2+}. The application of a 360 nm light dose (input$_1$) degrades **15** to **16** *via* an intramolecular hydrogen abstraction by the $n\pi^*$ triplet excited state of the nitroaryl group. This releases Zn^{2+}, which serves as the input to the second gate **17**. If EDTA is added, Zn^{2+} is competitively complexed to prevent interaction with **17**. Thus the first gate is a light dose, EDTA-driven INHIBIT(EDTA) gate where EDTA is the disabling input$_2$. The emission output at 510 nm from YES gate **17** only emerges if free Zn^{2+} is available because of the Zn^{2+}-induced blocking of PET from the dipyridylamine unit to the borodipyrromethene fluorophore (Figure 9.1).

9.3 Two-Input Two-Output Logic

A 1:2 demultiplexer (Figure 9.2) is needed to recover and re-route data bits which were previously sent down one transmission line from a 2:1 multiplexer.[84,85] Such a device is able to respond to a single input in two different ways depending on an address input (Table 9.1). A good number of molecular examples are known,[33] but Akkaya's laboratory now provides a pointer to theranostic applications.[86]

A fluorescent module (**18**) and a photosensitive module (**19**) are joined by click chemistry to produce the demultiplexer, and H^{+} is deployed as the address input. The data input is considered to be light intensity or light dose at 625 nm (see later). **18** in isolation shows fluorescence in both acidic and neutral conditions but crucially protonation of its dimethylaminophenyl group decreases its ICT nature and shifts its λ_{abs} from 681 to 647 nm. **19** on its own absorbs in the same region but, rather than fluorescence, its excitation produces $^1O_2$. The heavy atom nature of the iodo substituents induces intersystem crossing so that the triplet excited state of **19** is obtained. This is what leads to $^1O_2$ production from the commonplace $^3O_2$. Furthermore, protonation of its pyridyl group increases its ICT nature and shifts its λ_{abs} from 647 to 671 nm. When **19** and **18** are coupled there is electronic energy transfer (EET) between them but importantly the direction of the transfer can be

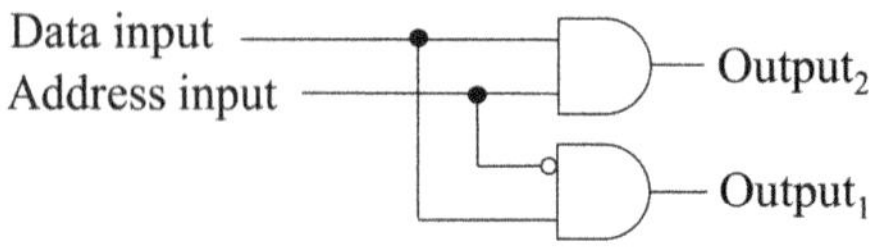

Figure 9.2 Logic gate array for a 1 : 2 demultiplexer.

Table 9.1 Truth table for a 1 : 2 demultiplexer.

Address input$_1$	Data input$_2$	Output$_1$	Output$_2$
0	0	0	0
0	1	1	0
1	0	0	0
1	1	0	1

controlled by controlling the pH. Under neutral conditions, the excited state energy is transferred from **19** to **18** resulting in fluorescence at 715 nm (output$_1$), however the reverse is true under acidic conditions; the energy is transferred from **18** to **19** resulting in $^1O_2$ production (output$_2$) which is absent under neutral conditions.

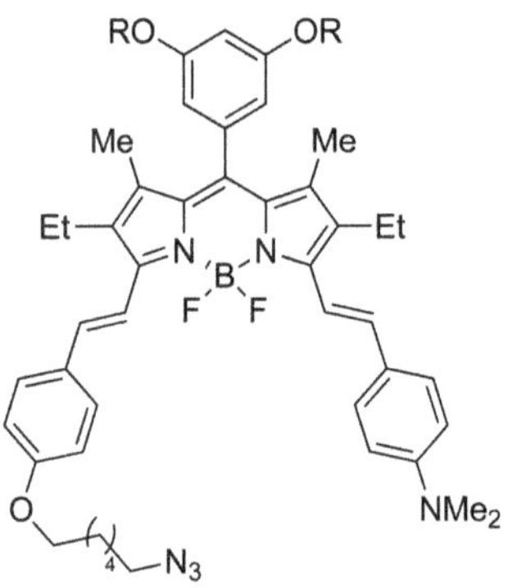

R = -CH$_2$CH(Et)(CH$_2$)$_2$Et

18

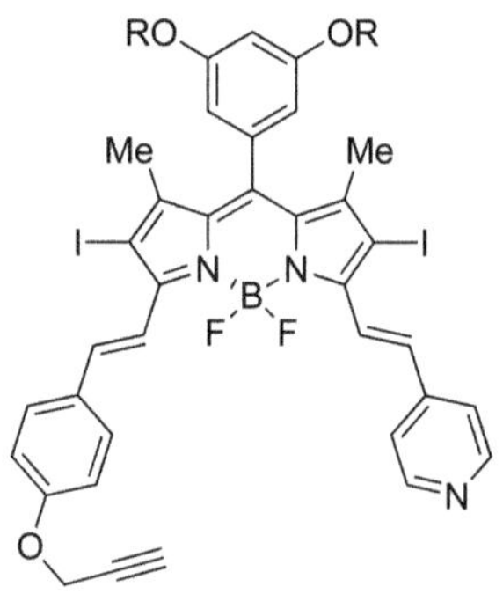

R = -CH$_2$CH(Et)(CH$_2$)$_2$Et

19

These results set up the following scenario with more hydrophilic versions of **18–19**. Cancer cells are known to have reduced pH[87] so the condition itself provides the address input and photodynamic therapy[88] provides the output of most interest. The advantage of **18–19**, however, is that it not only affects the tumor cell but it also identifies the cells in question. Normal cells, with normal pH will fluoresce while the tumor cells will be killed by $^1O_2$ produced upon irradiation with light. However, it needs to be pointed out that the use of light intensity and light dose interchangeably as the data input depending on the need leads to an inconsistency from a general molecular logical viewpoint. Furthermore, the use of excitation light intensity as an input to fluorescent devices is particularly problematic. In most logic studies concerning fluorescence,[33] excitation light intensity serves as the power supply only.

9.4 Three-Input One-Output Logic

Glass and his team examined the potential for advanced molecular sensors to monitor the release of neurotransmitters into the synaptic cleft under realistic conditions.[89] Glutamate is the most abundant neurotransmitter in the brain and in certain regions, glutamate release is accompanied with that of Zn^{2+}. Glutamate is stored along with Zn^{2+} in the synaptic vesicles at pH 5, however the pH in the synaptic cleft is much higher: 7.4. So a glutamate, Zn^{2+}-driven AND gate feeding into a H^+-driven INHIBIT(H^+) logic gate **20** is required, where H^+ is the disabling input (Figure 9.3). This means that whilst in the vesicle, the sensor fluorescence is low but upon release into the cleft and deprotonation, the fluorescence should increase. Previous efforts towards sensing for neurotransmitters such as glutamate[90] and γ-aminobutyric acid (GABA)[91] were not as clear as the Glass system.

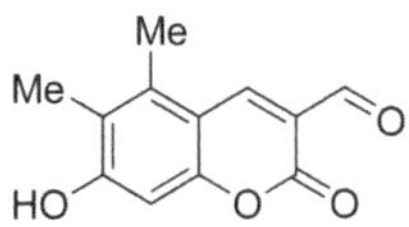

20

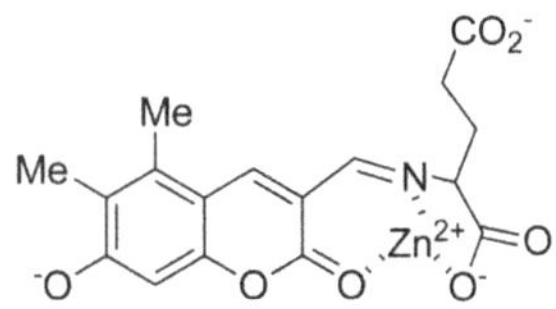

21

Glass' previous work laid the foundation by developing an iminium ion by reversible reaction of a coumarin aldehyde with a biogenic amine. The iminium ion, by virtue of its positive charge, enhances the ICT nature of the excited state, *cf.* the aldehyde. The resulting red-shift in the absorption spectrum permits selective excitation of the iminium ion so that a fluorescence emission signal can be observed.[92] Dopamine and norepinephrine could be visualized in this way.[93] Systems of increasing complexity could selectively target catecholamines,[94] phosphoserine[95] and glutamate in the cleft (as opposed to that in the vesicle).[96] However, we focus on sensor **20**. Here, Zn^{2+} can be chelated by the lactone oxygen, imine nitrogen and the α-carboxylate of glutamate. While an iminium is absent in this instance, the binding of the Zn^{2+} accomplishes the same result. Further, the phenolic group of the

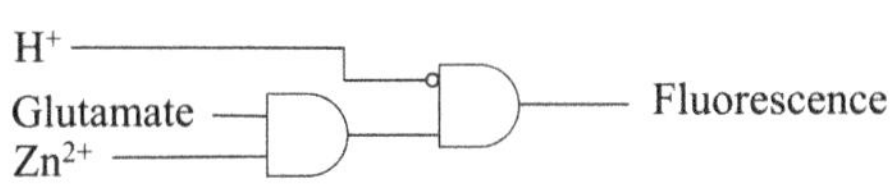

Figure 9.3 Logic gate array arising from **20**.

coumarin deprotonates to enrich the ICT nature of the excited state even more, as seen in complex **21**. The 'high' output fluorescence is an enhancement factor of 32 whereas the seven 'low' outputs are all below a factor of 11.5. The binding constants (β) for H^+, Zn^{2+} and glutamate are 6.1, 30 and 2.8 respectively. The relatively small β values for Zn^{2+} and glutamate could cause a collapse of **21** in the synaptic cleft where the concentration of both species would be decreased. Disassembly of designed sensing ensembles in multicompartmental bioenvironments is an ever-present danger. However, the success of precursors of **20** suggests that further structural adjustments may alleviate these issues.

Coordination polymer nanoparticles are produced in neutral solution from nucleotides, *e.g.* guanosine-5'-monophosphate (GMP) and lanthanide ions such as Eu^{3+}, which are oppositely-charged building blocks. These nanoparticles encapsulate the porphyrin derivative **22**, aided by its anionic nature.[97] Such encapsulation also isolates **22** from other copies so that aggregation-induced quenching of its fluorescence (at 612 nm) is avoided when pumped at 399 nm. Aggregation-induced quenching of fluorescence has been exploited nicely by this laboratory previously.[98] Hereby, the fluorescence is the output while Eu^{3+} serves as $input_1$. However, if the general metal complexant EDTA (considered as $input_2$) is applied, the Eu^{3+} ions are stripped from the nanoparticles and the scaffold collapses to release **22** into the solvent water where it aggregates, followed by fluorescence loss. Another input ($input_3$) can be applied in the form of H^+. The carboxylate groups of **22** are protonated to trigger another channel of aggregation, so that fluorescence is quenched again. In acidic solution, the nanoparticles dissociate as well. So, fluorescence output is 'high' only if $input_1$ (Eu^{3+}) is 'high' while $input_2$ (EDTA) and H^+ ($input_3$) are 'low'. The corresponding logic gate array is shown in Figure 9.4, where two INHIBIT gates are connected in series with the two disabling inputs being EDTA and H^+.

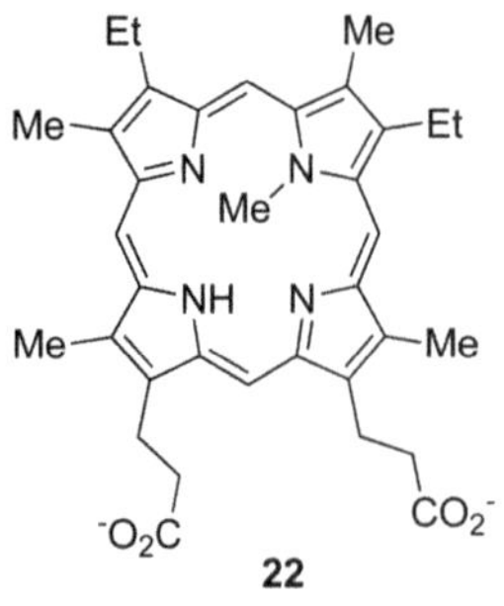

22

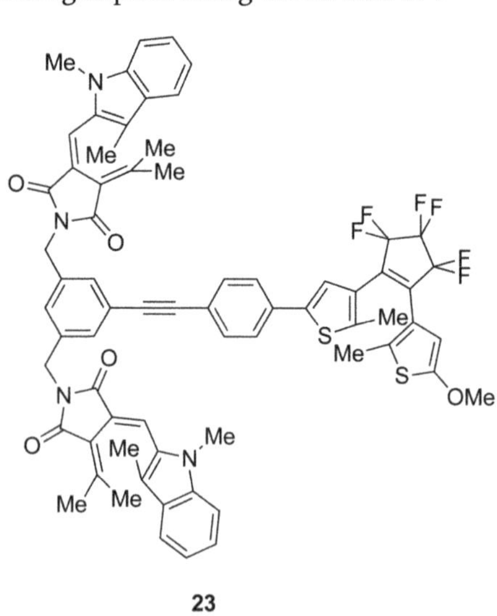

23

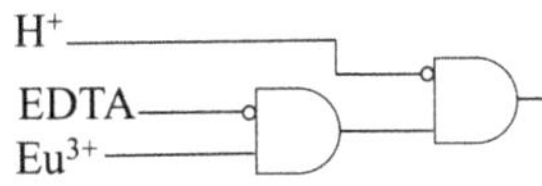

Figure 9.4 Logic gate array arising from **22** and GMP.

Table 9.2 Truth table for a 2-bit even parity generator and the corresponding sum column.

$Input_1$	$Input_2$	Output	Sum
0	0	0	0
0	1	1	2
1	0	1	2
1	1	0	2

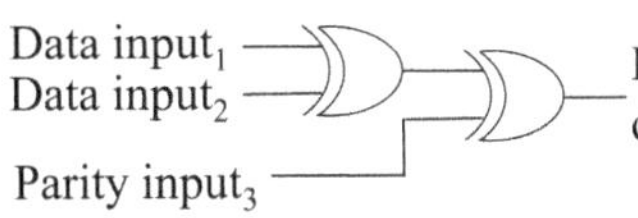

Figure 9.5 Logic gate array for a 3-bit parity checker.

Now we describe a three-input logic array which nests another two-input device, both of which concern 'parity'. In a mathematical context, 'parity' describes the odd/even nature of an integer. This idea is used in computer science to introduce the simplest way of checking the fidelity of data transmission over noisy lines. The line could be inter-computer or even intra-computer. For instance, the parity of a data string (as judged from the summed number of 1's) is made up to an even number by adding a parity bit 1 or 0 as necessary before data transmission. Such an action is performed by an even parity generator.[84,85] If the transmitted data arrives at its destination with an odd parity due to corruption, a separate parity checker[84,85] within the system recognises an error. In the simplest case of a 2-bit even parity generator, a two-input XOR logic gate does the trick. The first three columns of Table 9.2 represent a straightforward truth table of an XOR gate, but the final column gives the sum of the string '$Input_1Input_2$Output' which is of even parity independent from the input combination. The 'Output' provides the parity bit which is added by the generator. A 3-bit parity checker uses the output from an XOR gate (fed by the two data bits) and the parity bit input to feed a second XOR gate (Figure 9.5). The output is the parity error check because a '1' signals an error whereas a '0' says that all is well with the data transmission. The final column of Table 9.3 gives the sum of the string '$Input_1Input_2$Parity $input_3$'. When this has even parity, the parity error check output reads '0'.

The laboratories of Andréasson and Pischel have joined forces to emulate the devices of the previous paragraph with molecules containing separate fulgimide (FG) and dithienylethene (DTE) photochromic units,[99] *e.g.* **23**. The structure shown has both photochromics in their open (colourless) forms. Examples of the open (colourless) and closed (coloured) forms of DTE were previously encountered with **5** and **6**, respectively. When exposed to 380 nm light, the FG unit cyclizes much faster than DTE does. So the states, in order of appearance, are: FG_o–DTE_o, FG_c–DTE_o and FG_c–DTE_c, where the subscripts 'o' and 'c' refer to open (colourless) and closed (coloured) forms respectively. The closed form of FG is fluorescent at 630 nm, albeit weakly (quantum yield = 0.005), because it has a donor–acceptor π electron system with internal charge transfer (ICT) character.[34,41] So the FG_c–DTE_o state is fluorescent but the FG_c–DTE_c state is not fluorescent because of electronic energy transfer (EET) from FG_c to the coloured DTE_c. The upshot is that when the 380 nm light dose is gradually increased, the 630 nm emission intensity shows 'off–on–off' behaviour.[100,101] Older examples of this type which are driven by light dose[102,103] are available. Such 'off–on–off' actions can be understood as XOR logic[32,104] since two 380 nm light doses can be arranged as degenerate $input_1$ and $input_2$ as follows. Application of either one of these doses takes nonfluorescent FG_o–DTE_o to fluorescent FG_c–DTE_o. Application of both doses produces nonfluorescent FG_c–DTE_c.

Thus we have a molecular implementation of a XOR gate in the context of an even parity generator. It would be good if future work will achieve the final sum column of Table 9.2 directly by molecules as well so that the analogy with the semiconductor version is fully followed. On the other hand, the parity check (fourth) column of Table 9.3 for the parity checker is directly achieved by **23** in the following way. The argument made earlier for the even parity generator is also applicable to the first four rows of Table 9.3, *i.e.* with parity $input_3$ = 0. The last four rows of Table 9.3, *i.e.* with parity $input_3$ = 1, are seen to fit a XNOR function, and require a redefinition of >540 nm light dose as $input_1$. The 380 nm light dose remains as $input_2$ and the 630 nm fluorescence remains as output while another 380 nm light dose serves as parity $input_3$. As in previous studies by Raymo and Giordani,[82,83,105–108] the choice of the actual values of the ultraviolet and visible light doses, and even their order of application, is critical. However, it must be noted that the re-coding of $input_1$ between the first four rows of Table 9.3 and the last four rows is an inconsistency that can hopefully be removed in future studies.

Table 9.3 Truth table for a 3-bit parity checker and the corresponding sum column.

$Input_1$	$Input_2$	Parity $input_3$	Check output	Sum
0	0	0	0	0
0	1	0	1	1
1	0	0	1	1
1	1	0	0	2
0	0	1	1	1
0	1	1	0	2
1	0	1	0	2
1	1	1	1	3

9.5 Four-Input One-Output Logic

Inspired by the selective binding of Hg^{2+} and Ag^+ with thymine and cytosine, Lin's laboratory used the pyrene-based imide **24** in a similar way but to produce a logical outcome.[109] The non-planarity of **24** mitigates against aggregation in aqueous alcohol solution. Hg^{2+} or Ag^+ bind to **24** and cause strong quenching of fluorescence. Indeed, they are the only metal ions to do so. Also, they both display 2:1 (**24**–metal ion) stoichiometry as the DNA bases do. However, Ag^+ succeeds in binding only under alkaline conditions, *i.e.* H^+ inhibits the Ag^+ action. On the other hand, the introduction of S^{2-} as a metal scavenger would obviously inhibit the action of both metal ions. Thus, the logic gate array in Figure 9.6a suggests itself if metal binding to **24** is considered as the output. Since the observed output is the fluorescence, we need to add a NOT gate at the end to the array in Figure 9.6a to account for the fluorescence quenching produced upon heavy metal binding (Figure 9.6b). The heavy, redox-active metal ions which are bound to the deprotonated **24** naturally create up to three causes for the fluorescence quenching. The application of De Morgan's Law[51] to the right-most three gates gives us the simplified array in Figure 9.6c.

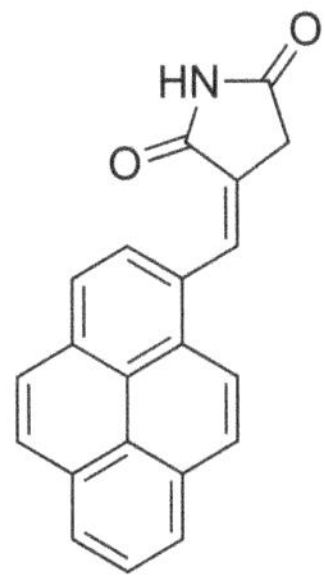

24

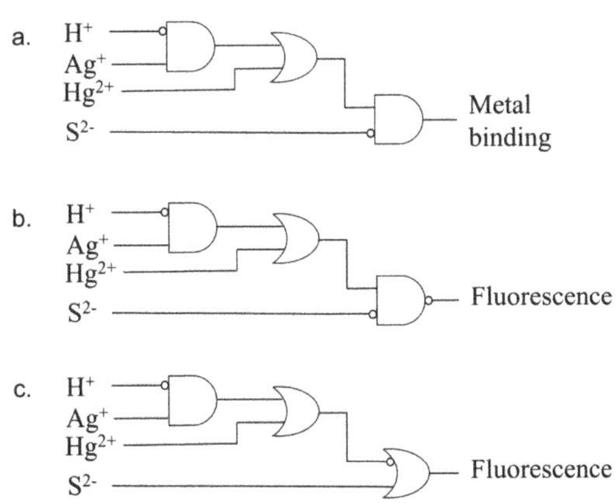

Figure 9.6 Logic gate arrays arising from **24**, (a) when metal binding is the output, (b) when fluorescence is the output and (c) when the gate array in part (b) is minimized.

Lin's team also produced a code for cryptography but this has several issues. The ordered input string 'H^+, Ag^+, Hg^{2+}, S^{2-}' is arbitrarily chosen and then read as hexadecimal numbers because the 4-input gate **24** naturally has 16 entries. The keys for cryptography can be arbitrary as long as they are shared between the sender and the receiver only. The output of **24** has only two possibilities (0 or 1) so that much information is lost during this mapping of input to output. To compensate for this loss, Lin's team re-coded the outputs as '.' and '–' instead of 0 and 1 so that the common Morse code combinations of '.' and '–' could be used from then onwards. Being common, Morse code is not ideal for secret codes, and perhaps more importantly, the final message that is read could have arisen from several starting messages.

9.6 Complex Computation

Common image-handling software such as Photoshop™ or Paintshop™ can directly detect the edges of objects. Molecules can solve the same problem just as the molecular systems in animal eyes[110] tackle this situation all the time. Indeed, animal survival depends on it. Picture the scene: a fish is swimming serenely in a stream when a hungry bird swoops looking for lunch. The fish spots the silhouette overhead and rapidly identifies the perimeter of the beak, head, wings and tail as something to evade. A quick flick of its fins and tail takes the fish deep below the surface, leaving the bird to find its meal elsewhere.

In particular, it is important to note that animal eyes, rather than brains, achieve this result. The ganglion cells in the retina run the edge detection algorithm so that time is saved instead of sending the visual signals to the brain. The brain is engaged when more detail about the object is required at leisure.

Following previous success with genetically engineered bacteria,[111] Ellington, Chen and co-workers used reactive DNA networks[112] to detect the edge of a binary (black and white) object, *e.g.* a hole in an opaque mask, through which UV light is shone. Basically, the network is composed of A and B, where UV irradiation activates A but deactivates B. So activated A and active B are not coincident either in irradiated or unirradiated areas. However, when activated A diffuses across the edge from irradiated to unirradiated areas, it meets active B so that a reaction takes place to create a fluorescent product. A is a small stem–loop construct which, when photo-decomposed, gives a smaller oligonucleotide. Through strand-displacement,[113–116] the latter unfolds B which is also a stem–loop construct. This unfolding initiates a process which finally switches 'on' fluorescence of a type of molecular beacon[117] based on electronic energy transfer.[34,41] Ellington and Chen's engineering of the reactive DNA network is exquisite since they arrange various fluorescent colour combinations/reversals of the irradiated and unirradiated areas. The highlight is a similar visualization of the part of the edge adjacent to the irradiated or unirradiated areas.

9.7 Conclusion

This short review of current trends in molecular Boolean computation has shown how fluorescent, photochromic, electrochemical and phosphorescent phenomena of designed molecules yield a range of logic outcomes. Even techniques such as calorimetry and the polymerase chain reaction were brought into play. The driving inputs covered a spectrum of ionic species, temperature and light dose. The molecular devices themselves ranged from simple inorganic complexes to multi-component organic supermolecules. One of the aims of this review was to illustrate the breadth of ideas, methods and materials involved in the field, so that some cross-fertilization would occur in the mind of the reader.

It is clear that creative research pours out of laboratories determined to demonstrate the commonality between chemistry and computation at the level of concepts and components. It is our hope that the out-pouring will continue well into the future.

Acknowledgements

We thank the Department of Employment and Learning, Northern Ireland, for support.

References

1. A. P. de Silva, H. Q. N. Gunaratne and C. P. McCoy, *Nature*, 1993, **364**, 42.
2. F. M. Raymo, *Adv. Mater.*, 2002, **14**, 401–414.
3. V. Balzani, A. Credi and M. Venturi, *ChemPhysChem*, 2003, **4**, 49.
4. A. P. de Silva and N. D. McClenaghan, *Chem.–Eur. J.*, 2004, **10**, 574.
5. D. Gust, T. A. Moore and A. L. Moore, *Chem. Commun.*, 2006, 1169.
6. V. Balzani, A. Credi and M. Venturi, *Nano Today*, 2007, **2**(2), 18.
7. A. P. de Silva and S. Uchiyama, *Nat. Nanotechnol.*, 2007, **2**, 399.
8. R. Ballardini, P. Ceroni, A. Credi, M. T. Gandolfi, M. Maestri, M. Semeraro, M. Venturi and V. Balzani, *Adv. Funct. Mater.*, 2007, **17**, 740.
9. U. Pischel, *Angew. Chem., Int. Ed.*, 2007, **46**, 4026.
10. A. Credi, *Angew. Chem., Int. Ed.*, 2007, **46**, 5472.
11. V. Balzani, A. Credi and M. Venturi, *Chem.–Eur. J.*, 2008, **14**, 26.
12. K. Szacilowski, *Chem. Rev.*, 2008, **108**, 3481.
13. Y. Benenson, *Mol. Biosyst.*, 2009, **5**, 675.
14. E. Katz and V. Privman, *Chem. Soc. Rev.*, 2010, **39**, 1835.
15. J. Andreasson and U. Pischel, *Chem. Soc. Rev.*, 2010, **39**, 174.
16. M. Amelia, L. Zou and A. Credi, *Coord. Chem. Rev.*, 2010, **254**, 2267.
17. U. Pischel, *Aust. J. Chem.*, 2010, **63**, 148.
18. P. Ceroni, A. Credi, M. Venturi and V. Balzani, *Photochem. Photobiol. Sci.*, 2010, **9**, 1561.
19. H. Tian, *Angew. Chem., Int. Ed.*, 2010, **49**, 4710.

20. S. Minko, E. Katz, M. Motornov, I. Tokarev and M. Pita, *J. Comput. Theor. Nanosci.*, 2011, **8**, 356.
21. A. P. de Silva, *Chem.–Asian J.*, 2011, **6**, 750.
22. G. de Ruiter and M. van der Boom, *Acc. Chem. Res.*, 2011, **44**, 563.
23. U. Pischel, J. Andreasson, D. Gust and V. F. Pais, *ChemPhysChem*, 2013, **14**, 28.
24. P. L. Gentili, *RSC Adv.*, 2013, **3**, 25523–25549.
25. B. Daly, J. Ling and A. P. de Silva, *Front. Chem. Sci. Eng.*, 2014, **8**, 240.
26. A. P. de Silva, N. D. McClenaghan and C. P. McCoy, in *Electron Transfer in Chemistry*, ed. V. Balzani, Wiley-VCH, Weinheim, 2001, vol. 5, p. 156.
27. A. P. de Silva, N. D. McClenaghan and C. P. McCoy, in *Molecular Switches*, ed. B. L. Feringa, Wiley-VCH, Weinheim, 2001, p. 339.
28. V. Balzani, M. Venturi and A. Credi, in *Molecular Devices and Machines*, Wiley-VCH, Weinheim, 2nd edn, 2008.
29. A. P. de Silva, T. P. Vance, B. Wannalerse and M. E. S. West, in *Molecular Switches*, ed. B. L. Feringa and W. R. Browne, Wiley-VCH, Weinheim, 2nd edn, 2011, p. 669.
30. *Molecular and Supramolecular Information Processing*, ed. E. Katz, Wiley-VCH, Weinheim, 2012.
31. *Biomolecular Information Processing*, ed. E. Katz, Wiley-VCH, Weinheim, 2012.
32. K. Szacilowski, *Infochemistry*, Wiley, Chichester, 2012.
33. A. P. de Silva, *Molecular Logic-based Computation*, Royal Society of Chemistry, Cambridge, 2013.
34. A. P. de Silva, H. Q. N. Gunaratne, T. Gunnlaugsson, A. J. M. Huxley, C. P. McCoy, J. T. Rademacher and T. E. Rice, *Chem. Rev.*, 1997, **97**, 1515.
35. A. P. de Silva, T. S. Moody and G. D. Wright, *Analyst*, 2009, **134**, 2385.
36. A. P. de Silva and S. Uchiyama, *Top. Curr. Chem.*, 2011, **300**, 1.
37. A. P. de Silva and K. R. A. S. Sandanayake, *J. Chem. Soc. Chem. Commun.*, 1989, 1183.
38. M. Vonlanthen and N. S. Finney, *J. Org. Chem.*, 2013, **78**, 3980.
39. R. S. Kathayat and N. S. Finney, *J. Am. Chem. Soc.*, 2013, **135**, 12612.
40. W. Lee and W. S. Jenks, *J. Org. Chem.*, 2001, **66**, 474.
41. B. Valeur and M. N. Berberan-Santos, *Molecular Fluorescence*, Wiley-VCH, Weinheim, 2nd edn, 2012.
42. K. Okabe, N. Inada, C. Gota, Y. Harada, T. Funatsu and S. Uchiyama, *Nat. Commun.*, 2012, **3**, 705.
43. S. Uchiyama, Y. Matsumura, A. P. de Silva and K. Iwai, *Anal. Chem.*, 2003, **75**, 5926.
44. S. Uchiyama, Y. Matsumura, A. P. de Silva and K. Iwai, *Anal. Chem.*, 2004, **76**, 1793.
45. C. Gota, K. Okabe, T. Funatsu, Y. Harada and S. Uchiyama, *J. Am. Chem. Soc.*, 2009, **131**, 2766.
46. T. Tsuji, S. Yoshida, A. Yoshida and S. Uchiyama, *Anal. Chem.*, 2013, **85**, 9815.

47. R. F. Service, *Science*, 2003, **302**, 556.
48. A. Coskun, J. M. Spruell, G. Barin, W. R. Dichtel, A. H. Flood, Y. Y. Botros and J. F. Stoddart, *Chem. Soc. Rev.*, 2012, **41**, 4827.
49. F. B. Meng, Y.-M. Hervault, Q. Shao, B. H. Hu, L. Norel, S. Rigaut and X. D. Chen, *Nat. Commun.*, 2014, **53**, 3023.
50. U. Pischel and J. Andreasson, *New J. Chem.*, 2010, **34**, 2701.
51. E. Hughes, *Electrical Technology*, Longman, Burnt Mill, 6th edn, 1990.
52. H. D. Abruna, P. Denisevich, M. Umana, T. J. Meyer and R. W. Murray, *J. Am. Chem. Soc.*, 1981, **103**, 1.
53. P. Denisevich, K. W. Willman and R. W. Murray, *J. Am. Chem. Soc.*, 1981, **103**, 4727.
54. Y. Q. Liu, A. Offenhausser and D. Mayer, *Angew. Chem., Int. Ed.*, 2010, **49**, 2595.
55. A. Credi, V. Balzani, S. J. Langford and J. F. Stoddart, *J. Am. Chem. Soc.*, 1997, **119**, 2679.
56. A. P. de Silva and N. D. McClenaghan, *J. Am. Chem. Soc.*, 2000, **122**, 3965.
57. A. P. de Silva and N. D. McClenaghan, *Chem.–Eur. J.*, 2002, **8**, 4935.
58. A. P. de Silva, I. M. Dixon, H. Q. N. Gunaratne, T. Gunnlaugsson, P. R. S. Maxwell and T. E. Rice, *J. Am. Chem. Soc.*, 1999, **121**, 1393.
59. L. X. Mu, Y. Wang, Z. Zhang and W. J. Jin, *Chin. Chem. Lett.*, 2004, **15**, 1131.
60. R. A. Bissell and A. P. de Silva, *J. Chem. Soc. Chem. Commun.*, 1991, 1148.
61. J. J. Cao, X. A. Ma, M. R. Min, T. T. Cao, S. F. Wu and H. Tian, *Chem. Commun.*, 2014, **50**, 3224.
62. *Photochromism. Molecules and Systems*, ed. H. Durr and H. Bouas-Laurent, Elsevier, Amsterdam, 1990.
63. J. N. Moorthy, S. Mandal, A. Mukhopadhyay and S. Samanta, *J. Am. Chem. Soc.*, 2013, **135**, 6872.
64. J. S. Park, E. Karnas, K. Ohkubo, P. Chen, K. M. Kadish, S. Fukuzumi, C. W. Bielawski, T. W. Hudnall, V. M. Lynch and J. L. Sessler, *Science*, 2010, **329**, 1324.
65. D. S. Kim, V. M. Lynch, J. S. Park and J. L. Sessler, *J. Am. Chem. Soc.*, 2013, **135**, 14889.
66. H. Bavireddi, P. Bharate and R. Kikkeri, *Chem. Commun.*, 2013, **49**, 9185–9187.
67. P. L. Gentili, *Chem. Phys.*, 2007, **336**, 64.
68. P. L. Gentili, *J. Phys. Chem. A*, 2008, **112**, 11992.
69. A. P. de Silva, M. R. James, B. O. F. McKinney, D. A. Pears and S. M. Weir, *Nat. Mater.*, 2006, **5**, 787.
70. G. de Ruiter, Y. H. Wijsboom, N. Oded and M. E. van der Boom, *ACS Appl. Mater. Interfaces*, 2010, **2**, 3578.
71. G. de Ruiter and M. E. van der Boom, *Acc. Chem. Res.*, 2011, **44**, 563.
72. G. de Ruiter and M. E. van der Boom, *J. Mater. Chem.*, 2011, **21**, 17575.
73. V. Balzani, M. Venturi and A. Credi, *Molecular Devices and Machines*, Wiley-VCH, Weinheim, 2nd edn, 2008.

74. *Molecular Switches*, ed. B. Feringa and W. S. Browne, Wiley-VCH, Weinheim, 2nd edn, 2012.
75. R. Saiki, D. Gelfand, S. Stoffel, S. Scharf, R. Higuchi, G. Horn, K. Mullis and H. Erlich, *Science*, 1988, **239**, 487.
76. K. B. Mullis, *Angew. Chem., Int. Ed. Engl.*, 1994, **33**, 1209.
77. K. Kleppe, E. Ohtsuka, R. Kleppe, I. Molineux and H. G. Khorana, *J. Mol. Biol.*, 1971, **56**, 341.
78. K. S. Park, C. Jung and H. G. Park, *Angew. Chem., Int. Ed.*, 2010, **49**, 9757–9760.
79. T. Carrell, *Nature*, 2011, **469**, 46.
80. E. T. Ecik, A. Atilgan, R. Guliyev, T. B. Uyar, A. Gumusa and E. U. Akkaya, *Dalton Trans.*, 2014, **43**, 67.
81. J. H. Kaplan and G. C. R. Ellis-Davies, *Proc. Natl. Acad. Sci. U. S. A.*, 1988, **85**, 6571.
82. F. M. Raymo and S. Giordani, *Org. Lett.*, 2001, **3**, 3475.
83. S. Giordani, M. A. Cejas and F. M. Raymo, *Tetrahedron*, 2004, **60**, 10973.
84. A. P. Malvino and J. A. Brown, *Digital Computer Electronics*, Glencoe, New York, 3rd edn, 1993.
85. A. L. Sedra and K. C. Smith, *Microelectronic Circuits*, Oxford University press, Oxford, 5th edn, 2003.
86. S. Erbas-Cakmak, O. A. Bozdemir, Y. Cakmak and E. U. Akkaya, *Chem. Sci.*, 2013, **4**, 858.
87. J. A. Goode and D. J. Chadwick, *The Tumor Microenvironment*, Wiley, New York, 2001.
88. *Photodynamic Therapy*, ed. T. Patrice, Royal Society of Chemistry, Cambridge, 2003.
89. K. S. Hettie, J. L. Klockow and T. E. Glass, *J. Am. Chem. Soc.*, 2014, **136**, 4877.
90. M. Bonizzoni, L. Fabbrizzi, G. Piovani and A. Taglietti, *Tetrahedron*, 2004, **60**, 11159.
91. A. P. de Silva, H. Q. N. Gunaratne, C. McVeigh, G. E. M. Maguire, P. R. S. Maxwell and E. O'Hanlon, *Chem. Commun.*, 1996, 2191.
92. E. K. Feuster and T. E. Glass, *J. Am. Chem. Soc.*, 2003, **125**, 16174.
93. K. S. Hettie, X. Liu, K. D. Gillis and T. E. Glass, *ACS Chem. Neurosci.*, 2013, **4**, 918.
94. K. E. Secor and T. E. Glass, *Org. Lett.*, 2004, **6**, 3727.
95. C. M. Cooley, K. S. Hettie, J. L. Klockow, S. Garrison and T. E. Glass, *Org. Biomol. Chem.*, 2013, **11**, 7387.
96. J. L. Klockow, K. S. Hettie and T. E. Glass, *ACS Chem. Neurosci.*, 2013, **4**, 1334.
97. F. Pu, E. G. Ju, J. S. Ren and X. G. Qu, *Adv. Mater.*, 2014, **26**, 1111.
98. E. G. Ju, X. J. Yang, Y. H. Lin, F. Pu, J. S. Ren and X. G. Qu, *Chem. Commun.*, 2012, **48**, 11662.
99. M. Bälter, S. M. Li, J. R. Nilsson, J. Andréasson and U. Pischel, *J. Am. Chem. Soc.*, 2013, **135**, 10230.
100. A. P. de Silva, H. Q. N. Gunaratne and C. P. McCoy, *Chem. Commun.*, 1996, 2399.

101. S. A. de Silva, A. Zavaleta, D. E. Baron, O. Allam, E. Isidor, N. Kashimura and J. M. Percarpio, *Tetrahedron Lett.*, 1997, **38**, 2237.
102. M. P. O'Neil, M. P. Niemczyk, W. A. Svec, D. Gosztola, G. L. Gaines III and M. R. Wasielewski, *Science*, 1992, **257**, 63.
103. F. Pina, M. Maestri and V. Balzani, *J. Am. Chem. Soc.*, 2000, **122**, 4496.
104. K. Szacilowski, *Chem. Rev.*, 2008, **108**, 3481.
105. F. M. Raymo and S. Giordani, *Org. Lett.*, 2001, **3**, 1833.
106. F. M. Raymo and S. Giordani, *J. Am. Chem. Soc.*, 2002, **124**, 2004.
107. F. M. Raymo and S. Giordani, *Proc. Natl. Acad. Sci. U. S. A.*, 2002, **99**, 4941.
108. F. M. Raymo and S. Giordani, *J. Am. Chem. Soc.*, 2001, **123**, 4651.
109. D. Y. Tong, H. F. Duan, H. J. Zhuang, J. G. Cao, Z. L. Wei and Y. J. Lin, *RSC Adv.*, 2014, **4**, 5363.
110. *How Animals See the World*, ed. O. F. Lazareva, T. Shimizu and E. A. Wasserman, Oxford University Press, Oxford, 2012.
111. J. J. Tabor, H. M. Salis, Z. B. Simpson, A. A. Chevalier, A. Levskaya, E. M. Marcotte, C. A. Voigt and A. D. Ellington, *Cell*, 2009, **137**, 1272.
112. S. M. Chirieleison, P. B. Allen, Z. B. Simpson, A. D. Ellington and X. Chen, *Nat. Chem.*, 2013, **5**, 1000.
113. G. Seelig, D. Soloveichik, D. Y. Zhang and E. Winfree, *Science*, 2006, **314**, 1585–1588.
114. W. Fontana, *Science*, 2006, **314**, 1552.
115. G. Seelig, B. Yurke and E. Winfree, *J. Am. Chem. Soc.*, 2006, **128**, 12211.
116. B. M. Frezza, S. L. Cockroft and M. R. Ghadiri, *J. Am. Chem. Soc.*, 2007, **129**, 14875.
117. S. Tyagi and F. R. Kramer, *Nat. Biotechnol.*, 1996, **14**, 303.

CHAPTER 10

Metallosupramolecular Materials for Magnetic Applications: Spin-Crossover

SUZANNE NEVILLE*[a]

[a]F11, The University of Sydney, Eastern Ave, Camperdown, NSW 2006, Australia
*E-mail: suzanne.neville@sydney.edu.au

10.1 Introduction

A broad range of research groups are dedicated to understanding the spin crossover (SCO) phenomenon, designing and synthesising new SCO materials and advancing this field for application purposes.[1] The attraction of SCO materials research is not only its visual appeal but the vast range of observable magnetic features, allowing assessment *via* many theoretical and experimental approaches. There is a vast gap, however, between conceptually understanding the SCO process and the preparation of SCO materials with targeted properties owing to the diverse factors that govern the presence or absence of a spin transition (ST) and the overall magnetic character. Whilst a range of metal ions and respective coordination environments conducive to SCO are known, the presence of a ST is not guaranteed as solid state interactions, such as hydrogen-bonding and π-stacking, also influence the presence or absence of SCO. Thus, the successful preparation of metallosupramolecular

RSC Smart Materials No. 15
Functional Metallosupramolecular Materials
Edited by John G. Hardy and Felix H. Schacher

Published by the Royal Society of Chemistry, www.rsc.org

materials that show SCO presents a significant challenge, in particular as intra- and intermolecular interactions play a pivotal role in their design and realisation. Further complicating matters is that encapsulated molecules such as anions and solvents are critical in defining SCO character, either through direct contacts or indirect electrostatic influence as such molecules are to some extent present in the majority of metallosupramolecular materials. Thus it is not surprising that the number of discrete SCO species of nuclearity two or above is far fewer than those of mononuclear materials. In this chapter we explore the design, structure and magnetic features of the fairly exclusive class of discrete metallosupramolecular species that display SCO properties with a brief look at those of infinite polymeric nature for comparative purposes.

10.1.1 Brief Introduction to SCO

Depending on the nature of the ligand field of octahedral d^4–d^7 metals, the d-orbital energy gap between t_{2g} and e_g sets can vary. The relative magnitude of this orbital splitting (as governed by ligand field strength, Δ_o) compared to the energy required to pair electrons (mean spin-pairing energy, P) defines two potential d-electron distributions, namely, high spin (HS) and low spin (LS) (Figure 10.1(a)). Weak ligand fields stabilise the HS state and strong ligand fields stabilise the LS state. The process termed SCO (or a spin transition, ST) is facilitated at intermediate ligand field strengths where external perturbation by temperature, pressure or light irradiation, can result in gradual, abrupt, incomplete or stepwise switching between the two states.[1–4]

Importantly, owing to the difference in occupancy of the bonding and anti-bonding orbitals and the relative number of unpaired electrons between these two states, there are a number of distinct and measureable consequences of this switching. For example, in the case of iron(II), distinct optical changes occur between HS (white) and LS (purple) states, allowing such

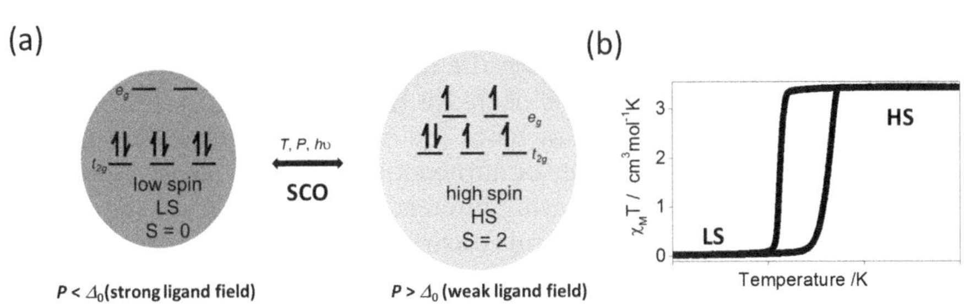

Figure 10.1 (a) Illustration of the d-orbital electron distribution (d_6) for the high spin (HS) and low spin (LS) state and the conditions for spin crossover (SCO). (b) Example of a temperature-dependent magnetic susceptibility profile for a hysteretic SCO.

materials to be envisaged for display devices and sensing applications.[1–5] Furthermore, in a HS iron(II) complex the electrons are distributed over both orbital sets ($t_{2g}^4e_g^2$), consequently this state shows a magnetic signal characteristic of four un-paired electrons, whereas the electrons in a LS iron(II) complex are all paired ($t_{2g}^6e_g^0$) resulting in no magnetic signal. Thus variable temperature magnetic susceptibility measurements readily map the ST behaviour between these two states. Potential hysteresis in the cooling and heating curves provides memory storage capability for these materials.[1–5] Finally, iron(II) HS complexes show longer metal–ligand bond lengths than LS complexes, owing to the electronic occupancy of the anti-bonding orbitals (typically a 0.2 Å contraction), so the spin state may be indirectly assessed by structural methods.

A clear aim in the field of SCO research is to develop materials with highly cooperative (*i.e.*, hysteretic) ST behaviours, thus many early SCO examples focused on gaining an understanding of structural features that result in such properties.[1,4–10] Indeed, cooperativity between SCO sites facilitated by structural linkages such as hydrogen-bonding and π-stacking, were found to be important for enhancing solid state communication by allowing a more effective spin state propagation through the lattice.[5,10–15] Another demonstrated approach to providing enhanced solid state communication is through directly coordinating spin switching sites through bridging ligands, *i.e.*, polynuclear materials.[5,10–15] In this respect, dinuclear materials provide the simplest examples of the effect of coordinating spin switching sites. Larger and more complex metallosupramolecular systems are thus targeted to provide further information and to uncover novel spin switching phenomena.

10.1.2 Metallosupramolecular SCO Materials

There are a number of considerations for designing metallosupramolecular materials with SCO properties that differ from those with other or no targeted functionality, in particular, owing to the specific geometric and ligand field requirements that must be met in order for SCO to occur, *i.e.* intermediate ligand field strength. Further to this, the linked SCO sites must be in an arrangement that allows for the structural variation over the spin transition (*i.e.*, for iron(II) 5–10% decrease HS to LS). If SCO sites are too tightly joined they may prevent or block the other sites from undergoing a transition (referred to as negative- or anti-cooperativity). Although a number of grids and metallocycles displaying interesting SCO properties have been developed, the design and successful construction of these systems still represents a significant challenge. By far the most examples are dinuclear and polymeric metallosupramolecular SCO materials. Examples representing those between dinuclear and polymeric are known and show very elegant methods of supramolecular design which should enable the emergence of many more examples in the future.

10.2 Dinuclear

Of all the discrete SCO-type materials reported, those representing dinuclear species are the most abundant. The reason for this is, firstly, that while synthetic design and realisation is still a significant challenge, there are only two metal sites to contend with compared to say a hexanuclear material. Secondly, focus has been strongly directed at their synthesis as the fundamental insight into SCO properties that dinuclear materials provide is of significant importance. Foremost, dinuclear materials represent the most basic prototype of a polymeric material and thus offer significant understanding towards assessing the relative outcome of directly coordinating SCO sites. Furthermore, dinuclear materials can show direct ([HS–HS] to [LS–LS]) or step-wise ([HS–HS] to [HS–LS] to [LS–LS]) SCO processes and associated one- or two-step (or partial) STs. Isolation of distinct [HS–LS] or mixed 1 : 1 [HS–HS] and [LS–LS] discrete units (as observed by structural or spectroscopic methods) has been a significant driving force in the production of novel dinuclear SCO materials. Finally, where short bridging ligands are employed in linking SCO sites, an interplay between magnetic coupling and SCO may result.

There are too many dinuclear SCO materials reported to assess them all individually here, thus focus will be directed towards general synthetic design approaches and materials that show a significant degree of supramolecular design ingenuity and distinction. Dinuclear materials are classified here based on (1) short (2–3 atoms) or (2) long bridges between metal centres. There are many excellent reviews of dinuclear SCO materials for which to otherwise refer.[16–18]

10.2.1 Short Bridging Ligands

10.2.1.1 One Compartment Ligands

The thermal, pressure and light-induced SCO properties of a family of structurally similar dinuclear materials, which are formed by the bridging of iron(II) sites by bis-bidentate ligands, have been intensively investigated. The $[Fe^{II}N_6]$ coordination environments are completed by chalcogen anions and terminal bidentate ligands. This family is particularly versatile as the SCO properties can be tuned and rationalised by the relative ligand field strength of the peripheral ligand and coordinated anions. For example, $[Fe(bpym)(NCS)_2]_2bpym$ (bpym = 2,2′-bipyrimidine), also known as bpym-S, where bpym is both the bridging and terminal ligand, represents a HS analogue in this series,[19] though, inclusion of the stronger field ligand NCSe (bpym-Se) results in a half SCO around 100 K which is relatively abrupt.[20] The application of pressure, which also effectively increases ligand field strength, induces a ST in bpym-S with features similar to that of bpym-Se under ambient pressure, and increases the ST temperature in bpym-Se.[21] The bpym-Se analogue also provided early evidence of an ordered [HS–LS] intermediate state species.[22] Further ligand field optimisation within this series utilising, for example, 2,2′-bithiazoline (bt) in place of bpym, sees complete STs with subtle two-step character for both S and Se analogues.[20] The transition temperatures of both the bt-S and bt-Se analogues are significantly increased to that of bpym-Se analogous material.

Bidentate ligands with short bridging moieties have also been exploited in the dinuclear species [(pypzH)(NCSe)Fe(μ-pypz)$_2$Fe(NCSe)(pypzH)]·2H$_2$O] (pypz = 2-pyrazolylpyridine) (Figure 10.2(a)).[23] This particular dinuclear material is interesting in that in contains both protonated and de-protonated pypz ligands each with distinct binding modes. The deprotonated ligands bridge the iron(II) sites through the pyrazolyl nitrogens and furthermore each bind to an iron(II) site through the pyridine nitrogen. The protonated pypz ligands act as terminal bidentate groups, with the octahedral coordination environment completed by *cis*-NCSe anions. The dinuclear exists in the [HS–HS] state at RT and converts to a [LS–LS] state in one abrupt step at 225 K and represents the first structurally characterised dinuclear example of this type where a [HS–LS] phase does not exist (Figure 10.2(a)).

In a distinct approach, a family of dinuclear species have been reported which are representative of the well-known 1-D chain triazole systems [Fe^{II}(R-1,2,4-trz)$_3$](anion)$_2$.[24–26] The iron(II) sites in these discrete species are linked by triply bridged $\mu_{1,2}$-triazole ligands of various 4-functionalisation, as per their 1-D chain counterparts. The formation of a polynuclear species is inhibited by the presence of terminal *cis*-NCS anions and mono-dentate R-1,2,4-trz ligand ligands. The salen functionalised analogue (Figure 10.2(b)) is of particular note as the one step SCO occurring around 150 K, as followed by magnetic susceptibility, can be additionally mapped using temperature-dependent emission spectroscopy owing to an intra-ligand electronic transition.[24] Further to this, a HS analogue was reported containing a different solvent combination,

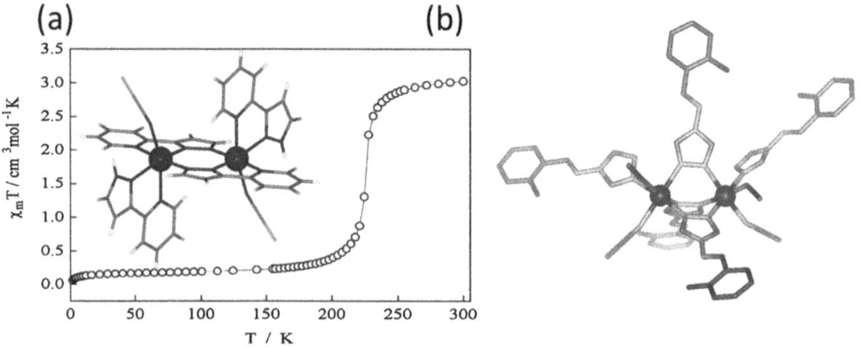

Figure 10.2 (a) Structure of the dinuclear material [(pypzH)(NCSe)Fe(μ-pypz)$_2$ Fe(NCSe)(pypzH)]·2H$_2$O with short pyrazolyl bridges and the one-step SCO proceeding directly from [HS–HS] to [LS–LS] states. Adapted with permission from ref. 23. (b) Structural representation of the triazole-based dinuclear material [Fe$_2$(R-1,2,4-trz)$_5$(NCS)$_4$]·(guest) (R = *N*-salicylidene-4-amino).

indicating prospects of guest-related SCO phenomena.[26] Most recently in this series, with an appended benzene group, a half gradual one-step SCO is observed.[25] Further subtle ligand modification within this family may thus be exploited in the future to enhance solid state interactions towards achieving higher transition temperatures and more cooperative ST behaviours.

10.2.1.2 Two Compartment Ligands

Short bridges between metal sites have also been very successful in producing dinuclear SCO materials where two pocket, fairly rigid, ligands have been employed. For example, an extensive family of pyrazolate-containing double binding pocket ligands have been utilised to produce dinuclear materials of the type *trans*-[Fe_2(μ-bpypz)$_2$(NCX)$_2$(R-py)$_2$].[27–30] The structures feature double bpypz$^-$ ligand bridges, occupying the equatorial iron(II) sites, with the axial sites filled by NCX and R-py groups in a *trans*-disposition with respect to iron(II) centres (Figure 10.3(a)). This series is particularly important as it has allowed a broad systematic investigation of the relative effect of electron withdrawing and donating substituents on SCO properties. Such studies are often challenging, as structural modulations may occur over the series, and fundamentally very important as they lend to the possibility of predicting SCO behaviour 'pre-synthesis'. In particular, the R-py group has been modified over a range of simple R groups (*i.e.* R = H, 4-Me, 4-Me_2N, 3-Me, 3-Cl, 3-Br) and the relative thermal and light-induced SCO behaviours assessed for both the X = S and BH_3 analogues. As expected with respect to ligand field strength considerations, the NCS$^-$ series shows lower transition temperatures than the $NCBH_3^-$ series. Significantly, within each series, the

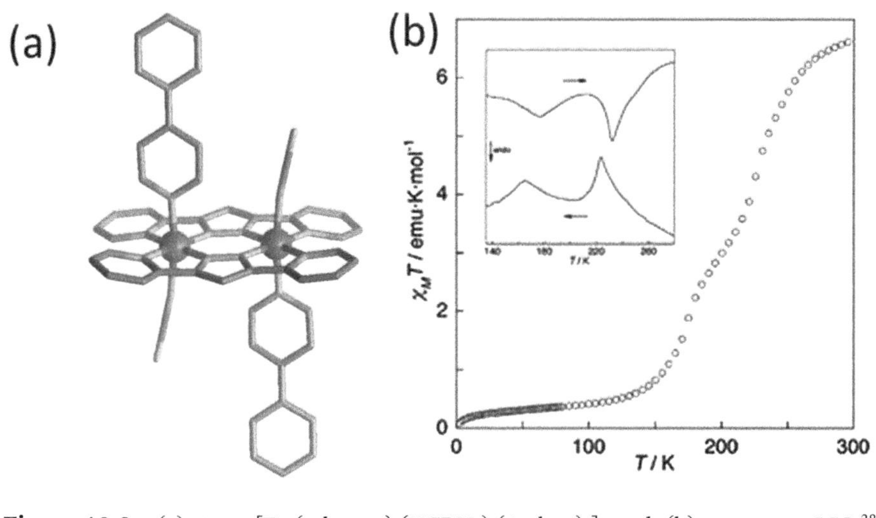

Figure 10.3 (a) *trans*-[Fe_2(μ-bpypz)$_2$($NCBH_3$)$_2$(4-phpy)$_2$] and (b) two-step SCO.[28] Reproduced with permission from ref. 28.

transition temperature is seen to increase with R-py ligands of greater electron withdrawing potential, following a particularly convincing linear trend of Hammett constant (quantifying electron donating and withdrawing capacity) *versus* transition temperature.[27,29–32] Larger R group substituents have also been investigated within this family, for example, where R = 4-phenyl and X = BH_3, a two-step SCO results where a mixed 1:1 ratio of distinct [LS–LS] and [HS–HS] species exist at the intermediate plateau—the first of its type (Figure 10.3).[28] Furthermore, in a logical extension, bridging bis-pyridyl type ligands have been utilised to form one-dimensional polymeric SCO materials; an approach that has significant scope for future expansion.[33]

Subtle ligand modification, where the pyrazolate bridging core is replaced with triazolate results in SCO where R-py is pyridine and X = BH_3. Interestingly, detailed studies revealed different SCO behaviour of a powder *versus* single crystalline samples, such that the powder shows a two-step SCO and the crystals a half one-step SCO.[34,35] This highlights that varied sample preparation conditions should be considered for all SCO materials.

Along a similar line, dual metal binding pocket ligands have been exploited in dinuclear species of the general formula $[Fe^{II}_2(PMRT)_2](X)_4$, where PMRT represents 4-substituted (R) 3,5-bis[((2-pyridylmethyl)-amino)methyl]4*H*-1,2,4-triazole, and X a range of anions. The general structure of these dinuclear species comprises 1,2,4-triazole bridged iron(II) sites from two ligands with the pendant pyridyl groups from each ligand completing the axial coordination. The original member of this family $[Fe^{II}_2(PMAT)_2](BF)_4\cdot(DMF)$ (M = NH_2) revealed an abrupt one-step SCO where structural analysis, for the first time, confirmed an isolated [HS–LS] species which remains trapped to low temperature.[36] This material additionally provided the first evidence of a [HS–LS] species using variable temperature Mössbauer spectroscopy in the absence of a magnetic field.[37] Having established a successful design approach for attaining SCO in dinuclear species, systematic modification of both the ligand 4-substituent and anion were carried out and revealed a range of SCO and non-SCO behaviours.[38,39] Significant variation in complex packing and solvation, as is the case in many such studies, precluded extraction of structure *versus* magnetic character parameters. Importantly though, where R = Ph^{tBu}, in the material $[Fe^{II}_2(PMPh^{tBu}T)_2](BF)_4\cdot3.5(H_2O)$, a hysteretic SCO results with the largest hysteresis loop size for a dinuclear species.[40] Dependent on scan rate, the hysteresis loop varies between 22 and 41 K—highlighting, most notably, that reports on SCO materials should clearly detail experimental conditions, in addition to measurements under a variety of situations.

10.2.2 Longer Bridging Ligands

10.2.2.1 Monodentate Ligands

A range of dinuclear iron(II) complexes where capped metal sites are bridged by monodentate bis-pyridyl type ligands are emerging with various strategies employed for terminating the dinuclear unit. These materials are, in general, not expected to reveal significant cooperativity between metal sites owing to the long and flexible bridging component, however, are of fundamental interest for understanding SCO outcomes of discrete species. For example, the material $[Fe^{II}_2(3\text{-bpp})_2(NCS)_4(4,4'\text{-bpy})]\cdot 2(MeOH)$, comprised of 3-bpp (2,6-bis(pyrazole-3-yl)pyridine) and *trans*-NCS groups terminate the dinuclear and 4,4′-bpy (4,4′-bipyridine) bridging the iron(II) sites.[41] This material shows a partial SCO where at low temperatures, a mixed [HS–LS] exists. Detailed crystallographic studies revealed a series of thermally-dependent SCO and non-SCO related phase transitions associated with this material.[42] The [HS–LS] species can additionally be switched to the [HS–HS] state by light-irradiation and interestingly, a complete LS dinuclear species can be induced by pressure.[41,43] Similarly, the capping ligand di(2-picolyl)amine has been used with a range of bridging ligands of varying length and flexibility, to produce a family of SCO dinuclear complexes where ligand strain is thought to play a large role in the magnetic properties.[44–46]

Other dinuclear SCO materials have also been reported which contain weakly linked metal ion sites, for example using dicyanamide, capped by N-donor end groups.[47,48]

10.2.2.2 Two Compartment Ligands

Ligands that contain two pockets, pre-defined to house metal ions, are a common feature within dinuclear materials synthesis. This is largely because of the reduced synthetic complexity of these materials, thus allowing more broad systematic studies to be conducted. There are two general classes of such materials: (1) those that contain two of these two-compartment ligands and (2) those that contain a single two-compartment ligand.

The primary example of the former is the dinuclear SCO material $[Fe_2(NCS)_4(ddpp)_2]\cdot$guest where guest manipulation results in a range of structural and magnetic features of interest (Figure 10.4(a)).[49,50] Where

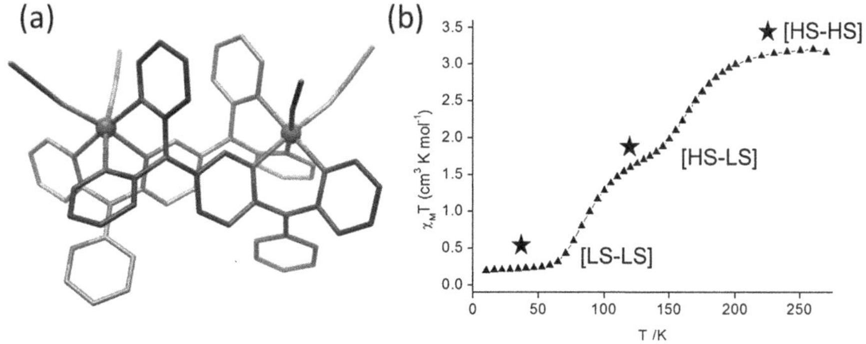

Figure 10.4 (a) The dinuclear material $[Fe_2(NCS)_4(ddpp)_2]\cdot$guest with head-to-tail bridging ligands and (b) where guest = $4(CH_2Cl_2)$ at two-step SCO with distinct HS and LS sites. Adapted from ref. 49.

guest = 4(CH_2Cl_2), a complete two-step SCO occurs and parallel structural analysis resolves unique [HS–HS], [HS–LS] and [LS–LS] states at the respective plateau regions (Figure 10.4(b)). The challenge to identify materials where distinct [HS–LS] species exist is assisted in this material by the presence of two crystallographically unique iron(II) sites at all temperatures. Subtle distortive deviation between these sites, owing to solid state influences, is enough to stabilise this [HS–LS] state. With guest variation to chloroform, the intermediate [HS–LS] state is no longer stabilised as both iron(II) sites in the dinuclear are now structurally and magnetically equivalent, resulting in a one-step ST. These studies reveal that not only is guest type important for SCO character but so is the relative quantity of guest molecules. Indeed, for both the dichloromethane and chloroform solvates, step-wise removal of solvent molecules can be achieved by specific heating profiles and results in a variety of one-step SCO behaviours highlighting that these systems act as diverse molecular sensors.

Di-2,2′-pyridyl metal binding groups also feature in the dinuclear material $[Fe_2(dpyatriz)_2(H_2O)_2(CH_3CN)_2](ClO_4)$ (dpyatriz = 2,4,6-tris(dipyridin-2-ylamino)-1,3,5-triazine) where iron(II) sites are bridged by bis-bidentate dpyatriz ligands.[51] The octahedral coordination environments are completed by water and acetonitrile molecules resulting in an atypical SCO $[FeN_5O]$ configuration. Indeed, the relatively high transition temperature of this material (*ca.* 260 K), coupled with fairly abrupt transition character, suggests that this type of coordination environment could be explored more extensively.

The use of two-compartment ligands that only require a single ligand for complex formation is a facile approach to the generation of dinuclear species.[52] For example, tetradentate N_2O_2 binding pockets are formed in a family of Schiff-base-like ligands with back-to-back binding pockets (L, Figure 10.5(a)).[53,54] In this way, dinuclear materials of the type $[Fe^{II}_2(L)(X)_4]$ are formed and the ligand field adjusted by terminal ligands (X) in the axial positions, such as pyridine and methyl-imidazole (meim), and Schiff-base

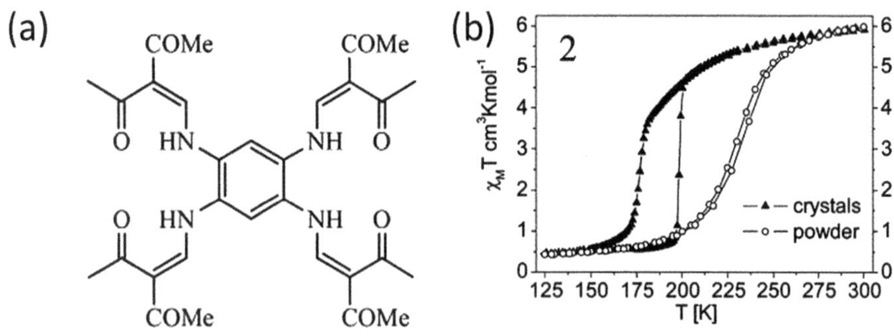

Figure 10.5 (a) The octadentate, dinucleating $N_2O_2^{2-}$ coordinating Schiff-base-like ligand (L) used in the formation of $[Fe^{II}_2L(meim)_4](meim)_4$, (b) displaying a hysteretic SCO which is sample preparation dependent. Reproduced with permission from ref. 53.

ligand external functionalisation. Indeed, the analogue $[Fe_2L(meim)_4](meim)_4$ (where L is a methoxy-functionalised Schiff-base ligand) reveals a cooperative SCO with a 21 K hysteresis loop (Figure 10.5).[53] Detailed structural assessment indicates that the extent of cooperative interactions dictates the SCO behaviour in these materials. As such it is not surprising that a rapidly precipitated powder shows a more gradual SCO behaviour with no hysteresis compared to a pristine single crystal sample (Figure 10.5(b)).

10.2.2.3 *Triple Helicates*

Triple helicate dinuclear species typify the supramolecular approach to designing multi-nuclear species. These materials are generally constructed by twisted bis-bidentate ligands linking metal sites.[55] For example, triple helicate dinuclear species using short azine metal ion bridging components have been reported.[56] The two-pocket ligand design where imidazole rings are separated by an azine group enables a targeted binding configuration and substitution on the imidazole ring defines the relative HS or LS character of the materials at RT. Further anion variation within the SCO-active un-substituted imidazole ring ligand leads to subtle transition temperature shifts of the overall abrupt ST characters of this family. In each case a [HS–LS] species is isolated and a [LS–LS] state not observed—as is often the case where long and flexible ligands are utilised. However, the combination of flexible binding ligand and anion and ligand hydrogen-bonding interactions leads to a versatile SCO character including the presence of polymorphs of the BF_4^- anion analogue.

Imidazole-based bis-bidentate Schiff base ligands have also been utilised, where anion variation results in a range of SCO behaviours.[57,58] Extensive ligand, anion and solvent hydrogen bonding interactions facilitate this anion sensing behaviour. Highlighting furthermore that only minor variations need to be made to modulate SCO behaviour, a similar bis-bidentate ligand (where the ligand spacer and imidazole group have been modified, see Figure 10.6(a)) incorporated into a helicate dinuclear (acetonitrile solvate) results in a somewhat abrupt and complete SCO around 150 K.[59] Structurally, distinct [HS–HS]

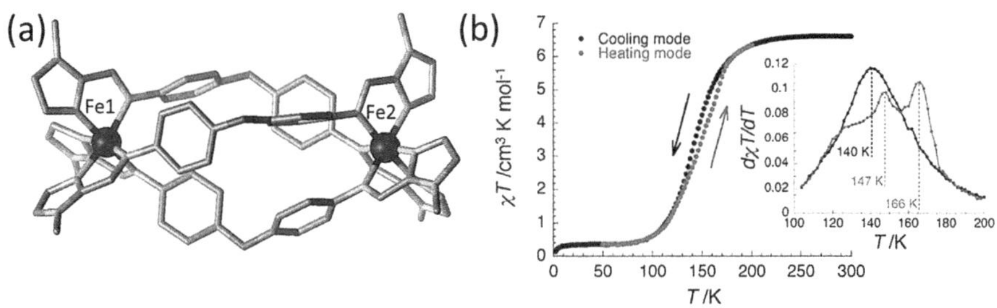

Figure 10.6 (a) The imidazole-based triple helicate $[Fe_2(L)_3]^{4+}$ which shows (b) a reproducible one-step on cooling and two-step on heating SCO process. Reproduced with permission from ref. 59.

and [LS–LS] dinuclear species are isolated at the high and low temperature plateau regions, respectively. Some anomalous SCO behaviour is observed in this material: firstly, while on cooling, a smooth one-step SCO occurs, with heating a subtle two-step character appears (Figure 10.6(b)) and secondly, while a complete LS to metastable HS state is attained with irradiation at 10 K, the relaxation process follows a complex series of steps. Rationalisation of each of these points is complex but the underlying feature is that a [HS–LS] species can be stabilised under specific conditions. Further studies on the analogous hydrate species revealed a 'half' one-step SCO where a [HS–LS] species is trapped at low temperatures.[60] This guest-sensitive ST occurs at a much higher temperature than the acetonitrile solvate suggesting supramolecular interactions are intrinsic to the SCO character of these helical dinuclear species.

10.3 Trinuclear

A number of trinuclear SCO materials are known; common to all examples are incomplete spin transition behaviours such that there are no trinuclear SCO materials where all three metal sites switch from HS to LS states. This is mostly due to the diverse coordination environments in some, or all, of the metal centres—as this is a design feature that promotes the formation of trinuclear subunits.

10.3.1 Linear Trinuclear Materials

The most extensively studied family of SCO trinuclear materials consist of a linear array of three iron(II) centres where a central $Fe^{II}N_6$ core bridges the peripheral metal sites through 1,2,4-triazole ligands of various functionalisation (R-trz, Figure 10.7(a)).[61] This central iron(II) site displays an equivalent

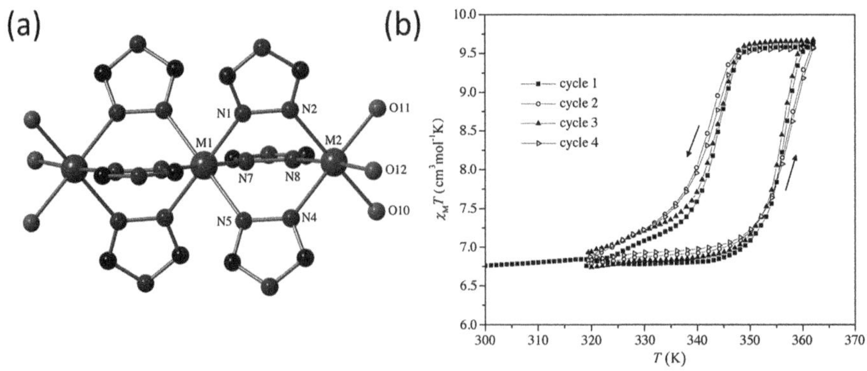

Figure 10.7 (a) Trinuclear subunit of the type $[Fe_3(\mu\text{-R-trz})_6(H_2O)_6](A)_6{\cdot}nH_2O$, where R-trz is neutral, or $[Fe_3(\mu\text{-R-trz})_6(H_2O)_6]{\cdot}nH_2O$, where R-trz is anionic. (b) Thermal SCO behaviour of $[Fe_3(\mu\text{-L})_6(H_2O)_6]{\cdot}5H_2O$ over successive cycles above RT. Reproduced with permission from ref. 64.

coordination to those found in 1-D chain systems of the type $[Fe^{II}(R\text{-}trz)_3](A)_2$ (A = anion), which have been extensively studied for their cooperative and ambient SCO transitions. As such, this inner iron(II) site undergoes a spin transition whereas the outer iron(II) sites, capped with bound water molecules ($Fe^{II}N_3O_3$), remain HS (Figure 10.7(a)). Depending on the functional group of the 1,2,4-triazole ligand, the charge balancing anion and degree of lattice solvation, different transition temperatures result. For example, $[Fe_3(hyetrz)_6(H_2O)_6](CF_3SO_3)_6$ (hyetrz = 4-(2′-hydroxyethyl)-1,2,4-triazole) displays a gradual SCO centred at room temperature[62] and for this reason has been assessed for nanoscale spin state domain imaging.[63] Exemplifying the versatility of this family, the material $[Fe_3(\mu\text{-}L)_6(H_2O)_6]\cdot 5H_2O$ (L = 4-(1,2,4-triazol-4-yl)ethanesulfonate), obtained by partial dehydration of the parent 'as-made' material, shows a hysteretic transition above room temperature ($T_{1/2}\uparrow\downarrow$ 357, 343 K, Figure 10.7(b))—the highest reported for a non-polymeric structure.[64]

Rather different to other linear-style trimers, a 1,2,3-benzotriazolate bridging ligand has been used to form the bimetallic cluster $[Fe^{II}{}_3Zn_6Cl_6(L_{12})]$ (L = 5,6-dimethoxy-1,2,3-benzotriazolate).[65] In distilling this rather complex structure, there are three approximately linear iron(II) sites bridged by 1,2,3-triazolate groups (in the 1,2-position), as per the 1,2,4-triazole derivatives described earlier. Neighbouring 3-position nitrogen atoms of the 1,2,3-triazole are linked furthermore by zinc(II) ions, to provide a rigid cluster-like nona-nuclear material. The short μ-1,2 bridge between iron(II) sites provides an overall $[FeN_6]$ coordination environment and the central site undergoes a HS to LS transition below 160 K (when the material is desolvated).

10.3.2 Non-Linear Trinuclear Materials

Other successful strategies that have resulted in trinuclear SCO materials have relied on metal-containing units that comprise a combination of fixed and labile M–L bonds to direct formation. For this reason, cluster-like arrays of metal ions are commonplace. For example, using a pentadentate blocking ligand, the bimetallic trinuclear material $[(NC)_3Cr^{III}(CNFe^{III}(saldpt))_3]$ ((saldpt) H_2 = *N*,*N*′-bis(1-hydroxy-2-benzylidene)-1,7-diamino-4-azaheptane) results.[66] Structurally, three cyanide ligands of a central hexacyanato-Cr^{III} unit are capped by Fe^{III}(saldpt) groups which show a gradual SCO. Also using a pentadentate Schiff base ligand, the triangular SCO material $[Fe_3L_2(NCS)_4(H_2O)]$ (LH = 1,3-bis[(2-pyridylmethyl)imino]propan-2-ol) forms through the bridging of iron(II) sites by both the ligand and NCS anions. The unique triangular arrangement of iron(II) sites and assortment of ligand and anion binding modes result in distinct iron(II) sites, some SCO-active and some inactive, and a gradual and incomplete spin transition.[67]

Also reliant on exploiting directionally defined coordination environments are a family of polynuclear SCO materials where tris(pyrazolyl)methane capped iron(II) sites are bridged by functionalised tris(pyrazolyl)methane ligands (Figure 10.8).[68] In this way, dinuclear, trinuclear and

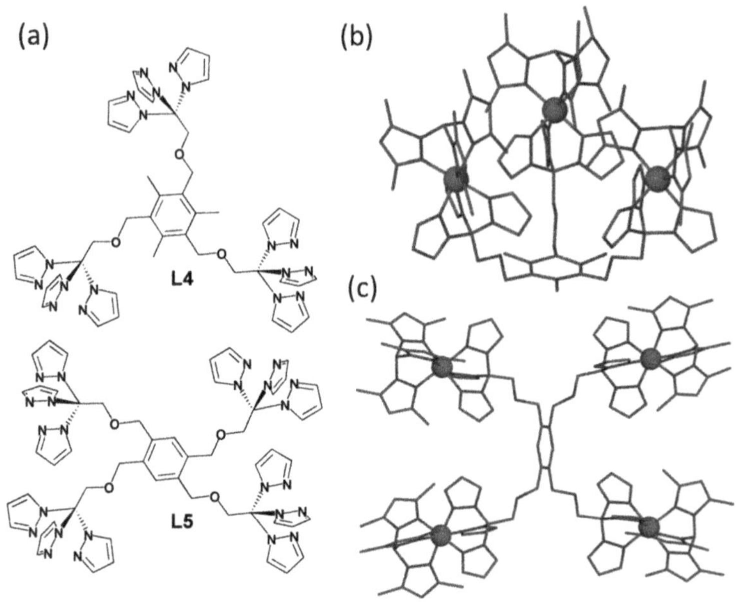

Figure 10.8 (a) Semi-rigid tri- and tetra-podal tris(pyrazolyl)methane bridging ligands. Example of a (b) trinuclear and (c) tetranuclear mixed ligand iron(II) SCO material. Reproduced with permission from ref. 68.

tetranuclear materials result where the bridging ligands, such as those depicted in Figure 10.8(a), are used to define the nuclearity. The SCO properties of these materials range from gradual to abrupt and all show LS iron(II) environments at RT with SCO resultant above RT.

10.4 Tetranuclear

10.4.1 Poly-Aromatic Square Grids

Compared to the intrinsic complication of designing trinuclear materials, squares are conceptually more readily accessible. Indeed, achieving the archetypal [2 × 2]-array of metal ions in a square grid is feasible by exploiting the defined directionality of polypyridyl compartmental ligands. In this way metal sites are linked by bridging tris-chelate groups where the combination of defined binding mode and relatively rigid ligands results in interesting structures and magnetic properties. A variety of grid sizes have been reported and SCO properties have been modulated by subtle ligand functionalisation, and anion and metal variation (iron(II) and cobalt(II)).

The primary example, involving ligand directed self-assembly of $[Fe_4L_4]^{8+}$ grids, utilises terpyridine-type coordination environments where the SCO properties depend on the nature of the substituent in the 2-position on the central pyrimidine group.[69] Irrespective of functionalization, gradual and incomplete transitions without hysteresis are observed—although subtle thermal hysteresis in the light-activated regime suggests underlying cooperative effects.[70] The most important facet of this family is the presence of multiple stepped switching, where for example, three magnetic levels can be isolated (3HS:1LS, 2HS:2LS, 1HS:3LS) and can be addressed by temperature, pressure and light-irradiation.[70] An innovative further step has been taken within this family, whereby pendant pyridyl-functionalised ligands are utilised to link squares into 1-D or 2-D superstructures by a secondary metal ion.[71] While the structural consequence is intriguing, the SCO behaviour in these higher dimensional materials is diminished. This approach could conceivably be exploited in the future to combine multiple metal ion physical and chemical properties into the one material for a distinct approach to multi-functionality.

Similarly, terpy-type compartmental polypyridyl ligand design also includes examples where the central aromatic ring is a pyrazole group. The anionic nature of the ligand results in a lower overall charge of the iron(II) grid complexes, which has been exploited to induce sequential oxidation. A tetranuclear complex of the type $[Fe_4L_4]^{4+}$ shows a ST over a broad temperature range that over a series of steps sees the isolation of four species (4HS, 3HS:1LS, 2HS:2LS and 1HS:3LS).[72] This, coupled with the structurally characterised two-step oxidation process, evidencing the preservation of the grids, results in a dual switchable system with many accessible spin and oxidation state derivatives. Slight synthetic variation results in the formation of a unique trinuclear analogue, $[Fe_3(HL)_2L_2]^{4+}$, where one of the square vertices does not contain a metal ion (Figure 10.9(a)).[73] This structural deviation has a large impact on the overall magnetic properties where the solvated (acetonitrile) species remains locked in a mixed spin [1HS–1LS] state over all temperatures and the desolvated material undergoes an abrupt and hysteretic one-step SCO above RT from [2HS:1LS] to [1HS:2LS] iron(II) states (Figure 10.9(b)). The broad hysteresis loop of 26 K is unusual for such grid systems and is likely due to the combination of internal hydrogen-bonding

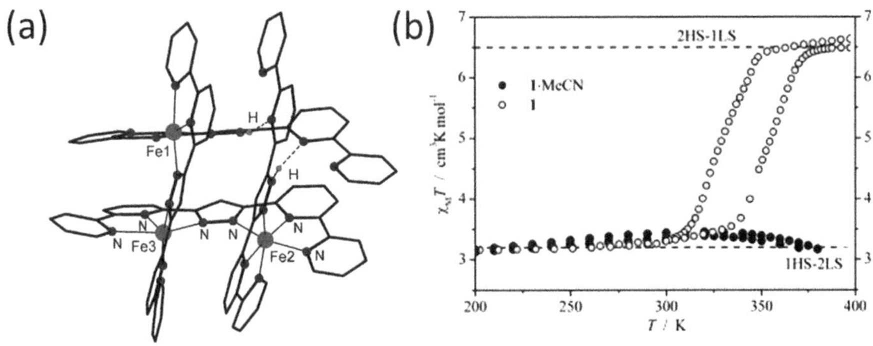

Figure 10.9 (a) Grid-type trinuclear material generated through metal site vacancy resulting in (b) guest-sensitive abrupt and hysteretic ST properties. Reproduced with permission from ref. 73.

interactions at the vacant vortex, which are closely associated with the SCO sites, and those of the unbound pyridyl groups, available for host–host or host–guest interactions. This material highlights, importantly, that future careful ligand design to include specific hydrogen-bonding interaction sites may allow more materials of such cooperative nature to be developed.

Variations to this approach have thus been presented which show unique high temperature or multi-step SCO properties. The most comprehensive studies comprise a range of bis(tridentate) Schiff base ligands where [2 × 2] grids are constructed using either iron(II) or cobalt(II). Indeed, the cobalt(II) analogue represents the highest nuclearity discrete cobalt(II)-containing SCO material reported and displays a multi-step transition where three states can be isolated: 2HS:2LS, 1HS:3LS and 4LS.[74] While SCO cobalt(II) complexes are relatively rare, the SCO process here is driven by partial ligand deprotonation which distorts the metal centre sufficiently for a ST to occur. The analogous iron(II) material also undergoes a ST, although gradual, and occurs just above RT.[75] The SCO properties of this system can be modulated by ligand bulk, anion variation and the degree of hydration. Notably, the dehydrated material shows a distinctly different SCO process than the hydrated analogue consistent with ligand field and crystal packing influences, providing an avenue for molecular sensing, which is more commonly investigated in nanoporous polymeric systems.[76] A further addition to this Schiff-base-type compartmental ligand series includes a bridging oxygen atom in the ligand design. In this way iron(II) square grids are formed which contain oxo-bridges and resultant FeN_4O_2 coordination environments. The somewhat abrupt ST here is similar in character to the previous example but occurs at a much lower temperature (*i.e.* <200 K). Interestingly, the isolated 2HS:2LS state in this material occurs in pairs within each grid rather than the more common checker-board arrangement.[77]

The final example in this tetranuclear grid series is a material where selective switching is possible by using different laser stimuli.[78] The incorporation of an imidazole-based two-compartment ligand into a $[Fe_4L_4]^{4+}$ array leads to a multi-step SCO where iron(II) sites switch from 4HS to 3HS:1LS to 2HS:2LS states. Chemical oxidation of the $[Fe^{II}_4]$ species leads to a mixed valent $[Fe^{II}_2Fe^{III}_3]$ grid where each iron site is in the LS state. Most importantly, comprehensive photo-excitation studies (magnetic susceptibility, Mössbauer and structural characterisation) revealed the $[Fe_4L_4]^{4+}$ in the low temperature 2HS:2LS state can be switched using either green or red light to the 4HS state. Furthermore, the LS iron sites in the thermal SCO-inactive $[Fe^{II}_2Fe^{III}_3]$ grid can both be converted to the HS state using green or red light irradiation, where notably, it is unusual to see a thermal ST inactive but photo-switchable SCO system.

10.4.2 Cyano-Bridged Tetranuclear Materials

Cyano-based bridging groups incorporated into multi-ligand squares provide a short link between SCO sites which may be beneficial for enhanced cooperativity. That being said, relatively few squares of this type have been reported, likely owing to the significant challenge in realising their conceptual design but providing significant scope for their further exploration. The similarity of the ST temperature (*i.e.* near or above RT) and gradual SCO character of the following cyano-bridged square materials implies an intrinsic behaviour which is governed largely by the rigid cyano-bridges rather than the capping ligands. Overall, irrespective of transition temperature, the appearance of one- *versus* multi-step SCO in these materials is driven by symmetry considerations (*i.e.* the number of crystallographically distinct sites).

The first of this type is the mixed ligand material $[Fe^{II}_4(\mu\text{-}CN)_4(L)_4(tpa)_2](PF_6)_4$ (L = 2,2′-bipyridine (bpy) or 1,10-phenanthroline (phen), tpa = tris(2-pyridylmethyl)amine). When L = bpy, a two-step SCO occurs where 1HS:3LS and 4LS grids are isolated where both distinct states are stable over a relatively large temperature range for a square material.[79] The 1HS:3LS state is also achieved with red light irradiation to the 4LS material at low temperatures.[80] When L = phen, owing to symmetry considerations, a single-step SCO is observed just above RT.[81] The changes in SCO behaviour achieved in this series enabled by minor ligand variation suggest the potential to further expand and optimise ST character within this family.

The capping tpa ligand also features in a unique $[Fe^{II}_4]$-type square where iron(II) sites are linked by bridging dicyanamide units.[82] A complete two-step SCO is observed where 2HS:2LS and 4LS grids are present at the respective steps. Dehydration of the material occurs *via* a reversible single-crystal transformation and the two-step SCO character is retained, although the steps are less pronounced.

Similarly, the capping ligand tris-pyrazolylborate (Tp) has been used in [2 × 2] square grid SCO materials. Indeed, a mixed oxidation state square was reported, $[Fe^{II}_2Fe^{III}_2]$, which comprises the capping ligand Tp on the iron(III) sites and the bidentate ligand bis(1-methylimidazol-2-yl)ketone on the iron(II) sites. Their linkage through cyano groups leads to a [2 × 2] grid where the iron(II) sites undergo a HS to LS conversion above RT (Figure 10.10).[83] Guest dependency was observed where the transition temperature of the dehydrated sample was reduced. Related to this is an angular trinuclear mixed valency material, comprising both tpa and Tp groups where there is a partial SCO of the iron(II) sites above RT.[84]

Of distinction to other SCO cyano-bridged squares where at least one tris-chelating ligand is employed, the material $[Fe^{II}_4(CN)_4(bpy)_4(bpym)_4](PF_6)_4$ contains two sets of bis-bidentate ligands.[84] Despite the added synthetic complexity of this approach, a cyano-bridged square results where half of the iron(II) sites undergo a LS to HS transition above RT.

10.4.3 Tetranuclear Cages

We look now at the use of bidentate organic components to direct the self-assembly of tetranuclear molecular cages—an approach that has seen significant success in the formation of SCO dinuclear helicate materials. The benefit in using such ligands is that the bidentate binding mode provides a pre-defined coordination which has fewer degrees of freedom than

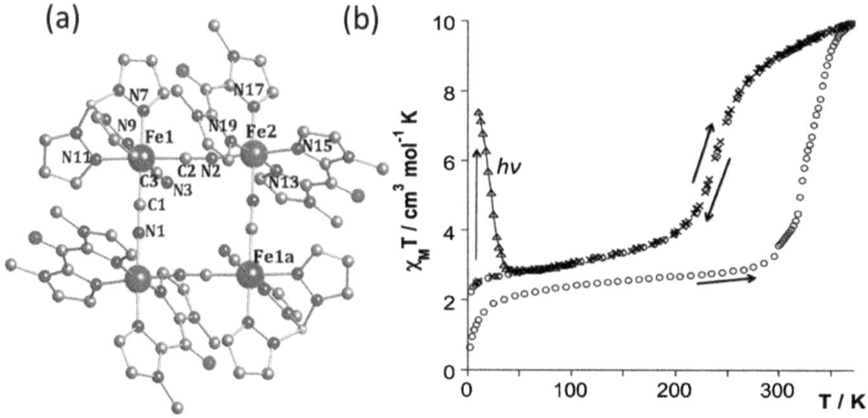

Figure 10.10 (a) A mixed valency cyano-bridged $[Fe^{II}{}_2Fe^{III}{}_2]$ square and (b) guest-dependent SCO properties (o = hydrated, × = dehydrated). Reproduced with permission from ref. 83.

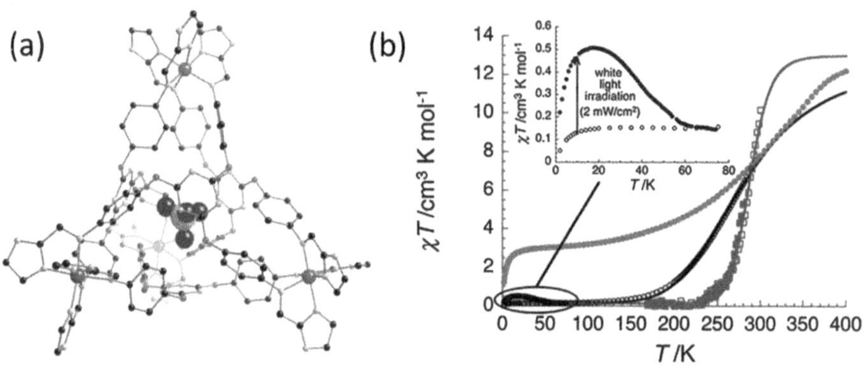

Figure 10.11 (a) Tetrahedral $[Fe^{II}{}_4L_4]^{8+}$ cage with face-capped ligands, corner occupied iron(II) sites and encapsulated BF_4 anion. (b) Thermal-, light- and solvent-induced SCO properties. Reproduced with permission from ref. 85.

monodentate ligands, thus reducing synthetic complexity as per the use of metallo-ligands in trinuclear materials synthesis. In this way, $[Fe^{II}{}_4L_4]^{8+}$-type nano-capsule materials have successfully incorporated SCO switching sites.[85,86] These materials consist of tris-imino-imidazole type ligands where the iron(II) sites lie on the corners and the ligands face-cap a tetrahedron (Figure 10.11(a)). The size of the tetrahedron is modulated by the ligand core and can host a range of anions. In the solid state, gradual one-step STs result which can be induced by a range of external stimuli, such as temperature and light (Figure 10.11(b)). The retention, and indeed enhancement, of SCO properties in dilute acetone solution shows the potential of these materials as control capture and release sensing materials (Figure 10.11(b)).[85]

10.5 Higher Nuclearity Discrete

There are few examples of higher nuclearity discrete materials that undergo a spin transition. To design very complex supramolecular metal–organic materials is a synthetic challenge in itself. An extra level of synthetic design complexity is required to enable externally addressable SCO active sites within such materials, in particular with respect to the subtle balance required between inner and outer coordination sphere solid state environments for SCO to occur. Two synthetic design approaches have thus far been successful in producing cages and clusters of nuclearity ranging from four to seven, namely the use of pre-formed metallo-ligand structural building units or the use of bidentate ligands. Overall, these approaches rely on reducing synthetic variability to allow both a targeted supramolecular design and further incorporation of SCO sites into considerably more complicated materials.

10.5.1 Clusters

A heptanuclear mixed valence SCO material was prepared, whereby complex formation is directed by the initial formation of a mononuclear Schiff-base iron(III) containing precursor. This metallo-ligand acts as both a capping ligand to block the formation of a polymeric species and provides a single accessible binding site for targeted complex formation. Henceforth, with the addition of iron(II) hexacyanide, a heptanuclear complex is formed composed of a central iron(II) hexacyanide environment capped by six iron(III) Schiff-base units.[87] The peripheral iron(III) sites of the mixed valence complex undergo a gradual SCO and the central iron(II) sites remain LS. A related trinuclear material also displays similar structural and magnetic features.[66]

In a distinct approach, but still reliant on terminal and bridging ligands, a family of pentanuclear SCO materials were prepared by the use of mononuclear bis-bidentate complexes with two labile binding sites.[88] A trigonal bipyramidal core is resultant by capping with $[M^{III}(CN)_6]^{3-}$ anions where three of the cyano groups are terminal. Where M^{III} = Fe and Co and the mononuclear contains Fe^{II}, gradual STs occur.

10.5.2 Cages

A family of supramolecular cages, termed 'nanoballs', have been prepared by the initial formation of a scorpionate-based copper(I) metallo-ligand, which, importantly, contain peripheral 4-pyridyl groups with rigid predefined directionality that direct the subsequent complex formation.[87,89,90] Upon addition of $Fe^{II}(NCX)_2$ (X = CH_3, S, Se, BH_3, Figure 10.12(a)), within hours, a pure crystalline product is produced, comprised of discrete rigid $[Fe^{II}_6(Cu^I(Tp^{4\text{-}py}))_8]$ 3 nm diameter shells. One-step gradual STs occur, where the ligand field strength of the iron(II) bound anion (*i.e.*, NCX) defines the transition temperature (Figure 10.12(b)). The LS to HS transition can additionally be accessed by light irradiation at 10 K (Figure 10.12(b)). Of further interest in this family, owing to the inefficient packing of nanoball

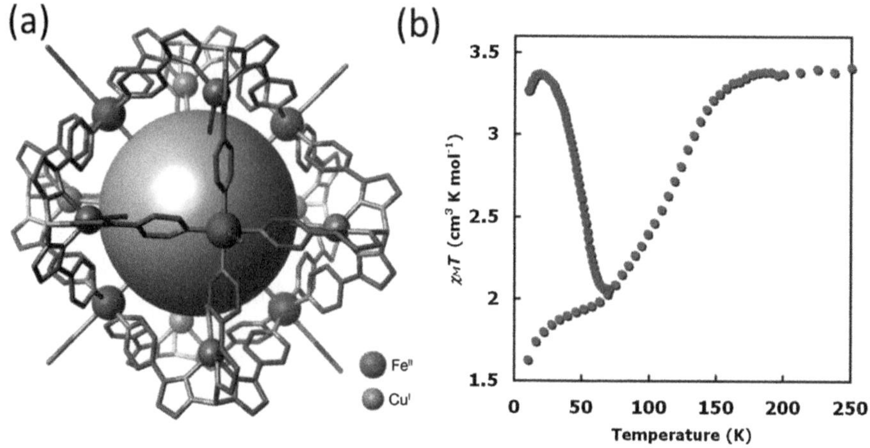

Figure 10.12 (a) The nanoball rigid $[Fe^{II}_6(Cu^I(Tp^{4\text{-}py}))_8]$ shell of 3 nm diameter. (b) The iron(II) sites undergo a SCO *via* thermal variation or light-irradiation. Adapted from ref. 89.

species in the solid state, large solvent-filled cavities result and the spin switching behavior is modulated by the guest species present, thus producing host–guest active switching materials reminiscent of porous framework materials.

10.6 Polymeric Materials

There are significantly more examples of SCO infinite polymeric coordination materials than those of discrete owing to their somewhat reduced synthetic and design complexity. In particular, as there is no need for complex capping ligand utilisation or otherwise ligands with pre-defined metal binding pockets, focus can be directed towards systematic ligand/anion/solvent modifications using established framework platforms to tune SCO properties. In this respect, there are clusters of distinct families representing 1-D, 2-D and 3-D framework materials where extensive studies have been carried out. While the majority of this chapter is devoted to exploring discrete SCO materials, we now have a brief look at those of polymeric origin. Focus will be directed towards examples of metallosupramolecular prominence and/or those exemplifying important SCO characteristics, such as multi-stepped behaviours. Attention should otherwise be directed towards the many comprehensive reviews on SCO polymeric materials.[14,17]

10.6.1 1-D Materials

One-dimensional chain-like materials are a very well-known and extensively explored branch of SCO materials. In particular, this is owing to the disposition of the $[Fe^{II}(R\text{-}1,2,4\text{-triazole})_3](anion)_2$ family to show highly cooperative and hysteretic STs, which can be centred on or near RT. These classic SCO materials are not only readily synthesised by precipitation with the respective reagents, but ligand R-substitution and anion variation are fairly straightforward to achieve. In this respect, many such materials have been reported, where initial investigations were focused on ascertaining structural features relevant to achieving high temperature and cooperative STs and further investigations have expanded to more advanced facets such as nanoparticle preparations, gels, thin films, dendrimers and liquid crystals.[91] A notable advancement in this family is the single crystal structural confirmation of the analogue $[Fe(NH_2trz)_3](NO_3)_2{\cdot}nH_2O$,[92] a longstanding challenge with these materials. It was found that while the individual chains are fairly isolated, they are interconnected *via* a dense hydrogen-bonding network through water and anions. These results are significant as they may lead to further advancements within this family as these detailed structure–function correlations may now be utilised for future material design.

One-dimensional chain materials with longer linking bridges are also of significant interest for their SCO properties. Indeed, in the same light as dinuclear materials, their study is of fundamental interest as there are several structural consequences of a mixed 1 : 1 HS : LS species. The most common case in multi-step SCO materials is a random distribution of HS and LS iron(II) sites such that 'averaged' HS/LS metal–ligand bond lengths are observed. When long-range HS and LS ordering occurs in 1-D chain materials, alternating [–HS–LS–HS–LS–] species may result or distinct [–HS–HS–HS–HS–] and [–LS–LS–LS–LS–] chains. An example of this is presented in the material $[Fe(NCS)_2(cddt)]{\cdot}2(CH_3OH)$ (cddt = 2-chloro-4,6-bis(dipyrid-2-ylamino)-1,3,5-triazine) where $[Fe^{II}$-*trans*$(NCS)_2]$ environments are bridged by cddt ligands through the bis(2-pyridyl)amine groups. This material shows an incomplete one-step SCO where structurally resolved HS and LS iron(II) sites alternate along each chain, *i.e.*, [–HS–LS–HS–LS–], at low temperatures.[93] This scenario is facilitated by the presence of two unique iron(II) sites over the entire temperature regime, where the magnetic distinction between the two iron(II) sites results from ligand–ligand π-interactions at one of these sites. In the related material $[Fe(NCSe)_2(bdpp)]$ (bdpp = 4,6-bis(2′,2″-pyridyl)pyrazine), an 'intermediate' case is revealed over the full two-step SCO.[94] At intermediate plateau temperatures, 'averaged' HS/LS Fe–N bond lengths are observed, however, diffuse scattering indicates that long-range ordering of the type [–HS–LS–HS–LS–], occurs along the chains but with no 3-D correlation. Finally, an example of the alternative mixed 1 : 1 HS–LS distinction, *i.e.*, [–HS–HS–HS–HS–] and [–LS–LS–LS–LS–] chains, has been reported in the material $[Fe^{III}(acen)(bpyp)](BPh_4)$ (acen = *N*,*N*′-bis(acetylacetonato)ethylenediamine, bpyp = 1,3-bis(4-pyridyl)propane).[95] Zig-zag 1-D chains are formed by the bridging of $[Fe^{III}(acen)]$ N_2O_2 environments by bpyp ligands, where notably there are two crystallographically distinct chains. While a one-step SCO is observed, a structure at the midpoint between complete HS and LS states revealed alternating [–HS–HS–HS–HS–] and [–LS–LS–LS–LS–] chains.

10.6.2 2-D and 3-D Frameworks

Higher dimensional (2- and 3-D) polymeric materials are a significant focus in the field of SCO materials owing to their potential enhancement of SCO-site communication pathways *via* multi-dimensional coordinative linkages. Furthermore, such materials often contain significant solvent-accessible void space where the inclusion, removal or exchange of guest molecules may be exploited to optimise SCO behaviour and for molecular sensing properties. In this respect, there are two main classes of 2-D and 3-D framework materials which have been extensively explored for host–guest sensing applications: (1) 2-D and 3-D Hofmann-type materials and (2) so-called SCOFs (SCO frameworks). While there are many other 2-D and 3-D materials that display SCO, the remainder of this chapter will be devoted to the structural design features and guest-sensitive SCO properties of these particular families. In particular, as the synthetic protocols for each are well established, rigorous systematic investigations have been enabled. Many reviews of Hofmann-type SCO materials exist, in particular for 3-D analogues, thus a brief look at several novel examples follows.[96–99]

10.6.2.1 Hofmann-Type Frameworks

Hofmann-type materials are characterised by heterometallic layers containing distorted octahedral iron(II) environments bridged or separated by aromatic N-donor ligands (although some distorted variations on this topology have been reported). The general formula of this family is $([Fe^{II}L)_nM^{II}(CN)_4]\cdot$guest ($n = 2$ for unidentate ligands, $n = 1$ for bis-unidentate ligands; M^{II} = Ni, Pd, Pt), and $[Fe^{II}(L)_n(M^{I}(CN)_2)_2]\cdot$guest ($n = 2$ for unidentate ligands, $n = 1$ for bis-unidentate ligands; M^{I} = Cu, Au, Ag). Thus the main distinction between these variations is that terminal axial ligands result in 2-D networks and bridging ligands result in 3-D frameworks (sometimes present as interpenetrated nets). The use of di- or tetra-cyano-metallate species does not affect the overall topology but does lead to a broad range of SCO behaviours.

With respect to 2-D Hofmann materials, many examples exist incorporating mono-functionalised pyridine ligands (X-py), where cooperative and hysteretic SCO behaviours exist and multi-step STs are also a feature. For example $[Fe(Clpy)_2Pd(CN)_4]$ displays a 're-entrant' two-step SCO, where a structural transformation is coincident with the HS–LS plateau. As we have seen in dinuclear and 1-D chain species, such studies are of fundamental importance. More complex, larger ligands have to a lesser extent been explored, however they are advantageous towards increasing solid state interactions and creating pore volume. For example, the use of rigid (stpy) *versus* flexible (pep) ligands in analogous 2-D Hofmann materials reveals a dense crystal packing and extended π–π interactions in the pep analogue enable a more cooperative ST.[100]

In a distinct approach, a functionalised 1,2,4-triazole ligand (thtrz) was incorporated into the 2-D Hofmann material $[FePd(CN)_4(thtrz)_2](EtOH, H_2O)$ (Figure 10.13(a)).[101] Unlike the 1-D chain 1,2,4-triazole materials

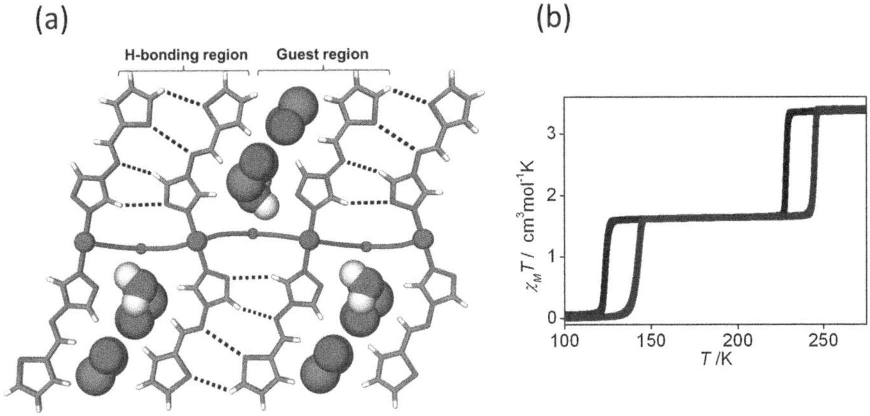

Figure 10.13 [$FePd(CN)_4(thtrz)_2$]·(EtOH, H_2O): (a) one 2-D layer showing guests and interactions and (b) two-step SCO. Reproduced with permission from ref. 101.

highlighted above, the ligand binding is monodentate, leaving the unbound nitrogen available for host–host and host–guest interactions. This consequentially results in a two-step SCO behaviour through the generation of multiple crystallographically distinct iron(II) sites *via* a range of host–host and host–guest interactions. This network of interactions facilitates an intermediate plateau stabilisation over a 130 K temperature range through an anti-cooperative effect where the transition of one SCO site induces a HS stabilisation of neighbouring sites (Figure 10.13(b)). A bis-1,2,4-triazole ligand has also been incorporated into the Hofmann-related material, ($Fe_3(tr_2ad)_4[Au(CN)_2]_2$) $[Au(CN)_2]_4$ revealing a novel 3-D topology, both mono and $\mu_{1,2}$ binding modes of the ligand and a multi-stepped SCO over a broad temperature range.[102] It is envisaged that many more Hofmann-type materials of this type will emerge with ever increasing complex ligand design and multi-functionality.

10.6.2.2 SCO Frameworks (SCOFs)

SCO frameworks (SCOFs) are defined as [$Fe^{II}(NCX)_2$(N-donor aromatic ligand)$_2$] coordination environments linked into two-dimensional grids *via* bridging N-donor aromatic ligands. Depending on the ligand size, flexibility and synthetic conditions, either stacked layer or interpenetrated grid motifs have been reported. Early examples of interpenetrated nets highlighted that these materials could sense the presence, absence or type of guest molecules and robust to flexible framework scaffolds could be generated through purposeful integration of ligand hydrogen bond sites.[7,103–106] Examples of stacked 2-D grid SCOF materials have more recently emerged with interesting guest sensing and magnetic behaviours. For example, [$Fe(NCS)_2(bpy)_2$]·*n*(guest) with all guests except chloroform (for example guest = toluene, nitrobenzene, acetone, methanol), a HS character is observed.[107] An abrupt and hysteretic two-step SCO in the chloroform analogue, while somewhat surprising, likely resides in local guest-related distortions.

Most recently, the ligand dpms was reported in $[Fe(NCX)_2(dpms)_2]$ (X = S, Se), where the flexible, sulfur-containing ligand results in parallel stacked pairs of self-interpenetrated undulating grids.[108] Interestingly, where X = S, a gradual and incomplete ST can be forced to completion under increasing pressure. On the other hand, the stronger ligand field of the Se analogue induces a complete one-step SCO at ambient pressures. Both materials show photo-excitation in the presence of light irradiation. Correlation of the thermal and light-induced characteristic transition temperatures revealed that their close proximity in the S analogue contributes to 'blocking' a complete thermal transition at ambient pressure.

SCOF materials offer vast potential for expansion, as per Hofmann-type systems, where creative ligand design (for example metallo-ligands) may result in advanced multi-functional behaviour and new SCO characters.

10.7 Outlook

The design and synthetic challenges that hamper the incorporation of SCO centres into metallosupramolecular materials are slowly being overcome with some very elegant examples of discrete SCO species now emerging. The use of features from these materials and those of infinite polymeric SCO species, which provide a useful 'template' for the construction of discrete materials, is likely to result in the expansion of this field. There is significant potential to utilise the elegant ligand design exploited in the vast field of metal–organic frameworks (MOFs) to drive the production of novel and more advanced features in discrete and polymeric SCO materials.

With respect to application advancement—the ultimate goal for SCO materials—the 'bottom-up' approach used in the preparation of discrete SCO materials presents a distinct advantage to the 'top-down' approach otherwise utilised in the preparation of, for example, nanoparticles. Whereas the result of particle size miniaturisation in nanoparticle preparations is somewhat disadvantageous (*i.e.*, conferring effective metal dilution), the realisation of 'nano-sized' molecular species sees an alternative route. Indeed, as 'nano-sized' metallosupramolecular materials with ST switching sites have already been generated (see Figure 10.12), the next challenge in this field is to isolate the discrete units, for example on a surface, and retain the SCO properties.

References

1. P. Gütlich and H. A. Goodwin, *Top. Curr. Chem.*, 2004, vol. 233–235.
2. M. A. Halcrow, *Spin-crossover materials: properties and applications*, John Wiley & Sons, Ltd., 2013.
3. P. Gütlich, A. B. Gaspar and Y. Garcia, *Beilstein J. Org. Chem.*, 2013, **9**, 342–391.

4. O. Kahn, *Molecular Magnetism*, VCH, New York, 1993.
5. O. Kahn and C. J. Martinez, *Science*, 1998, **279**, 44–48.
6. K. Madeja and E. König, *J. Inorg. Nucl. Chem.*, 1963, **25**, 377–385.
7. J. A. Real, E. Andres, M. C. Muñoz, M. Julve, T. Granier, A. Bousseksou and F. Varret, *Science*, 1995, **268**, 265–267.
8. L. Cambi and A. Cagnasso, *Atti R. Accad. Naz. Lincei*, 1931, **13**, 809.
9. E. König, *Coord. Chem. Rev.*, 1968, **3**, 471–495.
10. K. S. Murray, *Eur. J. Inorg. Chem.*, 2008, 3101.
11. M. Halcrow, *Chem. Soc. Rev.*, 2011, **40**, 4119.
12. J.-F. Létard, P. Guionneau and L. Goux-Capes, *Top. Curr. Chem.*, 2004, **235**, 221.
13. J. A. Real, A. B. Gaspar, V. Niel and M. C. Muñoz, *Coord. Chem. Rev.*, 2003, **236**, 121.
14. C. J. Kepert, Metal-organic framework materials, in *Porous Materials*, John Wiley & Sons, 2011.
15. Y. Garcia, V. Niel, M. C. Muñoz and J. A. Real, *Top. Curr. Chem.*, 2004, **233**, 229.
16. A. Bousseksou, G. Molnár, J. A. Real and K. Tanaka, *Coord. Chem. Rev.*, 2007, **251**, 1822.
17. K. S. Murray, *Eur. J. Inorg. Chem.*, 2008, 3101.
18. J. Olguín and S. Brooker, Spin-Crossover in discrete polynuclear complexes, in *Spin-Crossover Materials: Properties and Applications*, John Wiley & Sons, Ltd., Oxford, UK, 2013.
19. J. A. Real, J. Zarembowitch, O. Kahn and X. Solans, *Inorg. Chem.*, 1987, **26**, 2939.
20. J. A. Real, I. Castro, A. Bousseksou, M. Verdaguer, R. Burriel, M. Castro, J. Linares and F. Varret, *Inorg. Chem.*, 1997, **36**, 455.
21. V. Ksenofontov, A. B. Gaspar, J. A. Real and P. Gütlich, *J. Phys. Chem. B*, 2001, **105**, 12266.
22. V. Ksenofontov, H. Spiering, S. Reiman, Y. Garcia, A. B. Gaspar, N. Moliner, J. A. Real and P. Gütlich, *Chem. Phys. Lett.*, 2001, **348**, 381.
23. B. A. Leita, B. Moubaraki, K. S. Murray, J. P. Smith and J. D. Cashion, *Chem. Commun.*, 2004, 156.
24. Y. Garcia, F. Robert, A. D. Naik, G. Zhou, B. Tinant, K. Robeyns, S. Michotte and L. Piraux, *J. Am. Chem. Soc.*, 2011, **133**, 15850.
25. O. Roubeau, P. Gamez and S. J. Teat, *Eur. J. Inorg. Chem.*, 2013, 934.
26. H. S. Scott, T. M. Ross, B. Moubaraki, K. S. Murray and S. M. Neville, *Eur. J. Inorg. Chem.*, 2013, 803.
27. N. Suemura, M. Ohama and S. Kaizaki, *Chem. Commun.*, 2001, 1538.
28. K. Nakano, S. Kawata, K. Yoneda, A. Fuyuhiro, T. Yagi, S. Nasu, S. Morimoto and S. Kaizaki, *Chem. Commun.*, 2004, 2892.
29. K. Nakano, N. Suemura, S. Kawata, A. Fuyuhiro, T. Yagi, S. Nasu, S. Morimoto and S. Kaizaki, *Dalton Trans.*, 2004, 982.
30. K. Yoneda, K. Nakano, J. Fujioka, K. Yamada, T. Suzuki, A. Fuyuhiro, S. Kawata and S. Kaizaki, *Polyhedron*, 2005, **24**, 2437.
31. K. Nakano, N. Suemura, K. Yoneda, S. Kawata and S. Kaizaki, *Dalton Trans.*, 2005, 740.

32. J.-F. Létard, C. Cabonera, J. A. Real, S. Kawata and S. Kaizaki, *Chem.–Eur. J.*, 2009, **15**, 4146.
33. K. Yoneda, K. Adachi, S. Hayami, Y. Maeda, M. Katada, A. Fuyuhiro, S. Kawata and S. Kaizaki, *Chem. Commun.*, 2006, 45.
34. C. J. Schneider, J. D. Cashion, B. Moubaraki, S. M. Neville, S. R. Batten, D. R. Turner and K. S. Murray, *Polyhedron*, 2007, **26**, 1764.
35. C. J. Schneider, J. D. Cashion, N. F. Chilton, C. Etrillard, M. Fuentealba, J. A. K. Howard, J.-F. Létard, C. Milsmann, B. Moubaraki, H. A. Sparkes, S. R. Batten and K. S. Murray, *Eur. J. Inorg. Chem.*, 2013, 850.
36. M. H. Klingele, B. Moubaraki, J. D. Cashion, K. S. Murray and S. Brooker, *Chem. Commun.*, 2005, 987.
37. C. M. Grunert, S. Reiman, H. Spiering, J. A. Kitchen, S. Brooker and P. Gütlich, *Angew. Chem., Int. Ed.*, 2008, **47**, 2997.
38. J. A. Kitchen, N. G. White, G. N. L. Jameson, J. L. Tallon and S. Brooker, *Inorg. Chem.*, 2011, **50**, 4586.
39. J. A. Kitchen, J. Olguín, R. Kulmaczewski, N. G. White, V. A. Milway, G. N. L. Jameson, J. L. Tallon and S. Brooker, *Inorg. Chem.*, 2013, **52**, 11185.
40. R. Kulmaczewski, J. Olguín, J. A. Kitchen, H. L. C. Feltham, G. N. L. Jameson, J. L. Tallon and S. Brooker, *J. Am. Chem. Soc.*, 2014, **136**, 878.
41. D. Fedaoui, Y. Bouhadja, A. Kaiba, P. Guionneau, J.-F. Létard and P. Rosa, *Eur. J. Inorg. Chem.*, 2008, 1022.
42. A. Kaiba, H. J. Shepherd, D. Fedaoui, P. Rosa, A. E. Goeta, N. Rebbani, J.-F. Létard and P. Guionneau, *Dalton Trans.*, 2010, **39**, 2910.
43. H. J. Shepherd, P. Rosa, L. Vendier, N. Casati, J.-F. Létard, A. Bousseksou, P. Guionneau and G. Molnár, *Phys. Chem. Chem. Phys.*, 2012, **14**, 5265.
44. A. Y. Verat, N. Ould-Moussa, E. Jeanneau, B. Le Guennic, A. Bousseksou, S. A. Borshch and G. S. Matouzenko, *Chem.–Eur. J.*, 2009, **15**, 10070.
45. G. S. Matouzenko, E. Jeanneau, A. Y. Verat and A. Bousseksou, *Dalton Trans.*, 2011, **40**, 9608.
46. Y. de Gaetano, E. Jeanneau, A. Y. Verat, L. Rechignat, A. Bousseksou and G. S. Matouzenko, *Eur. J. Inorg. Chem.*, 2013, 1015.
47. S. R. Batten, J. Bjernemose, P. Jensen, B. A. Leita, K. S. Murray, B. Moubaraki, J. P. Smith and H. Toftlund, *Dalton Trans.*, 2004, 3370.
48. N. Ortega-Villar, A. L. Thompson, M. C. Muñoz, V. M. Ugalde-Saldivar, A. E. Goeta, R. Moreno-Esparza and J. A. Real, *Chem.–Eur. J.*, 2005, **11**, 5721.
49. J. J. M. Amoore, C. J. Kepert, J. D. Cashion, B. Moubaraki, S. M. Neville and K. S. Murray, *Chem.–Eur. J.*, 2006, **12**, 8220.
50. J. J. M. Amoore, S. M. Neville, B. Moubaraki, S. S. Iremonger, K. S. Murray, J.-F. Létard and C. J. Kepert, *Chem.–Eur. J.*, 2010, **16**, 1973.
51. M. Quesada, P. de Hoog, P. Gamez, O. Roubeau, G. Aromí, B. Donnadieu, C. Massera, M. Lutz, A. L. Spek and J. Reedijk, *Eur. J. Inorg. Chem.*, 2006, 1353.
52. S. Brooker, P. G. Plieger, B. Moubaraki and K. S. Murray, *Angew. Chem., Int. Ed.*, 1999, **38**, 408.
53. B. Weber, E. S. Kaps, J. Obel, K. Achterhold and F. G. Parak, *Inorg. Chem.*, 2008, **47**, 10779.

54. B. Weber and E. Kaps, *Heterocycl. Chem.*, 2005, **16**, 391.
55. S. G. Telfer, B. Bocquet and A. F. Williams, *Inorg. Chem.*, 2001, **40**, 4818.
56. Y. Sunatsuki, R. Kawamoto, K. Fujita, H. Maruyama, T. Suzuki, H. Ishida, M. Kojima, S. Iijima and N. Matsumoto, *Inorg. Chem.*, 2009, **48**, 8784.
57. F. Tuna, M. R. Lees, G. J. Clarkson and M. J. Hannon, *Chem.–Eur. J.*, 2004, **10**, 5737.
58. Y. Garcia, C. M. Grunert, S. Reiman, O. van Campenhoudt and P. Gütlich, *Eur. J. Inorg. Chem.*, 2006, 3333.
59. D. Pelleteret, R. Clérac, C. Mathonière, E. Harté, W. Schmitt and P. E. Kruger, *Chem. Commun.*, 2009, 221.
60. R. J. Archer, C. S. Hawes, G. N. L. Jameson, V. McKee, B. Moubaraki, N. F. Chilton, K. S. Murray, W. Schmitt and P. E. Kruger, *Dalton Trans.*, 2011, **40**, 12368.
61. G. Aromí, L. A. Barrios, O. Roubeau and P. Gamez, *Coord. Chem. Rev.*, 2011, **255**, 485.
62. Y. Garcia, P. Guionneau, G. Bravic, D. Chasseau, J. A. K. Howard, O. Kahn, V. Ksenofontov, S. Reiman and P. Gütlich, *Eur. J. Inorg. Chem.*, 2000, 1531.
63. C. Chong, B. Berini, K. Boukheddaden, E. Codjovi, J. Linarés, Y. Garcia, A. D. Naik and F. Varret, *Phys. Status Solidi A*, 2010, **207**, 1227.
64. V. Gómez, J. Benet-Buchholz, E. Martin and J. R. Galán-Mascarós, *Chem.–Eur. J.*, 2014, **20**, 5369.
65. S. Biswas, M. Tonigold, H. Kelm, H.-J. Krüger and D. Volkmer, *Dalton Trans.*, 2010, **39**, 9851.
66. R. Herchel, R. Boĉa, M. Gembicky, J. Kožíšek and F. Renz, *Inorg. Chem.*, 2004, **43**, 4103.
67. G. Psomas, N. Bréfuel, F. Dahan and J.-P. Tuchagues, *Inorg. Chem.*, 2004, **43**, 4590.
68. C. J. Schneider, B. Moubaraki, J. D. Cashion, D. R. Turner, B. A. Leita, S. R. Batten and K. S. Murray, *Dalton Trans.*, 2011, **40**, 6939.
69. M. Ruben, E. Breuning, J. M. Lehn, V. Ksenofontov, F. Renz, P. Gütlich and G. B. M. Vaughan, *Chem.–Eur. J.*, 2003, **9**, 4422.
70. E. Breuning, M. Ruben, J. M. Lehn, F. Renz, Y. Garcia, V. Ksenofontov, P. Gütlich, E. Wegelius and K. Rissanen, *Angew. Chem., Int. Ed.*, 2000, **39**, 2504.
71. M. Ruben, U. Ziener, J.-M. Lehn, V. Ksenofontov, P. Gütlich and G. B. M. Vaughan, *Chem.–Eur. J.*, 2005, **11**, 94.
72. B. Schneider, S. Demeshko, S. Dechert and F. Meyer, *Angew. Chem., Int. Ed.*, 2010, **49**, 9274.
73. M. Steinert, B. Schneider, S. Dechert, S. Demeshko and F. Meyer, *Angew. Chem., Int. Ed.*, 2014, **53**, 6135.
74. S.-Q. Wu, Y.-T. Wang, A.-L. Cui and H.-Z. Kou, *Inorg. Chem.*, 2014, **53**, 2613.
75. Y.-T. Wang, S.-T. Li, S.-Q. Wu, A.-L. Cui, D.-Z. Shen and H.-Z. Kou, *J. Am. Chem. Soc.*, 2013, **135**, 5942.
76. G. J. Halder, C. J. Kepert, B. Moubaraki, K. S. Murray and J. D. Cashion, *Science*, 2002, **298**, 1762.

77. D.-Y. Wu, O. Sato, Y. Einaga and C.-Y. Duan, *Angew. Chem., Int. Ed.*, 2009, **48**, 1475.
78. T. Matsumoto, G. N. Newton, T. Shiga, S. Hayami, Y. Matsui, H. Okamoto, R. Kumai, Y. Murakami and H. Oshio, *Nat. Commun.*, 2014, **5**, 3865.
79. M. Nihei, M. Ui, M. Yokota, L. Han, A. Maeda, H. Kishida, H. Okamoto and H. Oshio, *Angew. Chem., Int. Ed.*, 2005, **44**, 6484.
80. T. Nishihara, M. Nihei, H. Oshio and K. Tanaka, *J. Phys.: Conf. Ser.*, 2009, **148**, 012033.
81. I. Boldog, F. J. Muñoz-Lara, A. B. Gaspar, M. C. Muñoz, M. Seredyuk and J. A. Real, *Inorg. Chem.*, 2009, **48**, 3710.
82. R.-J. Wei, Q. Huo, J. Tao, R.-B. Huang and L.-S. Zheng, *Angew. Chem., Int. Ed.*, 2011, **50**, 8940.
83. A. Mondal, Y. Li, P. Herson, M. Seuleiman, M.-L. Boillot, E. Riviére, M. Julve, L. Rechignat, A. Bousseksou and R. Lescouëzec, *Chem. Commun.*, 2012, **48**, 5653.
84. M. Nihei, M. Ui and H. Oshio, *Polyhedron*, 2009, **28**, 1718.
85. A. Ferguson, M. A. Squire, D. Siretanu, D. Mitcov, C. Mathonière, R. Clérac and P. E. Kruger, *Chem. Commun.*, 2013, **49**, 1597.
86. R. A. Bilbeisi, S. Zarra, H. L. C. Feltham, G. N. L. Jameson, J. K. Clegg, S. Brooker and J. R. Nitschke, *Chem.–Eur. J.*, 2013, **19**, 8058.
87. R. Boĉa, I. Šalitroš, J. Kožíšek, J. Linares, J. Moncol and F. Renz, *Dalton Trans.*, 2010, **39**, 2198.
88. M. Shatruk, A. Dragulescu-Andrasi, K. E. Chambers, S. A. Stoian, E. L. Bominaar, C. Achim and K. R. Dunbar, *J. Am. Chem. Soc.*, 2007, **129**, 6104.
89. M. B. Duriska, S. M. Neville, B. Moubaraki, J. D. Cashion, G. J. Halder, K. W. Chapman, C. Baldé, J.-F. Létard, K. S. Murray, C. J. Kepert and S. R. Batten, *Angew. Chem., Int. Ed.*, 2009, **48**, 2549.
90. M. B. Duriska, S. M. Neville, B. Moubaraki, K. S. Murray, C. Baldé, J.-F. Létard, C. J. Kepert and S. R. Batten, *ChemPlusChem*, 2012, **77**, 616.
91. O. Roubeau, *Chem.–Eur. J.*, 2012, **18**, 15230.
92. A. Grosjean, N. Daro, B. Kauffman, A. Kaiba, J.-F. Létard and P. Guionneau, *Chem. Commun.*, 2011, **47**, 12382.
93. S. M. Neville, B. A. Leita, D. A. Offermann, M. B. Duriska, B. Moubaraki, K. W. Chapman, G. J. Halder and K. S. Murray, *Eur. J. Inorg. Chem.*, 2007, 1073.
94. S. M. Neville, B. A. Leita, G. J. Halder, C. J. Kepert, B. Moubaraki, J.-F. Létard and K. S. Murray, *Chem.–Eur. J.*, 2008, **14**, 10123.
95. S. Imatomi, R. Kitashima, T. Hamamatsu, M. Okeda, Y. Ogawa and N. Matsumoto, *Chem. Lett.*, 2006, **35**, 502.
96. J. A. Real, A. B. Gaspar, V. Niel and M. C. Muñoz, *Coord. Chem. Rev.*, 2003, 121.
97. Y. Garcia, V. Niel, M. C. Muñoz, and J. A. Real, Spin Crossover in 1D, 2D, 3D Polymeric Fe(II) Networks, in *Spin Crossover in Transition Metal Compounds I, Top. Curr. Chem.*, ed. P. Gütlich and H. A. Goodwin, Springer-Verlag, Berlin, Heidelberg, 2004, vol. 233, p. 229.

98. J. A. Real, A. B. Gaspar and M. C. Muñoz, *Dalton Trans.*, 2005, 2062.
99. M. C. Muñoz and J. A. Real, *Coord. Chem. Rev.*, 2011, **255**, 2068.
100. R. Otahani, M. Arai, H. Ohba, A. Hori, M. Takata, S. Kitagawa and M. Ohba, *Eur. J. Inorg. Chem.*, 2013, 738.
101. Y. M. Klein, N. F. Sciortino, F. Ragon, C. E. Housecroft, C. J. Kepert and S. M. Neville, *Chem. Commun.*, 2014, **50**, 3838.
102. F. J. Munoz-Lara, A. B. Gaspar, M. C. Munoz, A. B. Lysenko, K. V. Domasevitch and J. A. Real, *Inorg. Chem.*, 2012, **51**, 13078.
103. N. Moliner, M. C. Muñoz, S. Létard, X. Solans, N. Menendez, A. Goujon, F. Varret and J. A. Real, *Inorg. Chem.*, 2000, **39**, 5390.
104. S. M. Neville, B. Moubaraki, K. S. Murray and C. J. Kepert, *Angew. Chem., Int. Ed.*, 2007, **46**, 2059.
105. S. M. Neville, G. J. Halder, K. W. Chapman, M. B. Duriska, P. D. Southon, J. D. Cashion, J.-F. Létard, B. Moubaraki, K. S. Murray and C. J. Kepert, *J. Am. Chem. Soc.*, 2008, **130**, 2869.
106. S. M. Neville, G. J. Halder, K. W. Chapman, M. B. Duriska, B. Moubaraki, K. S. Murray and C. J. Kepert, *J. Am. Chem. Soc.*, 2009, **131**, 12106.
107. C. J. Adams, C. M. Muñoz, R. E. Waddington and J. A. Real, *Inorg. Chem.*, 2011, **50**, 10633.
108. N. F. Sciortino, S. M. Neville, C. Desplanches, J. F. Létard, V. Martinez, J. A. Real, B. Moubaraki, K. S. Murray and C. J. Kepert, *Chem.–Eur. J.*, 2014, **20**, 7448.

CHAPTER 11

Metallosupramolecular Materials for Energy Applications: Light Harvesting

VAKAYIL K. PRAVEEN[a] AND AYYAPPANPILLAI AJAYAGHOSH*[a]

[a]Photosciences and Photonics Group, Chemical Sciences and Technology Division, CSIR-National Institute for Interdisciplinary Science and Technology (CSIR-NIIST), Trivandrum-695019, India
*E-mail: ajayaghosh@niist.res.in

11.1 Introduction

One of the reasons for the existence of life on Earth relies on the ability of photosynthetic organisms such as plants, algae and photosynthetic bacteria to harvest energy from the Sun. These great harvesters of solar energy convert the absorbed light energy to electronic excitation with the help of organized ensembles of pigment molecules (chlorophylls and carotenoids).[1,2] The mesoscale organization of assemblies of pigment molecules in the photosynthetic membrane allows coherent transfer of excitation energy towards a reaction centre. The excitation energy reaching the reaction centre initiates a series of electron transfer events that facilitate charge separation across the photosynthetic membrane. The charge separation process prompts a proton transfer process to generate charge imbalance, necessary to power the synthesis of the high-energy biological

RSC Smart Materials No. 15
Functional Metallosupramolecular Materials
Edited by John G. Hardy and Felix H. Schacher

Published by the Royal Society of Chemistry, www.rsc.org

fuel adenosine triphosphate. Inspired by Nature, researchers have been attempting to mimic the mechanism of photosynthetic organisms with the aid of synthetic molecular and supramolecular systems.[3–11] Such attempts have unravelled the role of various photophysical processes such as exciton coupling, migration, electron transfer and charge separation in natural light-harvesting systems. Recent studies have demonstrated that Nature uses the principles of quantum mechanics and mesoscale self-assembly in the creation of light-harvesting systems.[1,2]

Knowledge about natural light-harvesting processes has helped researchers to design artificial systems that can be efficiently applied in nanoscale optoelectronic and photonic devices. However, an intricate task in developing artificial systems is the spatial organization of chromophores within an optimum distance and ideal orientation that facilitates efficient homo- and hetero-energy transfer processes. Several molecular and supramolecular strategies have been adopted for this purpose. Among them, chromophoric assemblies have generated considerable interest because of their similarity with assemblies of pigments in natural light-harvesting systems. In this context, light harvesting and energy transfer in metallosupramolecular materials have attracted significant attention.[9–15] The assembly of metal ions and organic chromophores into a single hybrid system allows merging of the inherent properties of components in such a way to produce totally new properties, distinct from those of the building blocks.[15–20] Moreover, controlling the size, shape, and functions of these hybrid systems is considered to be a challenge due to the complexity of the coordination-induced multistep self-assembly process under dynamic conditions.[16–22] This chapter aims to put into perspective the developments in the design of metallo-supramolecular light-harvesting materials, particularly the use of crystalline metal–organic frameworks (MOFs, see also Chapter 7), self-assembly of nanoscale coordination polymers (NCPs, see also Chapters 4 and 5) and supramolecular metallogels (see also Chapter 6) as scaffolds for excitation energy transfer.

11.2 Light-Harvesting Metal–Organic Frameworks (MOFs)

Metal–organic frameworks MOFs are porous three-dimensional (3-D) coordination networks formed by metal ions and rigid organic multidentate ligands.[23–32] In recent years, MOFs have emerged as light-harvesting metallosupramolecular materials, because of the availability of a large number of chromophores as ligands that can act as light-harvesting units.[26–32] Depending on the spatial orientation of energy transfer donor and acceptor chromophores within the MOF structure, the energy transfer can take place from strut to strut or strut to metal nodes (antenna effect). Apart from this, the inherent porosity of MOFs allows encapsulation of either donor or acceptor chromophores and facilitates energy transfer either from the donor framework to a potential guest or in the reverse manner.[26–32]

The synthesis and characterization of near infrared (NIR)-emitting MOFs based on oligo(*p*-phenylenevinylene)-based bridging ligand **1** and ytterbium cations (Figure 11.1) have been reported by the research groups of Petoud and Rosi.[33] OPV ligands sensitize the Yb^{III} NIR emission through transfer of excitation energy *via* the antenna effect. Interestingly, the NIR emission quantum yields of MOFs were found to be high for Yb^{III}-based systems dispersed in solution (Figure 11.1b). This result demonstrates the importance of the rigid extended structure of the MOFs that protects the Yb^{III} ions from solvent vibrations, thereby leading to very good NIR emission.

Meyer, Lin and co-workers have described the synthesis and energy transfer properties of a phosphorescent MOF based on Ru^{II} complexed bridging ligand (**2**:L_{Ru}) and Zn^{2+} (Figure 11.2).[34] The Ru^{II}-based MOF building blocks were characterized with a strong absorption in the visible region and long-lived triplet metal-to-ligand charge transfer (^{3}MLCT). Doping with Os^{II}-complexed bridging ligand (**3**:L_{Os}) to the MOF structure has allowed studies of energy migration processes in the framework by monitoring the efficiency of the $Ru^{II} \rightarrow Os^{II}$ energy transfer. Time resolved emission studies gave clear evidence for $Ru^{II} \rightarrow Ru^{II}$ homo- and $Ru^{II} \rightarrow Os^{II}$ hetero-energy transfer processes. The results of photophysical studies revealed rapid and efficient energy migration in the MOF structure; ^{3}MLCT excited states of L_{Ru} are estimated to undergo approximately fifty $Ru^{II} \rightarrow Ru^{II}$ energy hops in their lifetimes.[34,35] Later, these systems were successfully utilized to demonstrate efficient energy migration assisted luminescence quenching (amplified quenching) processes with suitable quenchers such as 1,4-benzoquinone, *N*,*N*,*N*′,*N*′-tetramethylbenzidine, and methylene blue.[36,37]

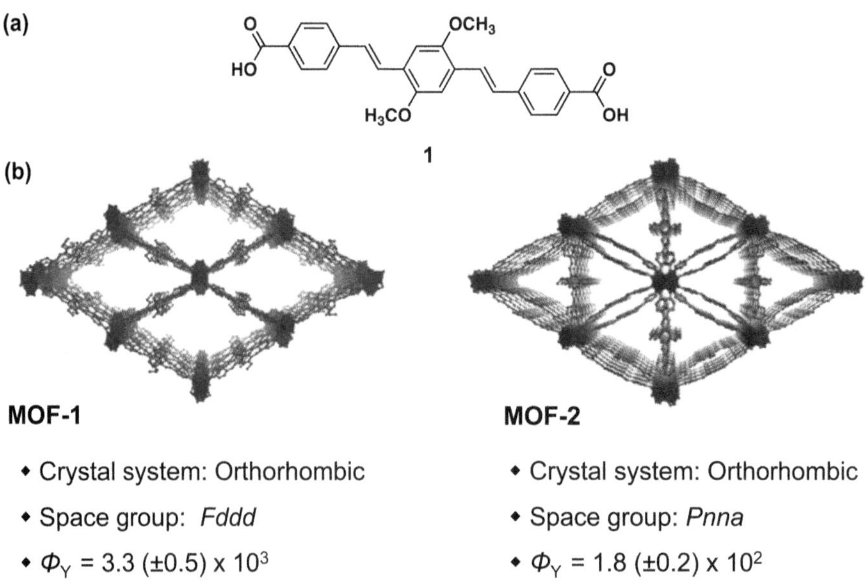

Figure 11.1 (a) Molecular structure of the OPV-based bridging ligand **1** and (b) projection of the MOFs viewed along the '*a*' crystallographic direction. Reprinted with permission from ref. 33.

Emission colour tuning of MOFs *via* ligand → metal and metal → metal energy transfer by co-doping of $GdCl_3$, and a 4,4′-bipyridine-based MOF with $EuCl_3$ and $TbCl_3$ has been demonstrated by Müller-Buschbaum and co-workers (Figure 11.3).[38,39] These MOFs exhibit emission colours spanning from green to red, depending on the content of the doped europium and terbium.

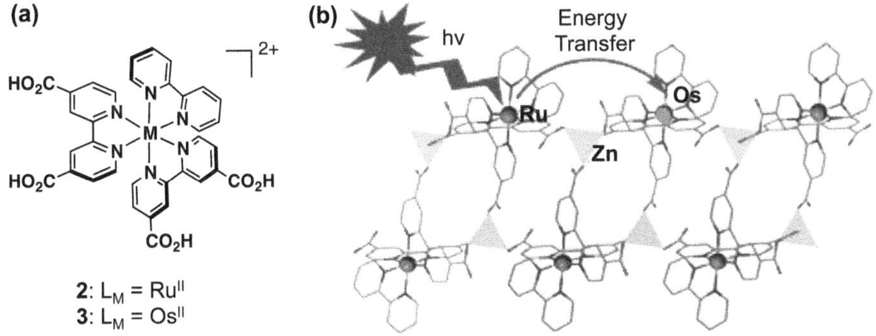

Figure 11.2 (a) Molecular structures of MOF building blocks, L_M. (b) A schematic representation of the Ru^{II} → Os^{II} energy transfer process in L_M–Zn MOF. Reprinted with permission from ref. 34. Copyright 2010, American Chemical Society.

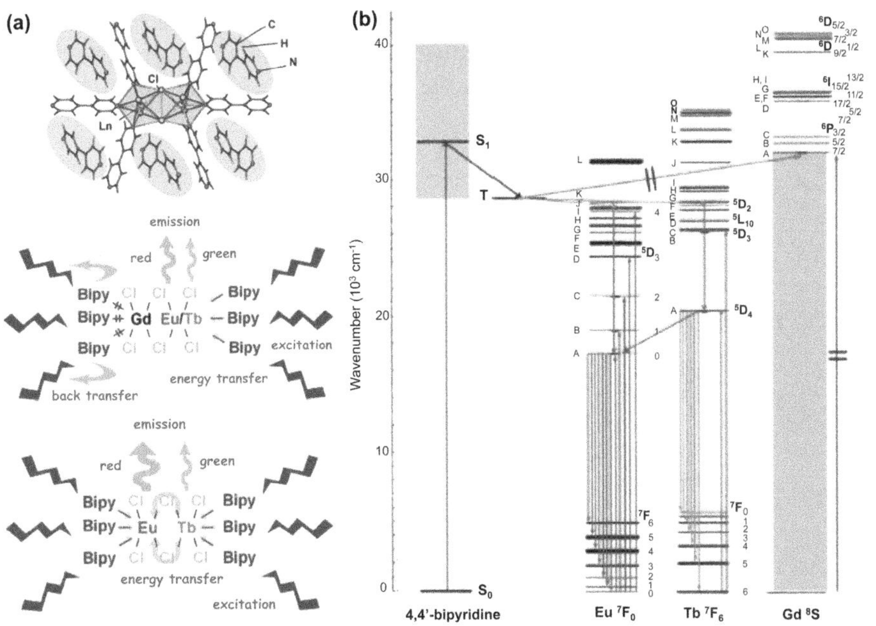

Figure 11.3 (a) Schematic depiction of the energy transfer processes between metal ions and ligands in comparison with the crystal structure and (b) excitation and emission processes between metal ions and ligands in $^{2}_{\infty}[Gd_{2-x-y}Eu_xTb_yCl_6(bipy)_3]\cdot 2bipy$. Reprinted with permission from ref. 39.

Photoluminescent studies revealed that, in addition to energy transfer from 4,4′-bipyridine to lanthanide ions, an energy transfer from the 4f-levels of Tb^{3+} to Eu^{3+} is responsible for the emission colour tuning.

The research group of Nenoff has introduced a family of luminescent MOFs composed of In, In–Eu, or Eu metal centres and trigonal bridging ligands **4–6** (Figure 11.4).[40,41] In these MOFs, the efficiency of energy transfer mediated luminescence from metal centres can be altered by varying the metal (Eu for In) doping levels and also the electronic properties of the ligands (electron rich or deficient). Systematic photophysical studies revealed that both the ligand and coordination environment of the metal centre are important in determining the quantum yield (QY) values of MOFs (Figure 11.4b). A combination of In, the ligand **6** and oxalic acid produced white-light emitting MOFs with very low QY value. However, doping of Eu enhanced the QY values of these MOFs. The purely red-emitting MOF made from Eu and **6** showed a 50% QY at room temperature and was marginally reduced (QY = 48%) at 150 °C due to reversible thermal quenching. This value is the highest reported for this class of materials at elevated temperatures.

A highly efficient resonance energy transfer from trigonal bridging ligand **7** to metal centres (Eu^{III} and/or Tb^{III}) was observed in microporous MOFs with a composition of $\{[Ln7(H_2O)_2]\cdot 3DMF\cdot H_2O\}_n$.[42] The efficiency of energy transfer from the ligand to Ln^{III} ions and the contribution from Eu^{III} and Tb^{III} in the emission spectra are found to be sensitive to induced-fit-type host–guest interactions. Thus, MOFs with a distinct pore size have the ability to discriminate volatile organic molecules by transducing the host–guest interactions into the clearly distinguishable and unique emission intensity ratio between Eu^{III} and Tb^{III}.

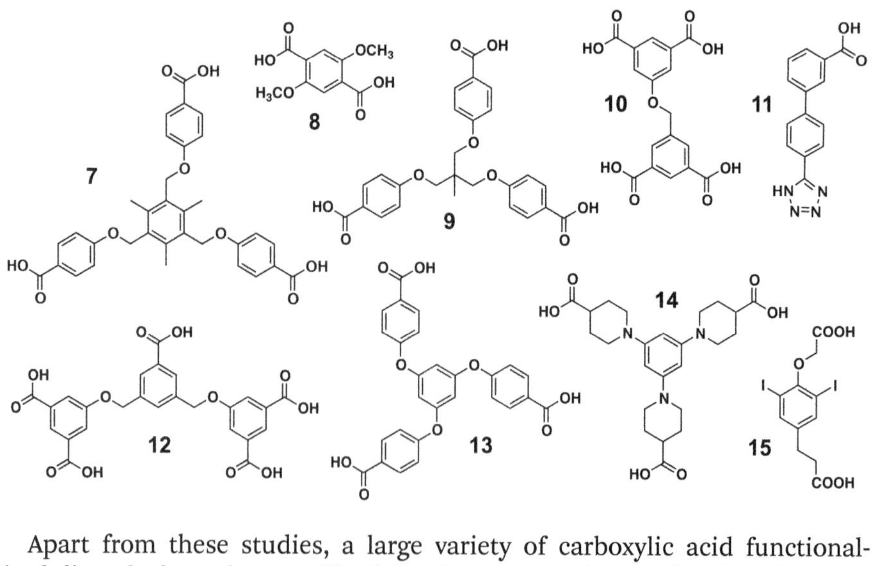

Apart from these studies, a large variety of carboxylic acid functionalized ligands have been utilized in the preparation of lanthanide-based

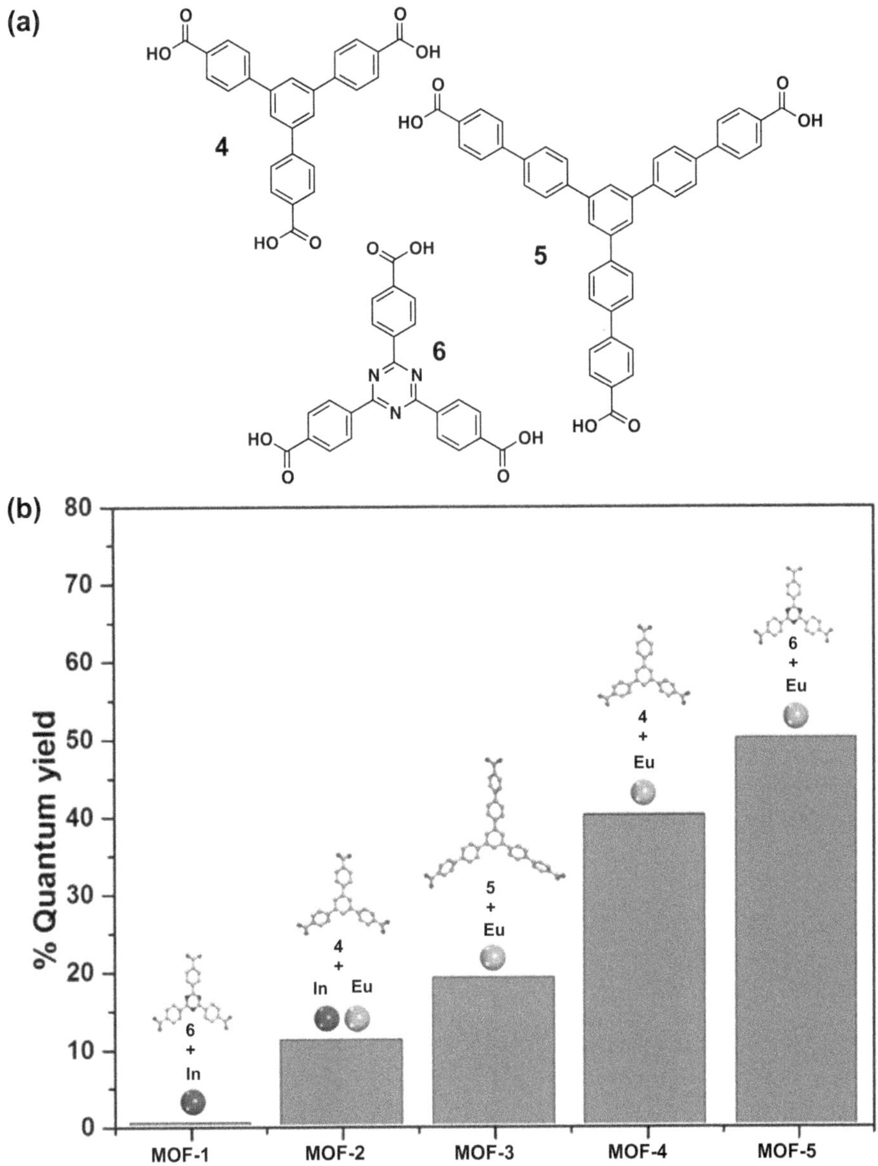

Figure 11.4 (a) Molecular structures of trigonal bridging ligands **4–6**. (b) A plot showing the percentage of quantum yield in MOF1-5. Reprinted with permission from ref. 41. Copyright 2014, American Chemical Society.

metal–organic frameworks showing ligand-sensitized (antenna effect) colour tunable luminescent properties, including white-light emission.[43–50] A luminescent thermometer based on mixed-lanthanide MOF $\{(Eu_xTb_{1-x})_2(\mathbf{8})_3(H_2O)_4{\cdot}DMF{\cdot}H_2O\}_n$ was reported by Qian, Chen, and co-workers.[43] MOFs with the following compositions: (i) $Ln_xGd_{1-x}\mathbf{9}$ in which Ln = Eu, Tb, Sm, and

Dy, $0 \leq x \leq 1$,[44] (ii) $\{[La_2(\mathbf{10})_2]\cdot(H_2O)_3\cdot(Me_2NH_2)_2\}_n$,[45] (iii) Zn^{II}–MOF of **11**,[46] (iv) $\{[Zn_2(\mathbf{12})\cdot H_2O]\cdot 3H_2O\cdot 3DMAc\cdot Me_2NH_2\}_n$,[47] (v) $\{[Zn_3(\mathbf{13})_2(H_2O)_2]\cdot 2H_2O\cdot 4DMF\}_n$,[48] and (vi) $\{[Ln(\mathbf{14})H_2O]\cdot 2DMF\cdot 3H_2O\}_n$[49] co-doped with Eu^{III} and Tb^{III} in different ratios emit tunable RGB colours towards white-light. The heavy atom substituted ligand **15** showed phosphorescence due to enhanced electronic population in the triplet excited state through intersystem crossing. The MOF made of **15** and Eu^{III} showed triplet ligand-to-metal energy transfer mediated characteristic lanthanide emission.[50]

The research groups of Farha, Wiederrecht, and Hupp have extensively investigated the light-harvesting and energy-transfer properties of porphyrin-based MOFs.[51–54] As an example, they have designed and synthesized a pillared-paddlewheel type MOF featuring a green emissive pyridine-functionalised boron dipyrromethene (BODIPY) **16** and a carboxylic acid functionalised Zn-porphyrin **17** showing red emission as struts and Zn^{II} with a d^{10} electronic configuration as metal nodes.[51] Investigation of the light-harvesting properties of the MOFs using confocal laser scanning microscopy revealed efficient transfer of excitation energy from the BODIPY-based ligands (pillar) to Zn-porphyrin-ligands (paddlewheel) (Figure 11.5a). The highlights of this study are the occurrence of energy transfer between the geometrically orthogonal struts owing to the slight deviation of their root-mean-square value of the angle from 90° and the non-participation of metal nodes in ligand-to-metal energy transfer, which eventually facilitated efficient strut-to-strut energy transfer processes upon excitation.

Anisotropic energy migration is reported in MOFs composed of pyridine-functionalised zincated-porphyrin **18** as pillars and **19** as a paddlewheel.[52]

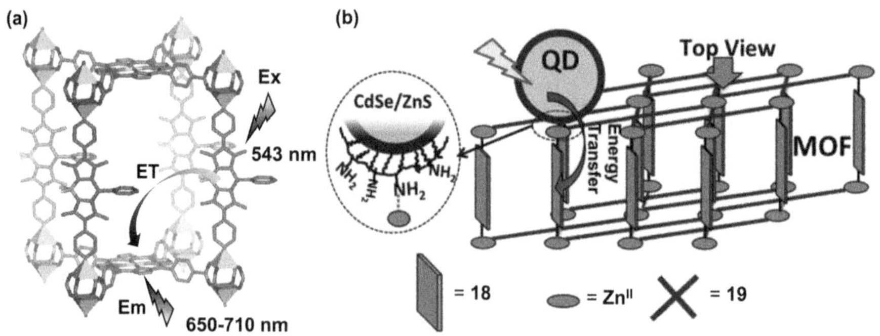

Figure 11.5 (a) Energy transfer (ET) process from ligand **16** to **17** in a pillared-paddlewheel-type MOF. Reprinted with permission from ref. 51. Copyright 2011, American Chemical Society. (b) Schematic representation showing energy transfer from CdSe/ZnS QDs functionalised on the MOF surface through the amine–Zn interaction to ligand **18**. The QD–MOF hybrid is able to harvest photons through energy transfer from the QDs to the MOF even in spectral regions where the MOF has little absorption. Reprinted with permission from ref. 53. Copyright 2013, American Chemical Society.

Fluorescence quenching studies and theoretical calculations revealed that exciton migration up to forty-five zincated-porphyrin struts occurred within the lifetime of the chromophore. Functionalization of the exterior surface of the Zn-porphyrin MOFs with size-tunable CdSe/ZnS core–shell quantum dots (QDs) was found to substantially improve the light-harvesting property of the material by facilitating an efficient energy transfer process from the QDs to the struts (>80%) (Figure 11.5b).[53] By taking into account the broad absorption feature of QDs in the visible region, sensitization by a monolayer of QDs can enhance the light-harvesting capacity of a single layer of MOF >50%.

The layer-by-layer (LbL) assembly of ligands **18** and **19** on an amine-functionalised silicon surface allowed the preparation of oriented porous thin films of pillared-paddlewheel-type MOFs.[54] The directional energy migration process in the MOF scaffold was demonstrated by functionalising an energy-accepting squaraine dye **20** on the periphery of the MOF film. Selective excitation of porphyrin units resulted in efficient energy hopping within the MOFs and irreversible trapping of excitation energy by the acceptor, followed by its far-red luminescence. This entire process was found to occur up to a 50-cycle LbL film of MOF.

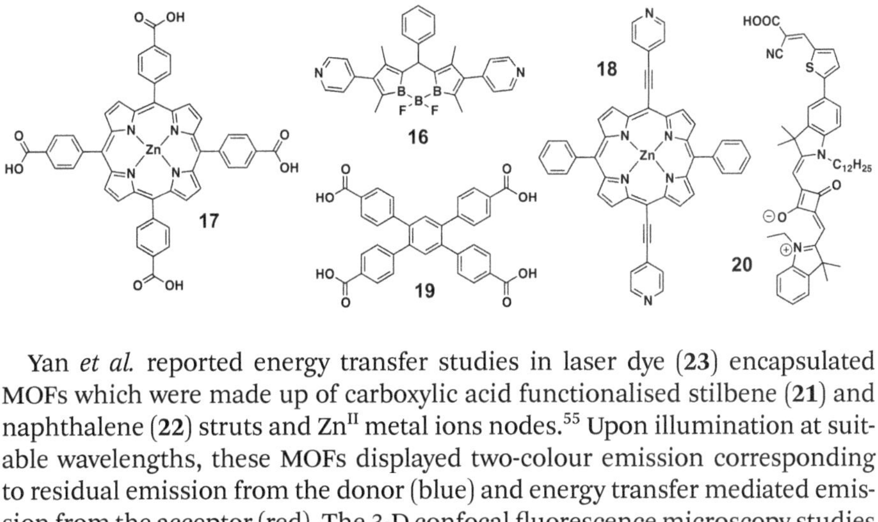

Yan *et al.* reported energy transfer studies in laser dye (**23**) encapsulated MOFs which were made up of carboxylic acid functionalised stilbene (**21**) and naphthalene (**22**) struts and Zn^{II} metal ions nodes.[55] Upon illumination at suitable wavelengths, these MOFs displayed two-colour emission corresponding to residual emission from the donor (blue) and energy transfer mediated emission from the acceptor (red). The 3-D confocal fluorescence microscopy studies revealed a ratiometric fluorescence change in different crystal planes within the MOF due to a variation in efficiency of energy transfer from the host to guest.

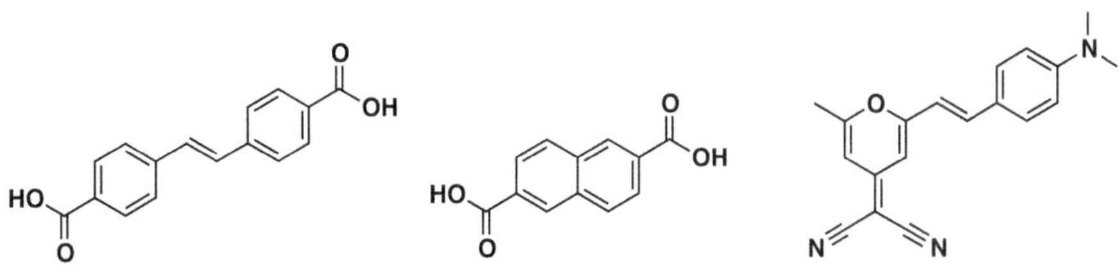

21 Energy donors **22** **23**: Energy acceptor

11.3 Energy Transfer in Self-Assembled Coordination Polymers

Nanoscale coordination polymers (NCPs) and their self-assembly have been developed as a scaling down technique of macroscopic crystalline MOFs to the nanoscale (nano-MOFs), thereby increasing the solvent dispersibility and processability to a certain extent.[56–58] Nanoscale coordination polymers or nano-MOFs find applications in various fields of chemistry and biology.[56–58] The first report on the use of self-assembled nanostructures of NCPs for light harvesting came from the research group of Loh.[59] For this purpose, they designed and synthesized an oligo(fluorene) molecule (**24**) functionalized with flexible alkyl chains and terminal carboxylic groups (Figure 11.6a). Upon addition of $Zn(OAc)_2$, the bridging ligands formed 1-D nanowires that show excellent stability in solvents (Figure 11.6b). The nanowires, upon encapsulation with red-emitting acceptor molecules **25**, showed efficient energy transfer from the blue-emitting ligand to the acceptor molecules. Energy transfer is facilitated by electrostatic interactions between negatively-charged nanowires owing to the presence of a carboxylate environment near the connecting Zn^{II} ions and the positively-charged acceptor molecules (Figure 11.6).

Uvdal and co-workers utilized a similar approach for the design of light-harvesting NCPs based on **26** and different lanthanide ions (Gd^{III}, Eu^{III} and Yb^{III}).[60] In this case, the self-assembly of NCPs resulted in the formation of solvent dispersible crystalline nanodisks with excellent photophysical properties such as efficient energy transfer from the ligand to the encapsulated acceptor molecules **26**. Temperature and method of encapsulation were found to influence the efficiency of the energy transfer process. The acceptor pre-loaded nanodisks showed better energy transfer properties when compared to those of the acceptor post-loaded nanodisks.

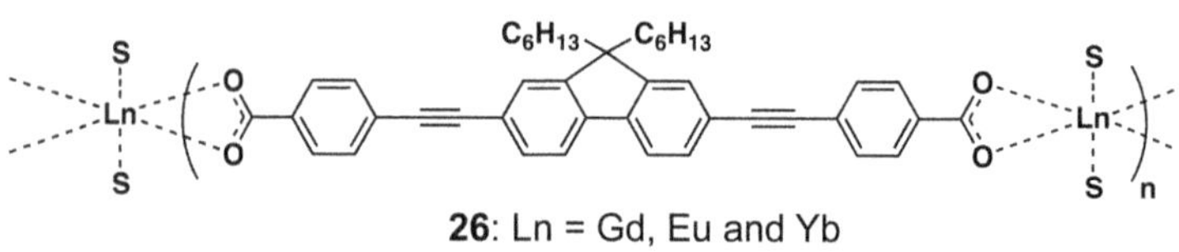

26: Ln = Gd, Eu and Yb

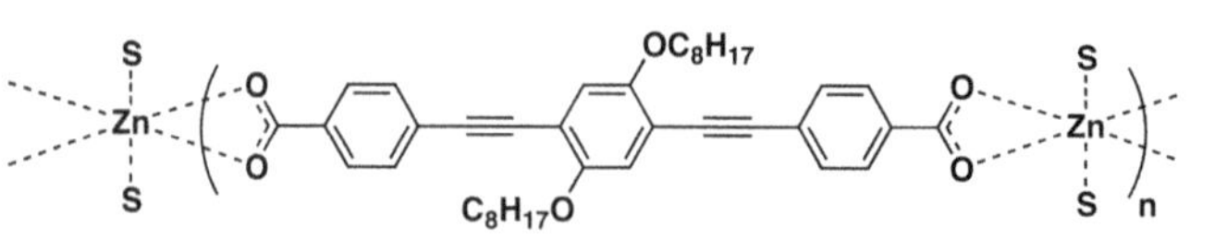

27

S = Solvent, Ligand or CH_3COO^-

Similarly, the self-assembly of amphiphilic NCPs (**27**) composed of carboxylic acid functionalised oligo(*p*-phenyleneethynylene) ligands and Zn^{II} ions to

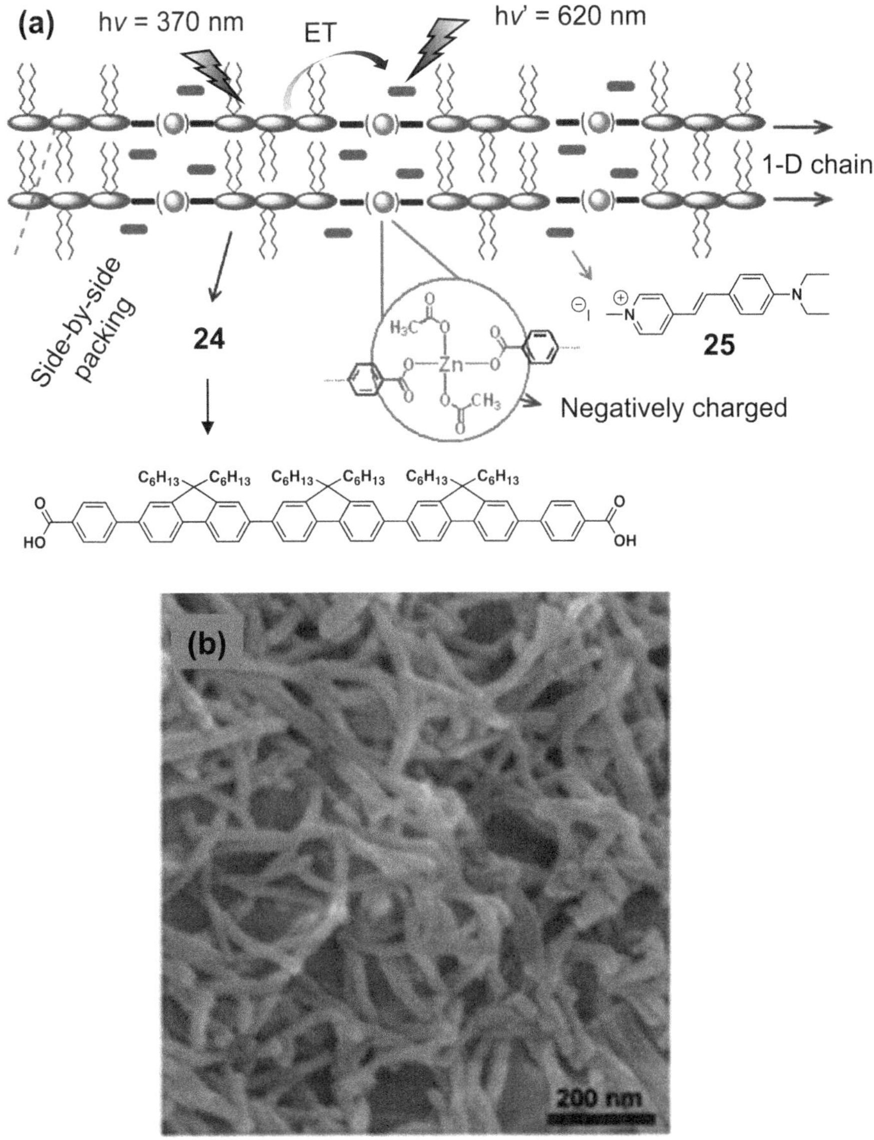

Figure 11.6 (a) A schematic representation showing energy transfer (ET) between **24** and **25** and in 1-D NCPs. (b) Scanning electron microscopy (SEM) image of nanowires formed by 1-D NCPs. Reprinted with permission from ref. 59. Copyright 2009, American Chemical Society.

soft vesicular and toroid nanostructures by controlling the reaction parameters has been reported.[61] The permanent porosity and dynamic nature of these nanostructures were utilized to encapsulate a suitable cationic dye (**25**) and to efficiently funnel excitation energy from the ligand to the acceptor dye. Strut-to-strut energy transfer in nano-MOFs made up of energy transfer donor (**24**) and acceptor (**28**) linkers of similar length (Figure 11.7a) has also been reported.[62] The disk-like nanostructure prepared from Gd^{III}, the blue-emitting **24** (98%), and a small percentage of the yellow-emitting **28** (2%) showed energy transfer from **24** → **28** upon selective excitation of the donor at 370 nm (Figure 11.7). Interestingly, the partial energy transfer from **24** to **28** in nano-MOFs produced a broad emission covering the entire visible region to give a white-light emission. The absence of interparticle energy transfer between homo ligand based nano-MOFs highlighted the importance of this multicomponent approach.

Kimizuka and co-workers reported the preparation and properties of water-dispersible amorphous nanoparticles from coordination polymer networks of nucleotides and lanthanides.[63] The nucleotides are selected as bridging ligands because of the chelating ability of the constituting nucleobases and phosphates in a bidentate manner. The size, properties, and functions

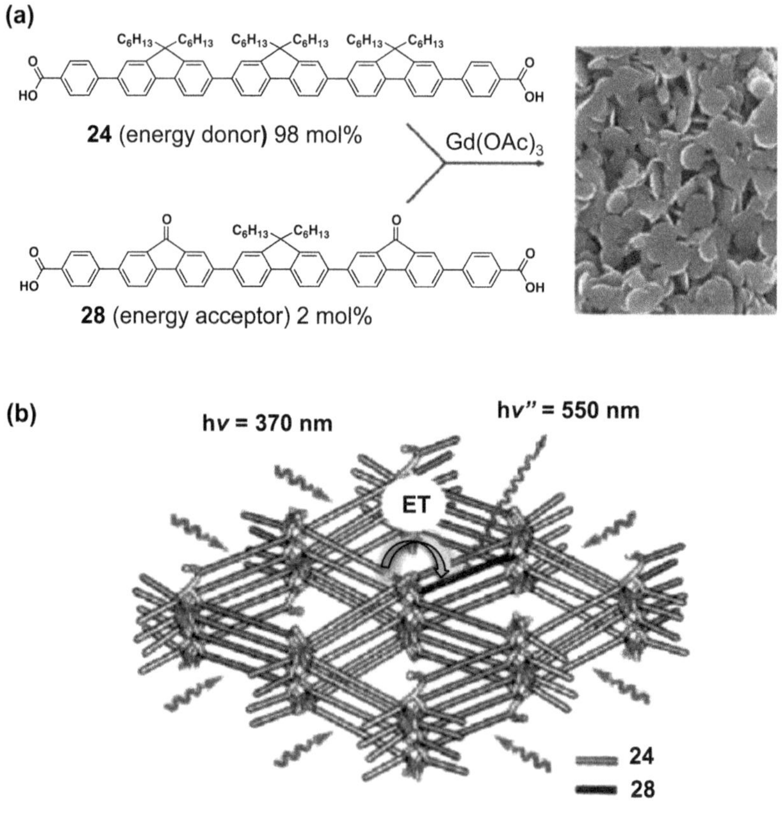

Figure 11.7 (a) Illustration showing the preparation of Gd–**24**–**28** multicomponent nanodisks. (b) Schematic representation of the energy transfer in self-assembled coordination polymers (nano-MOFs). Reprinted with permission from ref. 62. Copyright 2011, Wiley-VCH Verlag GmbH & Co. KGaA.

of coordination polymer nanoparticles (CPNs) were found to depend on the choice of nucleotides and lanthanide ions. For instance, the combination of guanosine 5′-monophosphate, **29** (Figure 11.8a) and Tb^{III} ions produced green luminescent CPNs. This luminescence property is attributed to energy transfer mediated sensitization of the emissive $^5D_4$ state of the Tb^{III} ion by the guanine base, facilitated by the coordination of two adjacent electron rich atoms (O and N) of the guanine to Tb^{III} ions. Furthermore, the observed luminescence from CPNs in water indicated that the interior of the nanoparticles is hydrophobic and thus reduces the possibility of the lanthanide

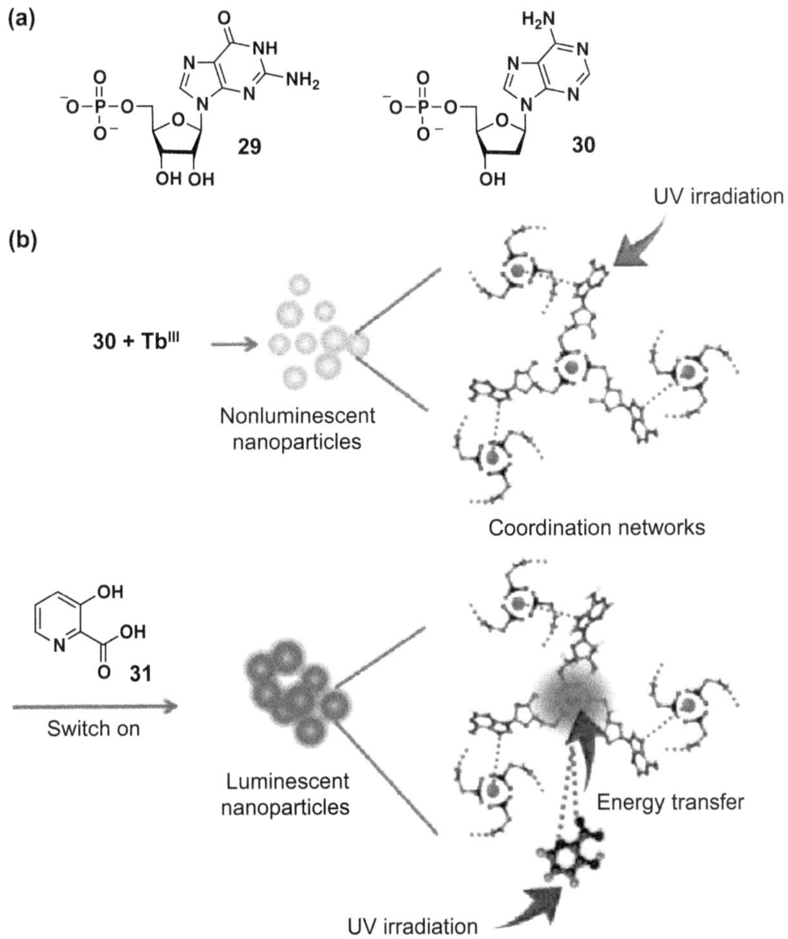

Figure 11.8 (a) Molecular structures of nucleotides (**29** and **30**) used to prepare CPNs in combination with lanthanide ions. (b) Schematic representation of nanoparticles formed in the absence (top) and presence (bottom) of ancillary ligand **31** from the coordination networks of **30** and Tb^{III} ions. Reprinted with permission from ref. 64. Copyright 2010, Wiley-VCH Verlag GmbH & Co. KGaA.

emission quenching *via* deactivation of excited states through coordination by water molecules.

While the nucleotide **29** and Tb^{III} ions produced luminescent nanoparticles, the combination of 2′-deoxyadenosine 5′-monophosphate, **30** (Figure 11.8a), and Tb^{III} ions resulted in nonluminescent nanoparticles because of the weak coordination ability of the adenine base of **30** and the competition of water molecules for the coordination sphere of Tb^{III} ions (Figure 11.8b).[63,64] However, by taking advantage of the adaptive nature of **30** and Tb^{III} CPNs, this issue has been resolved by incorporating a sensitizer hydroxypicolinic acid (**31**) as an ancillary ligand (Figure 11.8b).[64] The coordination polymeric nanoparticle of the ternary complex (**30**, **31** and Tb^{III}) showed enhanced luminescence due to the possible replacement of water molecules by this ancillary ligand. Consequently, changes in the local coordination environment of the Tb^{III} ions facilitate sensitized lanthanide emission. In general, the method of doping with an ancillary ligand can be useful to turn on the latent luminescent properties of coordination nanoparticles.

Ren, Qu, and co-workers have extended a similar concept to develop adaptive CPNs composed of **29** and Eu^{III} ions showing sensitized lanthanide emission that function as fluorescence-based switching logic gates.[65,66] Interestingly, encapsulation of the hydrophilic dye thioflavin T (**32**) and the thiazole orange (**33**) into **29**/Eu^{III} CPNs considerably enhanced the luminescent properties due to coordination network induced restriction of the conformational rotation of the dyes (Figure 11.9). In addition, a fluorescence

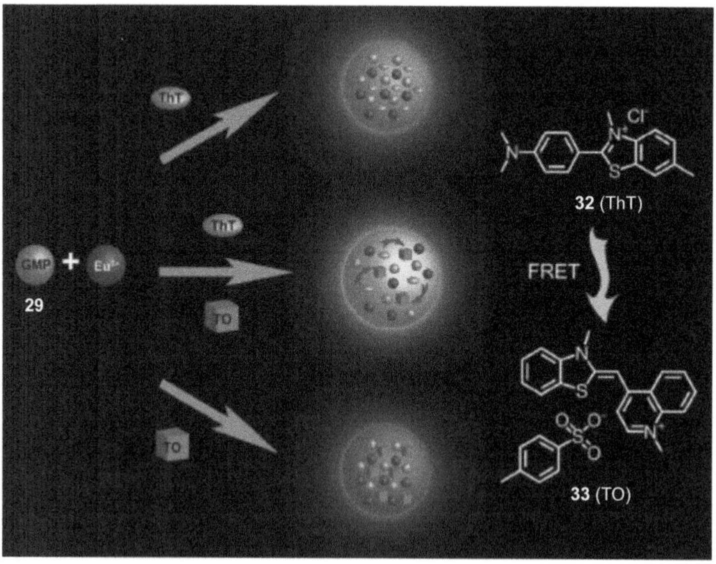

Figure 11.9 Construction of an artificial light-harvesting system by encapsulating two dyes into the CPNs through self-assembly. Reprinted with permission from ref. 66. Copyright 2014, Wiley-VCH Verlag GmbH & Co. KGaA.

resonance energy transfer mediated emission increase of the acceptor dye **33** and the corresponding quenching of emission from the donor dye **32** were observed upon simultaneously encapsulating the donor/acceptor dyes into **29**/Eu^{III} CPNs (Figure 11.9).

Droplet templated synthesis of discrete one-dimensional (1-D) nanostructures using an amphiphilic Zn–chlorophyll molecule (**34**) bearing a dendritic tetra(ethylene glycol) unit and an isonicotinic acid moiety at the 17th and 3rd positions of the chlorophyll ring, respectively, was reported by Numata and co-workers (Figure 11.10).[67] The amphiphilic molecules tend to organize perpendicularly at the interface between the 1,2-dichloroethane (DCE)–water emulsion guided by the 1-D Zn–pyridine coordination as well as the solvation of the hydrophilic ethylene glycol chains of the molecules by water. Interestingly, this self-assembly process displayed a concentration-dependent morphological transition from spheres to multilayered tubes. The hydrophobic nature of the interior of the tubular assembly has been effectively exploited for the extraction of a cyanine dye (**35**) from a nonpolar solvent (hexane). Photoluminescence studies revealed that excitation of the entrapped green fluorescent cyanine dyes within the interior space of the tubular assembly efficiently transfers excitation energy to red-emitting chlorophyll molecules located in the walls of the tubular assembly (Figure 11.10).

The mixed lanthanide (Ln = Gd^{III}, Tb^{III} and Eu^{III} ions) coordination polymers of 4-(dipyridin-2-yl)aminobenzoate (**36**) have been reported to form a nano-rod like morphology and emit white light (Figure 11.11).[68] The white-light emission is due to the ligand-sensitized red and green emission, respectively, from Eu^{III} and Tb^{III} complexes and the ligand-centred blue emission from the Gd^{III} complex. The emission properties of the mixed lanthanide coordination polymers can be tuned by varying the concentration of lanthanides and excitation wavelengths.

A multi-chromophoric system based on a tetra(thio)pyrene core functionalized with phenyl terpyridine units (**37**) formed nanoparticles upon complexation with Fe^{II}, Zn^{II}, and Nd^{III} salts owing to the formation of 'super' oligomeric coordination structures (Figure 11.12).[69] Interestingly, the nanoparticles prepared from **37** and Nd^{III} NCPs displayed very good emission in the infrared region due to a near complete transfer of the excitation energy from the multi-chromophoric units to the Nd^{III} ions.

Sun and co-workers reported the synthesis and two-photon absorbing properties of a 3-D metal–ligand complex (MLC) based on a tetraphenylmethane-based tetrahedral ligand (**38**) and Zn^{II} ions (Figure 11.13).[70,71] The reversible nature of the 3-D complex allowed the preparation of an extended 3-D coordination network by complexing with additional amounts of Zn^{II} ions and a linear bis-terpyridyl linker unit (**39**). Detailed steady state and time resolved studies revealed that the two-photon absorbing tetrahedral ligand efficiently transfers excitation energy to the linear bridging ligand. Thus, the extended 3-D coordination network showed enhanced two-photon activity and improved photostability.

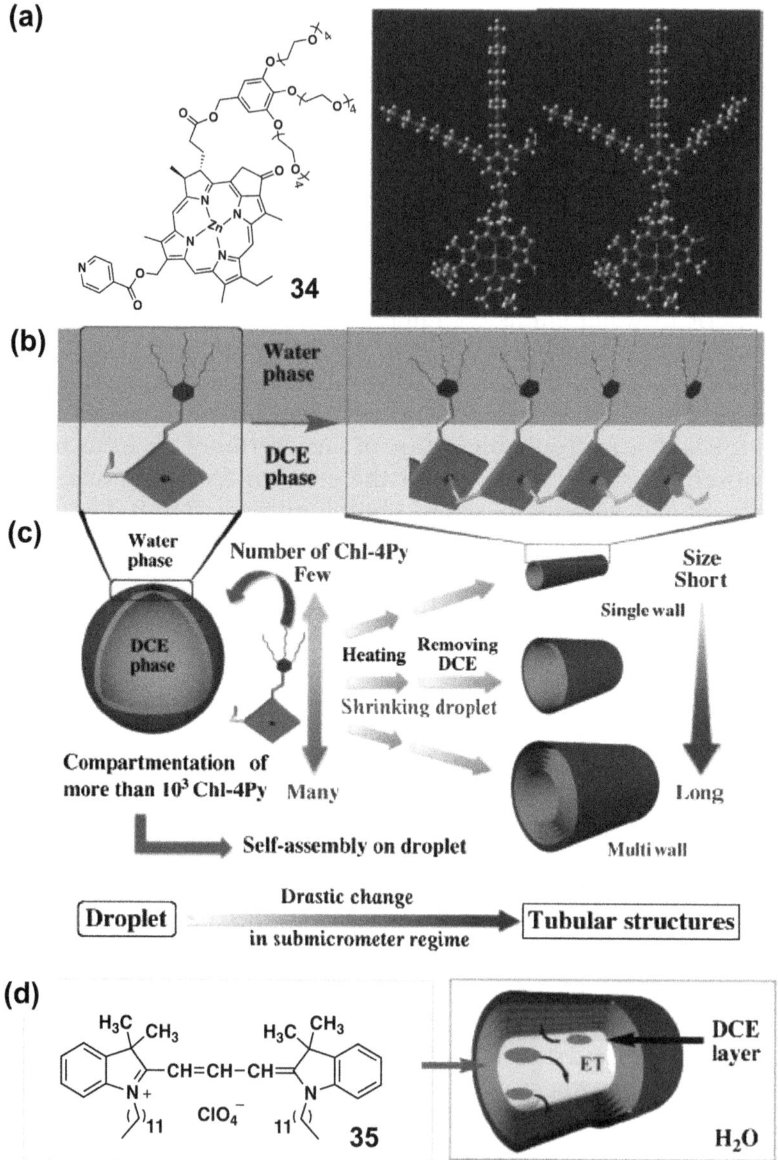

Figure 11.10 (a) Structure of the amphiphilic Zn–chlorophyll molecule **34**. (b) 1-D self-assembly of **34** at the DCE–water interface. (c) **34** compartmentalized on a DCE droplet and establishing its inherent 1-D assembling ability at the DCE–water interface through a dramatic change in the shape of the DCE droplet from spherical to tubular upon evaporating the DCE. (d) Schematic representation of energy transfer (ET) from entrapped **35** (donor) within the tubular assembly of **34** (acceptor). Reprinted with permission from ref. 67. Copyright 2012, Wiley-VCH Verlag GmbH & Co. KGaA.

Figure 11.11 (a) Molecular structure of ligand **36**. (b) Scanning electron microscopy and (c) transmission electron microscopy images of the nanorod formed by the mixed coordination polymers. Reprinted with permission from ref. 68.

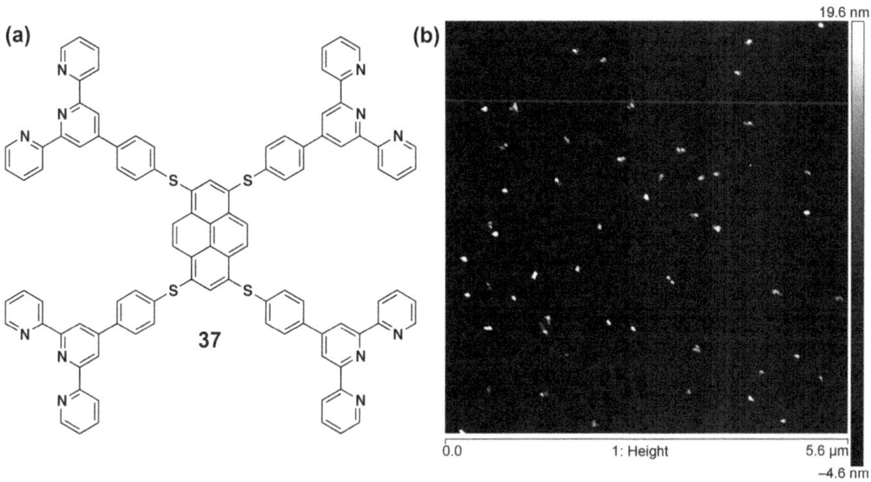

Figure 11.12 (a) Molecular structure of ligand **37**. (b) Atomic force microscopy peak force mode image from a solution of **37**·Zn_2 spin-coated on mica. Reprinted with permission from ref. 69. Copyright 2014, Wiley-VCH Verlag GmbH & Co. KGaA.

11.4 Metallo-Gels as Light-Harvesting Scaffolds

The spontaneous self-assembly of certain molecules upon changes in temperature is known to facilitate the entrapment of solvent molecules due to the formation of three-dimensional (3-D) nanostructures that may result in the gelation of solvents.[72–80] The gel phase, which combines the properties of both solution and solid phase, is considered as a medium to organize chromophores and favour efficient energy and charge migration processes.[76–80] The incorporation of suitable donor–acceptor molecules to such gel forming systems provides nanoarchitectures that undergo energy transfer processes.[76,81–86] Over the years, supramolecular gelation has evolved as a facile strategy for engineering light-harvesting assemblies with tunable properties.[76,81–86]

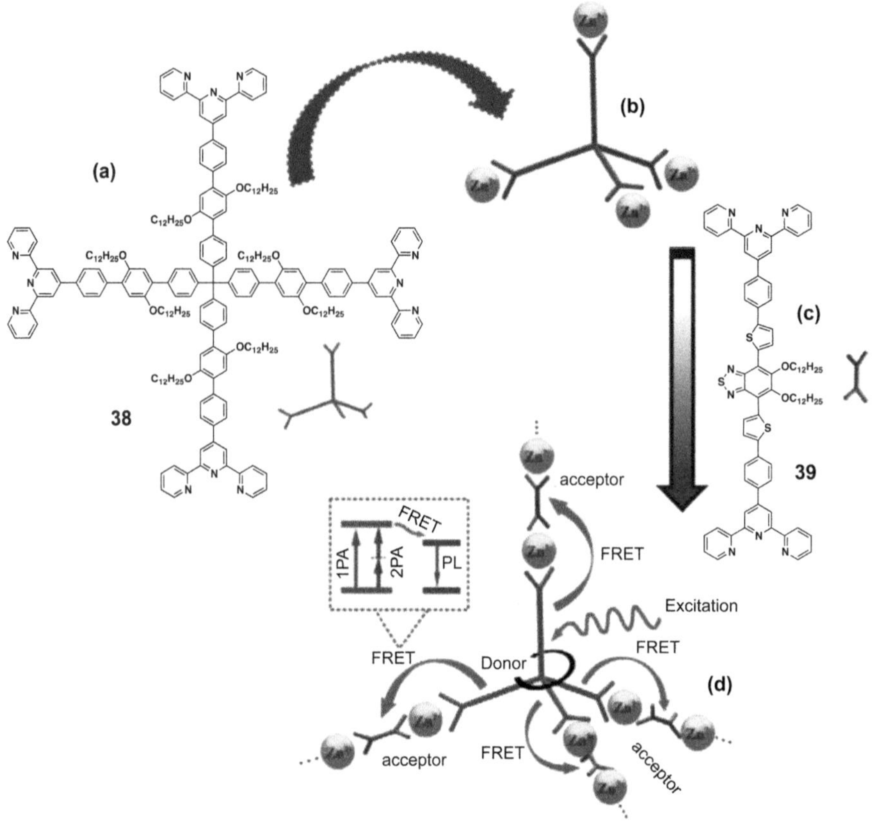

Figure 11.13 Schematic diagram of the synthesis of an extended 3-D metal ligand complex (MLC). (a) Tetrahedral molecule (**38**) used for the construction of donor. (b) Tetraphenylmethane-based 3-D MLC used as two-photon absorbing donor. (c) Uncoordinated monomer (**39**) used for the construction of the extended 3-D MLC. (d) The photoinduced energy transfer process in the extended 3-D MLC under one- and two-photon excitation. Reprinted with permission from ref. 71. Copyright 2013, Wiley-VCH Verlag GmbH & Co. KGaA.

The multiple binding modes of a tridentate ligand 2,6-bis(1′-methylbenzimidazolyl)-4-hydroxy-pyridine (BIP) with transition (2 : 1 complexation) and lanthanide (3 : 1 complexation) metal ions was utilized by Rowan and co-workers for the design of a stimuli-responsive supramolecular gel.[87–89] The compound **40** (Figure 11.14), which consists of a BIP moiety attached to either end of a pentaethylene glycol unit, formed a gel upon the addition of Eu(III)$(NO_3)_3$ (3 mol%), followed by Zn(II)$(ClO_4)_2$ (97 mol%) to a solution of **1** in $CHCl_3$–CH_3CN. Upon excitation at 340 nm, the **40**:Zn/Eu supramolecular gel system showed intense emission from Eu^{III} ions owing to efficient ligand-to-metal energy transfer processes. Interestingly, the addition of a small amount of formic acid to the luminescent gel of **40**:Zn/Eu resulted in the

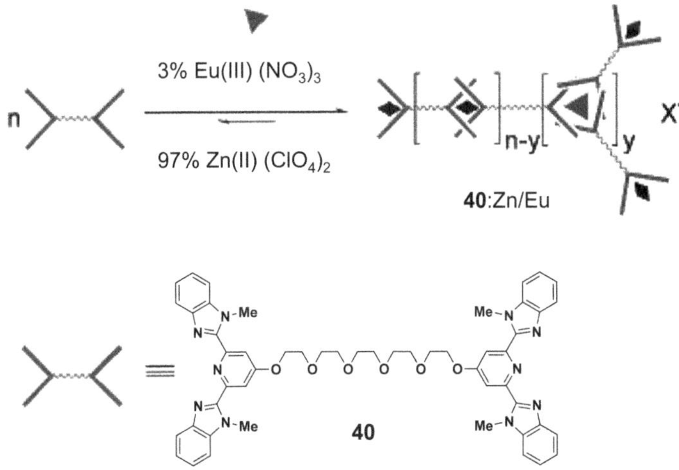

Figure 11.14 Schematic representation of the formation of a metallo-supramolecular gel using a combination of lanthanoid and transition metal ions mixed with the monomer **40**. Reprinted with permission from ref. 87. Copyright 2003, American Chemical Society.

disruption of the gel with strong quenching of the emission due to the displacement of BIP ligands by formate ions that switch off the antenna effect. The Eu^{III} emission was reverted upon removal of the solvent from the gel followed by re-swelling with CH_3CN.

Schanze and co-workers reported the self-assembling platinum acetylide oligomers **41** and **42** (Figure 11.15) forming aggregates and gels in hydrocarbon solvents.[90] These compounds displayed near-UV absorption and exhibited room temperature phosphorescence from a long-lived triplet excited state. They are good candidates to study triplet–triplet exciton migration and energy transfer processes in the aggregate/gel states. Compound **41** exhibited moderately intense phosphorescence from the platinum acetylide chromophore (λ_{max} = 500 nm) at ambient temperature in solution and in the aggregate/gel states. The mixed aggregates of **41** (energy donor) and small amounts of **42** (energy acceptor) exhibited phosphorescence emission predominantly from the latter ($\lambda_{max} \approx$ 580 nm) *via* triplet exciton diffusion among the donor aggregates followed by transfer of excitation energy to the low energy acceptor/trap. This process is observed only when the donor and the acceptor molecules are in the aggregate/gel state, which emphasizes the importance of the exciton diffusion within the donor scaffold in transporting the exciton to an acceptor/trap.[76,84–86]

The amphiphilic molecules **43** and **44** are good gelators of organic solvents such as methanol and ethanol. Doping of the gel either with Tb^{III} or Eu^{III} complexes, followed by excitation of the gelators resulted in green and red photoluminescence, respectively, due to excitation energy transfer from the gelator to the lanthanide complexes.[91,92] Fourier transform infrared spectroscopy studies revealed that the doped lanthanide complexes

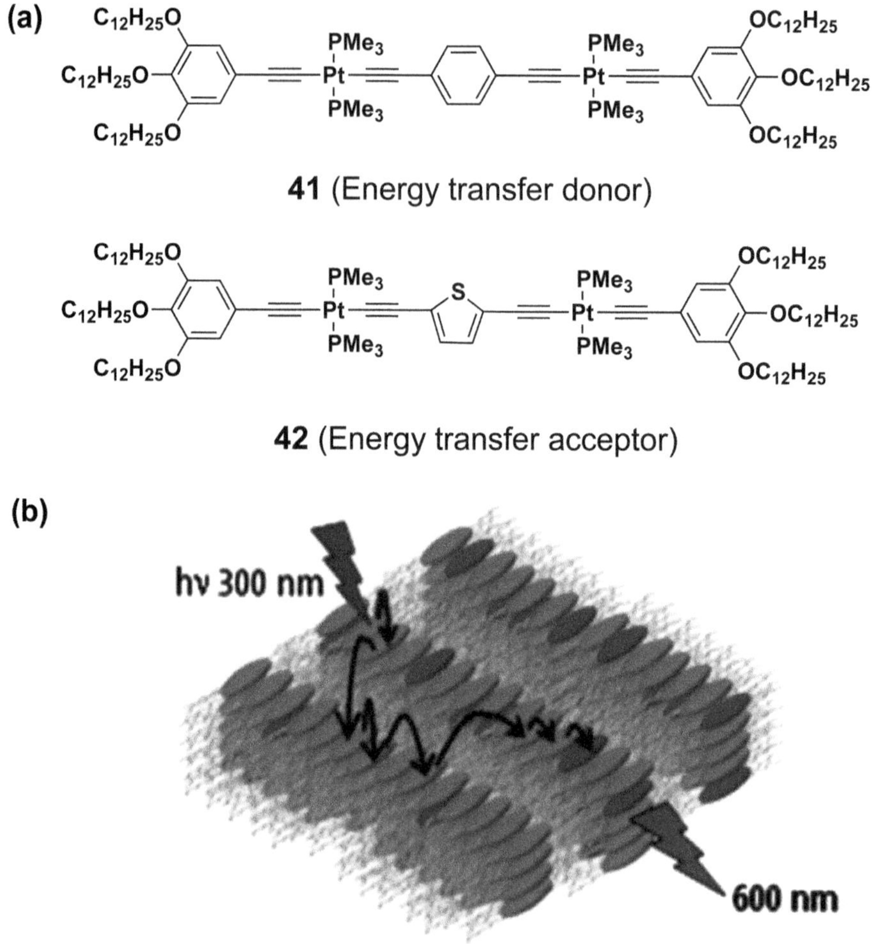

Figure 11.15 (a) Molecular structure of platinum acetylide oligomers **41** and **42**. (b) Schematic representation of the exciton diffusion assisted energy transfer within the mixed aggregates of **41** and **42**. Reprinted with permission from ref. 90. Copyright 2008, American Chemical Society.

coordinated to amide groups of the gelator, which in turn ensured close proximity of donor and acceptor moieties, and thereby efficient energy transfer processes.

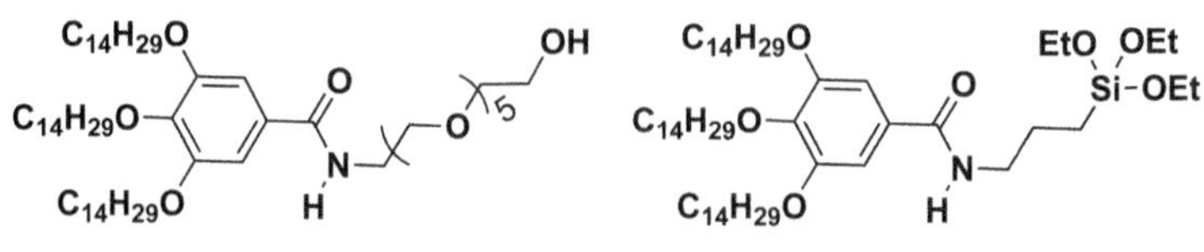

43 44

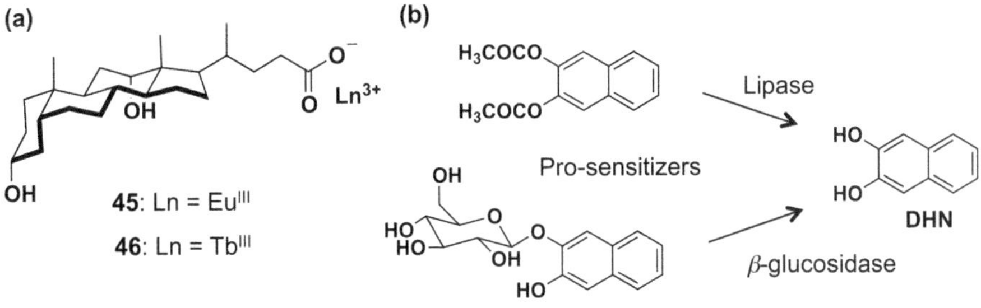

Figure 11.16 (a) Molecular structure of the lanthanide-cholate gelators **45** and **46**. (b) Enzyme activated generation of the sensitizer 2,3-dihydroxynaphthalene (DHN). Reprinted with permission from ref. 95.

Maitra and co-workers have demonstrated that sensitization of lanthanide photoluminescence can be achieved when a lanthanide-derived gel is doped with a suitable sensitizer.[93–97] Upon doping with pyrene, the gel derived from Eu-cholate **45** (Figure 11.16a) showed a 10-fold increase in Eu^{III} emission due to the antenna effect facilitated by the close vicinity of Eu^{III} and pyrene trapped in the hydrophobic pocket of the gel matrix.[93] This observation was unique for the gel matrix because no such emission enhancement was observed when the same experiment was conducted in a sodium dodecyl sulfate micelle. The antenna effect is considered to operate by the excitation of the donor to the first singlet state (S_1), intersystem crossing to the first triplet state (T_1), followed by triplet energy transfer to the lanthanide 4f electron, resulting in the rise of the lanthanide luminescence.

The energy transfer from Tb^{III} to Eu^{III} in mixed hydrogels has been exploited for the design of multicolour light-harvesting gels.[94] Different molar ratios of Tb^{III} and Eu^{III} ions were used to form cholate hydrogels and doped with 2,3-dihydroxynaphthalene (DHN) as a sensitizer. Tb^{III} emission was sensitized by DHN, followed by excitation energy transfer from Tb^{III} to Eu^{III}, which tuned the emission from green to yellow and reddish pink, depending on the molar ratios of lanthanides. Energy transfer mediated emission colour tuning was also observed in xerogels and the bulk state. In a different study, sensitization of Tb^{III} by DHN in the gel matrix was utilized to develop a 'pro-sensitizer'-based assay of enzymes such as lipase or β-glucosidase (Figure 11.16b).[95] Emission quenching of a hybrid metallogel from Tb^{III} and deoxycholate doped with DHN by 2,4,7-trinitrofluorenone as an electron-deficient aromatic compound has been reported,[96] which is an essential requirement to develop optoelectronic materials and sensors.

Ultrasound is reported to induce hydrogelation of a terpyridine derivative bearing a carboxylic group, **47** (Figure 11.17a), in its bis-protonated form **47**·2HBr.[98] Interestingly, hydrogelation of **47**·2HBr was observed with one equivalent of $EuCl_3·6H_2O$. Excitation of this gel at 350 nm showed complete quenching of the emission from the aggregated terpyridine chromophore

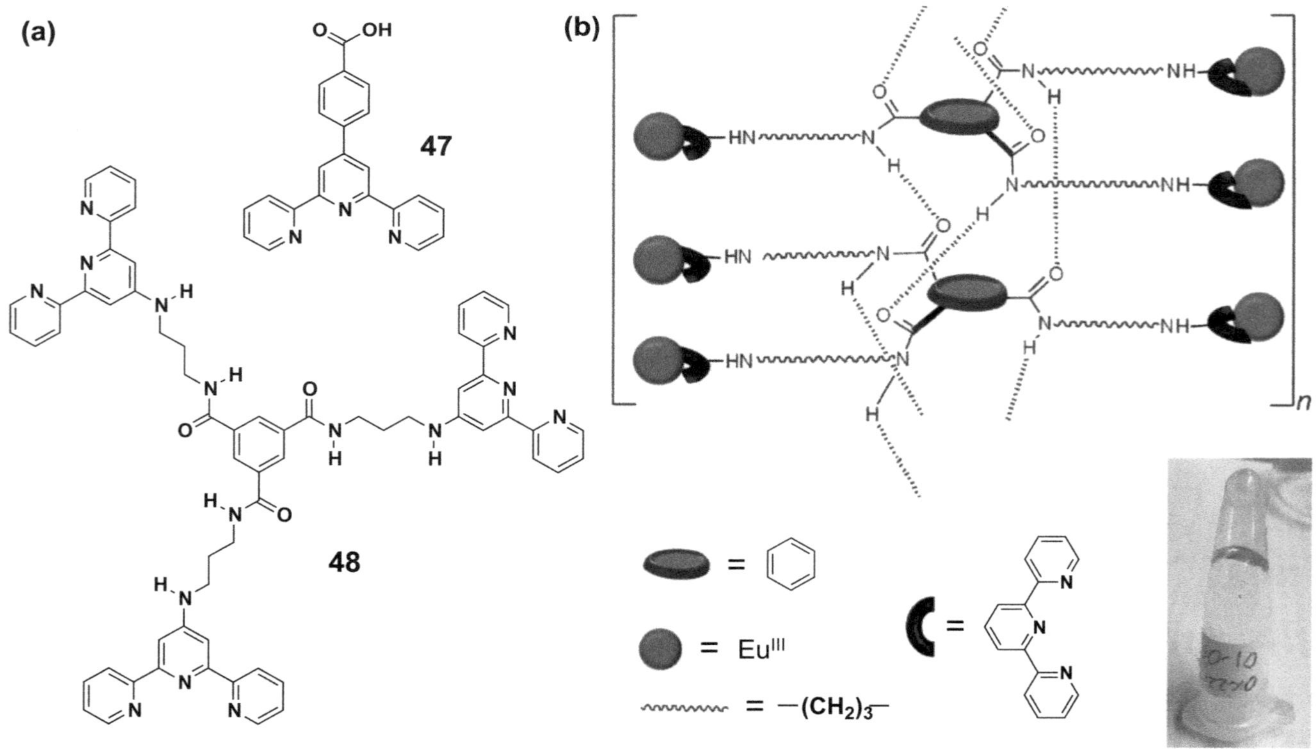

Figure 11.17 (a) Molecular structure of terpyridine-based ligands. (b) A schematic representation of the possible formation of three-fold hydrogen bonding in helices of the terpyridine-based tripodal ligand showing the europium binding sites on the outside of the helix. Inset shows the formation of a transparent $EuCl_3$–**48** gel in a H_2O–CH_3OH solvent mixture. Reprinted with permission from ref. 99. Copyright 2012, Wiley-VCH Verlag GmbH & Co. KGaA.

with the characteristic Eu^{III} emission having a lifetime value of 159.64 μs, indicating that lanthanide ions are sensitized by the antenna effect.

Boland, Gunnlaugsson and co-workers have demonstrated europium-directed supramolecular gelation of a tripodal ligand based on benzene-1,3,5-tricarboxyamide appended with three terpyridine moieties **48** (Figure 11.17a).[99] Morphological studies revealed initial self-assembly of the ligand, driven by three-fold hydrogen bonding. The supramolecular polymers thus formed were further connected by coordination of Eu^{III} ions to more than one terpyridine unit in the adjacent strands, **48** (Figure 11.17b). In the supramolecular gel state, terpyridine moieties functioned as a light-harvesting antenna for populating the excited state of Eu^{III}, and gave rise to characteristic Eu^{III}-centred red emission.

The interesting photoluminescence properties of lanthanide metal complexes make them attractive for use in soft materials.[74,75] Based on this idea, Kim and Chang designed a white-light emitting metallo-gel consisting of the Tb^{III} and Eu^{III} phenanthroline complexes derived from gelators **49** and **50**.[100,101] The similarity in the molecular structure allowed perfect mixing of **49** and **50** in gels of *n*-decane. At a suitable weight ratio of **49** and **50** (19 : 1), upon excitation of the ligand at 330 nm, white-light emission (Commission Internationale de l'Éclairage, CIE 1931 colour coordinates: $x = 0.36, y = 0.30$) was observed from the gel as a result of the balancing of ligand-centred bluish emission by the energy transfer mediated green and red emissions from the Tb^{III} and Eu^{III} ions, respectively.

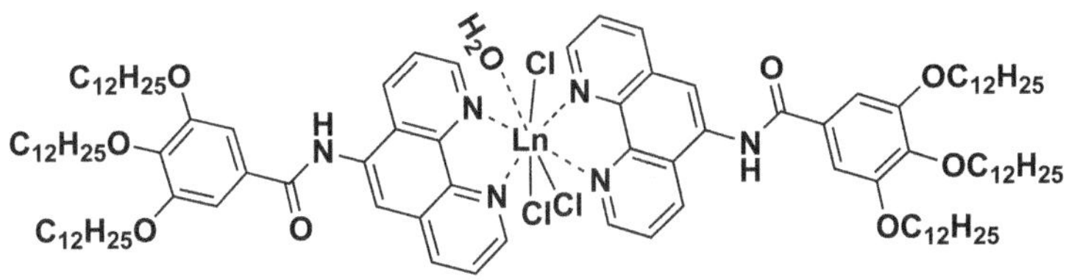

49: Ln = Tb^{III} **50**: Ln = Eu^{III}

11.5 Conclusions

The status of the current literature reveals that metallosupramolecular assemblies are one of the most promising materials for light energy harvesting applications. As demonstrated in this chapter, this alluring interest stems from several factors such as availability of a large spectrum of chromophore-based ligands and metal ions, ease of synthesis, processability (except highly crystalline MOFs), and the ability to spatially organize chromophores. The journey of metallosupramolecular materials is fascinating and the achievements in the area of light harvesting in terms of efficiency and emission colour tuning are excellent. Nevertheless, as in the case of any new energy-harvesting material, more challenges await us, such as controlling the direction of energy flow using an energy gradient involving a cascade energy transfer process, coupling of the energy transfer system with an electron transfer unit, and, ultimately, application of these materials in real devices, where durability and efficiency are major obstacles. Therefore, there is plenty of room for further improvements. In the coming years, we can expect significant improvements in the area of metallosupramolecular materials to make them a feasible choice for energy applications in terms of synthesis, processability, durability and efficacy.

Acknowledgements

We thank the Council of Scientific and Industrial Research (CSIR) and Department of Science and Technology (DST), Government of India, for financial support. A. A. is grateful to the Department of Atomic Energy (DAE), Government of India, for an Outstanding Researcher Award.

References

1. G. D. Scholes, G. R. Fleming, A. Olaya-Castro and R. van Grondelle, *Nat. Chem.*, 2011, **3**, 763.
2. J. Strümpfer, M. Şener and K. Schulten, *J. Phys. Chem. Lett.*, 2012, **3**, 536.
3. A. C. Benniston and A. Harriman, *Mater. Today*, 2008, **11**, 26.
4. G. Calzaferri and K. Lutkouskaya, *Photochem. Photobiol. Sci.*, 2008, **7**, 879.
5. F. Laquai, Y.-S. Park, J.-J. Kim and T. Basché, *Macromol. Rapid Commun.*, 2009, **30**, 1203.
6. M. R. Wasielewski, *Acc. Chem. Res.*, 2009, **42**, 1910.
7. K. V. Rao, K. K. R. Datta, M. Eswaramoorthy and S. J. George, *Chem.–Eur. J.*, 2012, **18**, 2184.
8. V. K. Praveen, C. Ranjith, E. Bandini, A. Ajayaghosh and N. Armaroli, *Chem. Soc. Rev.*, 2014, **43**, 4222.
9. A. Barbieri, B. Ventura and R. Ziessel, *Coord. Chem. Rev.*, 2012, **256**, 1732.
10. P. D. Frischmann, K. Mahata and F. Würthner, *Chem. Soc. Rev.*, 2013, **42**, 1847.
11. K. V. Rao, A. Jain and S. J. George, *J. Mater. Chem. C*, 2014, **2**, 3055.
12. C. Piguet and J.-C. G. Bünzli, *Chem. Soc. Rev.*, 1999, **28**, 347.
13. F. Barigelletti and L. Flamigni, *Chem. Soc. Rev.*, 2000, **29**, 1.
14. K. Binnemans, *Chem. Rev.*, 2009, **109**, 4283.
15. S. J. Bradberry, A. J. Savyasachi, M. Martinez-Calvo and T. Gunnlaugsson, *Coord. Chem. Rev.*, 2014, **273–274**, 226.
16. S. Kitagawa, R. Kitaura and S.-i. Noro, *Angew. Chem., Int. Ed.*, 2004, **43**, 2334.
17. R. Dobrawa and F. Würthner, *J. Polym. Sci., Part A: Polym. Chem.*, 2005, **43**, 4981.
18. M. Burnworth, D. Knapton, S. J. Rowan and C. Weder, *J. Inorg. Organomet. Polym.*, 2007, **17**, 91.
19. D. J. Tranchemontagne, Z. Ni, M. O'Keeffe and O. M. Yaghi, *Angew. Chem., Int. Ed.*, 2008, **47**, 5136.

20. V. A. Friese and D. G. Kurth, *Curr. Opin. Colloid Interface Sci.*, 2009, **14**, 81.
21. X. Wang and R. McHale, *Macromol. Rapid Commun.*, 2010, **31**, 331.
22. G. R. Whittell, M. D. Hager, U. S. Schubert and I. Manners, *Nat. Mater.*, 2011, **10**, 176.
23. S. R. Batten, N. R. Champness, X.-M. Chen, J. Garcia-Martinez, S. Kitagawa, L. Öhrström, M. O'Keeffe, M. P. Suh and J. Reedijk, *Pure Appl. Chem.*, 2013, **85**, 1715.
24. Y. Cui, Y. Yue, G. Qian and B. Chen, *Chem. Rev.*, 2011, **112**, 1126.
25. R. D. Mukhopadhyay, V. K. Praveen and A. Ajayaghosh, *Mater. Horiz.*, 2014, **1**, 572.
26. L. E. Kreno, K. Leong, O. K. Farha, M. Allendorf, R. P. Van Duyne and J. T. Hupp, *Chem. Rev.*, 2012, **112**, 1105.
27. Y. Cui, B. Chen and G. Qian, *Coord. Chem. Rev.*, 2014, **273–274**, 76.
28. P. Falcaro and S. Furukawa, *Angew. Chem., Int. Ed.*, 2012, **51**, 8431.
29. J.-L. Wang, C. Wang and W. Lin, *ACS Catal.*, 2012, **2**, 2630.
30. D. Liu, K. Lu, C. Poon and W. Lin, *Inorg. Chem.*, 2014, **53**, 1916.
31. J. Heine and K. Müller-Buschbaum, *Chem. Soc. Rev.*, 2013, **42**, 9232.
32. L. V. Meyer, F. Schönfeld and K. Müller-Buschbaum, *Chem. Commun.*, 2014, **50**, 8093.
33. K. A. White, D. A. Chengelis, M. Zeller, S. J. Geib, J. Szakos, S. Petoud and N. L. Rosi, *Chem. Commun.*, 2009, 4506.
34. C. A. Kent, B. P. Mehl, L. Ma, J. M. Papanikolas, T. J. Meyer and W. Lin, *J. Am. Chem. Soc.*, 2010, **132**, 12767.
35. C. A. Kent, D. Liu, A. Ito, T. Zhang, M. K. Brennaman, T. J. Meyer and W. Lin, *J. Mater. Chem. A*, 2013, **1**, 14982.
36. C. A. Kent, D. Liu, L. Ma, J. M. Papanikolas, T. J. Meyer and W. Lin, *J. Am. Chem. Soc.*, 2011, **133**, 12940.
37. C. A. Kent, D. Liu, T. J. Meyer and W. Lin, *J. Am. Chem. Soc.*, 2012, **134**, 3991.
38. C. J. Höller, M. Mai, C. Feldmann and K. Müller-Buschbaum, *Dalton Trans.*, 2009, **39**, 461.
39. P. R. Matthes, C. J. Höller, M. Mai, J. Heck, S. J. Sedlmaier, S. Schmiechen, C. Feldmann, W. Schnick and K. Müller-Buschbaum, *J. Mater. Chem.*, 2012, **22**, 10179.
40. D. F. Sava, L. E. S. Rohwer, M. A. Rodriguez and T. M. Nenoff, *J. Am. Chem. Soc.*, 2012, **134**, 3983.
41. D. F. Sava Gallis, L. E. S. Rohwer, M. A. Rodriguez and T. M. Nenoff, *Chem. Mater.*, 2014, **26**, 2943.
42. C. Zhan, S. Ou, C. Zou, M. Zhao and C.-D. Wu, *Anal. Chem.*, 2014, **86**, 6648.
43. Y. Cui, H. Xu, Y. Yue, Z. Guo, J. Yu, Z. Chen, J. Gao, Y. Yang, G. Qian and B. Chen, *J. Am. Chem. Soc.*, 2012, **134**, 3979.
44. S. Dang, J.-H. Zhang and Z.-M. Sun, *J. Mater. Chem.*, 2012, **22**, 8868.
45. M.-L. Ma, C. Ji and S.-Q. Zang, *Dalton Trans.*, 2013, **42**, 10579.
46. Z.-F. Liu, M.-F. Wu, S.-H. Wang, F.-K. Zheng, G.-E. Wang, J. Chen, Y. Xiao, A.-Q. Wu, G.-C. Guo and J.-S. Huang, *J. Mater. Chem. C*, 2013, **1**, 4634.

47. M.-L. Ma, J.-H. Qin, C. Ji, H. Xu, R. Wang, B.-J. Li, S.-Q. Zang, H.-W. Hou and S. R. Batten, *J. Mater. Chem. C*, 2014, **2**, 1085.
48. H. He, F. Sun, T. Borjigin, N. Zhao and G. Zhu, *Dalton Trans.*, 2014, **43**, 3716.
49. Q. Tang, S. Liu, Y. Liu, D. He, J. Miao, X. Wang, Y. Ji and Z. Zheng, *Inorg. Chem.*, 2014, **53**, 289.
50. A. Balamurugan, A. K. Gupta, R. Boomishankar, M. L. Reddy and M. Jayakannan, *ChemPlusChem*, 2013, **78**, 737.
51. C. Y. Lee, O. K. Farha, B. J. Hong, A. A. Sarjeant, S. T. Nguyen and J. T. Hupp, *J. Am. Chem. Soc.*, 2011, **133**, 15858.
52. H.-J. Son, S. Jin, S. Patwardhan, S. J. Wezenberg, N. C. Jeong, M. So, C. E. Wilmer, A. A. Sarjeant, G. C. Schatz, R. Q. Snurr, O. K. Farha, G. P. Wiederrecht and J. T. Hupp, *J. Am. Chem. Soc.*, 2013, **135**, 862.
53. S. Jin, H.-J. Son, O. K. Farha, G. P. Wiederrecht and J. T. Hupp, *J. Am. Chem. Soc.*, 2013, **135**, 955.
54. M. C. So, S. Jin, H.-J. Son, G. P. Wiederrecht, O. K. Farha and J. T. Hupp, *J. Am. Chem. Soc.*, 2013, **135**, 15698.
55. D. Yan, Y. Tang, H. Lin and D. Wang, *Sci. Rep.*, 2014, **4**, 4337, DOI: 10.1038/srep04337.
56. A. M. Spokoyny, D. Kim, A. Sumrein and C. A. Mirkin, *Chem. Soc. Rev.*, 2009, **38**, 1218.
57. W. Lin, W. J. Rieter and K. M. L. Taylor, *Angew. Chem., Int. Ed.*, 2009, **48**, 650.
58. W. L. Leong and J. J. Vittal, *Chem. Rev.*, 2011, **111**, 688.
59. X. Zhang, Z.-K. Chen and K. P. Loh, *J. Am. Chem. Soc.*, 2009, **131**, 7210.
60. X. Zhang, M. A. Ballem, M. Ahrén, A. Suska, P. Bergman and K. Uvdal, *J. Am. Chem. Soc.*, 2010, **132**, 10391.
61. V. M. Suresh, S. J. George and T. K. Maji, *Adv. Funct. Mater.*, 2013, **23**, 5585.
62. X. Zhang, M. A. Ballem, Z.-J. Hu, P. Bergman and K. Uvdal, *Angew. Chem., Int. Ed.*, 2011, **50**, 5729.
63. R. Nishiyabu, N. Hashimoto, T. Cho, K. Watanabe, T. Yasunaga, A. Endo, K. Kaneko, T. Niidome, M. Murata, C. Adachi, Y. Katayama, M. Hashizume and N. Kimizuka, *J. Am. Chem. Soc.*, 2009, **131**, 2151.
64. C. Aimé, R. Nishiyabu, R. Gondo and N. Kimizuka, *Chem.–Eur. J.*, 2010, **16**, 3604.
65. F. Pu, E. Ju, J. Ren and X. Qu, *Adv. Mater.*, 2014, **26**, 1111.
66. F. Pu, L. Wu, E. Ju, X. Ran, J. Ren and X. Qu, *Adv. Funct. Mater.*, 2014, **24**, 4549.
67. M. Numata, D. Kinoshita, N. Taniguchi, H. Tamiaki and A. Ohta, *Angew. Chem., Int. Ed.*, 2012, **51**, 1844.
68. A. R. Remya, S. Varughese and M. L. P. Reddy, *Dalton Trans.*, 2014, **43**, 10940.
69. A. Fermi, P. Ceroni, M. Roy, M. Gingras and G. Bergamini, *Chem.–Eur. J.*, 2014, **20**, 10661.

70. Z. B. Lim, H. Li, S. Sun, J. Y. Lek, A. Trewin, Y. M. Lam and A. C. Grimsdale, *J. Mater. Chem.*, 2012, **22**, 6218.
71. T. He, R. Chen, Z. B. Lim, D. Rajwar, L. Ma, Y. Wang, Y. Gao, A. C. Grimsdale and H. Sun, *Adv. Opt. Mater.*, 2014, **2**, 40.
72. F. Fages, *Angew. Chem., Int. Ed.*, 2006, **45**, 1680.
73. M.-O. M. Piepenbrock, G. O. Lloyd, N. Clarke and J. W. Steed, *Chem. Rev.*, 2010, **110**, 1960.
74. A. Y.-Y. Tam and V. W.-W. Yam, *Chem. Soc. Rev.*, 2013, **42**, 1540.
75. J. Zhang and C.-Y. Su, *Coord. Chem. Rev.*, 2013, **257**, 1373.
76. A. Ajayaghosh, V. K. Praveen and C. Vijayakumar, *Chem. Soc. Rev.*, 2008, **37**, 109.
77. S. S. Babu, K. K. Kartha and A. Ajayaghosh, *J. Phys. Chem. Lett.*, 2010, **1**, 3413.
78. S. S. Babu, S. Prasanthkumar and A. Ajayaghosh, *Angew. Chem., Int. Ed.*, 2012, **51**, 1766.
79. V. K. Praveen, C. Ranjith and N. Armaroli, *Angew. Chem., Int. Ed.*, 2014, **53**, 365.
80. S. S. Babu, V. K. Praveen and A. Ajayaghosh, *Chem. Rev.*, 2014, **114**, 1973.
81. V. K. Praveen, S. J. George, R. Varghese, C. Vijayakumar and A. Ajayaghosh, *J. Am. Chem. Soc.*, 2006, **128**, 7542.
82. A. Ajayaghosh, C. Vijayakumar, V. K. Praveen, S. S. Babu and R. Varghese, *J. Am. Chem. Soc.*, 2006, **128**, 7174.
83. A. Ajayaghosh, V. K. Praveen, S. Srinivasan and R. Varghese, *Adv. Mater.*, 2007, **19**, 411.
84. A. Ajayaghosh, V. K. Praveen, C. Vijayakumar and S. J. George, *Angew. Chem., Int. Ed.*, 2007, **46**, 6260.
85. C. Vijayakumar, V. K. Praveen and A. Ajayaghosh, *Adv. Mater.*, 2009, **21**, 2059.
86. C. Vijayakumar, V. K. Praveen, K. K. Kartha and A. Ajayaghosh, *Phys. Chem. Chem. Phys.*, 2011, **13**, 4942.
87. J. B. Beck and S. J. Rowan, *J. Am. Chem. Soc.*, 2003, **125**, 13922.
88. Y. Zhao, J. B. Beck, S. J. Rowan and A. M. Jamieson, *Macromolecules*, 2004, **37**, 3529.
89. W. Weng, J. B. Beck, A. M. Jamieson and S. J. Rowan, *J. Am. Chem. Soc.*, 2006, **128**, 11663.
90. T. Cardolaccia, Y. Li and K. S. Schanze, *J. Am. Chem. Soc.*, 2008, **130**, 2535.
91. Q. Wang, *J. Fluoresc.*, 2009, **19**, 793.
92. Q. Wang, K. Ogawa, Y. Li and H. Tamiaki, *Photochem. Photobiol. Sci.*, 2011, **10**, 60.
93. S. Bhowmik, S. Banerjee and U. Maitra, *Chem. Commun.*, 2010, **46**, 8642.
94. S. Banerjee, R. Kandanelli, S. Bhowmik and U. Maitra, *Soft Matter*, 2011, **7**, 8207.
95. S. Bhowmik and U. Maitra, *Chem. Commun.*, 2012, **48**, 4624.
96. R. Kandanelli, A. Sarkar and U. Maitra, *Dalton Trans.*, 2013, **42**, 15381.

97. V. S. Sajisha and U. Maitra, *Chimia*, 2013, **67**, 44.
98. L. Sambri, F. Cucinotta, G. De Paoli, S. Stagni and L. De Cola, *New J. Chem.*, 2010, **34**, 2093.
99. O. Kotova, R. Daly, C. M. G. dos Santos, M. Boese, P. E. Kruger, J. J. Boland and T. Gunnlaugsson, *Angew. Chem., Int. Ed.*, 2012, **51**, 7208.
100. H. Kim and J. Y. Chang, *Soft Matter*, 2011, **7**, 7952.
101. H. Kim and J. Y. Chang, *RSC Adv.*, 2013, **3**, 1774.

CHAPTER 12

Metallosupramolecular Assemblies for Application as Photocatalysts for the Production of Solar Fuels

DANILO DINI[a], MARY T. PRYCE[b], MARTIN SCHULZ*[c] AND JOHANNES G. VOS*[b]

[a]Department of Chemistry, University of Rome "La Sapienza", Rome, Italy; [b]SRC for Solar Energy Conversion, School of Chemical Sciences, Dublin City University, Dublin 9, Ireland; [c]Leibniz Institute of Photonic Technology (IPHT), Albert-Einstein-Straße 9, 07745 Jena, Germany
*E-mail: han.vos@dcu.ie

12.1 Introduction

Based on projections for the overall economic and population growth, worldwide energy consumption is expected to rise significantly in the future.[1,2] Global annual energy consumption is currently estimated at approximately 4.25×10^{20} J, which corresponds to an average energy consumption rate of 13.5 TW per annum.[3] At present, about 85% of our energy is produced from fossil fuels such as oil, gas and coal as well as nuclear power. The remainder comes from renewable energy sources such as wind power, solar power, biogas and the burning of biomass. The predicted increase in

RSC Smart Materials No. 15
Functional Metallosupramolecular Materials
Edited by John G. Hardy and Felix H. Schacher

Published by the Royal Society of Chemistry, www.rsc.org

the energy demand can, however, not be met by fossil fuels, the reserves of which are diminishing. While the development of fracking technologies will extend fossil fuel reserves, the introduction of these new techniques has given rise to further serious environmental concerns such as water pollution.

However, the biggest problem associated with fossil fuels is not so much the dwindling reserves, but the fact that their use has been widely credited with the generation of increasing levels of CO_2 in the atmosphere. Recently, CO_2 concentrations of 400 ppm were reported, the highest value for thousands of years. It is now generally accepted that these high levels are man-made and are directly responsible for climate change. The development of a sustainable energy carbon neutral supply for a growing world population is therefore one of the largest challenges of the 21st century.[1,2]

These new energy technologies should be efficient and cheap so that they can replace currently used fossil fuels, they must be carbon neutral and thus reduce CO_2 levels in the atmosphere. Wind energy is at present a well-developed technology, but problems exist with the location of wind turbines and the constant variation of wind strength leads to an intermittent energy output. The potential of other alternatives such as wave and tidal options are under investigation. Solar energy is widely considered to be one of the most promising avenues towards the development of new environmentally friendly energy sources of the future. Its energy source is virtually unlimited and solar irradiation provides an average global energy of about 4.93 MJ m^{-2} (1.37 kW h m^{-2}).[4] This translates into a usable energy of *circa* 3×10^{24} J per year, more than 10 000 times the current annual world energy demand. Methods for harvesting the energy of the sun have been investigated for several decades. In particular, photovoltaic cells based on silicon have found applications and in recent years considerable improvements have been made in the efficiency and cost for silicon-based photovoltaic cells. An alternative molecule-based solar cell has been developed by Grätzel and co-workers and this invention has resulted in a considerable interest in dye-sensitised solar cells.[5] However, a problem with photovoltaic cells is that they produce electricity that must be used instantly, and cannot be stored easily. The same can be said for other carbon neutral energy sources such as wind and wave. An alternative (and complementary) approach is the development of a photocatalytic cell that produces fuels or feedstock chemicals from solar irradiation.[6] Such cells can be driven by electricity obtained from wind or solar energy, or directly by sunlight.

This chapter focuses on the present interest in the application of supramolecular photocatalytic assemblies for solar driven generation of fuels and useful chemical products using a carbon neutral approach. The ideal scenario is the sunlight-driven splitting of water into hydrogen and oxygen, since hydrogen can be used as a carbon neutral fuel that does not produce CO_2 since only water is released.[7] Similarly, the light-driven reduction of carbon dioxide to methane, methanol, formate or carbon monoxide offers the possibility of storing solar energy by conversion into these C_1-synthons. The combination of light-driven CO_2 reduction with a hydrogen technology could make a huge contribution to renewable energy sources.[8,9] It is important to note that the carbon neutral solar driven recycling of CO_2 into useful products will create novel industrial processes where the CO_2 emitted will be reused and in this way greatly reduce the concentration of CO_2 in the atmosphere. Potentially, this could be the ultimate way of fighting climate change. The supramolecular design principles required for the design of such molecular photocatalysts will be presented, and conclusions about structure–property relationships will be drawn. An overview on the importance of the experimental conditions used and the parameters to assess the efficiency of the photocatalytic systems are also included. In our discussion a series of compounds will be discussed to outline the general principles and illustrate important interesting results. A number of reviews on the application of molecular photocatalysts for the generation of solar fuels are listed.[10]

12.2 Background

Investigations on the potential of molecular photocatalysts for hydrogen generation and CO_2 reduction have been taking place since the late 1970s and since then there has been increasing interest in the use of molecular components as solar dyes absorbing in the visible part of the spectrum. In these early studies the intermolecular approach was taken where mixtures of a light absorbing moiety and a catalytic centre were utilised to produce solar fuels. In this approach the photosensitiser was irradiated and electron transfer from the excited state of the sensitiser to the catalytic centre occurred. Sacrificial electron donors (S) such as triethylamine (TEA) were added to regenerate the dye (see Figure 12.1).

The development of intramolecular photochemical molecular devices (PMDs) is inspired by natural photosynthesis where a chain of carefully chosen porphyrin-type components fulfils tasks such as light absorption in the visible part of the spectrum, vectorial electron transfer from the light absorber to the catalytic centre and most importantly, charge separation. A typical PMD consists of a light absorbing unit (**PS**), a bridging ligand (**B**) and a catalytic centre (**C**), as shown in Figure 12.1.[11–13] Since 2006, the intramolecular option has been considered and this has led to the development of a wide range of bimetallic complexes that act as intramolecular photocatalytic centres. A typical example of supramolecular assemblies that have been studied for light-driven hydrogen generation from water are shown in Figure 12.2. In

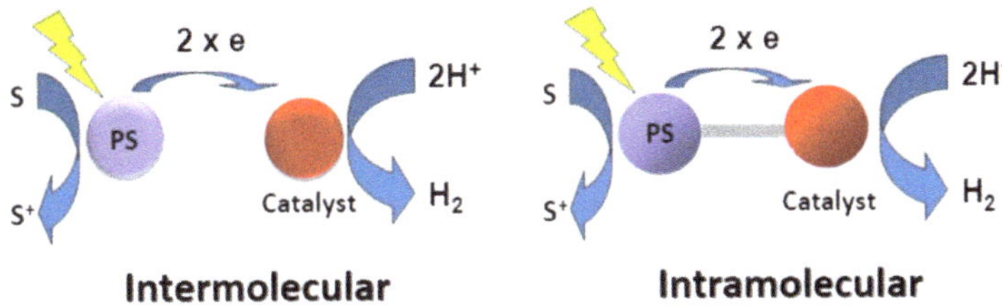

Figure 12.1 Intra- *vs.* intermolecular approach.

this case, the $Ru(bipy)_2$ (bipy = 2,2′-bipyridyl) centre acts as a photosensitiser and the covalently bound Pd unit as a hydrogen generating catalytic centre. Two bridging ligands differing only in the point of attachment of the second pyridine group in the tpy ligand are shown together with a typical monometallic precursor compound.

As already stated above, intermolecular systems were the first to be investigated. However, when comparing multi-component systems (*i.e.* a mixture of **PS** and **C**) with intramolecular analogues, the former seem to be disadvantageous, since they rely on both collision probability and diffusion processes. It is anticipated that this will limit the efficiency of the essential photo electron transfer. To overcome this drawback, a range of PMDs have been developed which connect the **PS** and **C** moieties *via* a suitable bridging ligand **B**. This bridging ligand should create the electronic interaction that exists between the various reaction partners in photosynthesis. It is worth nothing, however, that in the natural photosynthetic system, electron transfer occurs by a through-space process between carefully located reaction partners rather than *via* covalent bonds. The bridging ligand must, apart from providing a stable building block for holding the assembly together, also meet a number of different criteria such as:

(1) The bridge should allow electron transfer from the excited state to the catalytic centre.
(2) In this process the **PS** is oxidised and **C** is reduced; importantly, the bridge must stabilise this charge-separated state.
(3) For multi-electron processes, the bridge should act as a "charge pool".[1]

The **PS–B–C** approach as depicted in Figure 12.1 facilitates the control of the vectorial photoelectron transfer by allowing for a precise tuning of the molecular components and hence increasing the catalytic efficiency. Taking these considerations into account, the intramolecular assembly as shown in Figure 12.2 could be considered.

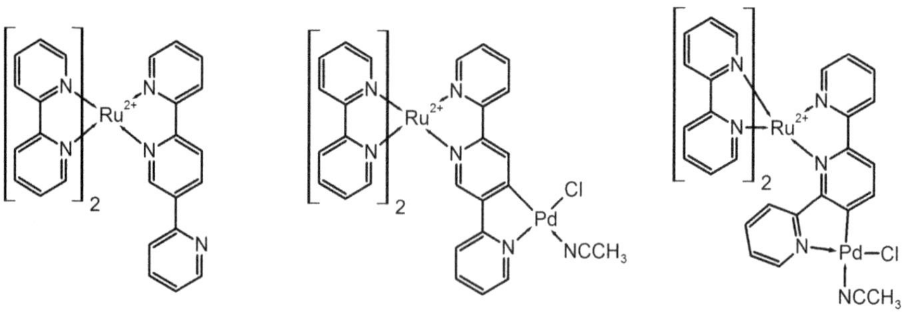

Ru-tpy **Ru-tpy-Pd** **Ru-tp-Pd**

Figure 12.2 Structures of monometallic precursor **Ru-tpy** and photocatalytic **Ru-tpy-Pd** compounds bridging ligand 2,2′:5′,2″-terpyridine (tpy) and **Ru-tp-Pd** based on 2,2′:6′,2″-terpyridine (**tp**).

A typical photoinduced catalytic process for hydrogen generation by a Ru/Pd-type assembly is shown in Scheme 12.1. After light absorption at the Ru-based photosensitiser, an electron is promoted from the Ru(II)-based ground state to a cluster of metal-to-ligand charge transfer states (MLCT). If the electronic properties of the bridging ligands are chosen correctly, electron transfer from bipy-based MLCT states to a bridge-based orbital will take place. This process will be discussed in more detail later. In subsequent steps, regeneration of the now formally Ru(III) centre by the sacrificial agent is observed. Subsequently, the electron will be transferred to the Pd moiety and hydrogen will be formed. Naturally, two photons are required to form a dihydrogen molecule. In the case shown in Scheme 12.1, release of a chloride ion from the catalytic centre is required to allow for the binding of protons at the catalytic centre with subsequent H_2 generation. This is generally observed for Pd-based hydrogen catalysts.

Scheme 12.1 shows that a sacrificial agent, *e.g.* triethylamine (TEA), is required to re-reduce the light-absorbing Ru centre so that the next photoexcitation can take place. It is important to realise that sacrificial agents such as TEA create radical species when acting as electron donors and these reaction products may participate in the photocatalytic system in dark reactions as shown in Scheme 12.2.

As outlined earlier, in intramolecular assemblies, the bridging ligand is expected to play an important role and does not only represent the chemical connection between a photosensitiser unit and a catalytically active centre but can also greatly affect the overall catalytic activity and stability of the

$$\text{Ru(II)-B-Pd(II)Cl}_2 \xrightarrow{h\nu} \text{Ru(II)}^*\text{-B-Pd(II)Cl}_2$$

$$\text{Ru(II)}^*\text{-B-Pd(II)Cl}_2 \rightarrow \text{Ru(III)-B}^{*\cdot}\text{-Pd(II)Cl}_2$$

$$\text{Ru(III)-B}^{*\cdot}\text{-Pd(II)Cl}_2 \xrightarrow{\text{TEA}} \text{Ru(II)- B}^{*\cdot}\text{-Pd(II)Cl}_2$$

$$\text{Ru(II)- B}^{*\cdot}\text{-Pd(II)Cl}_2 \rightarrow \text{Ru(II)-B-Pd(I)SCl} + \text{Cl}^-$$

$$\text{Ru(II)-B-Pd(I)SCl} + \text{H}^+ \rightarrow \text{Ru(II)-B-Pd(II)SCl} + 1/2\text{H}_2$$

Scheme 12.1 Proposed pathway for photocatalytic hydrogen generation. **B** = bridging ligand, S = solvent.

$$\text{TEA} \rightarrow \text{TEA}^{\cdot+} + e^-$$

$$\text{TEA}^{\cdot+} \rightarrow \text{Et}_2\text{N}^{\cdot}\text{CHCH}_3 + \text{H}^+$$

$$\text{Et}_2\text{N}^{\cdot}\text{CHCH}_3 \rightarrow \text{Et}_2\text{N}^+\text{=CHCH}_3 + e^-$$

Scheme 12.2 Decomposition pathway of TEA after electron donation to the sensitiser.

photocatalysts. Therefore, careful design protocols are required in the design of photosensitiser–bridge–catalyst assemblies, to yield intramolecular photocatalysts that are superior to intermolecular systems. The following sections will be concerned with the discussion of the coordination environment, photophysical, and photocatalytic properties of multinuclear metal complexes that show significant photocatalytic activity for hydrogen generation or CO_2 reduction. Of particular interest in the discussion will be the effect of the bridging ligand and the nature of both the photosensitiser and the catalyst on the photocatalytic properties of the compounds. Based on the results obtained, the possible design principles for the optimisation of the photocatalytic properties of this class of compounds will be suggested.

12.3 The Importance of Experimental Conditions and Parameters

As already pointed out in the Introduction, most studies so far have been carried out in solution and in these investigations many different approaches have been taken. Therefore, it is often difficult to directly compare one catalytic system with another. It is important to realise that when analysing different photocatalytic assemblies the conditions used are an important component and must be included. For example, for photocatalytic reduction of CO_2 and hydrogen generation, the amount of water present and the pH are important parameters that may strongly affect the results obtained. The formation of products such as methanol, formic acid and methane in photocatalytic CO_2 reduction are highly dependent on the presence of protons. On the other hand, while the presence of water during the photocatalytic hydrogen formation initially increases the efficiency of the process, at higher water concentration, lower product yields may be observed, most likely because of slower reaction rates as a result of changes in the polarity of the mixed solvent or due to decomposition of the compound.

A second issue to consider is that not all solvents are inert components in the photocatalytic reaction mixture; on occasion they can participate in the reaction and create erroneous results. A typical example is dimethylformamide (DMF). It is long known that DMF decomposes slowly under irradiation and at its boiling point with the formation of dimethylamine and carbon monoxide. The latter process has been observed in the preparation of compounds such as $[Ru(bipy)_2Cl_2]$ in DMF, when $[Ru(bipy)_2COCl]^+$ is also formed.[14] This decomposition is further catalysed by the presence of acids and bases such as the sacrificial agent TEA. TEA acts as a base and in water DMF decomposes forming CO, dimethylamine and formic acid. It is to be expected, therefore, that when the reaction mixture used for the photocatalytic studies on the reduction of CO_2 contains water, incorrect formate concentrations may be obtained. However, it is important to note that incorrect values may also be obtained when no water is used in the photocatalytic process but when samples taken for the measurement of formate are diluted with water for analysis. In that case the formate concentration depends on the timespan between the preparation of the analytic sample and the actual measurement.[15] This instability of DMF has important implications for the quantification of the CO_2 reduction products such as CO and formic acid. These inaccuracies in the quantification of the reaction products can be avoided by using $^{13}CO_2$ but this requires mass-sensitive analytical methods such as gas chromatography–mass spectrometry coupling (GC/MS). Importantly, Ishida and co-workers have recently reported that the related solvent *N*,*N*-dimethylacetamide does not show these problems.[16] These observations show that care has to be taken in the choice of the solvent used to assure that it does not actively contribute to the formation of products. In addition, in all homogeneous studies carried out, a sacrificial agent is added to back-reduce the photosensitiser after electron transfer to the catalytic centre. As already outlined above, TEA is such a sacrificial agent, but its involvement in the photocatalytic process is not limited to regeneration of the photosensitiser and potential decomposition of DMF. In addition, TEA forms radicals and protons (see Scheme 12.2) upon electron donation to the oxidised photocatalysts and these take part in the photocatalytic process. Similar processes are also likely to occur with other sacrificial agents. This is a particularly important issue since their concentrations are normally several orders of magnitude higher than that of the photocatalytic compounds that are normally present at concentrations of *ca.* 10^{-4} M.

An extensive range of different approaches to the design of efficient photocatalytic assemblies has been reported.[10] To further advance our understanding of molecularly driven photocatalytic solar fuel production and to propose further developments, the results reported need to be studied in detail taking into account issues such as their catalytic performances and stability. However, a comparison of the individual catalytic systems is a difficult task even if performance values such as turnover numbers (TONs), turnover frequencies (TOFs) or quantum efficiencies are given. For example, the TON is normally defined as the number of moles of product formed during the photocatalytic process divided by the number of moles of the photocatalyst present. However, sometimes a different approach is taken where the number of *electrons* rather than the number of *moles* produced per photocatalytic centre is used to calculate the TON. For hydrogen generation this results naturally in TON values that are twice as high. TON values should be related to the stability of the photocatalyst and values reported should be recorded before the formation of products has ceased. A TON value measured at an arbitrary time before termination of the reaction can also be related to the rate of the reaction and can be presented as the turnover frequency. TONs should be given together with the reaction time indicating when they are measured so that a reaction time profile can be established. Turnover numbers for intermolecular photocatalytic systems require particular attention. Such systems consist of separate molecular entities of photosensitiser and catalyst, however, not always in a 1 : 1 ratio. Thus, turnover numbers may be related to either the photosensitiser or catalyst but usually to the component with the lowest concentration. TOF values report the number of moles of product molecules divided by the number of moles of catalyst molecules and time. Since product formation *vs.* time is usually not linear, the initial TOF is given from a linear curve fit of the early stage of the reaction and thus corresponds to the initial turnover frequency; these values will often change later in the reaction. The quantum efficiency is an important parameter for photocatalytic processes and is defined as the number of moles of product molecules times the number of electrons involved and divided by the absorbed photons. Ferrioxalate actinometry is the method most often applied for the determination of photon flux.[17]

The comparison of photocatalytic results is further handicapped by the different experimental setups and conditions that are used by researchers, such as light sources and filter characteristics, reaction conditions (isobaric or isochoric), stirring, temperature control, product quantification *etc.* Among the light sources used for photocatalytic experiments, the irradiation wavelengths used range from the UV to visible light, using for example xenon or mercury arc lamps. Another approach is the use of light-emitting diodes (LED) that can emit over a wide wavelength range and are available as well-defined monochromatic light emitters in the UV or visible part of the spectrum to broad band white light sources. Besides, the spectral characteristics of the light source filters are often employed, but their spectral characteristics are usually not published. It is clear, therefore, that the spectral characteristics of the illumination set-up are of great importance for the photocatalytic performance and details of the set-up used should be reported in detail, especially when wavelength-dependent experiments are carried out.[18–23]

Product quantification is another critical parameter since, in particular for CO_2 reduction, a range of products may be expected. These include gaseous compounds such as CO and methane as well as liquids such as formic acid or methanol, and the solid oxalic acid. Depending on the pH of the reaction mixture, acids can be obtained in combination with their corresponding bases. Additionally, when investigating CO_2 reduction, hydrogen may also be formed as a side product in the presence of water and will be dependent on the pH used. Gaseous and volatile compounds are usually quantified by gas chromatography after headspace sampling, while liquids and solids require further sample preparation prior to analysis to remove compounds that potentially hamper the analytical procedure. This issue is particularly important since the photocatalytic reaction conditions contain generally high concentrations of sacrificial agents and their pH may also vary. This puts a high demand on the design of appropriate analytical methods. Generally, ionisable compounds such as organic acids may be quantified by capillary electrophoresis or ion chromatography but high performance liquid chromatography and colorimetric methods are also available. Therefore, a comprehensive analysis of the reduction products obtained may require a range of analytical techniques.

12.4 Photocatalytic Assemblies for Hydrogen Generation

12.4.1 The Supramolecular Concept

As already pointed out, the aim of this chapter is to consider the potential of supramolecular assemblies as photocatalysts for solar fuel production, and the focus of this section is hydrogen generation. In the approach taken (see Figures 12.1 and 12.2), a light-absorbing unit and a catalytic centre are covalently connected by a bridging ligand. The bridging ligand is chosen so that it allows efficient communication between the photosensitiser and the catalytic centre and to allow optimum electronic interaction between the different moieties in the assembly (see later). A simplified '1st generation' proposal for the photophysical properties required to allow for photoinduced vectorial electron transfer from the photosensitiser to the catalytic centre is shown in Figure 12.3.

In this section, the appropriateness of this scheme is considered by discussing the photophysical properties of **Ru-tpy-Pd** (see Figure 12.2), a compound that produces hydrogen with a turnover number of 130 in 18 hours upon 470 nm irradiation using an acetonitrile–water mixture of 90:10 and 2.3 M TEA as the sacrificial agent.[19] To assess the basic photophysical properties of **Ru-tpy-Pd** and its precursor **Ru-tpy** (see Figure 12.2), studies were carried out in acetonitrile in the femto- to the submicro-second time domains.[24]

Figure 12.3 suggests that excitation of **Ru-tpy-Pd** should lead to the population of singlet metal-to-ligand charge transfer states (^{1}MLCT), based on both the peripheral and the bridging ligands and that efficient intersystem

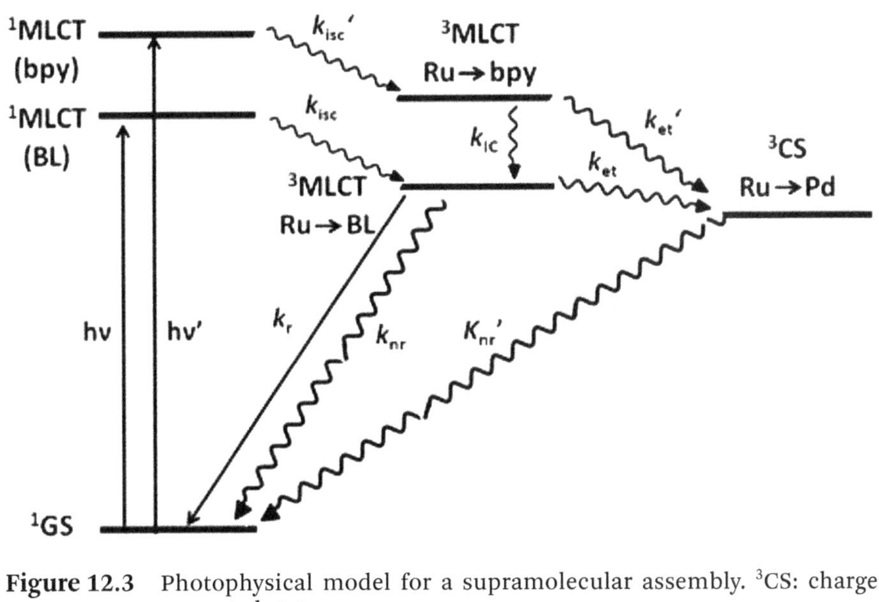

Figure 12.3 Photophysical model for a supramolecular assembly. ^{3}CS: charge separated state.

crossing from both the bipy and bridge-based ^{1}MCLT states to the ^{3}MLCT based on the bridging ligand is obtained. To investigate the accuracy of this model, detailed photophysical experiments involving time-resolved absorption (TA) in the femto–picosecond timescale were carried out on compounds **Ru-tpy** and **Ru-tpy-Pd**. The TA spectra observed after different time delays between pump and probe pulses are shown in Figure 12.4. In the spectra obtained for both compounds, an increase of the absorbance above 500 nm is observed which is associated with a triplet ligand-to-metal charge transfer (^{3}LMCT)-type transition. A ground state bleach is observed around 450 nm as expected. However, the most informative features are those found in the 420–350 nm region, that part of the spectrum where ligand-based transitions occur and this section will be analysed in more detail, since these data indicate a delicate interplay between the different MLCT states. In this range, two bands are observed for **Ru-tpy**: one at 370 nm, assigned to the population of an interligand state and one at 420 nm associated with the bridging ligand tpy. Note that intersystem crossing from the singlet to the triplet MLCT states is expected to occur within 100 fs and is outside the timescale of these experiments. Importantly, the time resolved data show that a decrease in the intensity of the 370 nm band is accompanied by an increase in the intensity of the 420 nm band on the same timescale. These observations can be explained in two different ways, first as an interligand electron transfer from the ^{3}MLCT$_{bipy}$ state to the ^{3}MLCT$_{tpy}$ state, so from the peripheral to the bridging ligand, or alternatively to an equilibration of these states. In the latter case, electron transfer can take place in both directions. For **Ru-tpy-Pd**, similar curves are obtained with some differences possibly related to the presence of the Pd(II) centre. Again, the 420 nm band increases but the brightness observed is less. The interpretation of this process is as stated for the precursor **Ru-tpy**. Importantly, as observed for **Ru-tpy**, the time-resolved spectrum remains the same between 20 and 500 ps. This suggests that electron transfer/delocalisation involving the bipy and tpy ligands is completed on the 20 ps timescale and is similar in both compounds. The decay of the state(s) present after 500 ps was investigated using ns time-resolved photoluminescence.

Upon excitation at 450 nm, the compounds show identical emission wavelengths but very different excited state lifetimes. For **Ru-tpy**, a mono-exponential decay was observed with a lifetime of 623 ns suggesting that both ^{3}MLCT$_{bipy}$ and ^{3}MLCT$_{tpy}$ states decay at the same rate, while for the photocatalyst **Ru-tpy-Pd**, a biexponential decay is observed with different time constants. The fast 88 ns component has been assigned to emission from the ^{3}MLCT$_{tpy}$ state to the ground state under the influence of the Pd centre; this is most likely related to the presence of enhanced spin–orbit coupling and/or an additional electron transfer pathway. Based on these data, a photophysical model for the light-induced processes observed as shown in Figure 12.5 can be obtained.

This study shows that when comparing Figures 12.3 and 12.5, the model description of a photocatalytic assembly as proposed in Figure 12.3, where full transfer of electrons upon excitation of both peripheral and bridging ligands is achieved, is not observed for **Ru-tpy-Pd**, mostly because intraligand

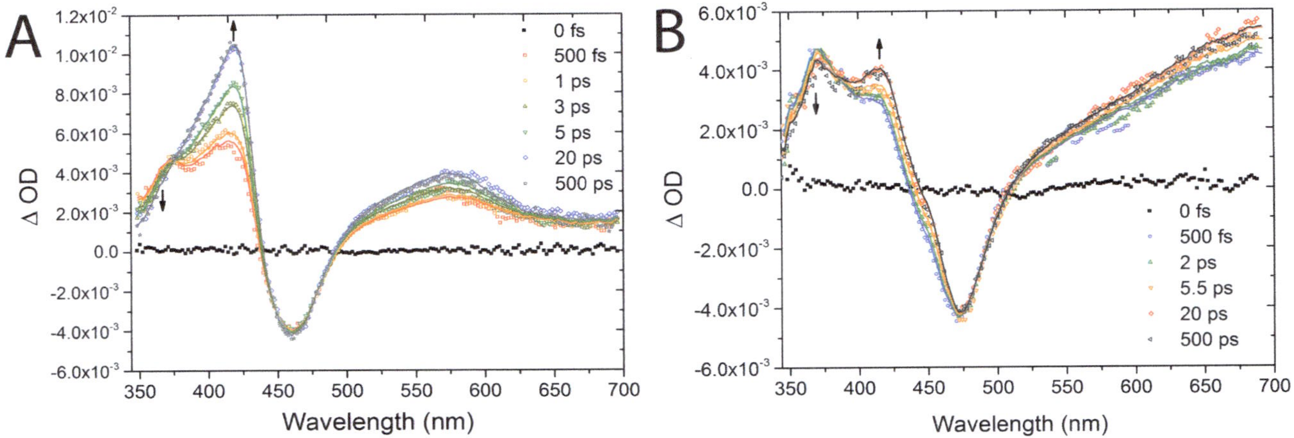

Figure 12.4 Time-resolved transition absorption spectra of the precursor **Ru-tpy** (A) and the photocatalytic compound **Ru-tpy-Pd** (B). Adapted from ref. 24.

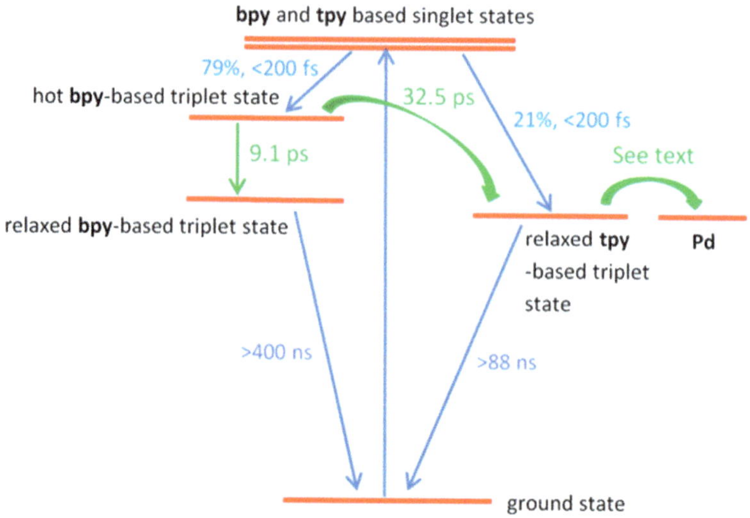

Figure 12.5 Simplified photophysical model for light-induced processes in **Ru-tpy-Pd**. Adapted from ref. 24.

electron transfer is not considered in Figure 12.3. The results shown for **Ru-tpy-Pd** indicate that a much more complicated electronic arrangement is obtained. After excitation into the ^{1}MLCT states of both types of ligands, only about 20% of electrons relax into the ^{3}MLCT state of the bridging ligand. The emission data also suggest that the electron transfer rate k_{et} from the bridging ligand to the Pd centre is in the order of 10^7 s^{-1}. The presence of two emission processes with different lifetimes suggests that the interaction between the photosensitiser and the Pd catalytic centre is weaker than expected from the model energy diagram shown in Figure 12.3. In the ideal situation outlined in Figure 12.3 where strong coupling between light absorber and catalytic centre is assumed, that should result in a complete quenching of the emission. This behaviour is observed for **Ru-tpphz-Pd**[25] discussed later. The emitting behaviour of **Ru-tpy-Pd** is surprising considering the short distance between the light absorbing Ru moiety and the Pd centre. The reason for this is at present not understood, but may be associated with cyclometallation of the bridging ligand, the effect of which on the electronic structure is not clear. Figure 12.5 indicates that 79% of the decay of the bipy-based ^{3}MLCT state, both radiative and non-radiative, is back to the ground state; this means that most of the energy absorbed initially by the compound decays back to the ground state, a process that clearly limits electron transfer from the photosensitiser to the catalytic centre. In the initial model shown in Figure 12.3, an effective electron transfer route from the higher energy peripheral ligands to the bridge was anticipated. This is clearly not the case for **Ru-tpy-Pd** and may be related to a relatively weak coupling between the ^{3}MLCT states of the peripheral and bridging ligands.

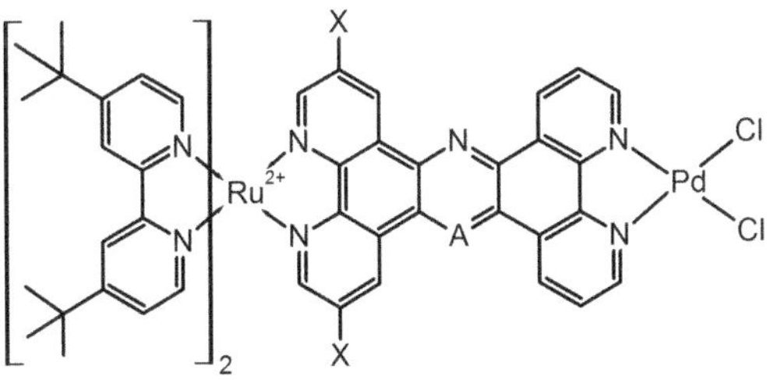

Figure 12.6 Structures of tetrapyrido[3,2-*a*:2′,3′-*c*:3″,2″-*h*:2‴,3‴-*j*]phenazine, tpphz and tetrapyrido[3,2-*a*:2′,3′-*c*:3″,2″-*h*:2‴,3‴-*j*] acridine, tpac type compounds.

This study indicates that for these supramolecular photocatalysts, both peripheral ligands and bridging ligands should be chosen with care to obtain efficient light-induced vectorial transfer from photosensitiser to catalytic centre. This would require that the excited states of the bridging ligands would be lower in energy than those of the peripheral ligands and that electronic coupling between both sets of states is efficient. This issue is further outlined below for **Ru-tpphz-Pd**.

The photophysical properties of **Ru-tpphz-Pd** (see Figure 12.6) were investigated in a similar manner.[26] These studies were carried out in CH_3CN and in the absence of a sacrificial agent in order to evaluate the photophysical properties of the assembly under non-catalytic conditions. An overview of the dynamics of electron-excitation transfer processes as evaluated by this study is given in Figure 12.7.

This scheme shows a comparison of the photophysical behaviour of the mononuclear precursor complex and the dinuclear photocatalyst **Ru-tpphz-Pd.** In both compounds, the initially populated singlet metal-to-ligand charge transfer excited states (^{1}MLCT) are based on the tbbipy/phenanthroline moieties. This first excited state successively undergoes relaxation to a ^{3}MLCT state, which is located on the phenanthroline part of the tpphz bridging ligand. For both compounds, the formation of the phenanthroline-centred ^{3}MLCT state from the initially populated mixture of singlet states takes place within 1.2 ps for the precursor and 0.8 ps for **Ru-tpphz-Pd**. However, there is a considerable difference in the decay of this triplet excited state. The decay time for **Ru-tpphz** is 240 ps, while for **Ru-tpphz-Pd** a decay time of 5 ps is observed. This decay is assigned to a transition from the equilibrated phenanthroline-centred ^{3}MLCT to an intra-ligand charge transfer (ILCT) state localised on the phenazine unit. These data show that the Pd(II) centre significantly influences the electronic structure of the bridging ligand by creating a significantly higher driving force for the formation of the phenazine-centred ILCT state which is formed after 5 ps, while for the precursor this state is populated after 240 ps. For **Ru-tpphz-Pd** this state further decays over 310 ps giving rise to the population of a long-lived ligand-to-metal charge transfer (LMCT) state on the Pd(II) centre, the decay of which was not observed on the time range accessible. The presence of the excited electron on the phenazine

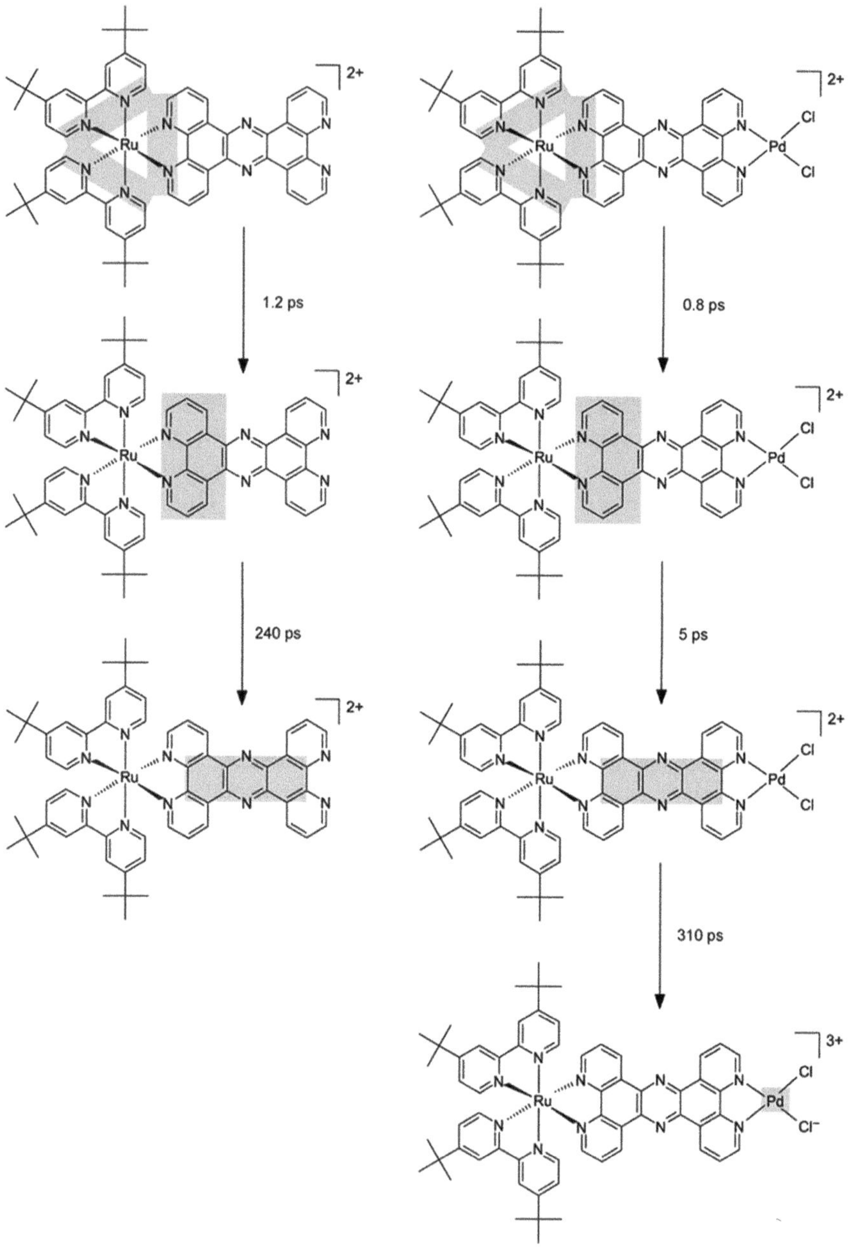

Figure 12.7 Electron transfer dynamics of **Ru-tpphz-Pd** and its precursor complex **Ru-tpphz**. Localisation of the photoexcited electron is illustrated by shaded areas. Reproduced with permission from ref. 26.

part of the bridging ligand is confirmed by electron paramagnetic resonance spectroscopy. The final reduction of the Pd(II) centre was expected to coincide with the dissociation of a Cl^- anion (see Scheme 12.1). This was supported by the observation that the catalytic activity of **Ru-tpphz-Pd** is completely inhibited by the addition of Cl^- to the catalytic mixture during catalysis and by DFT calculations.[25,18] Further confirmation for the interaction of the Pd centre with the excited state of the bridging ligand is the fact that while for the precursor complex emission is observed at 650 nm with a lifetime of 240 ns, for **Ru-tpphz-Pd** a relatively weaker emission at the same wavelength with a lifetime of <20 ns is noted. These results demonstrate that the compound behaves as expected on the basis of the principles of the molecular design chosen and the behaviour observed is closer to that sketched in Figure 12.3 than in Figure 12.5 indicating stronger coupling between the catalytic and light absorbing moieties in the compound.

An important observation from the previous section is that two different excited states, localised either on the bridge or on the peripheral ligands, are involved in the initial absorption processes. Figure 12.7 displays that upon initial light absorption of the peripheral ligands, a fast population of the phenanthroline bridge orbitals occurs. This behaviour is as expected if we consider the structural features of **Ru-tpphz-Pd** where the excited state of the bridging ligand is at a lower energy with respect to the excited state localised on the peripheral ligands. This leads to a number of questions: firstly, how is the photocatalytic efficiency of a particular compound affected by the wavelength of excitation? Secondly, how important is the nature of the peripheral ligands? These issues will be addressed in the following sections.

As outlined, the photophysical behaviour of **Ru-tpphz-Pd** (Figure 12.7) indicates that upon light absorption, the first populated excited state is located on the peripheral ligands. Moreover, the reaction pathway requires that population of the bridge takes place to allow for the photoinduced electron transfer to the catalytic centre. However, whether this intramolecular electron transfer process is efficient depends not only on the difference between the excited state energies of the peripheral and bridging moieties but also on the strength of their electronic coupling. The effect of the excitation wavelength on the hydrogen generation was assessed for **Ru-tpphz-Pd** and the irradiation wavelength was varied from 400 to 530 nm. The results obtained show that although the TON values obtained were highest when the irradiation wavelength approached the absorption maximum of the peripheral ligand, the catalytic efficiency obtained at lower energies of excitation is considerably higher than expected.[18] The catalytic activity of the compound does not strictly follow the absorption spectrum, but the absorption of photons with lower energy into the bridging-ligand-based ^{1}MLCT state leads to a higher catalytic activity than the absorption of high energy photons.

Considering that direct population of the bridge takes place at lower wavelengths, it appears that the catalytic efficiency is strongly correlated with the electron density located on the bridging ligand. Thus one way of optimising a photocatalyst is ensuring that light absorption leads to direct population of a high portion of bridging-ligand-localised excited states at all incident wavelengths, especially since polychromatic sunlight is regarded as our future energy source. These experiments show that by manipulation of the irradiation wavelength, different parts of the molecule can be excited, *i.e.* either bridging ligand or peripheral ligands, and that this can have important consequences for the capacity of the compound to generate hydrogen.

Another example of a wavelength-dependent photocatalytic process is observed for an analogue of **Ru-tpy-Pd** (see Figure 12.2), where in the photosensitiser moiety, the $Ru(bipy)_2$ centre is substituted by a $Ir(III)(ppy)_2$ moiety, where ppy = phenylpyridine, to form **Ir-tpy-Pd**.[27] The greater ligand-field stabilisation energy of the third row elements imparts an important advantage to Ir(III) complexes over Ru(II).[28] Consequently, Ir(III) compounds should exhibit greater stability as photosensitisers. This increased photostability is observed for a range of iridium complexes acting as photocatalysts for the photogeneration of hydrogen through an intermolecular mechanism of excitation transfer in the studies conducted by Bernhard and co-workers.[29,30] Surprisingly, few supramolecular systems based on iridium have been studied for the photocatalytic production of hydrogen through an intramolecular approach. Examples are the Ir–Co systems of Sakai[31] and Artero.[21,23] In the iridium studies carried out by Bernhard and co-workers,[29,30] (*vide infra*) visible light was used for the irradiation of the reaction mixtures. This is somewhat surprising since Ir(III) complexes generally exhibit only a weak absorption in the visible region, while in the UV region, much higher extinction coefficients are observed. To investigate whether irradiation into the stronger ligand-based UV transitions would yield higher TONs, **Ir-tpy-Pd** was irradiated at both 470 and 350 nm under the same experimental conditions.[27] Importantly, the TON obtained using 350 nm irradiation was only 16, whereas irradiation at 470 nm yielded a TON of 250. This is unexpected since the extinction coefficient of the intramolecular Ir–Pd compound at 470 nm is 1760 M^{-1} cm^{-1} while at 350 nm, the extinction coefficient is significantly larger at 36 312 M^{-1} cm^{-1}.

The results obtained may suggest that excitation around 350 nm into an intraligand or interligand π* transition does not lead to efficient electron transfer from the photosensitiser to the catalytic centre, while electron transfer is observed upon population of ^{3}MLCT states based on the bridge at lower energy. Time-dependent DFT studies indicate that 470 nm irradiation is likely to populate the lowest energy singlet and/or triplet excited states where electron density is redistributed from the Ir centre (including the cyclometallated ligands) towards the tpy (see Figure 12.2) bridging ligand and to a lesser extent towards the Pd centre. Thus, the Pd centre would be a better reducing agent following excitation at this wavelength. In contrast, irradiation at 350 nm can result in the population of both π*-type orbitals, while population of a state within the Pd MLCT set of states is also possible, the latter process resulting in electron transfer from the Pd centre towards the bridging ligand. Such an alteration of the electron distribution would diminish the reducing ability of the Pd centre and support the variation in TON with excitation wavelength. This observation shows that in order to achieve the most efficient photocatalytic efficiency, it is important to select a suitable excitation wavelength for photocatalytic studies; the choice should not simply be made on the basis of the transition with the highest extinction coefficient, but should also consider the nature of the electronic transition.

12.4.2 The Effect of Structural Parameters on Photocatalytic Behaviour

A detailed assessment of the photophysical properties of **Ru-tpy-Pd** shows that a large percentage of the light observed by peripheral ligands is deactivated by emission decay to the ground state and that the remainder transfers to the triplet state of the bridge. This indicates that the electronic coupling and/or energetics of the peripheral and bridging ligand are not optimised and that the nature of the peripheral ligands is more important than expected. Therefore, the photocatalytic behaviour of the compounds are dependent not only on the nature of the bridging ligand but also indirectly on the peripheral ligands. An example of such behaviour is observed for compounds **Ru-dpp-Pd** and **EtOOC-Ru-dpp-Pd**, see Figure 12.8.[32]

The absorption features of the two compounds are not significantly different. The dpp bridge, which contains a pyrazine unit, is expected to be the location of the lowest MLCT state and this is in agreement with the observation of the lowest MLCT transition observed at 530 nm.[33] Without ester-substituted bipyridine ligands, no hydrogen evolution was observed (**Ru-dpp-Pd**), however, upon the introduction of an ester group (**EtOOC-Ru-dpp-Pd**), a TON of 400 was obtained after 18 h irradiation with an irradiation wavelength of 470 nm. This observation suggests that esterification results in peripheral ligands, with better π-acceptor properties. This leads to a lowering of the π*-energy levels of these ligands and an enhanced electronic coupling

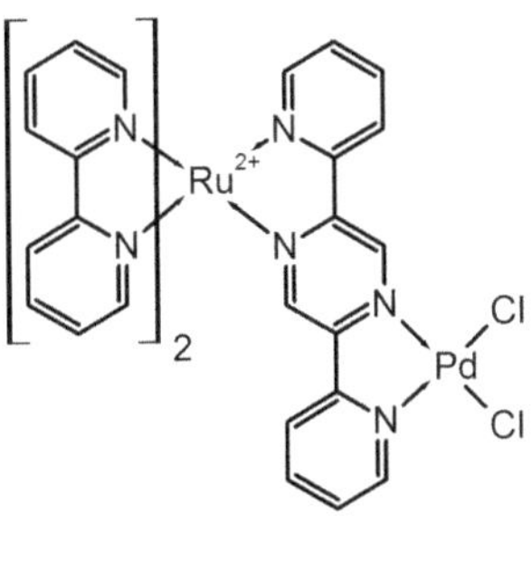

Ru-dpp-Pd

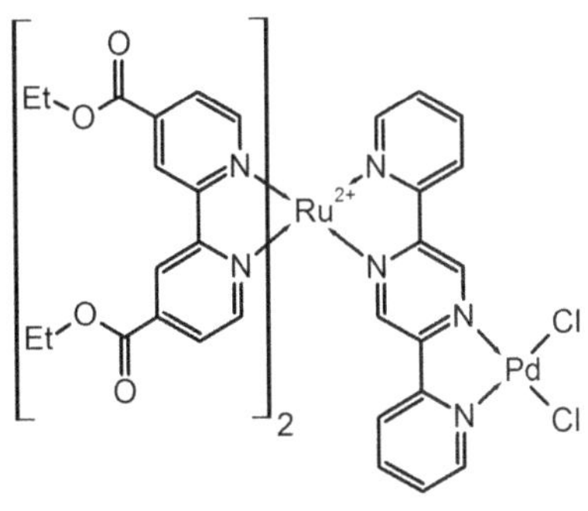

EtOOC-Ru-dpp-Pd

Figure 12.8 Structural features of 2,5-di(pyridin-2-yl)pyrazine-based compounds bridging dpp.

with the bridging ligand. It is important to note that because of the presence of a pyrazine moiety in dpp, the lowest ^{3}MLCT state will be located on the bridging ligand. However, while no hydrogen formation is observed for **Ru-dpp-Pd**, photocatalytic activity is evident for the ester analogue **EtOOC-Ru-dpp-Pd**. This indicates that the coupling between peripheral and bridging ligands is of prime importance.

Over the last number of years, the Brewer group has carried out detailed investigations on trimetallic assemblies. An example, compound **Ru-Rh-Ru-Cl**, is shown in Figure 12.9. The electrochemical and photocatalytic properties of these compounds were investigated and initially TONs in the range 36–270 were obtained after 20 h of irradiation depending on the nature of the peripheral ligands, bipy or phen, and the measurement conditions.[34] The authors also investigated the influence of the introduction of 4,7-diphenyl 1,10-phenanthroline (see Figure 12.9) and with this ligand a much improved TON value of 600 for this compound was reported, however these data were obtained under optimised conditions using a larger reaction volume (25 cm^3) and headspace (25 cm^3) during irradiation than used previously.[35]

In the supramolecular bimetallic assemblies under discussion, the choice of the bridging ligand is central to efficient hydrogen generation and careful consideration of its electronic properties and the distance between the light harvesting centre and the catalytic region is therefore necessary. The electronic properties of the bridging ligand are, for example, strongly dependent on its structural features as observed for **Ru-tpy-Pd** and **Ru-tp-Pd** (Figure 12.2). For **Ru-tpy-Pd**, a TON value of 130 after 18 hours' irradiation was obtained whereas **Ru-tp-Pd** produced no hydrogen.[19] The photophysical studies have shown transient absorption signals of the precursor [(bipy)$_2$Ru(tp)]$^{2+}$, compound **Ru-tp**, have almost disappeared within 200 ps confirming that, in contrast to **Ru-tpy-Pd**, the excited states in **Ru-tp** are short-lived.[24] The lack of photocatalytic hydrogen generation for the **Ru-tp-Pd** di-nuclear complex can be rationalised by the short excited state lifetime of the Ru(II) chromophore in the precursor. A metal-centred (^{3}MC) triplet excited state[36] is probably lying close in energy compared to the ^{3}MLCT states, providing a

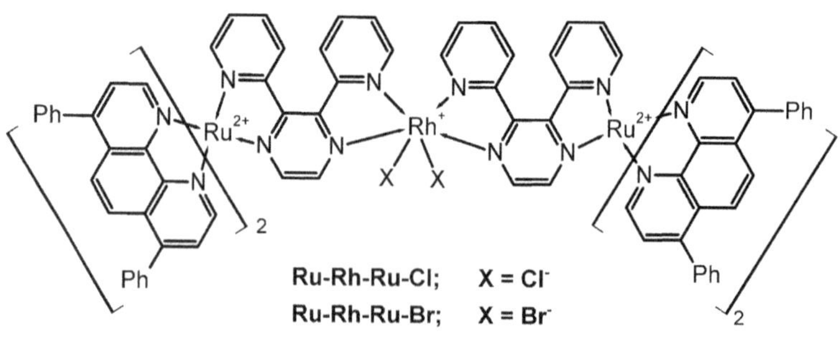

Figure 12.9 Structural features of typical Ru–Rh–Ru trinuclear photocatalysts, **Ru–Rh–Ru–X** photocatalytic assemblies. Bridging ligand: 2,3-di(pyridin-2-yl)pyrazine (2,3-dpp).

fast recombination channel to the ground state. The large steric hindrance induced by the **tp** ligand is probably the reason for such a ^{3}MC state to have relatively lower energy than that of the corresponding tpy-based complex. In addition, resonance Raman and DFT calculations indicate that for **Ru-tpy**, the lowest energy ^{3}MLCT orbital is based on the bridging ligand, while for **Ru-tp**, it is based mainly on the bipy ligands and the Ru centre. In addition to that, the considerably shorter lifetime of the excited state of **Ru-tp-Pd** compared to **Ru-tpy-Pd** (>10 *vs.* 100 ns) could represent another contributing factor.[19] This observation is in agreement with earlier reports that Ru–Pt diads coupled by the 2,3-dpp ligand do not show significant hydrogen generation.[10m]

Structural modifications of the bridging ligand of **Ru-tpphz-Pd** were carried out by Rau and co-workers and the photocatalytic properties of compounds **Ru-Br-tpphz-Pd** and **Ru-tpac-Pd** were investigated (Figure 12.6).[37] Introducing a brominated phenanthroline moiety as in **Ru-Br-tpphz-Pd** resulted in a TON value of 69 after 18 h of irradiation. By replacing the pyrazine ring of tpphz in **Ru-tpac-Pd** by an acridine moiety, a value of 139 was found. This compares with a value of 210 for **Ru-tpphz-Pd** under the same conditions (acetonitrile with 10% water and 2 M TEA). These observations show that the small variations in the chemical composition of the bridging ligand can have a significant effect on the efficiency of a supramolecular photocatalyst.

In the compounds discussed, catalytic centres based on Pd have been considered. In this section, some alternative options are considered. Brewer and co-workers have investigated a range of compounds of the type shown in Figure 12.9 using a trinuclear assembly containing two ruthenium centres and a Rh(III) catalytic centre capable of acting as a two-electron collector and as the catalytic centre.[34] In their studies they investigated the effect of changing the nature of the halide on the Rh catalytic centre. The photophysical properties of compounds **Ru-Rh-Ru-Cl** and **Ru-Rh-Ru-Br** indicate that the highest occupied molecular orbital (HOMO) is Ru-based and dπ in nature, while the lowest unoccupied molecular orbital has dσ* character and is located on the Rh-centred bridge of the assembly. This orientation of the molecular orbitals results in a low lying Ru → Rh metal-to-metal charge transfer (^{3}MMCT) excited state and the Rh centre takes on the role of electron acceptor in the molecule. Following electron transfer from the Ru to the Rh metal centre, Rh^+ is formed and it is this species that is responsible for hydrogen generation, halide loss being central to the catalytic process. The results obtained show that replacement of Cl^- (**Ru-Rh-Ru-Cl**) with Br^- (**Ru-Rh-Ru-Br**) did not result in significant photocatalytic activity as the TON value only increased from 28 for the chloride- to 38 for the bromide-containing compounds after 4 hours of irradiation. However, in later studies they observed greatly improved TONs for the bromide-containing compound after changing the catalytic conditions. By changing the solvent from acetonitrile to DMF, increasing the headspace, purging the system of hydrogen and by increasing the concentration of the sacrificial agents, the TONs for hydrogen formation increased to 280 after 19.5 h and a value of 820 was found after 50 h.[35] As discussed earlier, the presence of diphcnylphenanthroline as a peripheral ligand may also play a role.

Artero and co-workers have carried out detailed investigations on the potential of Co centres as catalysts for hydrogen generation.[21] The authors critically examined the effect of structural variation on the catalytic efficiency for hydrogen generation by systematic variation of both the bridging ligand and the nature of the photosensitiser (see Figure 12.10).

The Ru–Co complexes shown in Figure 12.10 are active as photocatalysts for the generation of hydrogen from water. In these complexes, the pyridine ring of the bridging ligand, (4-pyridine)imidazolo-[4,5-*f*]phenanthroline) (**Ru-PhenIm-Co**) or (4-pyridine)oxazolo-[4,5-*f*]phenanthroline (**Ru-PhenOx-Co**), is coordinated to a cobaloxime centre. A TON of 56 after 4 h (and 103 after 15 h when a 350 nm UV cut-off filter was used) was reported for **Ru-PhenOx-Co** in acetone using Et_3NHBF_4 as the proton source. Altering the structure of the bridge in **Ru-PhenOx-Co** by the replacement of the oxazolo moiety with an imidazolo unit to form compound **Ru-PhenIm-Co** led to a TON of 104 after 4 h under the same conditions. Thus, the nature of the bridge influences directly the capacity of the photocatalyst for generating hydrogen. This was further confirmed by examining the photocatalytic activity of individual Ru and cobaloxime units, which demonstrated that no hydrogen was detected in the absence of the bridging ligand. Therefore, the bridging ligand facilitates electron transfer from the Ru photoactive centre to the catalytic cobaloxime centre either through bonds or by an outer-sphere mechanism. An investigation of the iridium analogue (**Ir-PhenIm-Co**) of Ru (**Ru-PhenIm-Co**) under the same catalytic conditions yielded enhanced hydrogen production, with a TON of 210 after 15 h. This study elegantly demonstrates how catalytic centres based on first row transition metals can produce photocatalytic devices that are economically viable for use in the photocatalytic generation of molecular hydrogen.

An alternative Ir–Co assembly **Ir-Co-Ir** was considered by Sakai *et al.*[31] Using visible light irradiation for 8 hours, a TON value of 20 was reported for this compound. Based on NMR and electronic absorption spectroscopy, the authors suggested that compound **Ir-Co-Ir** forms by self-assembly (the Ir–Co ratio used was 3 : 1) in solution. The TON values are given with respect to the Co concentration. This value dropped to 10 for the corresponding multicomponent system using the model compounds $[Ir(ppy)_2(bipy)]^+$, and $[Co(bipy)_3]^{3+}$. It is noteworthy that the bridge utilised is not conjugated and thus is not expected to promote intramolecular electron transfer.

12.4.3 Intermolecular *vs.* Intramolecular Photocatalysis

As outlined in the Introduction, the intermolecular approach was initially considered when the first studies on the photochemical generation of H_2 were reported by Lehn *et al.* in the early 1980s.[38] These experiments focused on the use of $[Ru(bipy)_3]^{2+}$ as a photosensitiser in the presence of $CoCl_2$ in a $CH_3CN–H_2O–Et_3N$ mix, where the photosensitiser and catalyst were assessed for their potential to reduce CO_2 and/or generate hydrogen using visible light. Both H_2 and CO were detected, with a Co(I) species regarded as the 'reactive

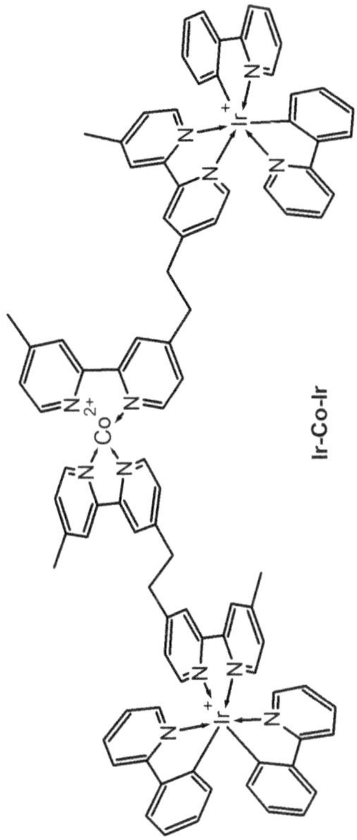

Ir-Co-Ir

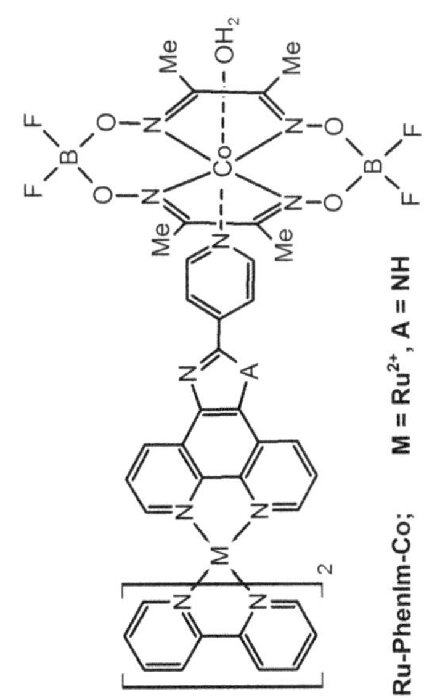

Ru-PhenIm-Co; M = Ru^{2+}, A = NH
Ru-PhenOx-Co; M = Ru^{2+}, A = O
Ir-PhenIm-Co; M = Ir^{3+}, A = NH

Figure 12.10 Structural features of cobalt-based photocatalysts.

intermediate' which formed upon reduction of Co(II) by the photoreduced ruthenium sensitiser. Subsequent studies demonstrated that the nature of the amine (*i.e.* sacrificial agent), solvent employed, photosensitiser, and ligands attached to the catalysts dramatically influence the catalytic activity of the system. In the subsequent thirty years, a multitude of photosensitisers, catalysts and reaction conditions were employed, with the amount of hydrogen formed ranging from very low TONs to values close to 9000. It is at this stage difficult to accurately determine the best catalyst/photosensitiser, or indeed the ideal conditions, as there are subtle differences in experimental conditions in individual research laboratories. Consequently, in this section, examples of intermolecular photocatalytic systems with some of the highest TONs reported to date will be discussed. Some examples comparing inter- *vs.* intramolecular approaches from our laboratories will be introduced, thereby allowing a comparison of the two approaches. For a comprehensive review on intermolecular approaches to hydrogen generation, there are a number of excellent reviews.[10]

In terms of photosensitisers, both organic (for example porphyrins, rhodamine) and inorganic (containing metals such as Re, Ru, or Ir) systems have been employed, with catalysts based on metals such as Pt, Pd, Rh and Co. A number of studies have focused on the efficiency of the photosensitiser. Four different photosensitisers were assessed using the same cobalt catalyst and the order of activity was $[Re(CO)_3Br(diimine)] > [Ir(ppy)_2(diimine)]^+ > [Ru(dmphen)_2(diimine)]^{2+} > [Ru(bipy)_2(diimine)]^{2+}$.[21] The superior performance of the Re and Ir photosensitisers was attributed to the excited state undergoing faster reductive quenching when compared to the studies on Ru complexes.

Bernhard and co-workers have shown that the multicomponent systems containing $[Ir(ppy)_2(bipy)]^+$, together with K_2PtCl_4 as a catalyst in 9 : 3 : 1 acetonitrile–water–triethanolamine TEOA solutions (λ = 465 nm, 20 h), produce hydrogen without the need for an electron relay.[39] In these experiments, a TON of 63 was achieved after 18 hours of irradiation, which equates to a quantum yield for hydrogen production of 0.26. A structure–activity study followed, which focused on the bis-cyclometallated bipyridyl iridium photosensitiser and a platinum or palladium catalyst.[40] The palladium catalyst performed better than the platinum catalyst. Additionally, when experiments were performed in THF-based solvents, results were approximately 10 times those observed in the CH_3CN systems. In this study, TONs varied from 50 to 310 (following 6 hours' irradiation at 460 nm) depending on the nature of the groups attached to the bipy ligand as shown in Figure 12.11. Subsequently, this group extended their studies using iridium photosensitisers, and utilised a Pt or Rh catalyst in TEA–H_2O–THF mixtures (λ = 460 nm, 48 h).[41] In this instance, the TONs for hydrogen were significantly higher, and varied depending on the groups on both the cyclometallated and bipy ligand. For instance, the TONs for the dimethylbipy iridium complexes ranged from approximately 900–1600. Replacing the methyl groups with one or two vinyl groups led to a dramatic increase in TON ranging from 5700 to 8500 following irradiation. Notably, all the vinyl bipy complexes are characterised by lower luminescence quantum yields and shorter lifetimes than their mono- or dimethyl analogues. The authors have explained this five-fold increase on replacing a methyl with a vinyl group as arising from an interaction of the vinyl moieties on the bipy ligand with colloidal Pt formed *in situ*.

Mono- and dinuclear iridium-containing bibenzimidazole complexes shown in Figure 12.12 were recently reported for their ability to produce hydrogen in acetone–H_2O–TEOA mixtures together with a rhodium catalyst, where visible light (λ > 420 nm) was used as the excitation source.[42] The dinuclear complex yielded, after 72 hours of irradiation, a TON of 3750 for H_2, with a dramatically lower value of 1020 for the monomer analogue. The authors attributed this variation to differences in molar absorptivity, as the dinuclear species absorbed significantly more than the monomer in the visible region of the spectrum. This possibility was ruled out in the studies described earlier by Bernhard and co-workers, as in their case the absorption of both the dimethyl and vinyl iridium complexes are similar at the excitation wavelength employed.

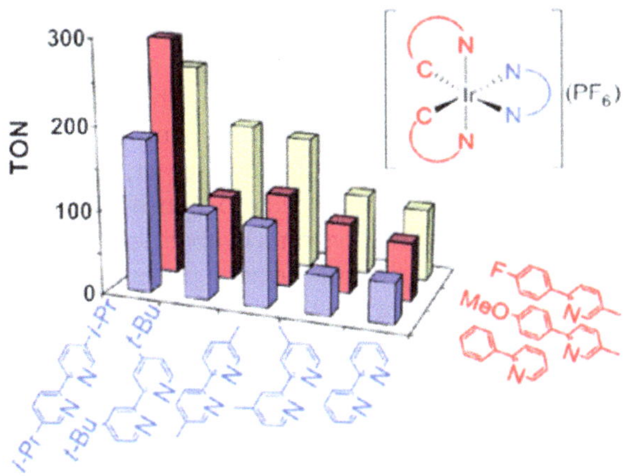

Figure 12.11 The effect of the substitution of Ir complexes on TON values obtained. Adapted from ref. 41*b*.

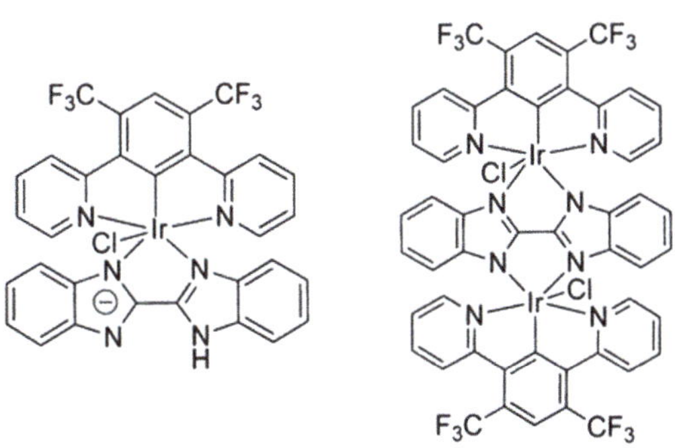

Figure 12.12 Mono- and dinuclear bibenzimidazole iridium complexes.

As already outlined in the Introduction, intramolecular approaches in designing molecular systems for the photocatalytic generation of hydrogen have been developed on the premise that electron transfer between the photosensitiser and the covalently attached catalytic centre will be significantly faster. Hence higher efficiencies for hydrogen production with respect to intermolecular cases are expected. In the intermolecular examples discussed earlier, physical diffusion of both catalyst and photosensitiser is required, and electron transfer between the photoexcited photosensitiser PS* and the catalyst requires a high frequency of reactive collisions.[43] In the next section, inter- and intramolecular options are compared for a series of compounds. In these studies, the hydrogen generation efficiencies of supramolecular assemblies are compared with those obtained for the precursors of these compounds under the same experimental conditions.

Intra- and intermolecular experiments were carried out by us using the $Ir(ppy)_2$ analogue **Ir-tpy-Pd** of **Ru-tpy-Pd** (see Figure 12.2) as the supramolecular assembly.[20] Intermolecular experiments were performed in CH_3CN–H_2O–TEA using the precursor **Ir-tpy** as a photosensitiser and $[Pt(CH_3CN)_2Cl_2]$ or $[Pd(CH_3CN)_2Cl_2]$ as the catalytic centres. Solutions containing 1 : 1 molar ratios of **Ir-tpy** and the catalyst yielded TONs of 245 and 166 for Pd and Pt catalysts, respectively, after 18 hours of irradiation with a wavelength of 470 nm. Under the same conditions, TON values of 254 and 308 were obtained for the supramolecular assemblies **Ir-tpy-Pd** and **Ir-tpy-Pt**, respectively. Although differences in TONs obtained are observed, no significant variation between intermolecular and intramolecular systems is observed.

Another example of inter- *vs.* intramolecular photocatalysis[19] is based on compound **Ru-tpy-Pd** (see Figure 12.2). Irradiation experiments were performed in CH_3CN–water (90 : 10) mixtures in the presence of 2.3 M TEA using 470 nm irradiation for 18 hours. Under these conditions, the intermolecular approach employing a 1 : 1 mixture of **Ru-tpy** and $[Pd(CH_3CN)_2Cl_2]$ yielded a TON of 70, while the intramolecular system produced a TON value of 130. The time dependent behaviour for this reaction was also studied for an 8 hour period. For **Ru-tpy-Pd**, hydrogen generation is greatest during the first 4 hours, whereas the intermolecular mixture shows little activity during this time period, therefore indicating an induction phase.

In a related study focusing on the compound **EtOOC-Ru-dpp-Pd**[32] (see Figure 12.8), a different type of behaviour was observed. In this instance, when a solution of the **EtOOC-Ru-dpp** precursor and $[Pd(CH_3CN)_2Cl_2]$ were irradiated, a TON of 450 was obtained, which is slightly higher than the corresponding TON of 400 for the intramolecular **EtOOC-Ru-dpp-Pd** complex. The data observed suggest that the **EtOOC-Ru-dpp-Pd** complex may be formed *in situ*, but further studies are required to fully understand the basis for this behaviour.

These studies comparing the inter- and intramolecular approach demonstrate that generally, the intramolecular photocatalytic process does generate higher TON values, but the difference is not that impressive and in most cases is less than an order of magnitude. Other groups using cobalt catalysts have also reported TONs for both the intra *versus* inter approach and have observed enhanced TONs for the intramolecular mechanism.[21,23]

A number of groups have probed the importance of the electron transfer rates between the photosensitiser and the catalyst in intermolecular systems. Such studies are of great importance as little is known about the mechanism of photocatalytic hydrogen generation and such data are needed to be able to compare and understand the differences between the inter- and intramolecular approach. Hamm and co-workers have eloquently demonstrated how time-resolved IR spectroscopy can be used to determine rates for electron transfer and also the quenching rate for a number of rhenium carbonyl photosensitisers where Co(III) glyoxime was used as the catalyst.[44] In these experiments, [Re(NCS)(CO)$_3$bipy] was used as the PS, [Co(dmgH)$_2$] (dmgH$_2$ = dimethylglyoxime) as the catalytic centre, and TEOA as the sacrificial agent, [HTEOA][BF$_4$] as the hydrogen donor, and 355 nm was employed as the excitation wavelength. Outlined in Scheme 12.3 are the reaction pathways and associated rate data discussed here, which were obtained in DMF.

A ^{3}MLCT-based rhenium excited state is generated following excitation, which in the presence of TEOA undergoes reductive quenching with $k_q = 9.7 \times 10^7$ M^{-1} s^{-1}. Subsequently, electron transfer ($k_1 = 1.3 \times 10^8$ M^{-1} s^{-1}) to the catalytic cobalt centre occurs yielding a cobalt(I) species ($Co^{II} \rightarrow Co^{I}$). The lifetime of the reduced PS$^-$ species varies (from 8.9 to 2.7 μs) as the concentration of the Co^{II} catalyst is increased (for instance from 1 to 3 mM). Following this electron transfer step, the cobalt centre is protonated ($Co^{I} \rightarrow Co^{III}$–H), and this cobalt hydride intermediate generates H$_2$ *via* a bimolecular reaction ($k_3 = 25$ M^{-1} s^{-1}). This reaction is regarded as the rate-determining step. Time-resolved infrared spectroscopy confirmed that photo-oxidised TEOA generates a second reducing equivalent which can be transferred to [Re(NCS)(CO)$_3$bipy] ($k_{2e}^- = 3.3 \times 10^8$ M^{-1} s^{-1}) if no Co(dmgH)$_2$ is present. TONs of 6000 for H$_2$ production with respect to rhenium were observed, while TON values

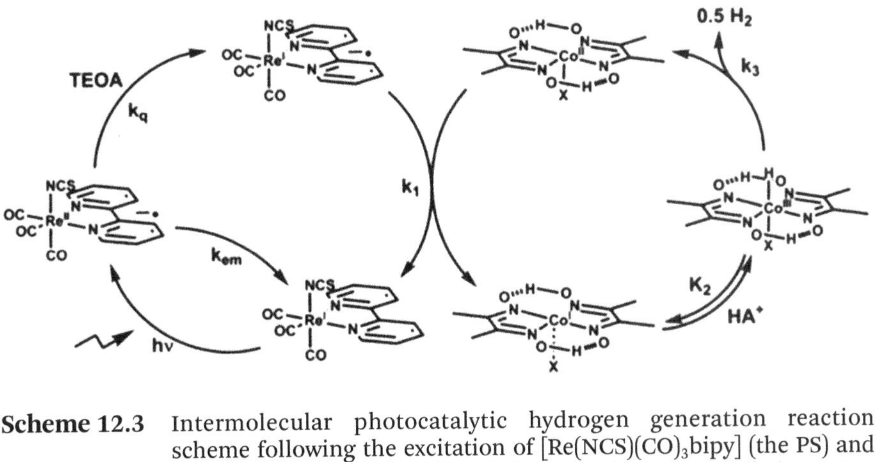

Scheme 12.3 Intermolecular photocatalytic hydrogen generation reaction scheme following the excitation of [Re(NCS)(CO)$_3$bipy] (the PS) and [Co(dmgH)$_2$] (dmgH$_2$ = dimethylglyoxime) as the catalytic centre. Reproduced with permission from ref. 44c. Copyright (2010) American Chemical Society.

with respect to cobalt never exceeded 1000. This indicates decomposition of the Co(dmgH)$_2$ catalyst. Similar kinetic data have been reported for other systems.[45] The kinetics obtained in these studies for the electron transfer process indicates a second order process that is dependent on the concentration of the catalyst. A concentration of 10^{-4} M for the catalyst translates to a first order rate constant for electron transfer between the two components of 10^4–10^5 s^{-1}.

Similar rates were previously reported by Espenson and Chao[46] when a hydridocobaloxime system was studied for its ability to generate hydrogen. Two different catalytic processes were observed when the concentration of the reactants was varied. A pseudo first order rate constant of 6×10^{-3} s^{-1} was observed at high H^+ and low cobaloxime concentrations. However, when the concentration of the hydridocobaloxime was increased, a second order process with a rate of 2×10^2 M^{-1} s^{-1} was observed. McCormick *et al.* reported the importance of the lifetime of the excited state for the intermolecular hydrogen generation using modified rhodamine as the PS and a cobalt-based glyoximate.[47] When the oxygen atom in rhodamine was substituted with S or Se, turnover frequencies (TOF) observed ranged from 0, 1700 and 5500 h^{-1}, respectively. The rhodamine ^{3}PS excited state is >1000 times longer than the singlet state (0.12–2.5 ns), thus facilitating bimolecular electron transfer which is necessary for the completion of H_2 formation as discussed in the earlier paragraphs. The Φ_{H2} yields also ranged from 0%, 12.2% and 32.8% respectively for these dyes (on replacing O with S, or Se) and correlate directly with the quantum yields for singlet oxygen generation, Φ_{O2}. The Φ_{O2} yields are 0.05, 0.17 to 0.67 for rhodamines with O, S and Se, respectively, as heteroatoms. The increase of TOF values with the increase of the atomic weight of the heteroatom of rhodamine is very impressive and determines the efficiency of the photosensitiser. In the case of the S and Se compounds, the PS* is located on a triplet excited state, while for the O compound, population of a singlet excited state is expected. It is worth noting that the lifetimes obtained vary between 1 and 2 ns. These values are still very short when compared to inorganic photosensitisers such as Ru and Ir complexes and it is somewhat surprising that such short lifetimes yield such high efficiencies. Previously, the group reported increased TONs for hydrogen formation in a xanthene dye when H was replaced with a halide. This was attributed to enhanced intersystem crossing (ISC) promoted by the heavy atom effect.[48]

12.5 Photocatalytic Assemblies for CO_2 Reduction

12.5.1 Introduction

The light-driven conversion of carbon dioxide (CO_2) into its closest reduction product, *i.e.* carbon monoxide (CO) involves the breaking of one C=O double bond. This process requires the absorption of 187 kcal mol^{-1} (or about 8 eV per single molecule),[49] which corresponds to a radiation wavelength of *ca.* 155 nm. This clearly shows that visible light (400–700 nm) cannot be used directly for the photo-induced transformation of CO_2 into CO or any other of its reduction products such as formaldehyde (CH_2O), methanol (CH_3OH) and methane (CH_4), the formation of which require at some stage the breaking of at least one C=O double bond present in the carbon dioxide substrate. When formate (HCO_2^-), oxalate [$(C_2O_4)^{2-}$] or their corresponding acidic forms are formed as reduction products of CO_2, the activation energy is reduced to approximately half (93 kcal mol^{-1}) with respect to CO formation. This is explained by the fact that their generation does not require the elimination of an oxygen atom from the starting material CO_2, but only the breaking of the weakest π-bonds between carbon and oxygen atoms. A light-induced activation of this second series of processes requires the absorption of UV radiation at about 300 nm (4 eV), which falls well outside the visible spectral range. Therefore, the development of a photocatalytic approach utilising visible light for the transformation of CO_2 requires the employment of light-absorbing mediators that are able to activate appropriate catalysts. This approach mimics the chlorophyll-activated photosynthesis for the transformation of water and carbon dioxide into cellulose. Besides light absorption *via* molecular photosensitisers, other options for CO_2 activation include electrochemical and photoelectrochemical routes[50] in which electrons are added to CO_2 *via* dark or photocatalytic electrolytic processes. The activation of CO_2 achieved through the electrochemical addition of an electron into the LUMO of CO_2, with formation of the radical anion $CO_2^{\bullet -}$, occurs at −1.9 V *vs.* the normal hydrogen electrode NHE.[49] The acquisition of the first electron by CO_2 occurs as described in eqn (12.1) at this very negative reduction potential due to the high stability of the CO_2 molecule as confirmed by its very negative free energy of formation being $\Delta G°f(CO_2) = -394$ kJ mol^{-1} [49] and the high position of the LUMO (−2.1 V *vs.* saturated calomel electrode as determined electrochemically in anhydrous dimethylsulfoxide).[51]

$$CO_2 + 1e^- \rightarrow CO_2^{\bullet -} \qquad (12.1)$$

Progress of the reaction described in eqn (12.1) is determined in part by the limited solubility of CO_2 in water and in organic solvents (in the order of a few grams per kilogram of solvent),[52] and by the tendency of the species to be converted into CO_3^{2-}, HCO_3^- and carbonic acid, H_2CO_3, in aqueous electrolytes in ratios that depend on the solution pH. These factors make the electrochemical reduction of CO_2 according to eqn (12.1) an extremely complicated process and this results generally in low current densities. In the absence of any light activation, these combined issues require large overpotentials (with respect to the range of electrochemical stability of water and several organic solvents) to realise the primary reduction of CO_2. Indeed, to carry out this process electrochemically, several important experimental conditions need to be controlled, such as the choice of the electrolyte and electrode material.[51] In particular, the electrolyte should not contain any water, and aprotic anhydrous solvents are required to prevent molecular hydrogen formation.[53] For the same reason, the materials constituting the cathode must present large overpotentials towards the reduction of hydrogen, *e.g.* Au, Pb, Hg, W or glassy carbon,[54,55] in order to afford the selective reduction of carbon dioxide (eqn (12.1)). If the activation of carbon dioxide through its monoelectronic reduction (eqn (12.1)) is crucial for the increase of its reactivity, the specificity of the products formed following this process is very limited as depicted in Scheme 12.4.

This is mainly a consequence of the generally high reactivity of the radical anion $CO_2^{\cdot -}$ that dimerises to form oxalate (reaction 2 in Scheme 12.4), or reacts with the parent substrate according to Lewis acid–base reactions 3 and 3′ in Scheme 12.4.[56] The products obtained from the various routes of $CO_2^{\cdot -}$ recombination are generally non-innocent in terms of electrochemical reactivity since they can undergo further reduction processes (see reactions 4, 4′, 5 and 5′ in Scheme 12.4) at the potential of $CO_2^{\cdot -}$ generation.[57] Besides the electrochemical route, which represents a typical heterogeneous process,[58–61] the activation of carbon dioxide *via* mono- (eqn (12.1)) or multi-electron[62] reduction can be carried out in two other ways: (i) chemically, by means of molecular catalysts as electron donating species that coordinate CO_2 efficiently (homogeneous process);[63,64] (ii) electrocatalytically, by applying a molecular catalyst either anchored on the cathode or dissolved in the electrolyte[65–69] with the aim of decreasing the activation energy for the electrochemical reduction of CO_2. The resulting reduction product $CO_2^{\cdot -}$ possesses a nucleophilic character,[70] and bears great synthetic importance since it can form organic compounds of interest upon the addition of appropriate electrophilic substrates.[71,72]

From simple energetic considerations, it can be shown that the monoelectronic reduction of CO_2 can be photoactivated with visible light with an energy higher that 1.9 eV ($\lambda < 630$ nm) (see eqn (12.2)).

$$CO_2 + M + h\nu \rightarrow [CO_2\text{–}M^*] \rightarrow CO_2^{\cdot -} + M^+ \qquad (12.2)$$

This requires the presence of a sensitiser[73] that acts as a photocatalyst with electron-donating properties. Upon light absorption, this photocatalyst transfers electron(s) to CO_2 either homogeneously[74] or heterogeneously[75] as a consequence of the initial formation of an excited state.[76] An alternative method of reducing CO_2 through visible light makes use of photoelectrochemical cells in which p-type semiconductor cathodes are activated by light and transfer electrons from the conduction band of the semiconductor to the LUMO of CO_2.[77] This approach is based on a process of photoelectrolysis[78] in which the electrical current accomplishes the same function as the sacrificial agent in the analogous process of photochemical reduction.[79] Photoelectrolysis presents the advantage of avoiding the need for a sacrificial chemical agent and, more importantly, this results generally in larger turnover numbers (TONs).[80,81] Despite the better suitability of the photoelectrochemical procedure, the most widely used approach for reducing CO_2 with visible light is in solution,[82,83] which aims to deliver the general relationships that correlate the molecular structure of the dye-sensitiser to its photocatalytic activity. Such relationships are valid when the mechanism of photo-assisted CO_2 reduction is known within a defined group of photocatalysts.[84] It is generally

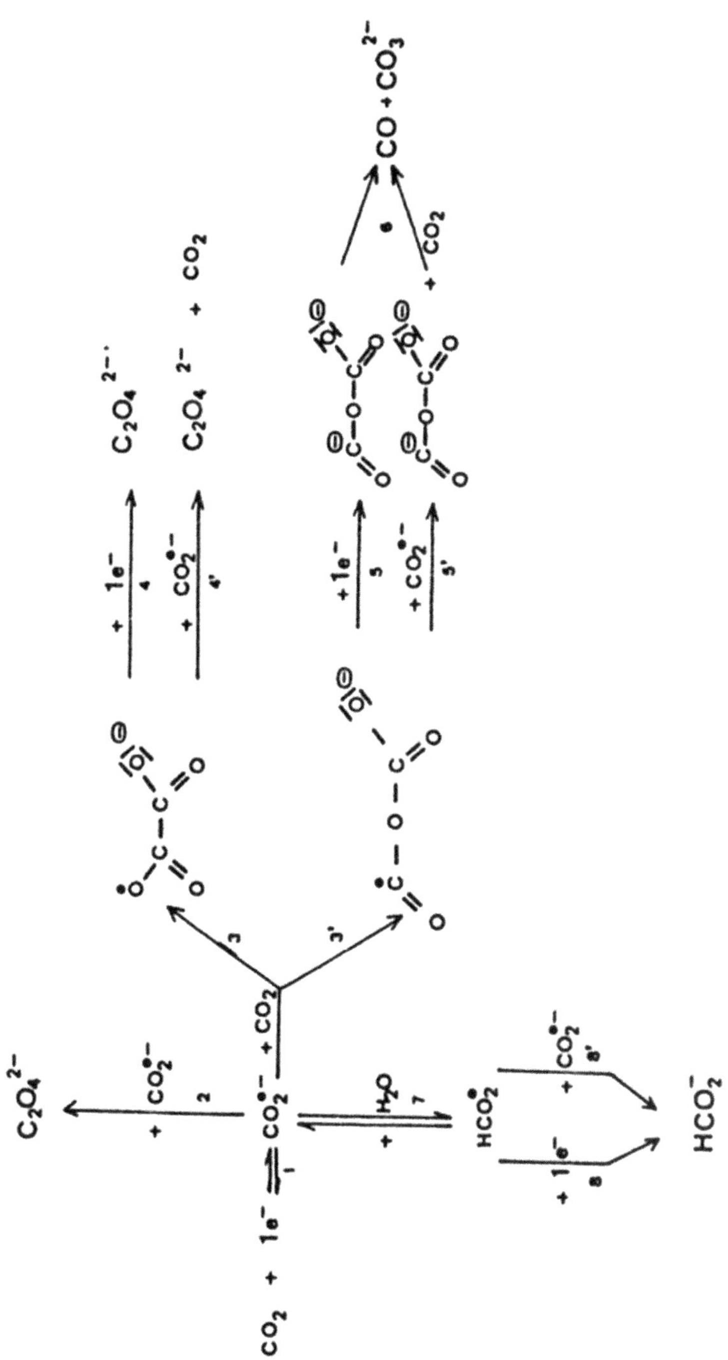

Scheme 12.4 Summary of the possible chemical and electrochemical processes following the mono-electronic reduction of carbon dioxide. Adapted from ref. 51.

difficult to obtain such data because of the high specificity of the interactions between the molecular photocatalyst and the CO_2 substrate.[85] Unfortunately, this limits the full rationalisation of the observed trends when different classes of sensitisers are employed.[80] In the next section, a number of examples of solution-based photocatalytic approaches for CO_2 reduction are discussed. In the literature, most studies are based on mixtures of photosensitisers and catalysts so a number of examples of such intermolecular light-driven CO_2 transformations are included. The application of supramolecular photocatalytic assemblies is also considered; particular attention is paid to the nature of the bridging ligand and its relationship to the photocatalytic behaviour observed.

12.5.2 Photocatalytic Assemblies for CO_2 Reduction

12.5.2.1 *Intermolecular Assemblies*

Lehn and Ziessel reported the first visible-light-driven CO_2 reduction by a molecular system as early as 1982.[86] The authors described an intermolecular homogeneous system comprising a $[Ru(bipy)_3]^{2+}$ photosensitiser and $CoCl_2$ as a catalyst precursor in acetonitrile–water using triethylamine as a sacrificial electron donor. Carbon monoxide and hydrogen were received as products with a low turnover number of 9 after 22 hours, which could later be enhanced to TON = 33 (15 h) by using DMF as a solvent and triethanolamine (TEOA) as a sacrificial electron donor.[38] In the same report, the authors also described the use of *fac*-$[Re(bipy)CO_3X]$ complexes (X = Cl^-, Br^-), which catalysed the reduction of CO_2 to CO with a turnover number of 48 after 4 hours in DMF, with triethanolamine as a sacrificial electron donor and $Et_4N^+Cl^-$ as an additive. CO and H_2 were produced simultaneously in the presence of the rhenium complex and $[Co(bipy)_3]^{2+}$. Interestingly, in the absence of the cobalt complex, CO was selectively produced and no hydrogen formation could be detected. In this context, the rhenium complex acts as a light-absorbing unit and as a catalyst at the same time. The addition of $Et_4N^+Cl^-$ increased the stability of the reaction components (*vide infra*). These results triggered further investigations on rhenium(I) photoreduction catalysts, which are now the most investigated complexes in the field of light-driven CO_2 reduction. One year later, in 1983, Lehn and co-workers also reported on the electrocatalytic reduction of CO_2 to CO by the *fac*-$[Re(bipy)CO_3Cl]$ catalyst with a Faradaic efficiency of 98%.[38,87] The electrocatalytic properties of similar Re(I) complexes were also investigated.[88,89]

While so far $[Ru(bipy)_3]Cl_2$ has served as a photosensitiser, it may also act as a precatalyst. In the absence of $CoCl_2$ (*vide supra*) and under visible light irradiation in DMF–TEOA, $[Ru(bipy)_3]Cl_2$ reduces CO_2 to formate with TONs up to 27.[38] Photo-induced cleavage of one bipyridine ligand is necessary to enable the formation of a bis(bipyridine)ruthenium species which can form hydrides that in turn insert CO_2 and CO. Carbonyl and hydride complexes that are likely intermediates in these processes, such as $[Ru(bipy)_2(CO)Cl]^+$,

$[Ru(bipy)_2(CO)_2]^{2+}$ and $[Ru(bipy)_2(CO)H]^+$, were synthesised by Kelly and co-workers.[90–94] The complex $[Ru(bipy)(tp)(CO)]^{2+}$ catalysed the electrochemical reduction of CO_2 to CO and small amounts of H_2, in the absence of water.[95,96] In the presence of water, formate became the major product and CO and H_2 were found as minor products. Moreover, a 1:1 mixture of $[Ru(bipy)_3]^{2+}/[Ru(bipy)_2(CO)_2]^{2+}$ in TEOA–DMF photochemically reduced CO_2 to formate under illumination at $\lambda > 320$ nm for 20 hours and furnished a maximum TON of 158 and a quantum yield for formate formation of 14%.[97] Interestingly, the use of 1-benzyl-1,4-dihydronicotinamide (BNAH) as a sacrificial electron donor instead of TEOA yielded lower amounts of formate but appreciable amounts of CO.

The discovery of the Re-catalysed photoreduction of CO_2 triggered extensive research effort in this field, leading to a wide range of inter- and intramolecular systems. These developments are discussed later. However, we shall bear in mind that rhenium, like ruthenium, is among the rarest metals on earth. Although catalysts based on, for example, rhenium or ruthenium, have played an important role in pioneering developments in light-driven carbon dioxide reduction, their scarcity precludes their large-scale technical use. Research efforts are thus increasingly aimed at developing photocatalytic systems based on earth-abundant metal centres.

Manganese, the lighter rhenium homologue is much more earth abundant. Electrocatalytic studies found the manganese(I) species *fac*-$[Mn(bipy)(CO)_3Br]$ to be active in the reduction of CO_2 to CO.[98,99] Interestingly, in an intermolecular photoreduction process, formic acid was found as the main product when *fac*-$[Mn(bipy)(CO)_3Br]$ was photosensitised by $[Ru(dmb)_3]^{2+}$.[100] Under illumination with visible light (≥480 nm), besides small amounts of CO and H_2, formic acid was produced with a TON = 149 (12 hours) and a quantum efficiency of 6%. The formation of different products using the same pre-catalyst $[Mn(bipy)(CO)_3Br]$ is a consequence of different reaction mechanisms in the electro- or photocatalytic processes. Formation of the soluble dimer $[Mn(bipy)(CO)_3]_2$ is an important reaction step and occurs electrochemically by one-electron reduction of $[Mn(bipy)(CO)_3Br]$ to $[Mn(bipy)(CO)_3Br]^-$ and subsequent formation of $[Mn(bipy)(CO)_3]^{\cdot}$ by loss of $Br^{\cdot}$, followed by dimerisation of $[Mn(bipy)(CO)_3]^{\cdot}$ to $[Mn(bipy)(CO)_3]_2$.[101] The photochemical dimerisation process (in the absence of sacrificial donors or photosensitisers) proceeds *via* photoinduced rearrangement of the *fac*-$[Mn(bipy)(CO)_3Br]$ isomer to the *mer*-$[Mn(bipy)(CO)_3Br]$ and subsequent photo-induced loss of $Br^{\cdot}$, followed by dimerisation of $[Mn(bipy)(CO)_3]$ to $[Mn(bipy)(CO)_3]_2$.[102] The formation of $[Mn(bipy)(CO)_3]_2$, important for the reduction of CO_2 to CO, can be followed by UV-vis and IR-spectroscopy. In the photocatalytic CO_2 reduction to formic acid by $[Mn(bipy)(CO)_3Br]/[Ru(dmb)_3]^{2+}$, the occurrence of dimer $[Mn(bipy)(CO)_3]_2$ signals was directly related to the formation of CO as a by-product. However, CO formation ceased within the first 10 min alongside the disappearance of the dimer $[Mn(bipy)(CO)_3]_2$. In the course of the reaction, hitherto unknown compound(s) formed, which led to the formation of formic acid as a major product.

The majority of CO_2-reducing photocatalysts yield the two-electron reduction products CO or HCOOH. Reduction products requiring the addition of several electrons, such as H_2CO ($4e^-$), H_3COH ($6e^-$), CH_4 ($8e^-$), are only rarely obtained. A surprisingly simple non-metal containing catalyst, namely pyridine, for the reduction of CO_2 was recently proposed. Methanol was selectively produced from CO_2 in a photoelectrochemical experiment using a p-GaP electrode and pyridine as a catalyst, with high Faradaic efficiencies (88–100%).[103] The first mechanistic investigations proposed the formation of a pyridine-H$^{\cdot}$ species as the key intermediate by one-electron reduction of a pyridine-H$^+$ (pyridinium) and subsequent formation of a pyridinium carbamate species by addition of CO_2.[104] Theoretical studies suggest that the one-electron reduction of pyridine-H$^+$ to pyridine-H$^{\cdot}$ is an electrode-surface-assisted process based on thermodynamic considerations.[105–108] However, the identification of the complete reaction mechanism remains a matter of research. Other experiments showed that a metal or semiconductor surface is not necessarily required. Only very recently, the homogeneous, intermolecular, pyridine-catalysed photoreduction of CO_2 to formate and methanol was reported. This process occurs in homogeneous aqueous solution using $[Ru(phenanthroline)_3]^{2+}$ as the photosensitiser and ascorbic acid as the sacrificial electron donor.[109] Formate was found as the main product with a maximum TON = 76, after 6 hours on irradiation with 470 nm LED light (TON referenced to the molar amount of photosensitiser). The yield of the photocatalytic process critically depends on a variety of parameters. Examination of the catalyst to photosensitiser ratio showed that an excess of catalyst was needed while an excess of the photosensitiser yielded no products. The pH value was found to be critical and the best results were obtained with a pH of around 5. A significant decrease of product formation was observed at pH 4 or 6. Addition of KCl significantly increased the product formation while addition of LiCl, NaCl, RbCl, CsCl, $MgCl_2$ had little or no effect. In an intramolecular approach, $[Ru(phen)_2(ptpb\beta)]^{2+}$ was employed both as photo- and electrocatalyst (ptpbβ = pyrido[3′,4′:5,6]pyrazino[2,3-*f*][1,10]phenanthroline). Both processes yield formate, formaldehyde and methanol.[110] The latter examples clearly demonstrate that CO_2 reduction catalysts are not limited to metal centres, metal or semiconductor surfaces but may also comprise organic moieties such as pyridine or analogous compounds.

Cobalt and nickel tetraazamacrocyclic complexes are active electrocatalysts for the reduction of CO_2 to CO.[111–113] Especially, Ni-cyclam (cyclam = 1,4,8,11-tetraazacyclotetradecane), shown in Figure 12.13, was found to be highly stable and selective for CO_2 reduction over proton reduction with TONs for CO formation exceeding 8000 after 8 h and a TOF of 1000 h^{-1}.[114–116] Cobalt macrocycles did not receive as much attention as electro- and photocatalysts with respect to their nickel-containing congeners although several mechanistic studies were published, especially on the cobalt macrocycle Co–HMD (HMD = 5,7,7,12,14,14-hexamethyl-1,4,8,11-tetraazacyclotetradeca-4,11-diene), see Figure 12.13.[111,117,118] Co–HMD and other cobalt

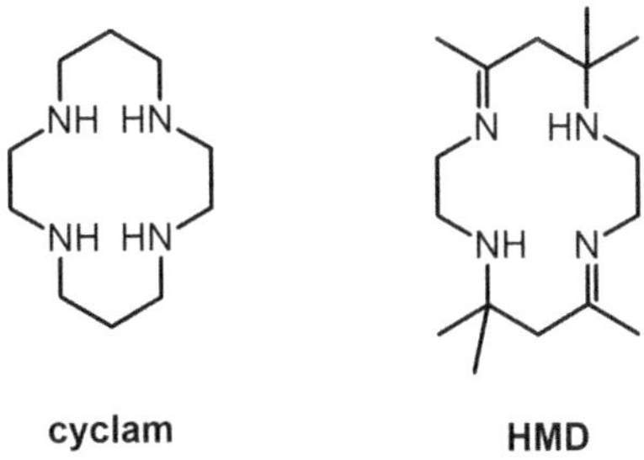

Figure 12.13 Cyclam and HMD ligands used for Co and Ni complexes.

macrocycles were successfully employed as photocatalysts in CO_2 to CO reduction using *p*-terphenyl as a photosensitiser and triethylamine as a sacrificial electron donor in acetonitrile–methanol and $\lambda > 290$ nm light for irradiation.[119–123] The reaction proceeds *via* a Co(I) species that binds CO_2 and a solvent molecule followed by reduction to CO and a Co(II) species. However, TONs of CO formation never exceeded 20 and small amounts of hydrogen were produced. Intramolecular Ni–cyclam/ruthenium complexes and an intermolecular system using a $[Ru(bipy)_3]^{2+}$ photosensitiser showed similar low turnover numbers.[124–127]

It is noteworthy that secondary amines such as cyclams are able to hydrogen bond to CO_2.[114] Hydrogen bonding interactions were also investigated in detail for Co–HMD by IR spectroscopy.[128] A similar bonding situation was found to be responsible for iridium-pincer-complex-catalysed CO_2 hydrogenation.[129]

In addition, porphyrin and corrole macrocycle complexes were tested for light-driven CO_2 reduction. Both compound classes have interesting absorption and electronic features that can be controlled by the choice of central metal and the *meso*-substituents. For instance, photosensitising Pd-tetraphenylporphyrins were coupled to a CO_2-reducing $[Re(bipy)(CO)_3(3\text{-picoline})]^+$ moiety. The intra- and intermolecular photoreduction processes of CO_2 to CO were low (TON = 2 and TON = 3 after 4 h, respectively).[130] Using a zinc porphyrin, however, increased the catalytic performance with a TON = 102 (2 h, $\lambda > 520$ nm) for the intermolecular process.[131]

Porphyrins have dominated the area of photosynthetic solar energy conversion owing to their resemblance to the natural photosynthetic chlorophyll pigment.[132–134] Recently, corroles were intensively investigated for their photophysical and photochemical properties and they have emerged as a promising alternative to porphyrins as components of photoactive arrays.[135,136] Free-base corroles were successfully used in assemblies for photoinduced charge separation, employing them as electron donors in combination with several electron acceptors.[137–139] Metallo-corroles can act as electron acceptors in the presence of photosensitising *p*-terphenyl.[140] Namely, Co(III)- and Fe(III)-corroles catalysed the photoreduction of CO_2 to CO with a TON between 50 and 100 in 10 hours, upon irradiation with $\lambda > 320$ nm light.

The catalytic cycle of mononuclear Re catalysts of the type [Re(bipy)$(CO)_3$X], where X = Cl or Br, is similar for both photo- and electrocatalysis as shown in Figure 12.14. In both catalytic cycles, a one-electron reduced species is observed as an important intermediate either by photoexcitation of the rhenium(I) complex and subsequent reductive quenching by the sacrificial electron donor or by electrochemical reduction. The negative charge of the one-electron reduced species resides on the bipyridine ligand.[88,141–146] This is followed by the loss of the halogen atom X^-, and the formation of the neutral $[Re(bipy)(CO)_3]^{\cdot}$ species, as evidenced by IR-spectroelectrochemical investigations, excited state lifetime experiments as well as HPLC and capillary electrophoresis.[88,147] Recently, the formation of the solvato species $[Re(bipy)(CO)_3(DMF)]^{\cdot}$ by cold spray ionisation mass spectrometry (CSI-MS) was reported.[148] Solvent/CO_2 ligand exchange affords the $[Re(bipy)(CO)_3(CO_2)]^{\cdot}$ species. Formation of CO from CO_2 requires the cleavage of one "oxide anion" from CO_2 under formation of water or carbonate, depending on the reaction conditions. CSI-MS experiments in DMF indicated the presence of $[Re(bipy)(CO)_3(CO_2)]^{\cdot}$ and $[Re(bipy)(CO)_3(COOH)]^{\cdot+}$. Subsequent transfer of a second electron to the metal centre and leaving of "O^{2-}" (*e.g.* as carbonate or water) is followed by cleavage of a Re–CO bond and the catalytic cycle can start again. In this way, the second electron can either be delivered by the sacrificial electron donor or by another one-electron reduced species.[87,147] Other experimental investigations proposed the presence of the dimeric complex $[Re(dmb)(CO)_3]_2(CO_2)$ as the actual catalytically active species, which is formed from CO_2 and two one-electron reduced Re species. This was also supported by the identification of a carbonate bridged product $[Re(dmb)(CO)_3]_2(CO_3)$.[149] Interestingly, the formation of the latter could be prevented by the addition of excess chloride.[150] Later studies were concerned with variations of the

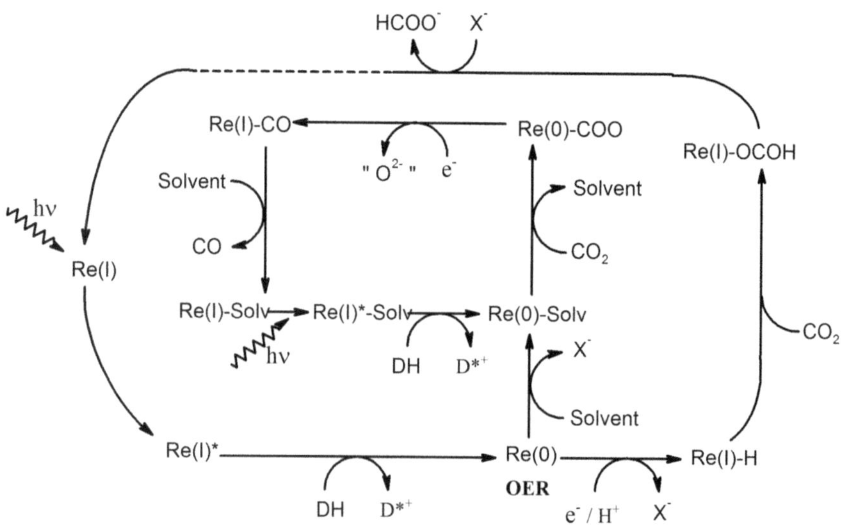

Figure 12.14 Catalytic cycle for photocatalytic CO_2 reduction with mononuclear [Re(bipy)$(CO)_3$X]-type complexes.

halide as well as the polypyridyl ligand.[62,81,145,147,151–154] A 25:1 molar ratio of $[Re(bipy)(CO)_3(CH_3CN)]^+$ and $[Re\{4,4'\text{-dimethoxy-bipy}\}(CO)_3\{P(OEt)_3\}]^+$ achieved a quantum yield of 59% for the photocatalytic reduction of CO_2 to CO. To the best of our knowledge, this is the highest value reported in the literature.[147] Electron transfer between these two reactants was found to be the rate-determining step,[147] where the former acts as the catalyst and the latter represents the photosensitiser. The $[Re(bipy)(CO)_3(CH_3CN)]^+$ catalyst alone, which also absorbs light, only achieved a quantum yield of 4%. This was attributed to the short excited state lifetime and thus fast return to the ground state which hampers an efficient reduction of the photoexcited complex by the sacrificial electron donor. The photosensitiser $[Re\{4,4'\text{-dimethoxy-bipy}\}(CO)_3\{P(OEt)_3\}]^+$, which by itself is also slightly catalytically active (TON 7.5 after 16 h, with $\lambda > 330$ nm irradiation), shows a much longer lifetime and yielded CO with a quantum yield of 33%. Higher quantum yields were prevented by an inner filter effect due to the stronger absorption of the one-electron reduced species which forms upon photoexcitation.

For the mononuclear ruthenium catalysts, the η^1-formate complex $[Ru(tp)(bipy)(OCHO)]^+$ is an important intermediate that could be isolated from the reaction of $[Ru(tp)(bipy)H]^+$ with CO_2. X-Ray diffraction experiments on $[Ru(tp)(bipy)(OCHO)]^+$ show the presence of a Ru–O bond while the hydridic H is bound to the COO carbon atom.[155] The nucleophilic attack of the hydrido ligand to the carbon atom of CO_2 is likely to be the rate-determining step in the formation of the formato complex. In line with these results are the investigations on the $[Ru(bipy)_2(CO)_2]^{2+}$ system. A hydride complex is formed from the neutral solvato complex $[Ru(bipy)_2(CO)(solvent)]$ followed by insertion of CO_2 into the Ru–H bond which yields the formate complex. After electron transfer, the formate anion is cleaved from $[Ru(bipy)_2(CO)(OC(O)H)]$ resulting in the formation of the solvato complex $[Ru(bipy)_2(CO)(solvent)]$.[156] DFT calculations confirmed the formation of a formate intermediate appearing after CO_2 insertion into the Ru–H bond in such a way that a C–H bond and a Ru–O bond are formed. It is noteworthy that hydrogen bonding to possibly present water molecules has a major influence on the CO_2 insertion rate (which reinforces the necessity of using anhydrous solvents in studies to obtain data that is comparable between laboratories).[157]

Despite the high selectivity of the $[Re(bipy)(CO)_3X]$-type photocatalysts to form CO and the high quantum yields, they suffer from a major drawback, which is the poor absorption of visible light. In order to improve the absorption properties, a series of supramolecular complexes has been developed, comprising one or more ruthenium- or osmium-polypyridyl photosensitisers and a $[Re(bipy)(CO)_3X]$-type catalyst.

12.5.3 Supramolecular Photocatalysts

As outlined in the Introduction of the chapter, in general supramolecular photocatalysts are comprised of a photosensitising unit (**PS**) that is connected *via* a bridging ligand (**B**) with a catalytically active centre (**C**).[10j] However, in the area of photocatalytic CO_2 reduction, a range of assemblies consisting of more than one photosensitiser or catalyst unit have also been reported. The **PS = C** approach, which has been reported for some Re(I) complexes, is limited by the lack of absorbance in the visible part of the spectrum while combining light-absorbing and catalytic properties at the same time. The more popular **PS–B–C** systems offer the possibility for optimisation of each unit individually. However, this naturally increases the synthetic workload. In the following section, a number of **PS–B–C** type assemblies will be discussed which may contain several photosensitisers. The effect of the nature of **B**, the bridging ligand, the peripheral ligands and the sacrificial agents will be discussed.

Among the series of compounds, the dinuclear Ru/Re complex **d_2Ru-Re** (for TON values and structures, see Figures 12.15 and 12.16) has become a benchmark compound. Using light with wavelengths longer than 500 nm, a turnover number of 170 was obtained after 16 h irradiation with $\lambda > 500$ nm, a quantum efficiency of 12% and a high selectivity for CO, while small amounts of hydrogen were obtained (formic acid was not detected). Improved TON numbers were obtained with the tetranuclear **$RuRe_3$** system for which a TON of 240 (16 h with respect to the added supramolecule) and a quantum efficiency for CO production of 9% were reported.[158] Modifications of the bipy ligands in the photosensitiser had a major influence on the photophysical and photocatalytic properties of the complexes. For instance, electron-withdrawing substituents, such as CF_3, in the 4-position of the bipy ligand led to a stabilisation of ruthenium-site-based ^{3}MLCT states and hence hampered an efficient electron transfer to the catalytic rhenium site, leading to poor catalytic efficiencies.[158,159] Also, the bridge plays an important role, and the effect of its nature on the photocatalytic behaviour of the compound was investigated.

As shown in Figure 12.16, non-conjugated alkyl-chain bridges were used as well as a conjugated (4-methyl-4′-[1,10]phenanthroline-[5,6-*d*]imidazol-2-yl)

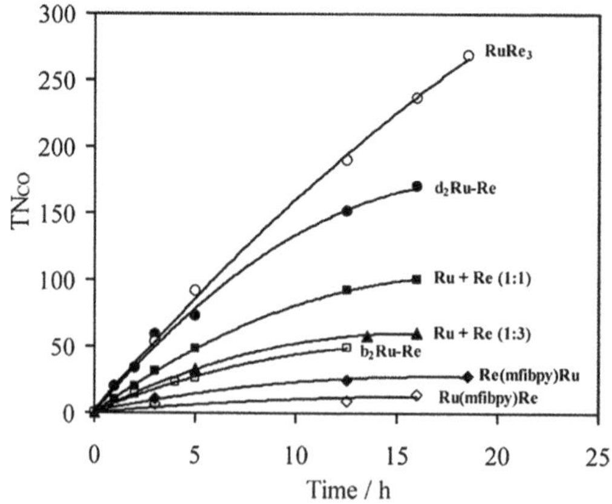

Figure 12.15 Turnover number for production of CO from CO_2 for the various compounds shown in Figure 12.16 as a function of irradiation time, $\lambda > 500$ nm. Reprinted with permission from ref. 158. Copyright (2005) American Chemical Society.

bipyridine (**mfibipy**) unit. It was found that the alkyl chain bridged Ru/Re complexes acted as much better catalysts than those with the conjugated **mfibipy** bridge. A comparison of their catalytic performance is shown in Figure 12.15. The more extended π-system on the phenanthroline side lowers the energy of the π*-orbitals and thus the photoexcited electron is mainly located on the phenanthroline side of the bridge. In the case of the **Ru-mfibipy-Re** complex, this suggests a lower likelihood of finding the photoexcited electron on the catalytic Re-side and hence a decreased catalytic activity. However, according to the reasoning above, for the **Re-mfibipy-Ru** complex, the photoexcited electron can be expected on the Re-side. The strong electronic communication across the bridge also lowered the reducing power of the one-electron reduced species to −1.1 V *vs.* Ag/AgCl, which is too positive to efficiently reduce CO_2.[158] A reduction potential of >−1.41 V *vs.* Ag/AgCl for mononuclear $[Re(diimine)(CO)_3(PR_3)]^+$ was found to be necessary for the efficient reduction of CO_2.[159] Although the presence of communication *via* the **mfibipy** bridge enhanced the electron transfer between the photosensitiser and catalyst units, other effects, such as redox properties and the localisation of the one-electron reduced species, are much more influential.

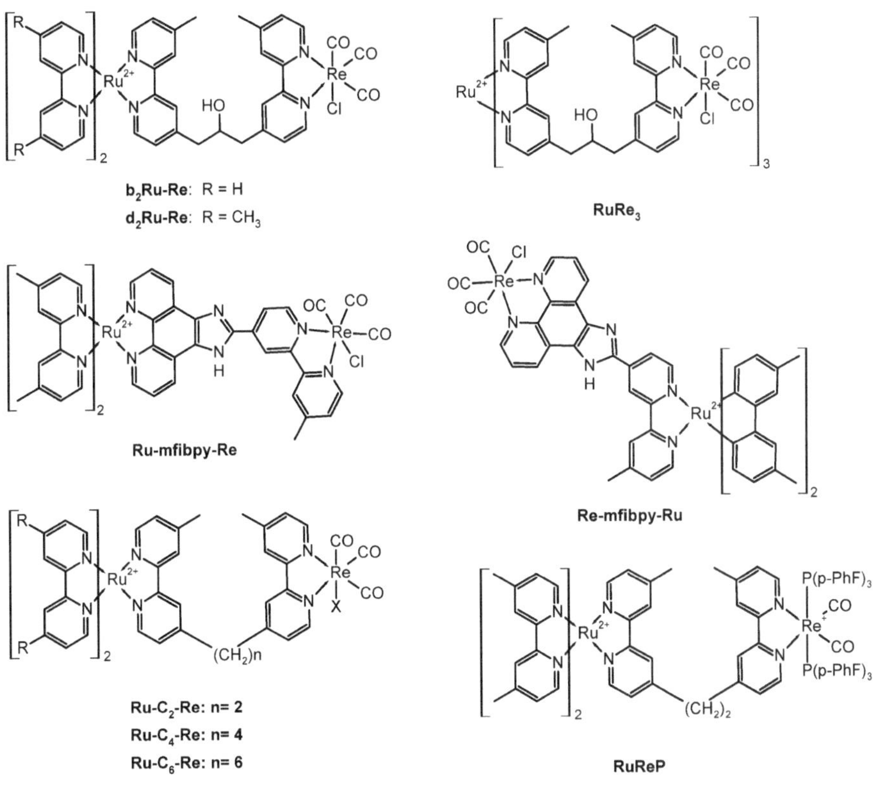

Figure 12.16 Structures of the ruthenium/rhenium complexes discussed in ref. 158.

The length of the alkyl chain also has an impact on the electron-transfer properties of the complexes. Besides the C_3-chain of the carbinol linker given in Figure 12.16, C_2-, C_4- and C_6-linkers were examined.[160] The photophysical and electrochemical properties of the complexes with the three different chain lengths are almost identical. However, the first oxidation wave and the first reduction wave of **Ru-C$_2$-Re** are 20–30 mV lower. This observation was discussed as a very weak interaction between the metal centres. Illumination with light with $\lambda > 500$ nm mainly afforded CO with very small amounts of hydrogen but without the formation of formic acid. With a quantum efficiency of 13% for the formation of CO from CO_2 (TON = 180 after 16 h), **Ru-C$_2$-Re** performed slightly better than **Ru-C$_4$-Re** and **Ru-C$_6$-Re**, both giving a quantum yield of 11% (TON = 120 after 16 h, $\lambda > 500$ nm).

On the basis of these studies, two basic design principles that should be applied to optimise the photocatalytic efficiency of the Ru/Re complexes can be deduced:

(1) Absorption of light and metal-to-ligand electron transfer should localise the excited electron on the bridging ligand rather than on the periphery.
(2) The reducing power of the catalytically active centre is higher when non-conjugated bridges are deployed, which are thus favoured over conjugated bridges such as **mfibipy**.

Introduction of a second phosphane ligand, namely tris-(*p*-fluorophenyl) phosphane, at the rhenium centre (compound **RuReP**, Figure 12.16) renders the driving force for the electron transfer from the excited Ru site to the Re site exergonic. Compound **RuReP** was found to catalyse the photoreduction of CO_2 to CO with TON = 207 (20 h, $\lambda > 500$ nm), TOF = 4.7 min^{-1} and a quantum efficiency of 15% using BNAH as the sacrificial electron donor.[161] As mentioned earlier, the choice of the sacrificial donor is an important factor for these homogeneous photocatalytic systems. By changing from BNAH to 1,3-dimethyl-2-phenyl-2,3-dihydro-1*H*-benzo[*d*]imidazole (BIH), the TON and TOF were increased to 2000 (10 h, $\lambda > 500$ nm) and TOF = 35.7 min^{-1} and the quantum efficiency increased to 45%.[162] All studies were carried out in DMF.

Based on the Ru/Re design principles, Os/Re complexes were developed (see Figure 12.17). Due to the Os-polypyridyl photosensitiser, they absorb a wider range of visible light than the respective ruthenium complexes.[163] These osmium compounds show ^{3}MLCT absorption between 540 and 730 nm, which was attributed to the stronger spin–orbit coupling of osmium with respect to ruthenium. The UV-vis absorption spectrum of **OsReP-X** is the sum of the UV-vis absorption comparisons of the mononuclear Re and Os precursor compounds. This indicates negligible coupling between the Os and Re moieties in the ground state. With BIH as a sacrificial electron donor instead of BNAH, a TON of 1138 (20 h) was obtained for the selective reduction of CO_2 to CO by **OsReP-Cl** (quantum yield for the formation of CO 10%). Hydrogen and formic acid were found as minor products. A mix of the mononuclear Os- and Re-precursor compounds gave less than half the amount

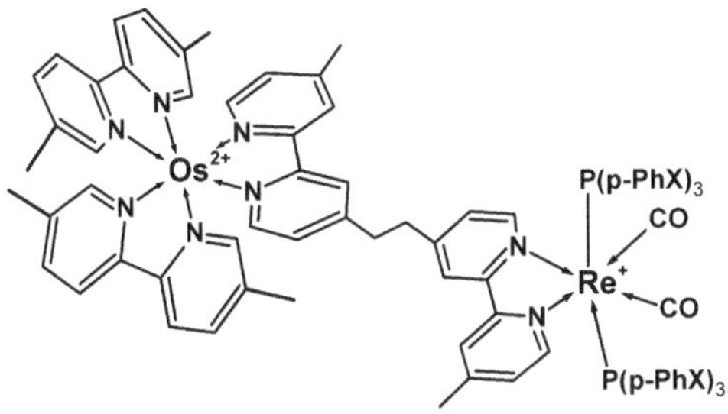

OsReP-X (X = F, Cl)

Figure 12.17 Structure of the osmium/rhenium complexes reported in ref. 163.

of CO. It is interesting to note that the formation of BI^+ (the oxidized product of BIH) agrees very well with the formation of CO. This indicates that BIH acts as a two-electron donor for the CO_2 reduction process. The 4-phenyl substituent X of the phosphine ligand in the Re unit has a marked effect on the photocatalysis. **OsReP-Cl** showed a better photocatalytic performance (TON_{CO} = 1138 (20 h), TOF_{CO} = 3.3 min^{-1}, quantum efficiency for CO = 0.12, λ > 620 nm) in comparison to **OsReP-F** (TON_{CO} = 762 (20 h), TOF_{CO} = 1.6 min^{-1}, quantum efficiency for CO = 0.10).

The efficiencies of the photochemical formation of the one-electron reduced species of the photocatalysts were very similar when comparing **OsReP-Cl** and **OsReP-F**. Thus the difference of the photocatalytic performance was attributed to intramolecular electron transfer processes from the one-electron reduced species of the Os unit to the Re unit. Based on electrochemical data, the intramolecular electron transfer process from the one-electron reduced Os species to the Re unit is exergonic in both cases. In the case of **OsReP-Cl**, the electron localisation on the Re unit should be more pronounced than in the case of **OsReP-F**, which is a possible reason for the better performance of **OsReP-Cl**.

Based on the Ru/Re design principles, as mentioned earlier, several Ru/Ru supramolecular photocatalysts were developed comprising a ruthenium trisbipy photosensitiser linked by a C_2-alkyl chain with a $[Ru(bipy)_2(CO)_2]^{2+}$ catalyst as shown in Figure 12.18.[164] These were the first catalysts reported for the visible light-driven photocatalytic reduction of CO_2 to formic acid. Besides CO and H_2 as side products, formic acid was formed with a high selectivity of about 89%. It should be mentioned that although the Ru(II) catalytic centre also absorbs light, it does not absorb at the irradiation wavelengths used for the photocatalysis (λ > 480 nm). The number of photosensitiser and catalyst units in the supramolecular systems strongly influenced the products of photocatalytic processes. A higher ratio of the photosensitiser with respect to the catalyst furnished a higher yield of formic acid. For instance, Ru_3 has two photosensitiser units and one catalyst unit producing formic acid with the highest selectivity and yield. In contrast, using Ru_4 with one photosensitiser unit and three catalyst units, CO was the main product, although the

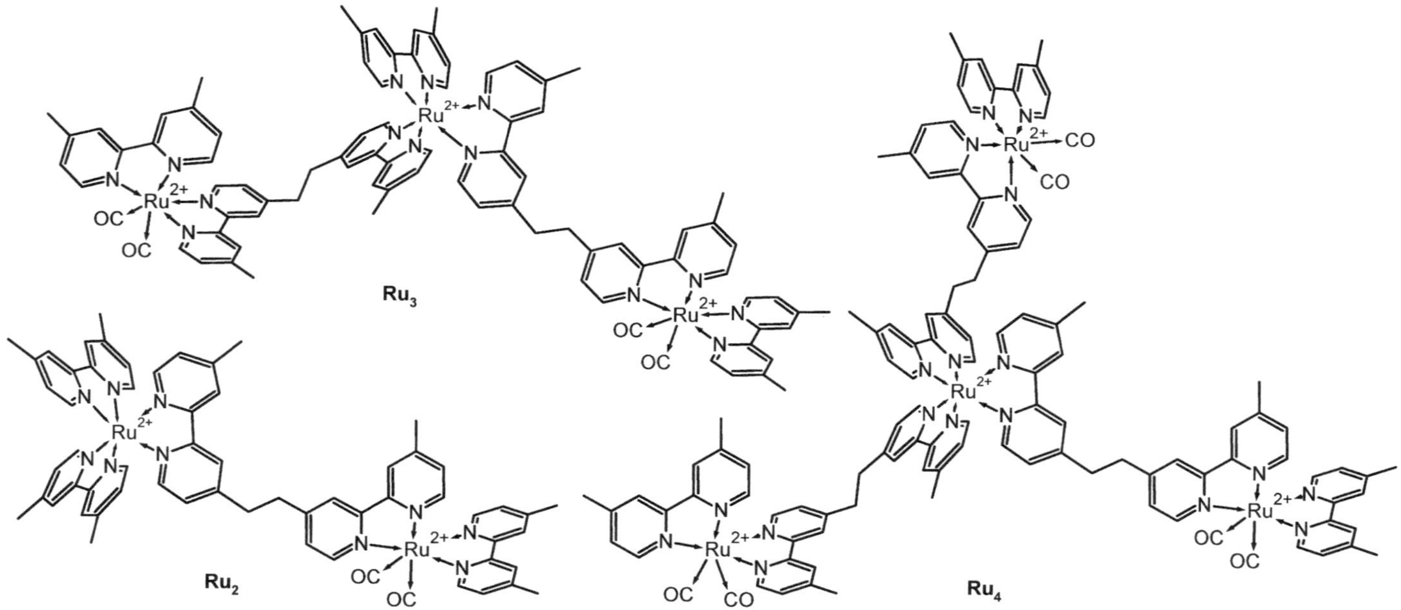

Figure 12.18 Structural features of the ruthenium/ruthenium complexes discussed in ref. 164.

photocatalytic activity of Ru_4 was much lower than of Ru_3. Interestingly, an intermolecular system of the mononuclear $[Ru(dmb)_3]^{2+}$ photosensitiser and $[Ru(dmb)_2(CO)_2]^{2+}$ catalyst only furnished small amounts of all three products. The intermolecular system rapidly decomposed concomitant with the formation of oligomeric ruthenium compounds. The same decomposition was observed for the catalyst consisting of one photosensitiser and two or three catalytically active units which also only furnished small amounts of all products. Dinuclear intramolecular Ru/Co– or Ru/Ni–bisphenanthroline complexes were found to be slightly active in the photoreduction of CO_2 to CO, HCOOH and H_2 in DMF–H_2O–TEOA (Figure 12.19).[165] The metal centres are bridged by ethylene and pentamethylene chains.

Expectedly, the absorption features of the complexes are dominated by the Ru-moiety (no shift in the absorption maximum at 451 nm) and no considerable contribution from the Co or Ni moieties was observed. This is supported by electrochemistry data showing no shift of the Ru-based oxidation potential at 0.86 V (in ACN *vs.* Fc/Fc^+). The excited Ru moiety photoreduces the catalytic centre as evidenced by a 30–50% luminescence quenching of the 615 nm emission of the dinuclear compounds with respect to the mononuclear Ru complex. The reduced catalytic centre subsequently transforms CO_2 into CO with the concomitant formation of HCOOH and H_2. The observed TONs for CO production never exceeded 10 after illumination with 400–750 nm light in DMF–H_2O–TEOA. The dinuclear ruthenium–nickel complex is less active than the ruthenium–cobalt analogue and produces a larger amount of hydrogen. The catalytic activity is almost equal for the intramolecular and intermolecular systems because the Ru light absorber and the Ni/Co catalytic centre are electronically insulated. Interestingly, the intramolecular system showed a slightly larger CO–H_2 ratio with respect to the intermolecular system. The mechanism very likely involves the coordination of CO_2 to a photoreduced cobalt(I) species, which is most likely the rate-determining step. Thus the overall efficiency is not significantly affected by the electron transfer rate from the photosensitiser to the catalytic centre and as a result the inter- and intramolecular systems show similar photocatalytic efficiencies.

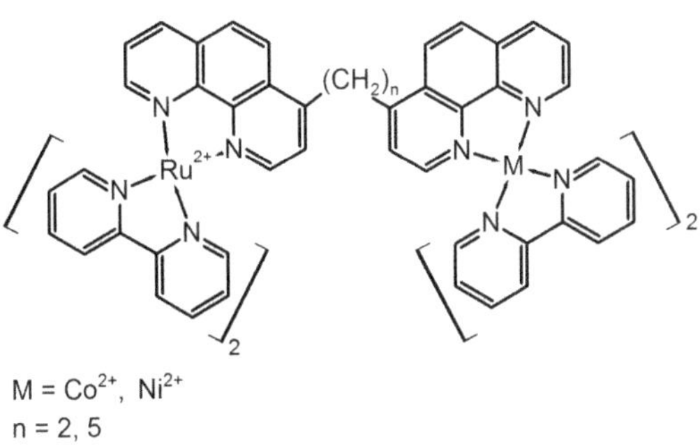

Figure 12.19 Dinuclear Ru/Co– and Ru/Ni–bisphenanthroline complexes.

12.6 Concluding Remarks

The overview given in this chapter on the study of homogeneous photocatalytic systems for solar fuel generation shows that the field is very diverse. A wide range of both inter- and intramolecular approaches have been studied. While intermolecular studies have been reported for decades, the supramolecular approach has only recently been developed. The latter approach was developed by the argument that by covalently connecting the photosensitiser and the catalytic centre to produce a supramolecular assembly, increased photocatalytic activity could be expected. In photocatalytic studies, the concept behind the design of intramolecular catalysts was the utilisation of bridging ligands that allow electronic interaction between the photosensitiser and the catalytic centre as shown in Figure 12.1. The emission properties of these bimetallic photocatalysts show a reduced emission lifetime or a complete quenching with respect to that observed for the precursors, as a function of the level of communication between the photosensitiser and the catalytic centre. Further evidence for interaction between the two reaction centres is obtained from changes in spectroscopic features such as absorption or emission wavelength, although it has to be pointed out that significant changes in these parameters are not always obtained. Importantly, photophysical studies of the design of intramolecular assemblies for hydrogen generation indicate (see Figures 12.3 and 12.5) that electronic coupling between peripheral and bridging ligands is an important factor that strongly affects the photoinduced electron transfer across the bridge. These observations are mirrored in the photocatalytic studies that show that TONs are dependent not only on the nature of the bridge, but also on the nature of the peripheral ligands. This shows that it is extremely important to tune the excited state energies of bridging ligands with those of the peripheral ligands in such a manner that their electronic coupling is optimised. This creates a range of additional parameters that should be considered in the optimisation of the photocatalytic behaviour of these supramolecular systems.

Apart from the supramolecular approach, a number of intermolecular options are discussed in this chapter. These studies show that if one compares the range of photocatalytic results obtained *i.e.* for example, looking at turnover numbers, the most efficient systems with the highest TONs are intermolecular in nature rather than supramolecular. Of particular interest is the application of iridium-based photosensitisers that show TONs of several thousand; such values are not obtained by intramolecular assemblies. Clearly, at this stage, the use of a bimetallic supramolecular assembly is less efficient than this particular intermolecular approach, although when comparable Ru-based systems are investigated under the same conditions (as described in this chapter), the intramolecular approach is somewhat more efficient. This is partly explained by the peripheral/bridging ligand coupling issue discussed earlier, but another important observation as shown in Scheme 12.3 is that for intermolecular hydrogen generation, the rate-determining step is not associated with electron transfer between the sensitiser and the catalyst, but with hydrogen formation at the catalytic centre, which has a second order rate constant of 25 M^{-1} s^{-1}. Therefore, for intramolecular systems such as those considered here, the electron transfer processes involving the photosensitisers may not be rate limiting. To allow further advances in this area, there is a need for detailed kinetic and photophysical studies on a wide range of assemblies to identify the rate-determining step of the intramolecular mechanism and further optimise intramolecular systems.

Another issue is that in dinuclear assemblies containing Pd-based catalysts for hydrogen generation, hydrogen may not be generated by the photocatalytic assembly but by Pd moieties as colloids no longer coordinated to the bridging ligand. In a similar manner, the weak coordination strength of first row transition metals, such as cobalt, raises questions. In some of the compounds discussed, Pd colloids do form, but the results obtained thus far suggest that their formation is based upon decomposition of the compounds, at that stage of the reaction where hydrogen formation is terminating. Further research on this issue is taking place at present but the results discussed earlier show that intramolecular photocatalysis is the main contributor to hydrogen generation.

Until now, many light absorbing "antenna" components in PMDs have contained rare, low abundant elements such as ruthenium, iridium, rhenium or rhodium.[10] In view of using the solar energy conversion technology discussed herein as a major part of the prospective global energy supply, the use of rare metals is not sustainable and hence research is also focusing on the utilisation of first row transition metals, such as cobalt and copper.

Clearly, the effectiveness of the photocatalytic system for hydrogen generation under investigation depends on a wide range of factors, and this at present prohibits a meaningful comparison between inter- and intramolecular photocatalytic mechanisms in this area. Therefore, more studies are needed where detailed kinetic analyses are carried out for well-defined photocatalytic cycles using systematic variations of the experimental conditions.

In the area of photocatalytic CO_2 reduction, non-conjugated bridging ligands have been suggested on the basis that a weak interaction between the two centres would prevent the fast back reaction expected in the strong coupling case. The literature discussed would suggest that for hydrogen generation, a direct coupling between the catalytic centre and the photosensitiser is to be recommended, however, the results obtained from photocatalytic CO_2 reduction indicate that little or no electronic coupling between the photosensitiser and the catalytic centre is to be preferred. This is maybe a somewhat unexpected but important observation. If one considers natural photosynthesis, however, no covalent bonds are used in the electron cascade, but electron transfer between the reaction partners takes place extremely fast, since they are carefully arranged in space, their location is both optimised and fixed. One may argue, therefore, that the approach taken in the design of supramolecular systems for CO_2 reduction is closer to natural synthesis than the strongly coupled assemblies proposed for hydrogen generation. A problem with optimising the photocatalytic efficiencies for CO_2 reduction is that the number of products obtained from carbon dioxide reduction are strongly dependent on the choice of the catalytic centre. Here the choice of the metal centre seems to have a larger impact than the choice of the respective ligands. At the same time, careful tuning of all three components of the **PS–B–C** system can lead to a significant increase in the catalytic performance and robustness of the catalysts. Interestingly enough, also, the sacrificial donor has a profound impact on stability, catalytic performance and even product ratios. Similar effects can be expected from choice of electrode material in the course of the necessary immobilisation of the catalytic systems and will require fundamental investigations on the function-determining electron transfer and reaction steps.

At present, the study of these photochemical molecular devices (PMDs) is mainly carried out in homogeneous solution, in which case, the addition of a sacrificial agent is required. The need for such sacrificial agents is clearly far from ideal and systems and devices will have to be developed to avoid this. The immobilisation of such photocatalytic assemblies on electrodes as part of a photoelectrochemical device has been proposed as a way forward. In that case, the electrode acts as an electron donor for the regeneration of the photosensitiser and no sacrificial agents are required. While the development of such photoelectrocatalytic electrode assemblies has so far not been considered in great detail, the development of electrochemical alternatives towards water splitting and CO_2 reduction have been considered. Figure 12.20 shows a potential photoelectrochemical cell, which consists of two different compartments—a photocathode based on p-type semiconductors for the reduction of either hydrogen or CO_2 and a TiO_2-based photoanode for the generation of oxygen. This mirrors the design of tandem photovoltaic dye-sensitised solar cells.

At present, an increasing number of investigations are involving the immobilisation of such catalysts on solid electrode surfaces. Prior to this, the two compartments shown in Figure 12.20 have been investigated separately. Studies at the reduction side of water splitting include the investigation of modified NiO-type sensitised photocathodes,[166] while TiO_2-based photoanodes for oxygen generation have been studied separately. In the study of

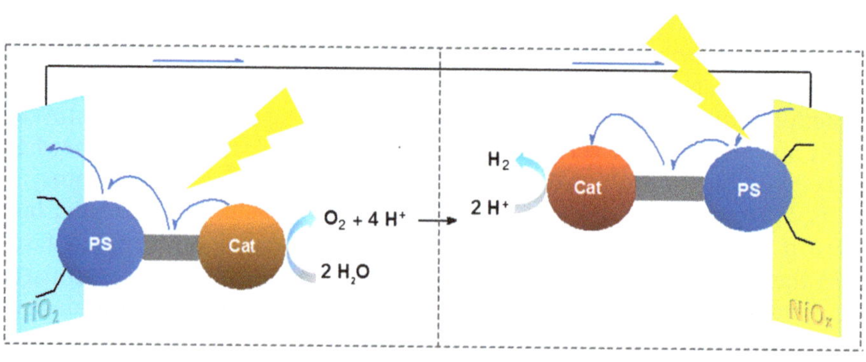

Figure 12.20 Schematic example of light-driven water splitting using surface immobilised supramolecular assemblies.

photocatalytic and electrocatalytic water oxidation, supramolecular systems have been reported both when immobilised on electrode surfaces[167] and in solution-based studies.[168] The photoanode development by Nocera and co-workers[169] has so far proved to be the most efficient oxygen generating system. The Nocera approach is not based on supramolecular catalysts, but deposition of cobalt catalysts on the electrode surfaces shows the formation of very intricate supramolecular structures. For the further development of this very interesting and environmentally important area of research, photocatalytic systems with higher efficiency and specificity (particularly in the case of CO_2 reduction), need to be developed. Furthermore, the study of photoelectrochemical cells as outlined in Figure 12.20 needs to be extended and investigations should be aimed at achieving higher photoelectrochemical efficiencies, higher current densities and increased stability.

Acknowledgements

The authors wish to thank the COST Action CM1202, PERSPECT-H2O for support.

References

1. V. Balzani, A. Credi and M. Venturi, *ChemSusChem*, 2008, **1**, 26.
2. *Annual Energy Outlook 2011 – With Projections to 2033*, U.S. Energy Information Administration, U.S. Department of Energy, DOE/EIA, Washington, DC 20585, 2011.
3. T. R. Cook, D. K. Dogutan, S. Y. Reece, Y. Surendranath, T. S. Teets and D. G. Nocera, *Chem. Rev.*, 2010, **110**, 6474.
4. N. S. Lewis and D. G. Nocera, *Proc. Natl. Acad. Sci. U. S. A.*, 2006, **103**, 15729.
5. A. Hagfeldt and M. Grätzel, *Acc. Chem. Res.*, 2000, **33**, 269.
6. L. Sun, L. Hammarström, B. Åkermark and S. Styring, *Chem. Soc. Rev.*, 2001, **30**, 36.
7. S. Ott, *Science*, 2011, **333**(6050), 1714–1715.
8. A. Boddien, D. Mellmann, F. Gärtner, R. Jackstell, H. Junge, P. J. Dyson, G. Laurenczy, R. Ludwig and M. Beller, *Science*, 2011, **333**, 1733.
9. C. Federsel, R. Jackstell and M. Beller, *Angew. Chem., Int. Ed.*, 2010, **49**, 6254.
10. (a) E. Fujita, *Coord. Chem. Rev.*, 1999, **185**, 373; (b) A. J. Morris, G. J. Meyer and E. Fujita, *Acc. Chem. Res.*, 2009, **42**, 1983; (c) T. S. Teets and D. G. Nocera, *Chem. Commun.*, 2011, **47**, 9268; (d) K. Sakai and H. Ozawa, *Coord. Chem. Rev.*, 2007, **251**, 2753; (e) S. Losse, J. G. Vos and S. Rau, *Coord. Chem. Rev.*, 2010, **254**, 2492; (f) V. Artero, M. Chavarot-Kerlidou and M. Fonecave, *Angew. Chem., Int. Ed.*, 2011, **50**, 7238; (g) H. Ozawa and K. Saki, *Chem. Commun.*, 2011, **47**, 2227; (h) A. Inagaki and M. Akita, *Coord. Chem. Rev.*, 2010, **254**, 1220; (i) W. T. Eckenhoff and R. Eisenberg, *Dalton Trans.*, 2012, **41**, 13004; (j) M. Schulz, M. Karnahl, M. Schwalbe and J. G. Vos, *Coord. Chem. Rev.*, 2012, **256**, 1682; (k) P. D. Frischmann,

K. Mahata and F. Wurther, *Chem. Soc. Rev.*, 2013, **42**, 1847; (l) S. Rau, D. Walther and J. G. Vos, *Dalton Trans.*, 2007, 915; (m) G. F. Manbeck and K. J. Brewer, *Coord. Chem. Rev.*, 2013, **257**, 1660.
11. H. Ozawa and K. Sakai, *Chem. Commun.*, 2011, **47**, 2227.
12. V. Artero, M. Chavarot-Kerlidou and M. Fontecave, *Angew. Chem., Int. Ed.*, 2011, **50**, 7238.
13. M. Schwalbe, M. Karnahl, H. Gorls, D. Chartrand, F. Laverdiere, G. S. Hanan, S. Tschierlei, B. Dietzek, M. Schmitt, J. Popp, J. G. Vos and S. Rau, *Dalton Trans.*, 2009, **20**, 4012.
14. J. M. Clear, J. M. Kelly, C. M. O'Connell, J. G. Vos, C. J. Cardin, S. R. Costa and A. J. Edwards, *J. Chem. Soc., Chem. Commun.*, 1980, 750.
15. A. Paul, D. Connolly, M. Schulz, M. T. Pryce and J. G. Vos, *Inorg. Chem.*, 2012, **51**, 1977.
16. Y. Kuramochi, M. Kamiya and H. Ishida, *Inorg. Chem.*, 2014, **53**, 3326.
17. H. J. Kuhn, S. E. Braslavsky and R. Schmidt, *Pure Appl. Chem.*, 2004, **76**, 2105.
18. S. Tschierlei, M. Karnahl, M. Presselt, B. Dietzek, J. Guthmuller, L. González, M. Schmitt, S. Rau and J. Popp, *Angew. Chem., Int. Ed.*, 2010, **49**, 3981.
19. G. Singh Bindra, M. Schulz, A. Paul, R. Groarke, S. Soman, J. L. Inglis, W. R. Browne, M. G. Pfeffer, S. Rau, B. J. MacLean, *et al.*, *Dalton Trans.*, 2012, **41**, 13050.
20. S. Soman, G. Singh Bindra, A. Paul, R. Groarke, J. C. Manton, F. M. Connaughton, M. Schulz, D. Dini, C. Long, M. T. Pryce, *et al.*, *Dalton Trans.*, 2012, **41**, 12678.
21. A. Fihri, V. Artero, A. Pereira and M. Fontecave, *Dalton Trans.*, 2008, 5567.
22. K. Kiyosawa, N. Shiraishi, T. Shimada, D. Masui, H. Tachibana, S. Takagi, O. Ishitani, D. A. Tryk and H. Inoue, *J. Phys. Chem. C*, 2009, **113**, 11667.
23. A. Fihri, V. Artero, M. Razavet, C. Baffert, W. Leibl and M. Fontecave, *Angew. Chem., Int. Ed.*, 2008, **47**, 564–567.
24. Q. Pan, F. Mecozzi, J. P. Korterik, D. Sharma, J. L. Herek, J. G. Vos, W. R. Browne and A. Huijser, *J. Phys. Chem. C*, 2014, **118**, 20799.
25. S. Rau, B. Schäfer, D. Gleich, E. Anders, M. Rudolph, M. Friedrich, H. Görls, W. Henry and J. G. Vos, *Angew. Chem., Int. Ed.*, 2006, **45**, 6215.
26. S. Tschierlei, M. Presselt, C. Kuhnt, A. Yartsev, T. Pascher, V. Sundström, M. Karnahl, M. Schwalbe, B. Schäfer, S. Rau, M. Schmitt, B. Dietzek and J. Popp, *Chem.–Eur. J.*, 2009, **15**, 7678.
27. S. Soman, G. Singh Bindra, A. Paul, R. Groarke, J. C. Manton, F. M. Connaughton, M. Schulz, D. Dini, C. Long, M. T. Pryce and J. G. Vos, *Dalton Trans.*, 2012, **41**, 12678.
28. R. J. Watts and J. van Houten, *J. Am. Chem. Soc.*, 1974, **96**, 4334.
29. J. I. Goldsmith, W. R. Hudson, M. S. Lowry, T. H. Anderson and S. Bernhard, *J. Am. Chem. Soc.*, 2005, **127**, 7502.
30. M. S. Lowry and S. Bernhard, *Chem.–Eur. J.*, 2006, **12**, 7970.
31. S. Jasimuddin, T. Yamada, K. Fukuju, J. Otsuki and K. Sakai, *Chem. Commun.*, 2012, **46**, 8466.

32. G. Singh Bindra, M. Schulz, A. Paul, S. Soman, R. Groarke, J. Inglis, M. T. Pryce, W. R. Browne, S. Rau, B. J. Maclean and J. G. Vos, *Dalton Trans.*, 2011, **40**, 10812.
33. J. Hirschmann, G. Singh Bindra, S. Soman, A. Paul, R. Groarke, M. Schulz, M. T. Pryce, W. R. Browne, S. Rau and J. G. Vos, *Dalton Trans.*, 2011, **40**, 10545.
34. (a) M. Elvington, J. Brown, S. M. Arachchige and K. J. Brewer, *J. Am. Chem. Soc.*, 2007, **129**, 10644; (b) M. Elvington and K. J. Brewer, *Inorg. Chem.*, 2006, **45**, 5242; (c) S. M. Arachchige, J. Brown and K. J. Brewer, *J. Photochem. Photobiol., A*, 2008, **197**, 13; (d) K. Rangan, S. M. Arachchige, J. R. Brown and K. J. Brewer, *Energy Environ. Sci.*, 2009, **2**, 410; (e) S. M. Arachchige, R. Shaw, T. A. White, V. Shenoy, H.-M. Tsui and K. J. Brewer, *ChemSusChem*, 2011, **4**, 514.
35. T. A. White, S. L. H. Higgins, S. M. Arachchige and K. J. Brewer, *Angew. Chem., Int. Ed.*, 2011, **50**, 12209.
36. A. Juris, V. Balzani, F. Barigelletti, S. Campagna, P. Belser and A. von Zelewsky, *Coord. Chem. Rev.*, 1988, **84**, 85.
37. (a) C. Kuhnt, M. Karnahl, S. Rau, M. Smitt, B. Dietzek and J. Popp, *ChemPhysChem*, 2011, **12**, 2101; (b) M. Karnahl, C. Kuhnt, F. W. Heinemann, M. Schmitt, S. Rau, J. Popp and B. Dietzek, *Chem. Phys.*, 2012, **393**, 65.
38. J. Hawecker, J.-M. Lehn and R. Ziessel, *J. Chem. Soc., Chem. Commun.*, 1983, 537.
39. L. L. Tinker, N. D. McDaniel, P. N. Curtin, C. K. Smith, M. J. Ireland and S. Bernhard, *Chem.–Eur. J.*, 2007, **13**, 8726.
40. P. N. Curtin, L. L. Tinker, C. M. Burgess, E. D. Cline and S. Bernhard, *Inorg. Chem.*, 2009, **48**, 10498.
41. (a) S. Metz and S. Bernhard, *Chem. Commun.*, 2010, **46**, 7551; (b) E. D. Cline, S. E. Adamson and S. Bernard, *Inorg. Chem.*, 2008, **47**, 10378.
42. J.-G. Cai, Z.-T. Yu, Y.-J. Yuan, F. Li and Z.-G. Zou, *ACS Catal.*, 2014, **4**, 1953.
43. P. Lei, M. Hedlund, R. Lomoth, H. Rensmo, O. Johansson and L. Hammarström, *J. Am. Chem. Soc.*, 2008, **130**, 26.
44. (a) M. Guttentag, A. Rodenburg, C. Bachmann, A. Senn, P. Hamm and R. Alberto, *Dalton Trans.*, 2013, **42**, 334; (b) M. Guttentag, A. Rodenberg, R. Kopelent, B. Probst, C. Buchwalder, M. Brandstätter, P. Hamm and R. Alberto, *Eur. J. Inorg. Chem.*, 2012, 59; (c) B. Probst, A. Rodenberg, M. Guttentag, P. Hamm and R. Alberto, *Inorg. Chem.*, 2010, **49**, 6453.
45. (a) C. V. Krishnan, B. S. Brunschwig, C. Creutz and N. J. Sutin, *J. Am. Chem. Soc.*, 1985, **107**, 2005; (b) E. D. Cline, S. E. Adamson and S. Bernhard, *Inorg. Chem.*, 2008, **47**, 10378; (c) P. Du, J. Schneider, G. Luo, W. W. Brennessel and R. Eisenberg, *Inorg. Chem.*, 2009, **48**, 4952.
46. T. Chao and J. H. Espenson, *J. Am. Chem. Soc.*, 1978, **100**, 129.
47. T. M. McCormick, B. D. Calitree, A. Orchard, N. D. Kraut, F. V. Bright, M. R. Detty and R. Eisenberg, *J. Am. Chem. Soc.*, 2010, **132**, 15480.

48. T. Lazarides, T. McCormick, P. Du, G. Luo, B. Lindley and R. Eisenberg, *J. Am. Chem. Soc.*, 2009, **131**, 9192.
49. (a) C. E. Housecroft and A. G. Sharpe, *Inorganic Chemistry*, Pearson, Harlow (UK), 4th edn, 2012, p. 460; (b) N. Sutin, C. Creutz and E. Fujita, *Comments Inorg. Chem.*, 1997, **19**, 67.
50. *Electrochemical and Electrocatalytic Reactions of Carbon Dioxide*, ed. B. P. Sullivan, K. Krist and H. E. Guard, Elsevier, Amsterdam, 1993.
51. P. A. Christensen and A. Hamnett, *Techniques and Mechanisms in Electrochemistry*, Blackie Academic and Professional, London, 1994, p. 286.
52. W. S. Dodds, L. F. Stutzman and B. J. Sollami, *Ind. Eng. Chem. Chem. Eng. Data Series*, 1956, **1**, 92.
53. A. Gennaro, A. A. Isse, M. G. Severin, E. Vianello, I. Bhugun and J. M. Saveant, *J. Chem. Soc., Faraday Trans.*, 1996, **92**, 3963.
54. B. R. Eggins and J. McNeill, *J. Electroanal. Chem.*, 1983, **148**, 17.
55. M. C. Massebleau, E. Dunach, M. Troupel and J. Perichon, *New J. Chem.*, 1990, **14**, 259.
56. *Carbon Dioxide as Chemical Feedstock*, ed. M. Aresta, Wiley, Weinheim, 2010.
57. J. C. Gressin, D. Michelet, L. Nadjo and J. M. Saveant, *Nouv. J. Chim.*, 1979, **3**, 345.
58. E. E. Benson, C. P. Kubiak, A. J. Sathrum and J. M. Smieja, *Chem. Soc. Rev.*, 2009, **38**, 89.
59. S. Kaneco, K. Iiba, S. Suzuki, K. Ohta and T. Mizuno, *J. Phys. Chem. B*, 1999, **103**, 7456.
60. R. P. S. Chaplin and A. A. Wragg, *J. Appl. Electrochem.*, 2003, **33**, 1107.
61. S. Kaneco, K. Iiba, H. Katsumata, T. Suzuki and K. Ohta, *Electrochim. Acta*, 2006, **51**, 4880.
62. K. Tanaka and D. Ooyama, *Coord. Chem. Rev.*, 2002, **226**, 211.
63. S. J. Archibald, S. P. Foxon, J. D. Freeman, J. E. Hobson, R. N. Perutz and P. H. Walton, *J. Chem. Soc., Dalton Trans.*, 2002, 2797.
64. W. Leitner, *Coord. Chem. Rev.*, 1996, **153**, 257.
65. J. W. Raebiger, J. W. Turner, B. C. Noll, C. J. Curtis, A. Miedaner, B. Cox and D. L. Du Bois, *Organometallics*, 2006, **25**, 3345.
66. A. Begum and P. G. Pickup, *Electrochem. Commun.*, 2007, **9**, 2525.
67. J. M. Saveant, *Chem. Rev.*, 2008, **108**, 2348.
68. R. Angamuthu, P. Byers, M. Lutz, A. L. Spek and E. Bouwman, *Science*, 2010, **327**, 313.
69. M. Shibata and N. Furuya, *Electrochim. Acta*, 2003, **48**, 3953.
70. M. Aresta and A. Di Benedetto, *Dalton Trans.*, 2007, 2975.
71. T. Sakakura, J. C. Choi and H. Yasuda, *Chem. Rev.*, 2007, **107**, 2365.
72. M. Mikkelsen, M. Jorgensen and F. C. Krebs, *Energy Environ. Sci.*, 2010, **3**, 43.
73. N. Roberston, *Angew. Chem., Int. Ed.*, 2006, **45**, 2338.
74. H. Tsubaki, A. Sekine, Y. Ohashi, K. Koike, H. Takeda and O. Ishitani, *J. Am. Chem. Soc.*, 2005, **127**, 15544.

75. J. S. Hwang, J. S. Chang, S. E. Park, K. Ikeue and M. Anpo, *Top. Catal.*, 2005, **35**, 311.
76. H. Gerischer and F. Willig, *Top. Curr. Chem.*, 1976, **61**, 31.
77. M. Halmann, *Nature*, 1978, **275**, 115.
78. L. A. Harris and R. H. Wilson, *Annu. Rev. Mater. Sci.*, 1978, **8**, 99.
79. C. A. Craig, L. O. Spreer, J. W. Otvos and M. Calvin, *J. Phys. Chem.*, 1990, **94**, 7957.
80. A. J. Morris, G. J. Meyer and E. Fujita, *Acc. Chem. Res.*, 2010, **42**, 1983.
81. H. Takeda and O. Ishitani, *Coord. Chem. Rev.*, 2010, **254**, 346.
82. A. M. Khenkin, I. Efremenko, L. Weiner, J. M. L. Martin and R. Neumann, *Chem.–Eur. J.*, 2010, **16**, 1356.
83. T. W. Woolerton, S. Sheard, E. Reisner, E. Pierce, S. W. Ragsdale and F. A. Armstrong, *J. Am. Chem. Soc.*, 2010, **132**, 2132.
84. Y. Hayashi, S. Kita, B. S. Brunschwig and E. Fujita, *J. Am. Chem. Soc.*, 2003, **125**, 11976.
85. T. Sakakura, J. C. Choi and H. Yasuda, *Chem. Rev.*, 2007, **107**, 2365.
86. J. M. Lehn and R. Ziessel, *Proc. Natl. Acad. Sci. U. S. A.*, 1982, **79**, 701.
87. J. Hawecker, J.-M. Lehn and R. Ziessel, *Helv. Chim. Acta*, 1986, **69**, 1990.
88. J. M. Smieja and C. P. Kubiak, *Inorg. Chem.*, 2010, **49**, 9283.
89. J. L. Inglis, B. J. MacLean, M. T. Pryce and J. G. Vos, *Coord. Chem. Rev.*, 2012, **256**, 2571.
90. J. M. Kelly and J. G. Vos, *Angew. Chem.*, 1982, **94**, 644.
91. J. M. Kelly, C. M. O'Connell and J. G. Vos, *Inorg. Chim. Acta*, 1982, **64**, L75.
92. J. M. Kelly, C. M. O'Connell and J. G. Vos, *Dalton Trans.*, 1986, 253.
93. J. M. Clear, J. M. Kelly, C. M. O'Connell, J. G. Vos, C. J. Cardin, S. R. Costa and A. J. Edwards, *Chem. Commun.*, 1980, 750.
94. J. G. Haasnoot, W. Hinrichs, O. Weir and J. G. Vos, *Inorg. Chem.*, 1986, **25**, 4140.
95. H. Ishida, K. Tanaka and T. Tanaka, *Organometallics*, 1987, **6**, 181.
96. K. Tanaka, T. Mizukawa and H. Nagao, *Inorg. Chem.*, 1994, **33**, 3415.
97. H. Ishida, T. Terada, K. Tanaka and T. Tanaka, *Inorg. Chem.*, 1990, **29**, 905.
98. M. Bourrez, F. Molton, S. Chardon-Noblat and A. Deronzier, *Angew. Chem., Int. Ed.*, 2011, **50**, 9903.
99. J. M. Smieja, M. D. Sampson, K. A. Grice, E. E. Benson, J. D. Froehlich and C. P. Kubiak, *Inorg. Chem.*, 2013, **52**, 2484.
100. H. Takeda, H. Koizumi, K. Okamoto and O. Ishitani, *Chem. Commun.*, 2014, **50**, 1491.
101. F. Hartl, B. D. Rossenaar, G. J. Stor and D. J. Stufkens, *Recl. Trav. Chim. Pays-Bas*, 2010, **114**, 565.
102. G. J. Stor, S. L. Morrison, D. J. Stufkens and A. Oskam, *Organometallics*, 1994, **13**, 2641.
103. E. E. Barton, D. M. Rampulla and A. B. Bocarsly, *J. Am. Chem. Soc.*, 2008, **130**, 6342.

104. E. B. Cole, P. S. Lakkaraju, D. M. Rampulla, A. J. Morris, E. Abelev and A. B. Bocarsly, *J. Am. Chem. Soc.*, 2010, **132**, 11539.
105. M. Z. Ertem, S. J. Konezny, C. M. Araujo and V. S. Batista, *J. Phys. Chem. Lett.*, 2013, **4**, 745.
106. J. A. Keith and E. A. Carter, *J. Phys. Chem. Lett.*, 2013, **4**, 4058.
107. J. A. Tossell, *Comput. Theor. Chem.*, 2011, **977**, 123.
108. M. Z. Kamrath, R. A. Relph and M. A. Johnson, *J. Am. Chem. Soc.*, 2010, **132**, 15508.
109. D. J. Boston, C. Xu, D. W. Armstrong and F. M. MacDonnell, *J. Am. Chem. Soc.*, 2013, **135**, 16252.
110. D. J. Boston, Y. M. F. Pachón, R. O. Lezna, N. R. de Tacconi and F. M. MacDonnell, *Inorg. Chem.*, 2014, **53**, 6544.
111. B. J. Fisher and R. Eisenberg, *J. Am. Chem. Soc.*, 1980, **102**, 7361.
112. M. Beley, J. P. Collin, R. Ruppert and J. P. Sauvage, *Chem. Commun.*, 1984, 1315.
113. A. H. A. Tinnemans, T. P. M. Koster, D. H. M. W. Thewissen and A. Mackor, *Recl. Trav. Chim. Pays-Bas*, 1984, **103**, 288.
114. M. Beley, J. P. Collin, R. Ruppert and J. P. Sauvage, *J. Am. Chem. Soc.*, 1986, **108**, 7461.
115. E. Kimura, M. Haruta, T. Koike, M. Shionoya, K. Takenouchi and Y. Iitaka, *Inorg. Chem.*, 1993, **32**, 2779.
116. M. Shionoya, E. Kimura and Y. Iitaka, *J. Am. Chem. Soc.*, 1990, **112**, 9237.
117. E. Fujita, D. J. Szalda, C. Creutz and N. Sutin, *J. Am. Chem. Soc.*, 1988, **110**, 4870.
118. E. Fujita, L. R. Furenlid and M. W. Renner, *J. Am. Chem. Soc.*, 1997, **119**, 4549.
119. T. Dhanasekaran, J. Grodkowski, P. Neta, P. Hambright and E. Fujita, *J. Phys. Chem. A*, 1999, **103**, 7742–7748.
120. D. Behar, T. Dhanasekaran, P. Neta, C. M. Hosten, D. Ejeh, P. Hambright and E. Fujita, *J. Phys. Chem. A*, 1998, **102**, 2870–2877.
121. J. Grodkowski, D. Behar, P. Neta and P. Hambright, *J. Phys. Chem. A*, 1997, **101**, 248.
122. S. Matsuoka, K. Yamamoto, T. Ogata, M. Kusaba, N. Nakashima, E. Fujita, S. Yanagida, D. A. Morgenstern, R. E. Wittrig, P. E. Fanwick, *et al.*, *J. Am. Chem. Soc.*, 1993, **115**, 601.
123. T. Ogata, S. Yanagida, B. S. Brunschwig and E. Fujita, *J. Am. Chem. Soc.*, 1995, **117**, 6708.
124. E. Kimura, S. Wada, M. Shionoya, T. Takahashi and Y. Litaka, *Chem. Commun.*, 1990, 397.
125. E. Fujita, S. J. Milder and B. S. Brunschwig, *Inorg. Chem.*, 1992, **31**, 2079.
126. E. Kimura, X. Bu, M. Shionoya, S. Wada and S. Maruyama, *Inorg. Chem.*, 1992, **31**, 4542.
127. E. K. Barefield, A. Bianchi, E. J. Billo, P. J. Connolly, P. Paoletti, J. S. Summers and D. G. Van Derveer, *Inorg. Chem.*, 1986, **25**, 4197.
128. E. Fujita, C. Creutz, N. Sutin and B. S. Brunschwig, *Inorg. Chem.*, 1993, **32**, 2657.

129. T. J. Schmeier, G. E. Dobereiner, R. H. Crabtree and N. Hazari, *J. Am. Chem. Soc.*, 2011, **133**, 9274.
130. J. Schneider, K. Q. Vuong, J. A. Calladine, X.-Z. Sun, A. C. Whitwood, M. W. George and R. N. Perutz, *Inorg. Chem.*, 2011, **50**, 11877.
131. C. D. Windle, M. V. Câmpian, A.-K. Duhme-Klair, E. A. Gibson, R. N. Perutz and J. Schneider, *Chem. Commun.*, 2012, **48**, 8189.
132. V. Balzani, A. Juris, M. Venturi, S. Campagna and S. Serroni, *Chem. Rev.*, 1996, **96**, 759.
133. D. Gust, T. A. Moore and A. L. Moore, *Acc. Chem. Res.*, 2001, **34**, 40.
134. M. R. Wasielewski, *Chem. Rev.*, 1992, **92**, 435.
135. L. Flamigni and D. T. Gryko, *Chem. Soc. Rev.*, 2009, **38**, 1635.
136. C. M. Lemon and P. J. Brothers, *J. Porphyrins Phthalocyanines*, 2011, **15**, 809.
137. L. Flamigni, A. I. Ciuciu, H. Langhals, B. Böck and D. T. Gryko, *Chem.–Asian J.*, 2012, **7**, 582.
138. F. D'Souza, R. Chitta, K. Ohkubo, M. Tasior, N. K. Subbaiyan, M. E. Zandler, M. K. Rogacki, D. T. Gryko and S. Fukuzumi, *J. Am. Chem. Soc.*, 2008, **130**, 14263.
139. M. Tasior, D. T. Gryko, M. Cembor, J. S. Jaworski, B. Ventura and L. Flamigni, *New J. Chem.*, 2007, **31**, 247.
140. J. Grodkowski, P. Neta, E. Fujita, A. Mahammed, L. Simkhovich and Z. Gross, *J. Phys. Chem. A*, 2002, **106**, 4772.
141. B. P. Sullivan, C. M. Bolinger, D. Conrad, W. J. Vining and T. J. Meyer, *J. Chem. Soc., Chem. Commun.*, 1985, 1414.
142. K. Kalyanasundaram, *J. Chem. Soc., Faraday Trans. 2*, 1986, **82**, 2401.
143. C. Kutal, M. A. Weber, G. Ferraudi and D. Geiger, *Organometallics*, 1985, **4**, 2161.
144. F. P. A. Johnson, M. W. George, F. Hartl and J. J. Turner, *Organometallics*, 1996, **15**, 3374.
145. G. J. Stor, F. Hartl, J. W. M. Van Outersterp and D. J. Stufkens, *Organometallics*, 1995, **14**, 1115.
146. E. Fujita, B. S. Brunschwig, S. Kita and Y. Hayashi, *J. Am. Chem. Soc.*, 2003, **125**, 11976.
147. H. Takeda, K. Koike, H. Inoue and O. Ishitani, *J. Am. Chem. Soc.*, 2008, **130**, 2023.
148. Y. Kou, Y. Nabetani, D. Masui, T. Shimada, S. Takagi, H. Tachibana and H. Inoue, *J. Am. Chem. Soc.*, 2014, **136**, 6021.
149. J. Agarwal, E. Fujita, H. F. Schaefer and J. T. Muckerman, *J. Am. Chem. Soc.*, 2012, **134**, 5180.
150. Y. Hayashi, S. Kita, B. S. Brunschwig and E. Fujita, *J. Am. Chem. Soc.*, 2003, **125**, 11976.
151. P. Kurz, B. Probst, B. Spingler and R. Alberto, *Eur. J. Inorg. Chem.*, 2006, 2966.
152. H. Hori, J. Ishihara, K. Koike, K. Takeuchi, T. Ibusuki and O. Ishitani, *J. Photochem. Photobiol., A*, 1999, **120**, 119.
153. G. Calzaferri, K. Hadener and J. W. Li, *J. Photochem. Photobiol., A*, 1992, **64**, 259.

154. H. Tsubaki, A. Sekine, Y. Ohashi, K. Koike, H. Takeda and O. Ishitani, *J. Am. Chem. Soc.*, 2005, **127**, 15544.
155. H. Konno, A. Kobayashi, K. Sakamoto, F. Fagalde, N. E. Katz, H. Saitoh and O. Ishitani, *Inorg. Chim. Acta*, 2000, **299**, 155.
156. J. R. Pugh, M. R. M. Bruce, B. P. Sullivan and T. J. Meyer, *Inorg. Chem.*, 1991, **30**, 86.
157. C. Creutz, M. H. Chou, H. Hou and J. T. Muckerman, *Inorg. Chem.*, 2010, **49**, 9809.
158. B. Gholamkhass, H. Mametsuka, K. Koike, T. Tanabe, M. Furue and O. Ishitani, *Inorg. Chem.*, 2005, **44**, 2326.
159. K. Koike, H. Hori, M. Ishizuka, J. R. Westwell, K. Takeuchi, T. Ibusuki, K. Enjouji, H. Konno, K. Sakamoto and O. Ishitani, *Organometallics*, 1997, **16**, 5724.
160. K. Koike, S. Naito, S. Sato, Y. Tamaki and O. Ishitani, *J. Photochem. Photobiol., A*, 2009, **207**, 109–114.
161. Y. Tamaki, K. Watanabe, K. Koike, H. Inoue, T. Morimoto and O. Ishitani, *Faraday Discuss.*, 2012, **155**, 115.
162. Y. Tamaki, K. Koike, T. Morimoto and O. Ishitani, *J. Catal.*, 2013, **304**, 22.
163. Y. Tamaki, K. Koike, T. Morimoto, Y. Yamazaki and O. Ishitani, *Inorg. Chem.*, 2013, **52**, 11902.
164. Y. Tamaki, T. Morimoto, K. Koike and O. Ishitani, *Proc. Natl. Acad. Sci. U. S. A.*, 2012, **109**, 15673.
165. N. Komatsuzaki, Y. Himeda, T. Hirose, H. Sugihara and K. Kasuga, *Bull. Chem. Soc. Jpn.*, 1999, **72**, 725–731.
166. (a) L. Li, L. Duan, F. fen, C. Li, M. Wang, A. Hagfeldt, L. Sun and L. Hammarström, *Chem. Commun.*, 2012, **48**, 988; (b) Z. Li, M. He, Z. Huang, U. Ozkan and Y. Wu, *J. Am. Chem. Soc.*, 2013, **135**, 11696; (c) F. Odobel and Y. Pellegrin, *J. Phys. Chem. Lett.*, 2013, **4**, 2551.
167. (a) J. R. Swierk and T. E. Mallouk, *Chem. Soc. Rev.*, 2013, **42**, 2357; (b) L. Alibabaei, M. K. Brennaman, M. R. Norris, B. Kalanyan, W. Song, M. D. Losego, J. J. Conception, R. A. Binstead, G. N. Parsons and T. J. Meyer, *Proc. Natl. Acad. Sci. U. S. A.*, 2013, **110**, 20008; (c) Y. Gao, X. Ding, J. Liu, L. Wang, Z. Lu, L. Li and L. Sun, *J. Am. Chem. Soc.*, 2013, **135**, 4219.
168. L. Kohler, N. Kaveevivitchai, R. Zong and R. P. Thummel, *Inorg. Chem.*, 2014, **53**, 912.
169. S. Y. Reece, J. A. Hamel, K. Sung, T. D. Jarvi, A. J. Esswein, J. J. H. Pijpers and D. G. Nocera, *Science*, 2011, **334**, 64648.

Subject Index